AF769533

Hen

Edited

HUBEF

Departn
College
Univers
Irvine, (

FRANC

Departn
USC Ca
Univers
School (
Los Ang

MARCE

Library of Congress Cataloging-in-Publication Data

Hemostasis and animal venoms / edited by Hubert Pirkle, Francis S.
Markland, Jr.
p. cm. -- (Hematology ; v. 7)
Papers presented at the Symposium on Animal Venoms and Hemostasis
held in San Diego, Calif. on July 20-21, 1985.
ISBN 0-8247-7806-5
1. Venom--Physiological effect--Congresses. 2. Hemostasis-
-Congresses. 3. Venom--Therapeutic use--Congresses. I. Pirkle,
Hubert. II. Markland, Francis S.
III. Symposium on Animal Venoms and Hemostasis (1985 : San Diego,
Calif.) IV. Series : Hematology (New York, N.Y.) ; v. 7.
QP632.V46H46 1988 87-32969
616.1'57061--dc19 CIP

Volum

Volum

Volume

Volume

Volume

Volume

MARCEL DEKKER, INC.
270 Madison Avenue, New York, New York 10016

Volume

Current printing (last digit):
10 9 8 7 6 5 4 3 2 1

PRINTED IN THE UNITED STATES OF AMERICA

Preface

It is the purpose of this book to bring together for the first time substantial coverage of our present knowledge of those components of animal venoms that affect the hemostatic mechanisms of man. The hemostatically active venom components are of interest both in their own right, especially from the standpoint of medical applications such as anticoagulant therapy or clinical laboratory testing, and also in relation to fundamental aspects of blood coagulation and platelet biology. Explosive recent progress in these latter fields has correspondingly heightened the importance of venom factors as probes, tools, and models.

A nucleus of material for this volume came from the Symposium on Animal Venoms and Hemostasis held in San Diego on July 20-21, 1985 in conjunction with the Xth Congress of the International Society of Thrombosis and Haemostasis. To this nucleus of new findings and perspectives presented at the symposium we have added further topics and critical reviews by active workers in the field in an effort to achieve a balanced presentation. We hope that the result is sufficiently detailed, current, and authoritative to meet the needs of specialists in the field while serving as a convenient reference source for outsiders looking in.

The editors gratefully acknowledge the help of the other members of the Symposium Organizing Committee, Patrick Gaffney, Zbigniew Latallo, and Neville Marsh, the encouragement and cooperation of Russ Doolittle as President of the Congress and his scientific program helpers John Griffin and Laurence Harker, the financial support of American Diagnostica Inc., Biotoxins, Inc., William Davis, M.D., George King Bio-Medical, Inc., Hi Chem, Inc., Latoxan, Miami Serpentarium Laboratories, Pentapharm Ltd, R.Q.P. Laboratories, Inc., The Upjohn Company, and a generous anonymous donor, the splendid contributions of our many authors, and the patient assistance of our publisher Marcel Dekker, Inc.

Hubert Pirkle
Francis S. Markland, Jr.

Contributors

Yuko Agishi, M.D. Professor, Department of Internal Medicine, Hokkaido University School of Medicine, Medical Hospital, Noboribetsu Branch Hospital, Noboribetsu City, Hokkaido, Japan

Alejandrô C. Alagon, M.D., Ph.D.* Assistant Professor, Department of Molecular Biology, Instituto de Investigaciones Biomedicas, Universidad Nacional Autonoma de Mexico, Mexico City, Mexico

H. Raymond Allen, Ph.D.† Graduate Student, Department of Chemistry and Biochemistry, University of Arkansas, Fayetteville, Arkansas

Miguel E. Alonso, Ph.D.‡ Associate Researcher, Centro de Quimica, Instituto Venezolano de Investigaciones Cientificas, Caracas, Venezuela

Carmen Luisa Arocha-Piñango, Full Researcher, Departamento de Medicina Experimental, Instituto Venezolano de Investigaciones Cientificas, Caracas, Venezuela

David L. Aronson, M.D.¶ Chief, Coagulation Branch, Division of Blood and Blood Products, Office of Biologics Research and Review, Bethesda, Maryland

---

Present affiliations:
*Assistant Professor, Department of Biochemistry, Centro de Investigacion sobre Ingenieria Genetica y Biotechnologia, Universidad Autonoma de Mexico, Cuernavaca, Mexico
†Postdoctoral Fellow, Department of Biochemistry and Molecular Biology, Louisiana State University Medical Center, New Orleans, Louisiana
‡Associate Professor, Departamento de Quimica, Facultad de Ciencias, Universidad de Los Andes, Merida, Venezuela
¶Research Professor, Department of Medicine, George Washington University Medical Center, Washington, D.C.

Graham S. Bailey, B.Sc., Ph.D., Department of Chemistry and Biological Chemistry, University of Essex, Colchester, England

Alan J. Barrett, Biochemistry Department, Strangeways Research Laboratory, Cambridge, England

William R. Bell, Jr., Ph.D., M.D., Professor of Medicine-Radiology, Department of Medicine, The Johns Hopkins University School of Medicine, Baltimore, Maryland

Jon B. Bjarnason, Ph.D., Professor, Department of Chemistry, Science Institute, University of Iceland, Reykjavik, Iceland

Kenneth M. Brinkhous, M.D., Alumni Distinguished Professor Emeritus, Department of Pathology, University of North Carolina School of Medicine, Chapel Hill, North Carolina

Andrei Z. Budzynski, Ph.D., Professor of Biochemistry, Department of Biochemistry, Temple University School of Medicine, Philadelphia, Pennsylvania

S. Campos, Laboratory of Molecular Biology, Faculdad de Ciencias Biologicas, Universidad Nacional Mayor de San Marcos, Lima, Peru

Zoila A. Carvajal, B.Sc., Research Assistant, Centro de Medicina Experimental, Instituto Venezolano de Investigaciones Cientificas, Caracas, Venezuela

Jui-yoa Chang, Ph.D., Scientific Specialist, Pharmaceuticals Research Laboratories, Ciba-Geigy Ltd., Basel, Switzerland

Jennifer Hong Chen, M.S.* Graduate Student and Research Assistant, Thrombosis and Haemostasis Research Unit, Suzhou Medical College, Suzhou, Republic of China

Chen-wu Chi, Ph.D., Professor, Shanghai Institute of Biochemistry, Chinese Academy of Sciences, Shanghai, Republic of China

W. Chitty, Centro de Medicina Experimental, Instituto Venezolano de Investigaciones Cientificas, Caracas, Venezuela

---

Present affiliation:
*Research Assistant, Department of Medicine, Stanford University, Stanford, California

Jacqueline Conard, Ph.D., Laboratoire Central d'Hematologie, Hospital Hôtel-Dieu, Paris, France

Gabriel J. Coscas, Professor of Clinical Ophthalmology and Chairman, Department of Ophthalmology-Eye University Clinic, Hospital of Creteil-University of Paris XII, Creteil, France

Lesley J. Creighton, B.Sc., Technician, Blood Products, National Institute of Biological Standards and Control, London, England

Norma Blumenfeld de Bosch, M.D., Head, Departamento de Investigaciones, Banco Municipal de Sangre del Distrito Federal, Caracas, Venezuela

Annie Derlon* Assistant Hemobiologist, Laboratoire Central de'Hematologie, Hospital Hôtel-Dieu, Paris, France

Emad W. Doaud, Ph.D., Biochemistry Department, Faculty of Medicine, Ain Shams University, Cairo, Egypt

Johannes Dodt,† Department of Clinical Chemistry and Clinical Biochemistry, University of Munich, Munich, Federal Republic of Germany

M. Farid El-Asmar, Ph.D., Professor, Biochemistry Department, Faculty of Medicine, Ain Shams University, Cairo, Egypt

E. Escobar, M.D., Laboratory of Molecular Biology, Faculdad de Ciencias Biologicas, Universidad Nacional Mayor de San Marcos, Lima, Peru

Herbert J. Evans, Ph.D., Associate Professor, Department of Biochemistry, Medical College of Virginia, Virginia Commonwealth University, Richmond, Virginia

Thomas Exner, M.Sc., Ph.D., Principal Scientific Officer, Hematology Department, Institute of Clinical Pathology and Medical Research, Westmead Hospital, Sydney, Australia

Ernst Fink, Plantorgan Arzneimittel, Bad Zwischenachn, Federal Republic of Germany

Elke Fortkamp, Gen-Bio-Tec, Heidelberg, Federal Republic of Germany

---

Present affiliations

*Hemobiologist, Laboratoire d'Hematologica, Centre Hospitals-Universitaire, Caen, France

†Institut fur Organische Chemie und Biochemie, Tednische Hochschule Darmstadt, Darmstadt, Federal Republic of Germany

Jay W. Fox, Ph.D., Associate Professor and Director, Protein and Nucleic Acid Research Facility, Department of Microbiology, University of Virginia Medical School, Charlottesville, Virginia

Hans Fritz, Ph.D., Professor and Head, Department of Clinical Chemistry and Clinical Biochemistry, University of Munich, Munich, Federal Republic of Germany

Michio Fujimaki, M.D., Department of Clinical Pathology, Tokyo Medical College, Tokyo, Japan

Katsuhiro Fukutake, M.D., Ph.D., Professor and Chairman, Department of Clinical Pathology, Tokyo Medical College, Tokyo, Japan

Kinichi Furukawa, M.D., Chairman, Second Department of Surgery, Thoracic Cardiovascular Surgery, Tokyo Medical College, Tokyo, Japan

P. J. Gaffney, Ph.D., National Institute of Biological Standards and Control, London, England

Alain Gaudric, M.D., Clinique Ophtalmologique, Hospital of Creteil-University of Paris XII, Creteil, France

Collis R. Geren, M.S., Ph.D., Professor, Department of Chemistry and Biochemistry, University of Arkansas, Fayetteville, Arkansas

José W. P. Govers-Riemslag, Department of Biochemistry, University of Limburg, Maastricht, The Netherlands

Pia Glas-Greenwalt, M.D., Research Assistant Professor of Medicine, Division of Nephrology, Department of Internal Medicine, University of Cincinnati Medical Center, Cincinnati, Ohio

Li-feng Guan, Ph.D.* Research Associate, Shanghai Institute of Biochemistry, Chinese Academy of Sciences, Shanghai, Republic of China

Franc Gubensek, Professor Dr., Senior Research Associate, Department of Biochemistry, "J. Stefan" Institute, E. Kardelj University, Ljubljana, Yugoslavia

---

Present affiliation:
*Research Fellow, Department of Biochemistry, University of Southern California, Los Angeles, California

Gudrun Heisterberg-Moutses, Gen-Bio-Tec, Heidelberg, Federal Republic of Germany

Wan-zhi Huang, Associate Professor, Department of Biology, University of Science and Technology of China, Hefei, Anhui, Republic of China

Shin Ishimaru, Assistant Professor, Department of Surgery, Tokyo Medical College, Tokyo, Japan

Sadaaki Iwanaga, Ph.D., Professor, Department of Biology, Faculty of Science, Kyushu University, Fukuoka, Japan

K. Shashi Kant, M.D., Associate Professor of Clinical Medicine, Department of Internal Medicine, University of Cincinnati Medical Center, Cincinnati, Ohio

Nancy E. Kirschbaum, Graduate Student, Department of Biochemistry, Temple University School of Medicine, Philadelphia, Pennsylvania

Darko Kocjan, Ph.D., Research and Development, Lek Works, Ljubljana, Yugoslavia

František Kornalík, M.D., Ph.D., Professor, Department of Pathophysiology, Medical Faculty, Charles University, Prague, Czechoslovakia

Lawrence F. Kress, Ph.D., Cancer Research Scientist V, Department of Experimental Biology, Roswell Park Memorial Institute, Buffalo, New York

F. Lazo, Laboratory of Molecular Biology, Faculdad de Ciencias Biologicas, Universidad Nacional Mayor de San Marcos, Lima, Peru

F. Lelong, Laboratoire Central d'Hematologie, Hospital of Creteil, University of Paris XII, Creteil, France

Yang Lu, M.S., Lecturer, Department of Biology, University of Science and Technology of China, Hefei, Anhui, Republic of China

Zi-xian Lu, Ph.D.* Adjunct Professor, Department of Biology, University of Science and Technology of China, Hefei, Anhui, Republic of China

---

Present affiliation:

*Professor, Shanghai Institute of Biochemistry, Chinese Academy of Sciences, Shanghai, Republic of China

Yunn-hwa Ma, Department of Pharmacology, National Taiwan University College of Medicine, Taipei, Taiwan

Francis S. Markland, Jr., Ph.D., Professor of Biochemistry and Acting Chairman, Department of Biochemistry and USC Cancer Center, University of Southern California School of Medicine, Los Angeles, California

Fritz Markwardt, Professor, Department of Biochemistry and Pharmacology of Hirudin, Institute of Pharmacology and Toxicology, Medical Academy Erfurt, Erfurt, Federal Republic of Germany

Neville A. Marsh, B.Sc., Ph.D.[†] Lecturer, Department of Physiology, Queen Elizabeth College, London, England

Hiroaki Matsumoto, Research Fellow, Department of Biology, Faculty of Science, Kyushu University, Fukuoka, Japan

Jürg Meier, Ph.D., Head, Biological Department, Pentapharm AG, Basel, Switzerland

Takashi Morita, Department of Biology, Faculty of Science, Kyushu University, Fukuoka, Japan

Amara L. Nouel, M.D., Professor, Chief, and Research Coordinator, Department of Hematology, Hospital Ruiz y Paez, Universidad de Oriente, Ciudad Bolivar, Venezuela

A. Ojeda, Departmento de Medicina Experimental, Instituto Venezolano de Investigaciones Cientificas, Caracas, Venezuela

J. Perales, Departmento de Medicina Experimental, Instituto Venezolano de Investigaciones Cientificas, Caracas, Venezuela

Paul M. Peyser, Research Assistant, Department of Psychology, Queen Elizabeth College, London, England

Hubert Pirkle, M.D., Professor, Department of Pathology, University of California, Irvine, California

---

Present affiliation:

[*]Senior Lecturer, Department of Physiology, King's College, London, England

Victor E. Pollak, M.D., F.R.C.P., F.A.C.P., Professor of Internal Medicine, Director, Division of Nephrology, Department of Internal Medicine, University of Cincinnati Medical Center, Cincinnati, Ohio

Marjorie S. Read, Ph.D., Research Assistant Professor of Experimental Pathology, Department of Pathology, University of North Carolina School of Medicine, Chapel Hill, North Carolina

Robert S. Reczkowski, Graduate Student, Department of Biochemistry, Temple University School of Medicine, Philadelphia, Pennsylvania

K.N.N. Reddy, Ph.D., Assistant Professor, Department of Biochemistry and Comprehensive Cancer Center, University of Southern California School of Medicine, Los Angeles, California

Michael Rieger,* German Cancer Research Center, Institute for Cell and Tumor Biology, Heidelberg, Federal Republic of Germany

A. Ritonja, Lek Works, Ljubljana, Yugoslavia

S. de Rodriguez, Lic. Bioanalysis Research Assistant, Centro de Medicina Experimental, Instituto Venezolano de Investigaciones Cientificas, Caracas, Venezuela

Jan Rosing, Department of Biochemistry, University of Limburg, Maastricht, The Netherlands

Changgeng Ruan, M.D., Professor, Thrombosis and Haemostasis Research Unit, Suzhou Medical College, Suzhou, Republic of China

Atsushi Sakai, Research Member, Department of Snake Venoms, The Japan Snake Institute, Nitta-gun, Japan

Nobuo Sakuragawa, M.D., Ph.D., Central Chemical Laboratory, Toyama Medical and Pharmaceutical University, Toyama City, Japan

Meyer Michel Samama, M.D., Ph.D., Professor, Laboratoire Central d' Hematologie, Hospital Hôtel-Dieu, Paris, France

Z. Z. Sapru, University of Essex, Colchester, England

Roy T. Sawyer, M.D., Biopharm (UK) Ltd., Swansea, Wales

**Present affiliation:*
German Cancer Research Center, Institute for Cell and Tumor Biology, Heidelberg, Federal Republic of Germany

Wolf-Dieter Schleuning, Dr. Med., Laboratoire Central et Division d'Hematologie, Centre Hospitalier Universitaire Vaudois, Lausanne, Switzerland

Silvia Schweitzer, Ph.D., Scientist, Department of Molecular Biology, Gen-Bio-Tec, Heidelberg, Federal Republic of Germany

Ursula Seemüller, M.D., Department of Clinical Chemistry and Clinical Biochemistry, Surgery Clinic City, University of Munich, Munich, Federal Republic of Germany

John R. Shainoff, Ph.D., Director, Thrombosis Section, Department of Brain and Vascular Research, Research Institute, The Cleveland Clinic Foundation, Cleveland, Ohio

Yu-yan Shu, Ph.D., Department of Biochemistry, Guangxi Medical College, Nanning, Republic of China

Scott V. Smith, Research Assistant, Department of Pathology, University of North Carolina School of Medicine, Chapel Hill, North Carolina

Reinhold Sommer, Managing Director, Gen-Bio-Tec, Heidelberg, Federal Republic of Germany

Han Speijer, Department of Biochemistry, University of Limburg, Maastricht, The Netherlands

Kurt F. Stocker, Ph.D., Director, Research and Development, Pentapharm AG, Basel, Switzerland

Eva Táborská, RNDr., Research Worker, Department of Pathophysiology, Medical Faculty, Charles University, Prague, Czechoslovakia

Kaoru Takahashi, M.D., Associate Professor, Central Clinical Laboratory, Toyama Medical and Pharmaceutical University, Toyama City, Japan

Guido Tans, Department of Biochemistry, University of Limburg, Maastricht, The Netherlands

M. L. Tasayco, Instituto Venezolano de Investigaciones Cientificas, Caracas, Venezuela

Che-ming Teng, Ph.D., Professor, Department of Pharmacology, National Taiwan University College of Medicine, Taipei, Taiwan

Ida Theodor, Department of Pathology, University of California, Irvine, California

A. Torres, M.D., Banco Municipal de Sangre del Distrito Federal, Caracas, Venezuela

Anthony T. Tu, Ph.D., Professor, Department of Biochemistry, Colorado State University, Fort Collins, Colorado

Minoru Ukita, Department of Clinical Pathology, Tokyo Medical College, Tokyo, Japan

Jan L. M. L. van Rijn, Ph.D, Department of Biochemistry, University of Limburg, Maastricht, The Netherlands

Haiying Wan, Assistant, Thrombosis and Haemostasis Research Unit, Suzhou Medical College, Suzhou, Republic of China

Chun Wang, Engineer, Department of Biology, University of Science and Technology of China, Hefei, Anhui, Republic of China

William R. Welches, Ph.D., Project Scientist, Brain and Vascular Research, The Cleveland Clinic Foundation, Cleveland, Ohio

B. C. Whaler, Department of Physiology, Queen Elizabeth College, London, England

A. Yarlequé, Laboratory of Molecular Biology, Faculdad de Ciencias Biologicas, Universidad Nacional Mayor de San Marcos, Lima, Peru

Xia Zhang, M.D.* Research Associate, Shanghai Institute of Biochemistry, Chinese Academy of Sciences, Shanghai, Republic of China

Robert F. A. Zwaal, Ph.D., Professor, Department of Biochemistry, University of Limburg, Maastricht, The Netherlands

---

**Present affiliation*:
Research Associate, VA Hospital, Division of Hematology, University of Miami, Miami, Florida

Contents

III. INHIBITORS

IV. PLATELET ACTIVATION AND AGGREGATION

V. HEMORRHAGIC FACTORS

VI. OTHER LABORATORY STUDIES

VII. CLINICAL APPLICATIONS

Hemostasis and Animal Venoms

I
COAGULANT FACTORS

1

Snake Venom Prothrombin Activators

Jan Rosing, Robert F. A. Zwaal, and Guido Tans
University of Limburg, Maastricht, The Netherlands

INTRODUCTION

Over the past 20 years, many snake venoms have been shown to contain proteins that are active in the hemostatic mechanism. These proteins may either have a procoagulant or an anticoagulant activity, since the literature gives examples of snake venoms that can promote or inhibit blood coagulation. Because such proteins may have potential therapeutic and research significance, it will be useful to gain more insight in their mode of action in the blood coagulation system. The objective of this chapter is to review those snake venoms that contain activators of the blood coagulation factor prothrombin. We became interested in venoms of this kind through our studies on the mechanism of prothrombin activation by its physiologic activator factor X_a. Therefore, the emphasis in this chapter will be on the structural and functional aspects of various venom prothrombin activators and how these relate and add to the current knowledge of the mechanism of prothrombin activation. To appreciate the mode of action of venom activators on prothrombin, it will be helpful to start with some background information on factor X_a-catalyzed prothrombin activation before we turn our attention to the snake venoms.

PROTHROMBIN ACTIVATION

Prothrombin is a single-chain zymogen with a molecular weight of 72,000 daltons that can be converted into the serine protease thrombin, the enzyme that catalyzes the conversion of fibrinogen into fibrin (for a review see Jackson and Nemerson, 1980). In vivo prothrombin activation is catalyzed by factor X_a. It appears that factor X_a has to cleave two peptide bonds in the prothrombin molecule to give rise to thrombin, and these are at position Arg_{274}-Thr_{275} (site 1) and Arg_{323}-Ile_{324} (site 2).* Both peptide bonds are available for proteolytic cleavage in the prothrombin molecule. Therefore, depending on the order of bond cleavage, two pathways of thrombin formation exist (Fig. 1). Cleavage at site 1 results in the loss of the activation peptide fragment 1.2 and in the formation of prethrombin 2, an intermediate that has the same molecular weight as thrombin (36,000 daltons), but which is not catalytically active. Prethrombin 2 can be further cleaved within a disulfide bridge to result in thrombin, a two-chain serine protease consisting of a light (A-chain) and a heavy chain (B-chain). The active site of thrombin is located on the B-chain region of the molecule. However, prothrombin can also be first cleaved at Arg_{323}-Ile_{324} (site 2) to result in so-called meizothrombin. This can be further cleaved at site 1, resulting in the formation of thrombin and the activation peptide fragment 1.2. During factor X_a-catalyzed prothrombin activation, both prethrombin 2 (Stenn and Blout, 1972; Hildebrant et al., 1973; Esmon and Jackson, 1974) and meizothrombin (Rosing et al., 1986) have been observed as transient reaction products. Therefore, it can be concluded that both peptide bonds of prothrombin are susceptible to proteolytic cleavage, and that the order of bond cleavage in prothrombin appears to be random. Cleavage at the Arg_{323}-Ile_{324} (site 2) position is a prerequisite for the exposure of the active site, since both meizothrombin and thrombin have esterolytic and amidolytic activity (Kornalik and Blombäck, 1975; Franza et al., 1975; Morita et al., 1976), whereas prethrombin 2 does not. However, both peptide bonds have to be cleaved to obtain full proteolytic activity towards fibrinogen because compared with thrombin, meizothrombin has a strongly reduced clotting activity (Kornalik and Blombäck, 1975; Franza et al., 1975).

---

*The nomenclature used for polypeptide fragments derived from prothrombin activation is that recommended by the International Committee on Thrombosis and Hemostasis (Jackson, C. M., 1977). The amino acid numbering used here is the one defined for the bovine molecule (Magnusson et al., 1975). The amino acid sequences of bovine and human prothrombin are very similar and share some 70 to 80% identical residues (Magnusson et al., 1975; Walz et al., 1977; Butkowski et al., 1977). However, the human molecule consists of 581 residues, whereas bovine prothrombin consists of 582 residues. This difference is the result of a deletion of the fourth residue in the human molecule. Consequently, the corresponding numbering of the cleavage sites in human prothrombin is Arg_{273}-Thr_{274} and Arg_{322}-Ile_{323}.

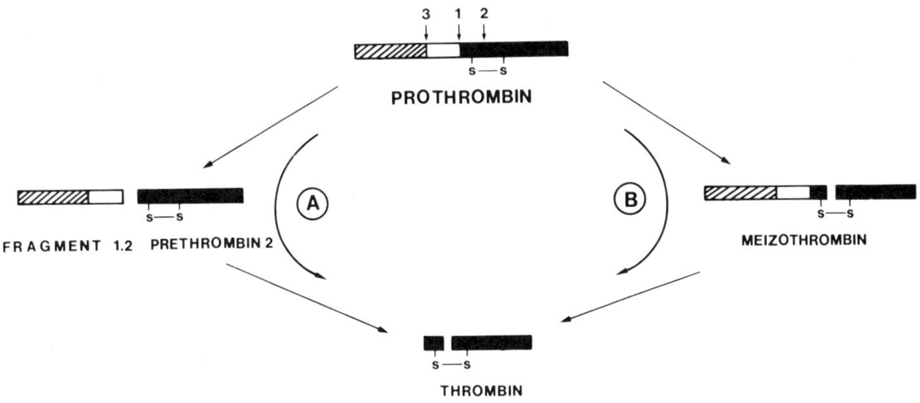

Figure 1 Possible pathways of prothrombin activation.

A few other peptide bonds in prothrombin and its activation products are susceptible to proteolytic cleavage by other serine proteases. Near the amino-terminal end of the prothrombin molecule, the Arg_{156}-Ser_{157} bond can be cleaved by thrombin. Cleavage of this peptide bond results in the formation of fragment 1 and prethrombin 1 (Jackson and Nemerson, 1980). When the corresponding bond in meizothrombin is cleaved, the product is meizothrombin des fragment 1. In human thrombin, there is also a rapid autocatalytic removal of a 13-residue peptide from the A-chain of thrombin, which has no effect on the catalytic properties of thrombin (Walz and Seegers, 1974; Downing et al., 1975; Walz et al., 1976). In the B-chain of human as well as bovine thrombin, there are at least two peptide bonds susceptible to autocatalytic cleavage. The proteolyzed forms of thrombin that result from these cleavages are called β-thrombin and γ-thrombin (Berliner et al., 1981; Mann et al., 1981; Novoa and Seegers, 1980; Workman and Lundblad, 1980). These autocatalytic cleavages have caused considerable confusion in the early literature concerning the pathways of prothrombin activation. It is now known, however, that these thrombin-catalyzed cleavages actually convert prothrombin and thrombin into products with a diminished procoagulant activity.

The activation of prothrombin by factor X_a shows a typical feature that is common for many other reactions of the blood coagulation cascade. Efficient prothrombin activation appears to be dependent on the presence of so-called accessory components, since prothrombin activation by factor X_a alone is a very slow process. The presence of calcium chloride ($CaCl_2$), phospholipid bilayers containing negatively charged phospholipids (in vivo provided by blood platelets), and a nonenzymatic protein cofactor factor V_a are necessary to obtain rates of prothrombin activation that are sufficiently high to have physiologic importance.

This is illustrated by an experiment from Jobin and Esnouf (1967) that is shown in Table 1. Jobin and Esnouf determined the effects of accessory components on the relative rate of thrombin formation by factor X_a. As can be seen, phospholipids plus $CaCl_2$ and factor V_a tremendously stimulate the reaction rate in a multiplicative way, and efficient prothrombin activation is only achieved with the complete prothrombinase complex consisting of factor X_a plus phospholipids, $CaCl_2$, and factor V_a.

Figure 2 shows a model for the assembly of the prothrombin-activating complex. It is generally accepted that factor X_a and factor V_a form a tight 1:1 stoichiometric complex that acts as the enzymatic unit in prothrombin activation. The phospholipid bilayer is thought to serve as a surface onto which the proteins bind and at which they can interact with each other. The stimulatory effect of phospholipid on prothrombin activation is presumably due to the fact that the substrate prothrombin has an affinity for both the enzyme factor X_a (or the factor X_a-factor V_a unit) and the phospholipid surface. The simultaneous binding of the enzyme (factor X_a) and the substrate prothrombin to the procoagulant phospholipid surface will facilitate the formation of productive enzyme-substrate complexes. This is reflected in a large decrease in K_m and a corresponding increase in the observed reaction rate (Rosing et al., 1980). For the binding of prothrombin and factor X_a to phospholipid bilayers, certain special amino acid residues are essential. These are the so-called γ-carboxyglutamic acid residues or Gla-residues, which are formed by a postribosomal carboxylation of specific glutamic acid residues in the amino-terminal portion of the protein molecules (Fig. 3). Without these Gla-residues, the binding of prothrombin and factor X_a to the lipid bilayers is greatly impaired, and no efficient thrombin formation can be obtained (Stenflo et al., 1974).

Table 1 The Effect of Accessory Components on the Relative Rate of Prothrombin Activation by Factor X_a

Activator	Relative rate
X_a, Ca^{2+}	1
X_a, Ca^{2+}, phospholipid	51
X_a, Ca^{2+}, V_a	318
X_a, Ca^{2+}, V_a, phospholipid	1640

Data taken from Jobin and Esnouf (1967).

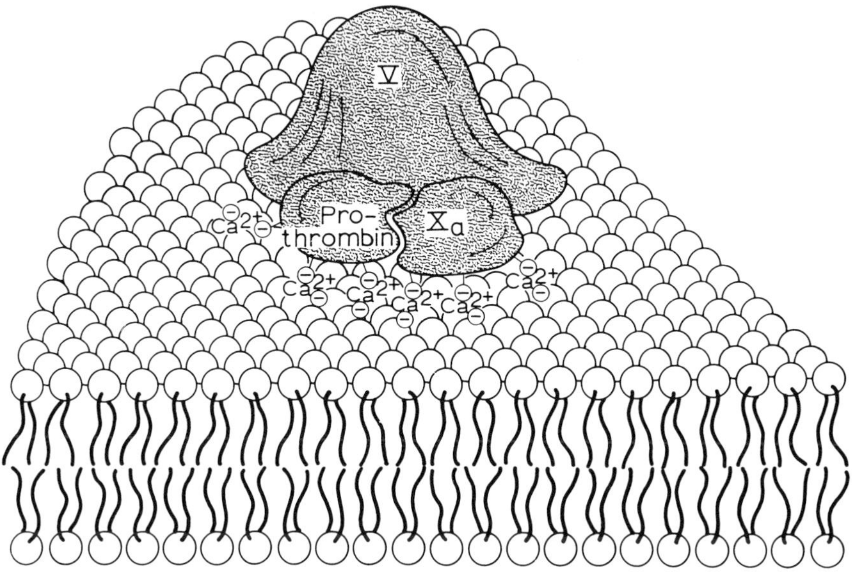

Figure 2 Schematic model for prothrombin-activating (prothrombinase) complex.

The role of the protein cofactor factor V_a in prothrombin activation is rather complex. The major function of factor V_a is to ensure a high V_{max} of thrombin formation (Nesheim et al., 1979; Rosing et al., 1980). This factor V_a-dependent increase in V_{max} appears to be caused by at least three different effects: (a) stimulation of the binding of factor X_a to negatively charged phospholipid surfaces, an effect which increases the amount of factor X_a that participates in prothrombin activation (Nesheim et al., 1979; Lindhout et al., 1982); (b) a factor V_a-induced shift in the pathway of prothrombin activation from a reaction sequence producing the catalytically inactive intermediate prethrombin 2 into one producing thrombin (Rosing et al., 1980); and (c) a factor V_a-dependent increase in a forward rate constant of the reaction pathway of prothrombin activation (Nesheim and Mann, 1983). In addition to these effects on the V_{max}, factor V_a is also able to stimulate the formation of enzyme-substrate complexes when phospholipids with a poor binding affinity for prothrombin are part of the prothrombinase complex (Pusey and Nelsestuen, 1983; van Rijn et al., 1984). In the absence of factor V_a, such phospholipids have a reduced effect on the K_m for prothrombin. Factor V_a stimulates thrombin formation on these procoagulant surfaces by decreasing the K_m for prothrombin, an effect that is presumably caused by promoting the interaction of

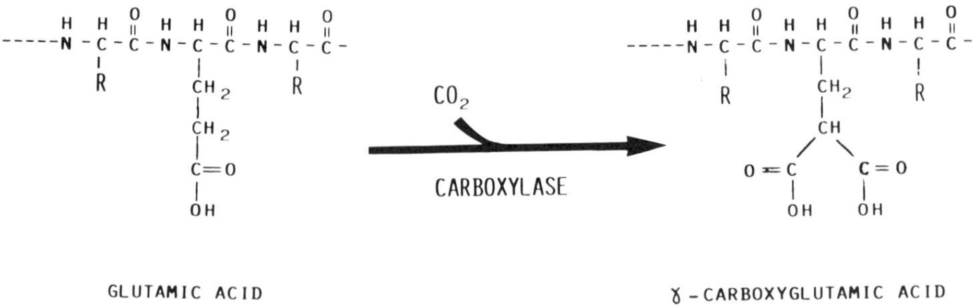

Figure 3 Conversion of glutamic acid into γ-carboxyglutamic acid.

prothrombin with the catalytic unit of the prothrombinase complex. The combined effects of factor V_a on the V_{max} of prothrombin activation and on the K_m for prothrombin cause a 1000 to 10,000-fold acceleration of factor X_a-catalyzed prothrombin activation.

PROTHROMBIN-ACTIVATING VENOMS

In view of the mechanism of prothrombin activation that has been outlined in the preceding section, it will not be surprising that a meaningful classification of prothrombin-activating snake venoms can be made by judging whether the venoms require the presence of the accessory components phospholipid and factor V_a to reach full prothrombin-activating capacity. Table 2 shows a list of venoms compiled in such a way. A number of venoms efficiently activate prothrombin irrespective of the presence of accessory components (e.g., *Echis carinatus* and *Dispholidus typus*). There are also several snake venoms (Australian Elapidae) that contain prothrombin activators that are accelerated by the presence of the accessory components of the prothrombinase complex. The latter venoms can be subdivided into a group that requires only phospholipids, which are the so-called "complete activators," and a group of venoms that are effective only in the presence of both phospholipids and factor V. These are the so-called "incomplete activators" (Denson, 1969, 1976). Finally, there are a number of venoms listed in Table 2 for which it is uncertain whether their prothrombin-converting principle requires the presence of accessory components. A large number of Bothrops venoms have been reported to contain prothrombin activators (Nahas et al., 1979; Sakuragawa and Takahashi, 1985). However, these venoms contain more than one procoagulant activity and, at present, not enough data are available to warrant classification of these venoms according to the criteria used in Table 2. For each group, venoms that have been most extensively

Table 2 Prothrombin-Activating Snake Venoms

Cofactor dependence	Snake venom	Venom activator with similar mode of action
None	*Echis carinatus* (saw-scaled viper)	*Echis coloratus*, *Dispholidus typus*, *Thelotornis kirtlandi capensis*, *Rhabdophis tigrinis tigrinis*
Phospholipids plus $CaCl_2$	*Oxyuranus scutellatus scutellatus* (Taipan snake)	*Demansia textilis textilis*, *Pseudonaja affinis*, *Pseudonaja nuchalis*, *Parademansia microlepidotus*
Phospholipids plus $CaCl_2$ plus factor V_a	*Notechis scutatus scutatus* (Tiger snake)	*Notechis ater niger*, *Notechis flinders*, *Notechis occidentalis*, *Notechis ater serventyi*, *Notechis ater humphreysi*, *Pseudechis porphyriacus*, *Hoplocephalus stephensii*, *Tropidechis carinatus*, *Cryptophis nigrescens*
Unknown		*Trimeresurus okinavensus*, *Bothrops atrox asper*, *Bothrops jararaca*, *Bothrops itapetiningae*, *Bothrops erythromelas*, *Bothrops castelnaudi*

studied and for which there is no doubt about these criteria are listed first. For classification of the other venoms listed in these groups, we have relied partly on purification data and direct prothrombin activation studies and partly on the fact that these venoms display similar effects in various clotting assays.

For the venom activators in the first group, there are no reports in the literature on stimulatory effects of phospholipids and/or factor V_a. The best known exponent of this group is the prothrombin activator from *Echis carinatus*, one of the most intensively studied venom activators. For this enzyme, a large amount of structural and functional data are available in the literature (Morita and Iwanaga, 1978; Briët et al., 1982; Rhee et al., 1982). The activator from *Dispholidus typus* has been shown to activate prothrombin by a mechanism similar to that displayed by *Echis carinatus* (Guillin et al., 1978; Bradlow et al., 1980). Apart from their capability to activate prothrombin, most of the venoms in the first group contain more than one procoagulant activity. Thus, in *Echis carinatus*, a factor X-activating principle has been reported by Denson (1976), an observation recently confirmed by Hemker et al. (1984). *Echis coloratus, Dispholidus typus*, as well as *Thelotornis kirtlandi* also contain factor X activators (Denson, 1976; Kornalik and Táborska, 1978; Bradlow et al., 1980).

The snake species listed in the groups that require phospholipids or phospholipids plus factor V_a all originate from Australia and belong to the Elapidae. The procoagulant action of these venoms appears to be primarily directed toward prothrombin. The actions of venom from *Oxyuranus scutellatus scutellatus* (Taipan snake) and *Notechis scutatus scutatus* (tiger snake) on prothrombin have been well characterized. Prothrombin activation by the activator from *Oxyuranus scutellatus scutellatus* is greatly stimulated by negatively charged phospholipids, but not by factor V_a (Pirkle et al., 1972; Lanchantin et al., 1973; Walker et al., 1980; Govers-Riemslag et al., 1987), whereas the venom activator from *Notechis scutatus scutatus* requires both phospholipids plus Ca^{2+} ions and factor V_a for efficient prothrombin activation (Jobin and Esnouf, 1966; Tans et al., 1985, 1987). The classification of the other venoms in these groups was made using data concerning the procoagulant activities in various clotting factor-deficient plasmas displayed by these venoms as compared with that displayed by Taipan snake and tiger snake venoms (Chester and Crawford, 1982; Marshall and Herzmann, 1983).

Structural Features of Purified Snake Venom Prothrombin Activators

Relatively little is known about the structure and function of the majority of the venom prothrombin activators listed in Table 2. Only three have been purified and characterized to a large extent. These are the activators of *Echis carinatus*, *Oxyuranus scutellatus scutellatus*, and *Notechis scutatus scutatus*. Fortunately, these venoms are representative of the different classes of prothrombin activators presented in Table 2. Therefore, throughout the rest of this chapter, these three venom activators will be discussed in more detail and will be used as examples of their respective class of prothrombin activators.

The best known and most extensively studied procoagulant snake venom is that of *Echis carinatus*. More than 20 years ago, it was reported that the venoms of *Echis carinatus* and *Echis coloratus* contained prothrombin activators (Gitter et al., 1960; Kornalik, 1963; Nahas et al., 1964). In 1972, Schieck and co-workers reported the first isolation of the prothrombin activator from the crude venom of *Echis carinatus*. Morita and Iwanaga (1978) purified this activator to homogeneity and obtained structural and functional data that were later confirmed by Rhee and associates (1982). The prothrombin activator from *Echis carinatus* appears to be a single-chain glycoprotein with an apparent molecular weight of 56,000 daltons as judged by SDS gel electrophoresis (Morita and Iwanaga, 1978; Rhee et al., 1982) and by gel permeation chromatography on Sepharose 6B (Morita and Iwanaga, 1978). The total carbohydrate content is approximately 16%, with about an equal distribution of hexose, hexosamine, and sialic acid (Rhee et al., 1982).

The activator of *Dispholidus typus* (boomslang) has also been purified and like the Echis activator was found to be a single-chain glycoprotein having a molecular weight of approximately 55,000 daltons (Guillin et al., 1978; Bradlow et al., 1980).

The purified activator from *Notechis scutatus scutatus* is a glycoprotein with an apparent molecular weight of 54,000 daltons and consists of two disulfide-linked polypeptide chains. The heavy and light chains have molecular masses of 32,000 and 23,000 daltons, respectively (Tans et al., 1985, 1987). *Notechis scutatus scutatus* can be inhibited by the active site-directed fluorescent dansyl-Glu-Gly-Arg-chloromethylketone (see also below). This inhibitor irreversibly alkylates the histidine of the catalytic triad of factor X_a-like enzymes (Kettner and Shaw, 1981) and, due to its fluorescent properties, it can be used to localize the position of the active site. Thus, it was found that the active site of the purified prothrombin activator of *Notechis scutatus scutatus* was located in the heavy-chain region of the molecule (Tans et al., 1985, 1987).

The structural features of the prothrombin activator isolated from the venom of *Oxyuranus scutellatus scutellatus* are of special interest. This activator does not require factor V_a for its procoagulant activity (Table 2). This appears to be due to the fact that the venom contains a factor V_a-like cofactor that is tightly complexed with the catalytic part of a prothrombin activator (Govers-Riemslag et al., 1987). These two parts can, however, be separated and isolated. The catalytic part containing the active site then appears to be a protein of approximately 57,000 daltons, which upon reduction, shows the fluorescent active site label to be bound to a subunit of approximately 30,000 daltons.

These data show that the structure of the elapid activators is similar to that of factor X_a, which also is a two-chain glycoprotein with its active site located in the heavy chain (cf. Jackson and Nemerson, 1980). Moreover, the purified activators of *Notechis scutatus scutatus* and *Oxyuranus scutellatus scutellatus* have been shown to contain the γ-carboxyglutamic acid or Gla-residues (Tans et al., 1985, 1987; Govers-Reimslag et al., 1987), which were first identified in the vitamin K-dependent blood coagulation factors. It is likely that these γ-carboxyglutamic acid residues facilitate the Ca^{2+}-dependent binding of these venom prothrombin activators to negatively charged phospholipid membranes. Therefore, phospholipids may promote venom-catalyzed prothrombin activation in a way similar to that reported for the factor X_a-dependent reaction. It is interesting that these snakes have developed the capability to synthesize γ-carboxyglutamic acid residues. Whether more snake species are capable of this is not known. In this respect, it is to note that the presence of Gla-residues was recently reported in a neurotoxin from the fish-hunting cone snail *Conus geographus* (McIntosh et al., 1984). At present, no information is available on the biochemistry of γ-carboxyglutamic acid synthesis in these snake or snail species. Since until now all γ-carboxyglutamic acid residues found in proteins result from a vitamin K-dependent carboxylation of glutamic acid, it is tempting to speculate that the Gla-residues found in the venom prothrombin activators are also synthesized by a vitamin K-dependent carboxylase.

Enzymatic Properties

It appears that the prothrombin activator from *Echis carinatus* has only a few properties in common with the elapid prothrombin activators and factor X_a. Echis appears to be remarkedly specific in cleaving just the one peptide bond that activates prothrombin (see next section). The activator from *Echis carinatus* has been found to be incapable of cleaving small oligopeptides such as S-2222, a chromogenic substrate that is readily hydrolyzed by factor X_a (Morita and Iwanaga, 1978; Rhee et al., 1982). Echis is also insensitive to serine proteinase inhibitors like DFP or the trypsin inhibitors of lima bean, soy bean, chicken ovomucoid, and bovine pancreas (Schieck et al., 1972; Morita and Iwanaga, 1978), and is also not inhibited by specific sulfhydryl reagents like parachloromercuribenzoate (PCMB), iodoacetate, and N-ethylamleimide (Morita and Iwanaga, 1978). In Table 3, we have summarized the inhibitors that have been reported to be effective. It can be seen that strong chelating agents such as EDTA or o-phenantroline effectively inhibit the Echis prothrombin activator. These data led

Table 3. Inhibitors of Purified Activators from Prothrombin Activating Snake Venoms

Venom activator	Inhibited by
*Echis carinatus**	EDTA o-phenantroline L-cysteine 2-mercaptoethanol
Oxyuranus scutellatus scutellatus†	Benzamidine STI Dansyl-GGACK
Notechis scutatus scutatus‡	DFP Benzamidine STI Dansyl-GGACK

*Morita and Iwanaga, 1978; Rhee et al., 1982.
†Govers-Riemslag et al., 1987.
‡Tans et al., 1985.

Morita and Iwanaga (1978) to propose that the activator isolated from *Echis carinatus* was not a serine proteinase like the coagulation factors found in mammalian plasma, but was a metalloproteinase. It should be mentioned, however, that Rhee and co-workers (1982), who isolated a prothrombin activator from *Echis carinatus* venom similar to the one described by Morita and Iwanaga (1978), reported that their venom activator was inhibited by iodoacetamide and that they were unable to detect Ca, Cr, Co, Cu, Fe, Mn, Ni, or Zn ions in their protein preparation. Not much is known about the other venom prothrombin activators that are listed in the same group as Echis (Table 2).

On the other hand, it is likely that the well-characterized elapid activators *Oxyuranus scutellatus scutellatus* and *Notechis scutatus scutatus* do belong to the class of serine proteinases, since both their structural features and their enzymatic properties appear to be similar to factor X_a. Table 3 shows that both venom activators are inhibited by the reversible serine proteinase inhibitor benzamidine, by soybean trypsin inhibitor, and by the active site-directed chloromethylketone dansyl-Glu-Gly-Arg-CH_2Cl. The purified activator from *Notechis scutatus scutatus* is inhibited by DFP in a similar manner as factor X_a (Tans et al., 1985, 1987).

Both venom activators also hydrolyze various peptide substrates designed for factor X_a, although the efficiency of substrate hydrolysis by these venom activators is one or two orders of magnitude lower than the catalytic efficiency of factor X_a (Walker et al., 1980; Tans et al., 1985, 1987; Govers-Riemslag, et al., 1987). Interestingly, the hydrolysis of these small substrates by the purified venom activator from *Notechis scutatus scutatus* is stimulated by factor V_a (Tans et al., 1985, 1987). Such a stimulation has not been observed for factor X_a. Recently, it was reported that the venom of *Demansia textilis textilis* contains a prothrombin activator that readily hydrolyzes the chromogenic substrate S-2222 (Masci et al., 1985).

Prothrombin Cleavage

As just discussed, two peptide bonds must be cleaved in prothrombin to complete its conversion into thrombin (Fig. 1). In the case of Echis-catalyzed prothrombin activation, the position of the cleavage site susceptible to the venom activator has been established by amino-terminal analysis of the products that are formed during the activation of bovine (Morita et al., 1976) and human (Briët et al., 1982) prothrombin. It appears that the Echis protein activates prothrombin by cleavage at the Arg_{323}-Ile_{324} position (site 2 in Fig. 1) and produces meizothrombin (Franza et al., 1975; Kornalik and Blombäck, 1975; Morita et al., 1976). Further peptide bond cleavage in meizothrombin, that results in thrombin formation is not due to the Echis activator, but must be attributed to autolysis of meizothrombin (Franza et al., 1975; Morita et al., 1976; Rhee et al., 1982). Thus, the Echis prothrombin activator appears to be incapable of cleaving off the fragment 1.2 region. For the sake of completeness, it should be mentioned that the venom of *Echis carinatus* can slowly cleave a second bond in prothrombin, producing an activation peptide that is almost identical to fragment 1 (Briët et al., 1982). The protolytic cleavage at this site (Gly_{159}-Ser_{160}) occurs, however, independent from the pathway by which Echis generates catalytically active products. A prothrombin activator with characteristics similar to the activator from *Echis carinatus* has been found in the venom of *Dispholidus typus*. This activator can only cleave prothrombin at site 2, giving rise to meizothrombin (Guillin et al., 1978). From the data reported by Kornalik and Táborská (1978) and later by Atkinson and associates (1980), it seems reasonable to assume that *Thelotornis kirtlandi capensis* (twig snake) venom will activate prothrombin in a similar manner. Finally, Morita and co-workers (1985) recently reported that the venom of *Rhabdophis tigrinus tigrinus* (Yamakagashi snake) contains an Echis-like prothrombin activator that also generates meizothrombin. Thus, the venoms classified in group I of Table 2 all appear to produce meizothrombin. This is, however, a product with strongly reduced clotting activity as compared with thrombin. The meizothrombin must, therefore, undergo a second peptide bond cleavage, either by autolysis or by other means, in order to obtain full procoagulant activity. In this respect, it may be important that the venoms of the majority of these snake species also contain factor X activators (Denson, 1976; Bradlow et al., 1980; Kornalik and Táborská, 1978; Hemker et al., 1984). In case of prothrombin activation by these venoms in plasma or whole blood, factor X_a may then also contribute to the cleavage of the Arg_{274}-Thr_{275} bond and, thus, aid in the production of thrombin.

The venoms from *Oxyuranus scutellatus scutellatus* and *Notechis scutatus scutatus* appear to be able to cleave prothrombin at both Arg_{274}-Thr_{275} and Arg_{323}-Ile_{324} (sites 1 and 2 in Fig. 1). However, the identity of cleavage products by determination of amino-terminal residues in the products formed has only been established in the case of human prothrombin activation by *Oxyuranus scutellatus scutellatus* (Lanchantin et al., 1973). In the other cases, the identity of cleavage sites has been inferred from the fact that the products formed during prothrombin activation appear identical to those observed during factor X_a-catalyzed thrombin formation as judged by polyacrylamide gel electrophoresis in the presence of SDS. During prothrombin activation by the activators from the venoms of *Notechis scutatus scutatus* and *Oxyuranus scutellatus scutellatus*, both prethrombin 2 and meizothrombin have been observed as intermediate products (Owen and Jackson, 1973; Tans et al., 1985, 1987; Govers-Riemslag et al., 1987). Therefore, the order of bond cleavage in prothrombin by these venoms appears to be random, as is the case in factor X_a-dependent prothrombin activation.

Kinetic Parameters of Prothrombin Activation by Snake Venom Activators

The kinetics of prothrombin activation by the purified activators from *Echis carinatus*, *Notechis scutatus scutatus*, and *Oxyuranus scutellatus scutellatus* have been studied in more detail. The kinetic parameters of venom-catalyzed prothrombin activation are summarized in Table 4. As expected, these data reflect the differences in cofactor dependence of the various venom prothrombin activators. *Echis carinatus* is a fairly efficient prothrombin activator that does not require the presence of accessory components. The K_m for prothrombin is approximately 1 μM and, from the data of Morita and Iwanaga (1978), the V_{max} can be calculated to be 280 turnovers per minutes. This means that at the plasma prothrombin concentration of 2 μM, this amounts to about 190 molecules activated per minute per molecule of venom activator. This is only 10- to 20-fold lower than the rate of factor X_a-catalyzed prothrombin activation under the most favorable reaction conditions (i.e., with phospholipids and factor V_a present) (Tans et al., 1980).

It can be easily seen why the Australian elapid venoms require the presence of phospholipids. In the absence of accessory compon-

Table 4 Kinetic Parameters of Prothrombin Activation by Purified Snake Venom Activators

Snake venom	K_m (μM)	V_{max} (turnovers/min)
Without accessory components		
Echis carinatus	1.03	280*
Oxyuranus s. Scutellatus	166	750†
Notechis s. scutatus	105	0.135‡
With phospholipid and $CaCl_2$		
Oxyuranus s. scutellatus	0.34	2137†
Notechis s. scutatus	0.20	0.302‡
With phospholipid, $CaCl_2$ and factor V_a		
Oxyuranus s. scutellatus	0.31	2198†
Notechis s. scutatus	0.16	1507‡

*Data from Morita and Iwanaga, 1978. V_{max} was calculated assuming a molecular weight of 56,000 daltons for the purified activator.
†Data from Govers-Riemslag et al., 1987.
‡Data from Tans et al., 1985.

ents, both venoms have a very high K_m, which means that it will be difficult to saturate the enzyme with substrate and achieve maximal rates of prothrombin activation. The presence of phospholipids plus $CaCl_2$ causes a large decrease in this K_m to levels of prothrombin that are sufficiently low to ensure that prothrombin activation can proceed at near V_{max}. In the case of *Notechis scutatus scutatus*, the V_{max} is, however, rather low and is increased enormously by the presence of factor V_a (Tans et al., 1985, 1987). *Oxyuranus scutellatus scutellatus* does not require factor V_a because this venom appears to contain its own factor V_a-like cofactor that is tightly complexed with the factor X_a-like catalytic unit of the prothrombin activator (Govers-Riemslag et al., 1987).

The kinetic parameters summarized in Table 4 explain the early reports of the effect of the accessory components on the rate of prothrombin activation, and give insight in the different requirements displayed by the different groups of venom activators. The Australian elapid venoms require phospholipids and factor V_a to obtain prothrombin activation rates equal to those found for factor X_a under comparable conditions. Apparently, the Echis activator is efficient enough and does not require further stimulation.

CLINICAL APPLICATIONS OF VENOM PROTHROMBIN ACTIVATORS

It will be clear that the snake venoms described in this chapter have been very useful in contributing to our understanding of the mechanism of prothrombin activaiton. They can, however, also be used as diagnostic tools in clinical assays of prothrombin. This is because the use of snake venoms allows a direct estimation of the amount of prothrombin in plasma samples, with minimal interference by other coagulation factors or inhibitors present in the plasma. The prothrombin activators from *Echis carinatus* and Taipan snake venom have especially attracted attention. Assays have been described that make use of these venoms to determine prothrombin levels in plasma (Denson et al., 1971; Bergström and Egberg, 1978; Kirchhof et al., 1978). Although the mechanism of action of these two activators is quite different, their diagnostic value appears similar since both can be used to determine the total level of prothrombin together with that of the degradation products prethrombin 1 and prethrombin 2. The assay principle is as follows. Plasma samples are incubated with the venom activator for a certain period of time to allow complete activation of prothrombin and, if present, also of its partial activation products prethrombin 1 and prethrombin 2. The thrombin formed as a result of this activation procedure can subsequently be determined in a clotting assay using fibrinogen or by measuring the amidolytic activity using thrombin-specific chromogenic substrates.

Potential problems that may arise in such venom-dependent prothrombin assays are: (a) the fact that the venom may convert part of the prothrombin into meizothrombin which, in case of a subsequent clotting assay, would result in an underestimation of the amount of prothrombin (meizothrombin has a low activity towards fibrinogen); and (b) the fact that the venoms activate prothrombin as well as prethrombin 1 and prethrombin 2, which makes it impossible to determine specifically the functionally active prothrombin. One can take advantage of the latter, however, when the outcome of a venom-dependent assay is compared with that of an assay in which only functionally active prothrombin is determined. From the difference between the two assay results, one can then calculate the amount of prothrombin activation intermediates present in plasma. Based on these principles, Taipan snake venom has been used to determine prethrombin 2 levels in human plasma (Aronson et al., 1984), and *Echis carinatus* has been used to aid in the diagnosis of disseminated intravascular coagulation (Sakuragawa et al., 1975, 1976).

It has often been suggested that another possible application of venom prothrombin activators would be in the determination of prothrombin during anticoagulant therapy. In the blood of patients using oral anticoagulants (vitamin K antagonists), part of the circulating prothrombin molecules lack γ-carboxyglutamic acid residues. These so-called descarboxyprothrombin molecules do not participate in the hemostatic mechanism because they lack the ability to interact with procoagulant phospholipids. Since the aim of a prothrombin determination during anticoagulant therapy is to distinguish between functionally active prothrombin and descarboxyprothrombin, Taipan snake venom and *Echis carinatus* venom cannot be used because these activators will rapidly activate both prothrombin and descarboxyprothrombin (Bergström and Egberg, 1978; Kirchhof et al., 1978). The venom of *Notechis scutatus scutatus* has so many properties in common that there will hardly be any advantage of using this venom over factor X_a. As yet, no reports are available on venom activators that specifically activate only prothrombin and not descarboxyprothrombin. In case of venom factor X_a activators, there is such an example since Russell's viper venom contains a factor X activator that rapidly activates factor X, but only slowly reacts with descarboxy-factor X.

CLINICAL SYMPTOMS OF SNAKE BITE ENVENOMATION

Snake venoms with prothrombin activors contain a wide variety of toxic compounds such as neurotoxins, cardiotoxins, and hemolytic factors (phospholipases) and hemorrhagic, myolytic, and proteolytic activities. Therefore, the clinical symptoms of envenomated patients are rather complex, and even show a wide variation for victims bitten by the same snake species. Since this chapter deals with snake venoms that contain proteins able to activate prothrombin, we will mainly focus on the clinical effects associated with the presence of the prothrombin activator.

Many centuries ago, it was already known that snake venoms had an effect on the circulating blood. In the 17th century, the phlogistic theory of van Helmont was published, in which he considered that snake venoms were "so cold that they coagulated the blood in the veins and arrested the circulation" (van Helmont, 1648). One century later, animal experiments and observations of victims of snake bite led to opposing views with respect to the coagulant actions of snake venoms. In 1737, Geoffry and Hunauld observed that blood from animals bitten by vipers was incoagulable, while in 1781, Fontana injected venom from the same viper in the jugular vein of a rabbit and observed that this caused its immediate death and coagulation of the blood in the vessels. At the end of the 19th century, it was realized that the coagulant or anticoagulant effect of snake venoms may depend on the doses of venom. Martin (1893) observed that injection of high doses of the venom of *Pseudechis porphyriacus* coagulated the blood in the vessels, while a low dose rendered the blood incoagulable. The clotting that occurs at high doses of venom is explained by a rapid activation of prothrombin by the venom prothrombin activator, which is followed by conversion of fibrinogen into fibrin and clot formation. In those cases where the blood became incoagulable, it is likely that defibrination occurred as a result of disseminated intravascular coagulation (DIC). Indeed, there are a number of reports in the literature describing an almost complete consumption of fibrinogen and prothrombin due to snake bite envenomation. Kornalik and Pudlak (1971) described a prolonged defibrination when rats were injected with sublethal doses of *Echis carinatus* venom. They showed that the prothrombin-activating principle of this venom caused a consumption coagulopathy that is the main mechanism of afibrinogenemia. Also, the prothrombin activators from *Echis carinatus* (Fainaru et al., 1974) and *Thelotornis kirtlandi capensis* (Kornalik and Táborská, 1978; Atkinson et al., 1980) are presumably responsible for a clinical case of DIC in patients being bitten by these snakes. In 1979, Herrmann and Bailey reported six cases of patients bitten by Australian elapidae (*Demansia nuchalis affinis* and *Notechis scutatus occidentalis*) with clinical and laboratory findings suggestive of DIC. All patients showed an increased plasma level of thrombin and a depletion of prothrombin, while fibrinogen was absent or very low. These abnormal coagulation parameters are likely caused by the efficient prothrombin activators that are present in the venom of these snakes.

It is not known to what extent the presence of a prothrombin activator in a venom contributes to the mortality from snakebite in man. Snakes containing venom prothrombin activators are among the most poisonous in the world, and in the literature, patients are described who died after being bitten by *Echis carinatus* (Warrel et al., 1974), *Echis coloratus* (Man, 1976), *Thelotornis kirtlandi* (Kornalik and Táborská, 1978), *Notechis scutatus* (Rollison, 1928), and *Oxyuranus scutellatus* (Flecker, 1944). Since there is a clear discrepancy between the severe alterations of the clotting system (consumption coagulopathy, DIC) and significant clinical manifestations in patients who recovered from snake bites, it is unlikely that the action of the prothrombin activator alone will cause death.

SNAKE VENOM CLEAVAGES OF PROTHROMBIN NOT RELATED TO ACTIVATION

From the discussion concerning the structure of prothrombin given in the section on prothrombin activation, it is clear that a few other peptide bonds in prothrombin have been identified as susceptible to proteolytic cleavage without giving rise to the production of thrombin. The most prominent one is the Arg_{156}-Ser_{157} bond that can be cleaved by thrombin to give rise to fragment 1 and prethrombin 1. Thus, it is possible that snake venoms may exist that can also cleave prothrombin at such sites without resulting in the activation of the molecule.

Such venoms may be useful tools in the production of fragments and derivatives of prothrombin that can be used for further scientific studies. At present, only a few venoms have been identified that can cleave additional peptide bonds in prothrombin. These are *Crotalus adamanteus* (Eastern diamondback rattlesnake), *Agkistrodon contortrix contortrix* (Southern copperhead), and *Agkistrodon acutus* (hundred pace snake) (Pirkle et al., 1976; Seegers et al., 1980) and *Bothrops atrox* (Niewiarowski et al., 1979). These venoms belong to the class of thromboserpentins since they contain thrombin-like enzymes. It will not be surprising that some of these venoms convert prothrombin into fragment 1 and prethrombin 1. However, not all thromboserpentins contain this activity. For example, the venom of *Agkistrodon rhodostoma* was reported to have no effect on prothrombin (Pirkle et al., 1976).

Agkistrodon acutus can also cleave prothrombin at an additional site near the amino-terminal end. This cleavage site has been tentatively identified as Lys_{44}-Tyr_{45} (Seegers et al., 1980). *Agkistrodon acutus* ultimately produces prethrombin 2 without giving rise to thrombin (Seegers et al., 1980) and this venom can, therefore, be very useful in the preparation and purification of this intermediate of the prothrombin activation pathway. It should be emphasized that the peptide bonds that are cleaved have not always been identified by amino-terminal analysis and, therefore, some caution must be exercised in the assignment of the reaction products that are formed. As an example of this, we can give the *Echis carinatus*-catalyzed cleavage of a second peptide bond in prothrombin that produces an activation peptide almost identical to fragment 1. However, Briët and co-workers (1982) identified this cleavage to be at Gly_{159}-Ser_{160}, rather than at Arg_{156}-Ser_{157}.

CONCLUDING REMARKS

In this chapter, we have reviewed the current state of knowledge of prothrombin-activating snake venoms. Based on their functional properties in the activation of prothrombin, it seems appropriate to classify the various venom prothrombin activators into two groups: one group being distinctly different from factor X_a (e.g., *Echis carinatus*), and one group sharing many properties with factor X_a (e.g., *Notechis scutatus scutatus* and *Oxyuranus scutellatus scutellatus*). We do not know whether this criterion will remain useful in the future when more data about other venom prothrombin activators become available. However, at present, the known venom prothrombin activators can be classified within these two characteristic groups. The catalytic properties of the *Echis carinatus* group of venom activators are entirely different from those of the Australian elapid venoms. Echis-like prothrombin activators are efficient enzymes that act independently of the accessory components phospholipid and factor V_a and rapidly cleave the Arg_{323}-Ile_{324} bond in prothrombin, a cleavage that results in the formation of meizothrombin. The enzymes of the Echis group probably do not belong to the class of serine proteinases. The factor X_a-like venom activators from the Australian elapidae are sometimes hard to distinguish from factor X_a in their ability to activate prothrombin. Like factor X_a, these venoms are strongly stimulated by phospholipids and $CaCl_2$. We feel that some of them will act analogous to *Notechis scutatus scutatus*, which is not only stimulated by phospholipids, but also by factor V_a, and that others will share the property of *Oxyuranus scutellatus scutellatus* in that they contain a factor that acts like factor V_a. Finally, the same inhibitors that inhibit factor X_a also inhibit these enzymes. Since there are reports that some of these venom activators are inhibited by DFP, it is likely that they belong to the class of serine proteinases.

Another feature separating these two groups of snake venoms, at least as it stands now, seems to be that the Australian elapid snakes effect their procoagulant action primarily through prothrombin activation, whereas the other venoms contain more than one procoagulant component.

We are well aware that the emphasis in this chapter has been on the biochemistry of venom-catalyzed prothrombin activation, rather than on the possible clinical features of venom activators. This is partly due to our interest in the use of snake venoms as a tool in the fundamental biochemical research of prothrombin activation, and partly because we felt that at present, sufficient knowledge exists about prothrombin activation to integrate the biochemical data about prothrombin activators available in the literature into one picture. We feel that a clear basic understanding of how the various venoms act will lead to new developments in prothrombin research and possibly to new clinical applications. Without any doubt, the studies of the venom activators from *Echis carinatus*, *Oxyuranus scutellatus scutellatus*, and *Notechis scutatus scutatus* have already significantly contributed to our understanding of the mechanism of prothrombin activation. Without these venoms, it would probably have taken much longer to solve the peptide bond cleavage patterns that lead to thrombin formation and to identify positively meizothrombin as an intermediate during factor X_a-catalyzed prothrombin activation. We feel, however, that much more remains to be learned from further studies of prothrombin-activating snake venoms.

REFERENCES

Aronson, D. L., Franza, B. R., and Bagley, J. (1984). The use of prothrombin activating snake venoms to measure human prethrombin 2: absence of prethrombin 2 in serum. Thromb. Res. 34:419-429.

Atkinson, P. M., Bradlow, B. A., White, J. A. M., Greig, H. B. W., and Gaillard, M. C. (1980). Clinical Features of twig snake (*Thelotornis capensis*) envenomation. S. Afr. Med. J. 58:1007-1011.

Bergström, K. and Egberg, N. (1978). Determination of vitamin K sensitive coagulation factors in plasma. Studies on three methods using synthetic chromogenic substrates. Thromb. Res. 12:531-547.

Berliner, L. J., Bauer, R. S., Change, T., Fenton, J. W., and Shen, Y. (1981). Active-site topography of human coagulant (α) and noncoagulant (γ) thrombins. Biochemistry 20:1831-1837.

Bradlow, B. A., Atkinson, P. M., Gomperts, E. D., and Gaillard, M. C. (1980). Studies on the coagulant effect of boomslang (*Dispholidus typus*) venom. Clin. Lab. Haematol. 2:317-331.

Briët, E., Noyes, C. M., Roberts, H. R., and Griffith, M. J. (1982). Cleavage and characterisation of human prothrombin by *Echis carinatus* venom. Thromb. Res. 27:591-600.

Butkowski, R. J., Elion, J., Downing, M. R., and Mann, K. G. (1977). Primary structure of human prethrombin 2 and α-thrombin. J. Biol. Chem. 252:4942-4957.

Chester, A. and Crawford, G. P. M. (1982). In vitro coagulant properties of venoms from Australian snakes. Toxicon 20:501-504.

Denson, K. W. E. (1969). Coagulant and anticoagulant action of snake venoms. Toxicon 7:5-11.

Denson, K. W. E., Borrett, R., and Biggs, R. (1971). The specific assay of prothrombin using the Taipan snake venom. Br. J. Haematol. 21:219-225.

Denson, K. W. E. (1976). Clot-inducing substances present in snake venoms with particular reference to *Echis carinatus*. Throm. Res. 8:351-360.

Downing, M. R., Butkowski, R. J., Clark, M. M., and Mann, K. G. (1975). Human prothrombin activation. J. Biol. Chem. 250:8897-8906.

Esmon, C. T. and Jackson, C. M. (1974). The conversion of prothrombin to thrombin. III. The Factor X_a-catalyzed activation of prothrombin. J. Biol. Chem. 249:7782-7790.

Fainaru, M., Eisenberg, S., Manny, N., and Hershko, C. (1974). The natural course of defibrination syndrome caused by *Echis colorata* in man. Thromb. Diaths. Hemorrh. 31:420-428.

Flecker, H. (1944). More fatal cases of bites of taipan (*Oxyuranus scutellatus*). Med. J. Aust. 2:383-384.

Fontana, F. (1781). Traité sur le venin de la vipère, Florence.

Franza, B. R., Aronson, D. L., and Finlayson, J. S. (1975). Activation of human prothrombin by a procoagulant fraction from the

the venom of *Echis carinatus*. J. Biol. Chem. 250:7057-7068.

Geoffry, E. F. and Hunauld, I. (1737). Mémoire dans lequel on examine si l'huile d'olive est un spécifique contre la morsure de vipères, Paris.

Gitter, S., Levi, G., Kochwa, S., de Vries, A., Rechnic, J., and Casper, J. (1960). Studies on the venom of *Echis coloratus*. Am. J. Trop. Med. Hyg. 9:391-395.

Govers-Riemslag, J. W. P., Speijer, H., Zwaal, R. F. A., and Rosing, J. (1987). Purification and characterization of the prothrombin activator from *Oxyuranus scutellatus* (Taipan snake). In Hemostasis and Animal Venoms, H. Pirkle and F. S. Markland (eds.), Marcel Dekker, New York, pp. 41-53.

Guillin, M. C., Bezeaud, A., and Menache, D. (1978). The mechanism of activation of human prothrombin by an activator isolated from *Dipholidus typus* venom. Biochim. Biophys. Acta 537:160-168.

Heldebrant, C. M., Butkowski, R. J., Bajaj, S. P., and Mann, K. G. (1973). The activation of prothrombin. II. Partial reactions, physical and chemical characterization of the intermediates. J. Biol. Chem. 248:7149-7163.

van Helmont, F. M. (1648). Discussion on the opinion that the virulence of viper venom is due to the animal being angry. Orthus Medicinae, Amsterdam.

Hemker, H. C., van Dam-Mieras, M. C. E., and Devilée, P. P. (1984). The action of *Echis carinatus* venom on the blood coagulation system. Demonstration of an activator of Factor X. Thromb. Res. 35:1-9.

Herrman, R. P. and Bailey, P. E. (1979). Plasma thrombin assay using a chromogenic substrate in disseminated intravascular coagulation due to snake bite envenomation. Thromb. Haemostas. 41:544-552.

Jackson, C. M. (1977). Thromb. Haemostas. 38:567.

Jackson, C. M. and Nemerson, Y. (1980). Blood coagulation. Ann. Rev. Biochem. 49:765-811.

Jobin, F. and Esnouf, M. P. (1966). Coagulant activity of tiger snake (*Notechis scutatus scutatus*) venom. Nature 211:873-875.

Jobin, F. and Esnouf, M. P. (1967). Studies on the formation of the prothrombin converting complex. Biochem. J. 102:666-674.

Kettner, C. A. and Shaw, E. N. (1981). Inactivation of trypsin-like enzymes with peptides of arginine chloromethyl ketone. Meth. Enzymol. 80:826-842.

Kirchhof, B. R. J., Vermeer, C., and Hemker, H. C. (1978). The determination of prothrombin using synthetic chromogenic substrates; choice of a suitable activator. Thromb. Res. 13:219-232.

Kornalik, F. (1963). Uber den einfluss von Echis carinata toxin auf die blutgerinnung in vitro. Folia Haemat. 80:73-78.

Kornalik, F. and Pudlák, P. (1971). A prolonged defibrination caused by *Echis carinatus* venom. Life Sci. 10 (II): 309-314.

Kornalik, F. and Blombäck, B. (1975). Prothrombin activation induced by ecarin- a prothrombin converting enzyme from *Echis carinatus* venom. Thromb. Res. 6:53-63.

Kornalik, F. and Táborská, E. (1978). Procoagulant and defibrinating potency of the venom gland extract of *Thelotornis kirtlandi*. Thromb. Res. 12:991-1001.

Lanchantin, G. F., Friedman, J. A., and Hart, D. W. (1973). Two forms of human thrombin. Isolation and characterization. J. Biol. Chem. 248:5956-5966.

Lindhout, T., Govers-Riemslag, J. W. P., van de Waart, P., Hemker, H. C., and Rosing, J. (1982). Factor V_a-Factor X_a interaction. Effects of phospholipid vesicles of varying composition. Biochemistry 21:5494-5502.

Magnusson, S., Sottrup-Jensen, L., Peterson, T. E., and Claeys, H. (1975). The primary structure of prothrombin, the role of vitamin K in blood coagulation and a thrombin catalyzed 'negative feedback' control mechanism for limiting the activation of prothrombin. In *Prothrombin and Related Coagulation Factors*, H. C. Hemker, and J. J. Veltkamp (eds.), Leiden Univ. Press, Leiden, pp. 25-46.

Man, G. (1976). Snake bites in Israel. M.D. thesis, University of Jerusalem.

Mann, K. G., Elion, J., Butkowski, R. J., Downing, M., and Nesheim, M. E. (1981). Prothrombin. Meth. Enzymol. 80:286-302.

Marshall, L. R. and Herzmann, R. P. (1983). Coagulant and anticoagulant actions of Australian snake venoms. Thromb. Haemostas. 50:707-711.

Martin, C. J. (1893). On some effects upon the blood produced by the injection of the Australian black snake (*Pseudechis porphyriacus*). J. Physiol (London) 15:380.

Masci, P. P., Whitaker, A. N., and de Jersey, J. (1985). Purification and characterization of the prothrombin activator of the common brown snake (*Pseudonaja textilis textilis*). Thromb. Haemostas. 54:285.

McIntosh, J. M., Olivera, B. M., Cruz, L. J., and Gray, W. R. (1984). γ-Carboxyglutamate in a neuroactive toxin. J. Biol. Chem. 259:14343-14346.

Morita, T., Iwanaga, S., and Suzuki, T. (1976). The mechanism of activation of bovine prothrombin by an activator isolated from *Echis carinatus* venom and characterization of the new active intermediate. J. Biochem. 79:1089-1108.

Morita, T. and Iwanaga, S. (1978). Purification and properties of prothrombin activator from the venom of *Echis carinatus*. J. Biochem. 83:559-570.

Morita, T., Iwanaga, S., and Sakai, A. (1985). A prothrombin activator found in the venom of *Rhabdophis tigrinus tigrinus* snake (Yamakagashi snake). Thromb. Haemostas. 54:312.

Nahas, L., Denson, K. W. E., and McFarlane, R. G. (1964). A study of the coagulant activity of eight snake venoms. Thromb. Diathes. Haemorrh. 12:355-367.

Nahas, L., Kamiguti, A. S., and Barros, M. A. R. (1979). Thrombin-like and Factor X-activator components of Bothrops snake venoms. Thromb. Haemostas. 41:314-328.

Nesheim, M. E., Taswell, J. B., and Mann, K. G. (1979). The contribution of bovine Factor V and Factor V_a to the activity of prothrombinase. J. Biol. Chem. 254:10952-10962.

Nesheim, M. E. and Mann, K. G. (1983). The kinetics and cofactor dependence of the two cleavages involved in prothrombin activation. J. Biol. Chem. 258:5386-5391.

Niewiarowski, S., Kirby, E. P., Brudzynski, T. M., and Stocker, K. (1979). Thrombocytin a serine protease from *Bothrop atrox* venom. 2. Interaction with platelets and plasma clotting factors. Biochemistry 18:3570-3577.

Novoa, E. and Seegers, W. H. (1980). Mechanisms of α-thrombin and β-thrombin-E formation: use of ecarin for isolation of meizothrombin 1. Thromb. Res. 18:657-668.

Owen, W. G. and Jackson, C. M. (1973). Activation of prothrombin with *Oxyuranus scutellatus scutellatus* (Taipan snake) venom. Thromb. Res. 3:705-714.

Pirkle, H., McIntosh, M., Theodor, I., and Vernon, S. (1972). Activation of prothrombin with Taipan snake venom. Thromb. Res. 1:559-568.

Pirkle, H., Markland, F. S., and Theodor, I. (1976). Thrombin-like enzymes of snake venoms: actions on prothrombin. Thromb. Res. 8:619-627.

Pusey, M. L. and Nelsestuen, G. L. (1983). The physical significance of K_m in the prothrombinase reaction. Biochem. Biophys. Res. Comm. 114:526-532.

Rhee, M., Morris, S., and Kosow, D. P. (1982). Role of meizothrombin and meizothrombin-(des F1) in the conversion of prothrombin to thrombin by the *Echis carinatus* venom coagulant. Biochemistry 21:3437-3443.

van Rijn, J. L. M. L., Govers-Riemslag, J. W. P., Zwaal, R. F. A., and Rosing, J. (1984). Kinetic studies of prothrombin activation: Effect of Factor V_a and phospholipids on the formation of the enzyme-substrate complex. Biochemistry 23:4557-4564.

Rollison, J. W. (1928). Fatal case of tiger-snake (*Notechis scutatus*) bite. Med. Sci. Arch. Adelaide Hosp. 7:18-20.

Rosing, J., Tans, G., Govers-Riemslag, J. W. P., Zwaal, R. F. A., and Hemker, H. C. (1980). The role of phospholipids and Factor

V_a in the prothrombinase complex. J. Biol. Chem. 255:274-283.

Rosing, J., Zwaal, R. F. A., and Tans, G. (1986). Formation of meizothrombin as an intermediate in Factor X_a-catalyzed prothrombin activation. J. Biol. Chem. 261:4224-4228.

Sakuragawa, N. and Takahashi, K. (1985). Coagulation studies on Habu snake (*Trimeresurus okinavensis*). Thromb. Haemostas. 54:313.

Sakuragawa, N., Takahashi, K., Hoshiyama, M., Jimbo, C., Matsuoka, M., and Ohnishi, Y. (1975). Significance of a prothrombin assay method using *Echis carinatus* venom for diagnostic information in disseminated intravascular coagulation syndrome. Thromb. Res. 7:643-653.

Sakuragawa, N., Takahashi, K., Hoshiyama, M., Jimbo, C., Niiya, K., Matsuoka, M., and Ohnishi, Y. (1976). Changes of prothrombin assayed by the method using *Echis carinatus* venom in experimental disseminated intravascular coagulation and a clinical case. Thromb. Res. 8:697-699.

Schieck, A., Kornalik, F., and Habermann, E. (1972). The prothrombin activating principle from *Echis carinatus* venom. I. Preparation and biochemical properties. Naunyn-Schmiedeberg's Arch. Pharmacol. 272:402-416.

Seegers, W. H., Teng, C., and Novoa, E. (1980). Preparation of bovine prethrombin 2: use of acutin and activation with prothrombinase or ecarin. Thromb. Res. 19:11-20.

Stenflow, J., Fernlund, P., Egan, W., and Roepstorff, P. (1974). Vitamin K-dependent modification of glutamic acid residues in prothrombin. Proc. Natl. Acad. Sci. USA 71:2730-2733.

Stenn, K. S. and Blout, E. R. (1972). Mechanism of bovine prothrombin activation by an insoluble preparation of bovine Factor X_a (thrombokinase). Biochemistry 11:4502-4515.

Tans, G., Rosing, J., van Dieijen, G., and Hemker, H. C. (1980). Conjectures on the mode of action of Factors V and VIII. In *The Regulation of Coagulation*, K. G. Mann and F. Taylor (eds.), pp. 173-185, Elsevier, North Holland (1980).

Tans, G., Govers-Riemslag, J. W. P., van Rijn, J. L. M. L., and Rosing, J. (1985). Purification and properties of a prothrombin activator from the venom of *Notechis scutatus scutatus*. J. Biol. Chem. 260:9366-9372.

Tans, G., Govers-Riemslag, J. W. P., van Rijn, J. L. M. L., and Rosing, J. (1987). Prothrombin activation by an activator purified from the venom of the mainland tiger snake (*Notechis scutatus scutatus*). In *Hemostasis and Animal Venoms*, H. Pirkle and F. S. Markland (eds.), Marcel Dekker, New York, pp. 29-40.

Walker, F. J., Owen, W. G., and Esmon, C. T. (1980). Characterization of the prothrombin activator from the venom of *Oxyuranus scutellatus scutellatus* (Taipan venom). Biochemistry 19:1020-1023

Walz, D. A. and Seegers, H. (1974). Amino acid sequence of human thrombin A chain. Biochem. Biophys. Res. Comm. 60:717-722.

Walz, D. A., Hewett-Emmett, D., Reuterby, J., and Seegers, W. H. (1976). Human thrombin: amino acid sequence of the autocatalytically removed 13 residue A1 peptide. Thromb. Res. 9:289-292.

Walz, D. A., Hewett-Emmett, D., and Seegers, W. H. (1977). Amino acid sequence of human prothrombin fragments 1 and 2. Proc. Natl. Acad. Sci. USA 74:1969-1972.

Warrel, D. A., Davidson, N. McD., Ormerod, L. D., Pope, H. M., Watkins, B. J., Greenwood, B. M., and Reid, H. A. (1974). Bites by the saw-scaled or carpet viper (*Echis carinatus*): Trial of two specific antivenoms. Br. Med. J. IV:437-440.

Workman, E. F. and Lundblad, R. L. (1980). Thrombin. In *CRC Handbook Series in Clinical Laboratory Science*, Sect. I: *Hematology*, R. M. Schmidt (ed.), CRC Press, Boca Raton, FL, 149-169.

2

Prothrombin Activation by an Activator Purified from the Venom of the Mainland Tiger Snake (*Notechis scutatus scutatus*)

Guido Tans, José W. P. Govers-Riemslag, Jan L. M. L. van Rijn, and Jan Rosing
University of Limburg, Maastricht, The Netherlands

The activation of the zymogen prothrombin into the active serine protease thrombin is one of the key events in blood coagulation. In vivo prothrombin activation is catalyzed by the serine proteinase factor X_a. However, factor X_a alone activates prothrombin at a very low rate. Therefore, calcium ions, negatively charged phospholipids, and the nonenzymatic protein cofactor factor V_a are required to sufficiently speed up reaction rates under physiologic conditions (Jackson and Nemerson, 1980).

Kinetic studies have shown that phospholipids promote prothrombin activation by decreasing the K_m for prothrombin, whereas the rate enhancement by factor V_a is mainly due to an increase in the V_{max} (k_{cat}) of factor X_a-catalyzed prothrombin activation (Nesheim et al., 1979; Rosing et al., 1980).

It is generally thought that the stimulatory effect of phospholipids is due to the fact that prothrombin, factor X_a, and factor V_a all have affinity for negatively charged phospholipid membranes. The simul-

---

Abbreviations used are: PL = phospholipid; PS = 1,2-dioleoyl-*sn*-glycero-3-phosphoserine; PC = 1,2-dioleoyl-*sn*-3-glycerophosphocholine; SDS = sodium dodecyl sulfate; S-2238 = H-D-phenylalanyl-L-pipecolyl-L-arginine-p-nitro-anilide; CBS 31.39 = CH_3SO_2-D-leucyl-glycyl-L-arginine-p-nitroanilide; S-2337 - N-benzoyl-L-isoleucyl-L-glytamyl-(piperidyl)-L-arginyl-p-nitroanilide hydrochloride; I2581 = N-dansyl-(*p*-guanidino)-phenylalanine-piperidide hydrochloride; DFP = diisopropylphosphofluoridate.

taneous binding of the coagulation factors will increase their local concentration at the membrane surface, facilitating the formation of the enzyme-substrate complex that will result in an increase in the rate of substrate (prothrombin) activation. The assembly of the proteins of the prothrombinase complex at the phospholipid surface also promotes the interaction of factor V_a with both the substrate prothrombin and the enzyme factor X_a. However, the mechanism by which factor V_a enhances the catalytic activity of factor X_a remains to be elucidated. It is unclear whether factor V_a stimulates prothrombin activation through an interaction with factor X_a (making factor X_a a better enzyme) or through an interaction with prothrombin (making prothrombin a better substrate).

It has been reported that the venom from the Mainland tiger snake (*Notechis scutatus scutatus*) contains a prothrombin activator that like factor X_a is stimulated by factor V_a and negatively charged phospholipids (Jobin and Esnouf, 1966). Since this venom prothrombin activator might lack one or more of the specific interactions with the accessory components of the prothrombinase complex (e.g., factor V_a), it can be a useful tool in the elucidation of the mechanism by which factor V_a promotes prothrombin activation. Therefore, we decided to purify this proteinase from the crude venom and characterize its catalytic and functional properties in prothrombin activation.

PURIFICATION

The isolation and purification of the activator are shown in Table 1. The main step is gel chromatography on Sephadex G-200. The activator eluted prior to the majority of protein present. Further removal of contaminants was achieved by chromatography over SP-Sephadex. The activator adhered to the resin and was eluted with a linear salt gradient. There was 10 mM benzamidine present during the isolation procedure to avoid degradation of the venom activator.

Gel electrophoresis in the presence of SDS showed that the purified activator migrated as a single band with an apparent molecular weight of approximately 54,000 daltons (Fig. 1). Upon reduction with 2-mercaptoethanol, two protein bands are seen, indicating that the activator consists of disulfide-linked heavy and light chains with apparent molecular weights of 32,000 and 23,000 daltons, respectively. The molecular weight of 54,000 daltons for the purified venom activator calculated from the relative migration distance on the SDS gel was confirmed by gel permeation chromatography on Sephadex G-200 (data not shown). The activator could be labeled with the active site-directed fluorescent chloromethylketone, dansyl-Glu-Gly-Arg-CH_2Cl, which resulted in a complete loss of enzymatic activity. The gels show that this fluorescent label was incorporated in the heavy chain, indicating that the active site is located within this domain of the molecule (Fig. 1).

Table 1 Purification of the Prothrombin Activator from the Venom of *Notechis scutatus scutatus*

	Total protein (mg)	Specific activity (nmol prothrombin activated/min/μg venom)	Yield (%)
Starting material	100	1.5	100
Sephadex G-200	7	12.0	60
SP-Sephadex	1.2	26.0	20

G-200 column chromatography (3.3 × 90 cm) was performed at 4°C at a flow rate of 15 ml/hr in 50 mM Tris-HCl (pH 7.6), 500 mM NaCl, and 10 mM benzamidine. The fractions containing activator were pooled and dialyzed against 25 mM Mes (pH 6.5) and 10 mM benzamidine, and applied to a SP-Sephadex column (2.2 × 20 cm). The activator was eluted with a linear salt gradient of 0 to 250 mM NaCl. Prothrombin activation was determined by the appearance of amidolytic activity towards the chromogenic substrate S-2238, as described earlier (Rosing et al., 1980). The activation mixture contained venom in appropriate dilutions, 0.4 μM prothrombin, 5 mM $CaCl_2$, 50 μM phospholipids (20% PS/80% PC), and 5 nM factor V_a at 37°C and pH 7.9.

PROTHROMBIN ACTIVATION

Kinetic Parameters of Prothrombin Activation by the Purified Activator

To gain more insight into the mechanism by which phospholipids and factor V_a stimulate venom-catalyzed prothrombin activation, we determined the kinetic parameters of prothrombin activation by the purified venom activator. Prothrombin activation was monitored through the appearance of amidolytic activity towards the chromogenic substrate S-2238. It was verified that the reaction was linear with time and proportional to the amount of enzyme present. In both the absence and presence of accessory components, the reaction obeyed Michaelis-Menten kinetics, and the kinetic parameters of venom-catalyzed prothrombin activation could be obtained from Lineweaver-Burk plots (Table 2). In the absence of accessory components, the rate of prothrombin activation is low due to very unfavorable kinetic parameters. The K_m is high (105 μM), and V_{max} is very low (0.002 nmol prothrombin activated per minute per μg venom). The presence of phospholipids stimulates the reaction by drastically decreasing K_m. Although factor V_a lowers K_m to about 10 μM, its dominant effect is a 3500-fold increase of the V_{max}. Finally, in the presence of both phospholipid and factor V_a, the most favorable kinetic parameters are observed. The K_m is low

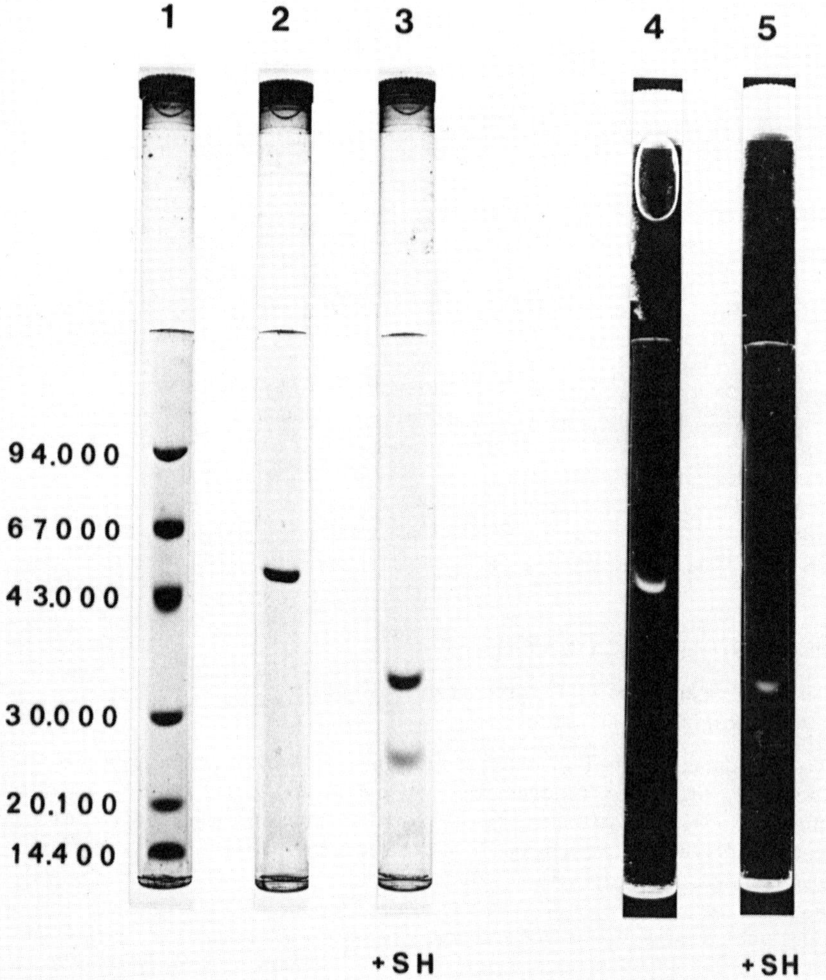

Figure 1 Gel electrophoresis of the purified prothrombin activator from *Notechis scutatus scutatus* venom in the presence of SDS was carried out on 10% gels according to Laemmli (1970). The gels contained molecular weight markers (1), 8 μg activator (2,3), or dansyl GGACK-labeled activator (4,5). Five percent 2-mercaptoethanol was present in reduced samples (gels 3 and 5).

Table 2 Kinetic Parameters of Prothrombin Activation by the Purified Activator from *Notechis scutatus scutatus*

Activator	Venom (μg/ml)	K_m (μM)	V_{max} (nmol prothrombin activated/min/μg venom)
Venom, Ca^{2+}	1.36	105	0.002
Venom, Ca^{2+}, PL	0.34	0.20	0.006
Venom, Ca^{2+}, V_a	3.41×10^{-4}	9.5	8.33
Venom, Ca^{2+}, PL, V_a	1.40×10^{-4}	0.16	27.7

The kinetic parameters of prothrombin activation were determined at 37°C at pH 7.9. Prothrombin activation was determined by the appearance of amidolytic activity towards the chromogenic substrate S-2238. The concentrations of reactants present were venom as indicated, varying amounts of prothrombin, 5 mM $CaCl_2$, 50 μM phospholipid vesicles (20% PS/80% PC), and/or 5 nM factor V_a.

due to the presence of phospholipid, and the high V_{max} is caused by the presence of factor V_a. Thus, the accessory components (phospholipid and factor V_a) stimulate prothrombin activation by the purified venom activator in a way that is remarkably similar to that by which they enhance factor X_a-dependent prothrombin activation (Rosing et al., 1980).

Peptide Bond Cleavage During Venom-Catalyzed Prothrombin Activation

Two peptide bonds must be cleaved in prothrombin to convert it into thrombin. These are the Arg_{274}-Thr_{275} (site 1) and the Arg_{323}-Ile_{324} (site 2) bonds. When only site 1 is cleaved, prethrombin 2, an intermediate without catalytic activity, is formed. Cleavage at site 2 gives rise to so-called meizothrombin, a catalytically active product with the same molecular weight as prothrombin, but which consists of two polypeptide chains linked together by a disulfide bridge.

The generation of activation products during venom-catalyzed prothrombin activation was followed by SDS gel electrophoresis (Figs. 2 and 3). Comparison of the nonreduced and reduced gels of a time course of prothrombin activation by the venom activator in the absence of factor V_a (Fig. 2) shows that prethrombin 2 is the major reaction product, and that small amounts of thrombin are formed. In the presence of factor V_a (Fig. 3), there are, however, only trace amounts of prethrombin 2, while thrombin and meizothrombin are the main reac-

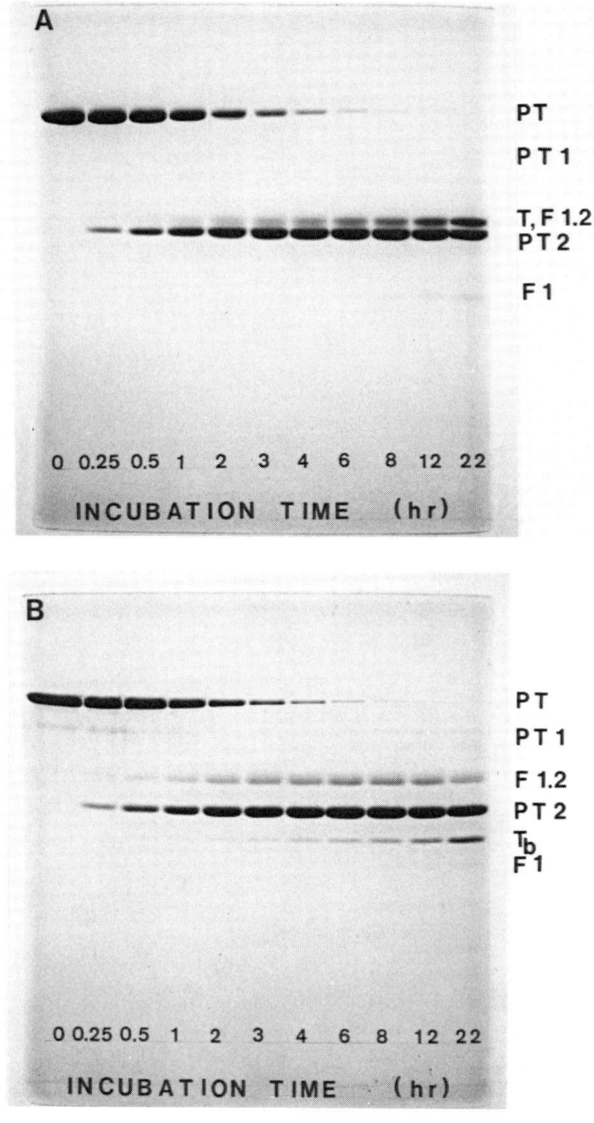

Figure 2 Prothrombin cleavage by the venom activator from *Notechis scutatus scutatus* in the absence of factor V_a. Prothrombin was activated at 37°C in a reaction mixture containing 50 mM Tris (pH 7.9), 175 mM NaCl, 5 mM $CaCl_2$, 200 μM I2581, 8.5 μg/ml venom activator, 50 μM PL (20% PS/80% PC), and 3 μM prothrombin. The reversible thrombin inhibitor I2581 was present to prevent feedback reactions of thrombin on prothrombin. At the indicated times, 50 μl samples were withdrawn for SDS gel electrophoresis on 10% slab gels according to Laemmli (1970). A = nonreduced samples; B = samples reduced with 5% 2-mercaptoethanol; PT = prothrombin; PT 1 = prethrombin 1; F 1.2 = fragment 1.2; T = thrombin; PT 2 = prethrombin 2; T_b = B-chain of thrombin; F 1 = fragment 1.

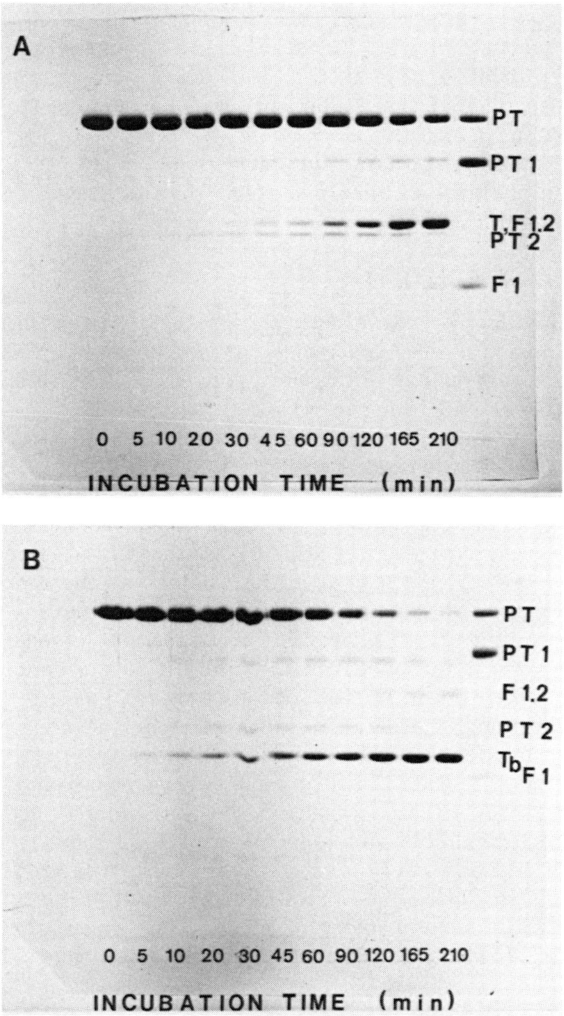

Figure 3 Prothrombin cleavage by the venom activator from *Notechis scutatus scutatus* in the presence of factor V_a. The reaction conditions were the same as described in Figure 2, except that the reaction mixture contained 8.5 ng/ml venom activator and 5 nM factor V_a.

tion products. The decreased intensity of the prothrombin band on the reduced gels corresponding with an increase in the intensity of the thrombin B-chain is indicative of meizothrombin formation under these conditions. This means that the venom activator can randomly cleave both the Arg_{274}-Thr_{275} and the Arg_{323}-Ile_{324} bonds. As in factor X_a-catalyzed prothrombin activation (Rosing et al., 1980, 1986), factor V_a also affects the preference by which these peptide bonds are cleaved by the venom activator.

γ-CARBOXYGLUTAMIC ACID CONTENT

Since the effect of phospholipid on the kinetic parameters of prothrombin activation by the venom activator resembles that observed for factor X_a, one would predict that venom-catalyzed prothrombin activation is a membrane-bound process and that the venom activator will, therefore, have an affinity for negatively charged phospholipids. In direct binding experiments, we were indeed able to show binding of the purified venom activator to phospholipid vesicles. Since binding occurred only in the presence of calcium, we determined whether the venom contained so-called γ-carboxyglutamic acid residues (Gla-residues). Such residues occur in vitamin K-dependent coagulation factors as the result of a postribosomal modification of glutamic acid residues and function in the Ca^{2+}-dependent binding of these proteins to negatively charged phospholipid membranes. We determined the Gla-content of the venom activator by high performance liquid chromatography (HPLC) using the method of Kuwada and Katayama (1983). As a control, the Gla-content in prothrombin, prothrombin fragment 1, and factor X was also determined (Table 3). In agreement with the literature, prothrombin and prothrombin fragment 1 contained approximately 10 Gla-residues, no Gla was detectable in prethrombin 1, and factor X contained some 11 Gla-residues. The purified venom activator contained eight Gla-residues per protein molecule, the presence of which explains its binding characteristics to negatively charged phospholipid and the stimulatory effect of phospholipids on venom-catalyzed prothrombin activation.

Table 3 γ-Carboxyglutamic Acid Content of the Prothrombin Activator from the Venom of *Notechis scutatus scutatus*

Protein	Gla-content (%)	Gla/protein (mole/mole)
Prothrombin	1.7 ± 0.4	10 ± 2
Prothrombin fragment 1	5.3 ± 1.4	8 ± 2
Factor X	2.4 ± 0.5	11 ± 3
Notechis scutatus scutatus	2.0 ± 0.5	8 ± 2

ENZYMATIC PROPERTIES

The Effect of Inhibitors on Venom-Catalyzed Prothrombin Activation

Differences between the venom activator and factor X_a were observed when the effect of various inhibitors on venom-dependent prothrombin activation was compared with their effect on the factor X_a-dependent reaction. Incubation with 50 mM DFP resulted in a progressive loss of enzymatic activity, until after 1 hour, about 30% of the activity of both factor X_a and *Notechis scutatus scutatus* was lost. This indicates that the venom activator also belongs to the class of serine proteinases. The reversible competitive inhibitor benzamidine showed similar inhibition of both venom activator and factor X_a. Dansyl-Glu-Gly-Arg-CH_2Cl, a chloromethylketone that rapidly alkylates the active site histidine of factor X_a (Kettner and Shaw, 1981), is a poor inhibitor of the venom activator, although complete inhibition of venom proteinase could be obtained at high chloromethylketone concentrations. Soybean trypsin inhibitor was required at 1000-fold higher concentrations to inhibit the venom than was required for factor X_a inhibition. We were not able to show any inhibition of venom-catalyzed prothrombin activation by an immunopurified antibody against bovine factor X_a. These data allow the conclusion that the activation of bovine prothrombin by the venom activator is solely due to the purified venom and is not caused by trace amounts of factors X or X_a that might be present in the prothrombin preparation.

Conversion of Chromogenic Substrates by the Venom Activator

One interesting difference between the purified activator and factor X_a was found upon testing the ability of the venom activator to cleave the factor X_a-specific chromogenic substrates CBS 31.39 and S-2337. Both substrates are converted at a high rate by factor X_a, but the conversion of these substrates by *Notechis scutatus scutatus* occurred at a 100- to 200-fold lower rate (data not shown). Factor V_a was, however, capable of stimulating the venom-catalyzed conversion of these substrates about eight-fold, whereas the factor X_a-dependent conversion was not affected by factor V_a. Figure 4 shows that this stimulation by factor V_a was concentration-dependent, and a plateau was reached at a 1:1 stoichiometry. This indicates that the venom activator and factor V_a form a tight 1:1 complex, which will act as the catalytic unit in either the conversion of small substrates (such as CBS 31.39) or in the activation of prothrombin.

CONCLUDING REMARKS

The data presented in this chapter concern the purification and characterization of a prothrombin activator from the venom of the Mainland

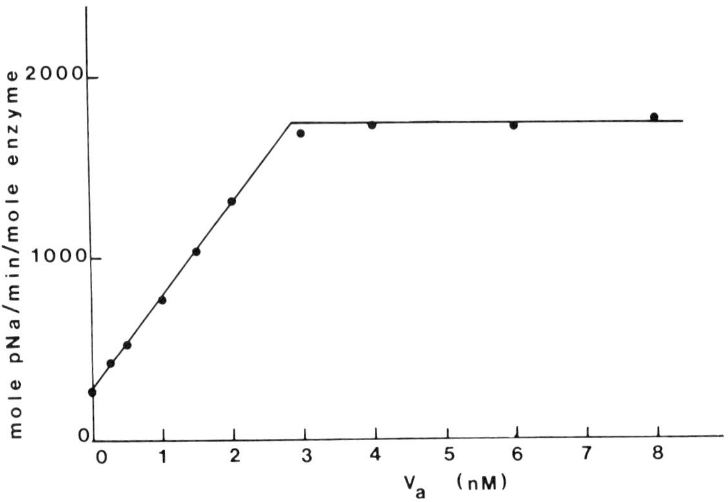

Figure 4 Effect of factor V_a on the amidolytic activity of *Notechis scutatus scutatus*. The rate of p-nitroanilide liberated from the chromogenic substrate by *Notechis scutatus scutatus* was determined at 37° C in 1 ml plastic cuvettes (1 cm path length) containing 50 mM Tris (pH 7.9), 175 mM NaCl, 0.5 mg/ml human serum albumin, 435 µM CBS 31.39, 136 ng venom/ml, and amounts of factor V_a as indicated in the figure. The rate of p-nitroanilide formation was calculated from the absorbance change at 405 nm using a molar extinction coefficient of 10,400.

tiger snake (*Notechis scutatus scutatus*). This venom was reported to contain a procoagulant activity that was attributed to a prothrombin activator, the action of which was dependent on the presence of phospholipids and factor V_a (Jobin and Esnouf, 1966).

The prothrombin activator that we purified from the crude venom of *Notechis scutatus scutatus* is remarkably similar to factor X_a. Its apparent molecular weight is approximately 54,000 daltons and it consists of a heavy and light chain of approximately 32,000 and 23,000 daltons, respectively, the heavy chain of which contains the active site. The activator is capable of activation of prothrombin at a very slow rate. The stimulatory effects of the accessory components phospholipid and factor V_a on the rate of prothrombin activation appear almost identical to those reported earlier for factor X_a-dependent activation (Rosing et al., 1980). Phospholipids cause a dramatic decrease in the K_m for prothrombin, whereas factor V_a increases the V_{max} of prothrombin activation. The decrease in the K_m is probably explained by the fact that phospholipids are able to bind both prothrombin and the venom activator. Simultaneous binding of the proteins to the phospholipid bilayer will promote the formation of the enzyme-substrate complex, decreasing the K_m. The purified activator contains eight γ-carboxyglutamic acid residues. These γ-carboxyglutamic acid residues likely play a role in the calcium-mediated binding of the venom activator to negatively charged phospholipids. To our knowledge, this is the first time that it is shown that a snake species is capable of synthesizing such amino acid residues.

From the effects of factor V_a on venom-catalyzed prothrombin activation and on the conversion of the chromogenic substrate CBS 31.39, it is obvious that factor V_a and the venom activator form a tight 1:1 complex with increased catalytic capacity. It would be interesting to compare the amino acid sequence of the venom activator with that of factor X_a in order to define the peptide regions in these protein molecules that might play a role in the interactions with phospholipids and factor V_a.

Both the Arg_{274}-Thr_{275} and Arg_{323}-Ile_{324} bonds of bovine prothrombin are susceptible to proteolytic cleavage by the venom activator. The order of bond cleavage appears to be random since prethrombin 2 (the major product in the absence of factor V_a) as well as thrombin and meizothrombin (the major products in the presence of factor V_a) are observed during venom-catalyzed prothrombin activation.

The fact that an activator with so many similarities to factor X_a shows a random order of bond cleavage in prothrombin activation prompted us to reinvestigate the possibility that meizothrombin may also occur as an intermediate in factor X_a-catalyzed prothrombin activation. Until now, there was no direct evidence for meizothrombin formation during the factor X_a-dependent reaction. Using techniques developed during our research of snake venom prothrombin activators (see also Govers-Riemslag et al., 1987), we were recently able to show the occurrence of meizothrombin as an intermediate during prothrombin activation by factor X_a (Rosing et al., 1986).

REFERENCES

Govers-Riemslag, J. W. P., Speijer, H., Zwaal, R. F. A., and Rosing, J. (1986). Purification and characterization of the prothrombin activator from *Oxyuranus scutellatus* (Taipan snake). In *Hemostasis and Animal Venoms*, H. Pirkle and F. S. Markland (eds.), Marcel Dekker, New York, pp. 41-53.

Jackson, C. M. and Nemerson, Y. (1980). Blood coagulation. Ann. Rev. Biochem. 49:765-811.

Jobin, F. and Esnouf, M. P. (1966). Coagulant activity of tiger snake (*Notechis scutatus scutatus*) venom. Nature 211:873-875.

Kettner, C. A. and Shaw, E. N. (1981). Inactivation of trypsin-like enzymes with peptides of arginine chloromethyl ketone. Methods Enzymol. 80:826-842.

Kuwada, M. and Katayama, K. (1983). An improved method for the determination of γ-carboxyglutamic acid in proteins, bone, and urine. Anal. Biochem. 131:173-179.

Laemmli, U. K. (1970). Cleavage of structural proteins during the assembly of the head of bacteriophage T4. Nature 227:680-685.

Nesheim, M. E., Taswell, J. B., and Mann, K. G. (1979). The contribution of bovine factor V and factor Va to the activity of prothrombinase. J. Biol. Chem. 254:10952-10962.

Rosing, J., Tans, G., Govers-Riemslag, J. W. P., Zwaal, R. F. A., and Hemker, H. C. (1980). The role of phospholipids and factor Va in the prothrombinase complex. J. Biol. Chem. 255:274-283.

Rosing, J., Zwaal, R. F. A., and Tans, G. (1986). Formation of meizothrombin as intermediate in factor Xa-catalyzed prothrombin activation. J. Biol. Chem. 261:4224-4228.

3

Purification and Characterization of the Prothrombin Activator from *Oxyuranus scutellatus* (Taipan Snake)

José W. P. Govers-Riemslag, Han Speijer, Robert F. A. Zwaal, and Jan Rosing
University of Limburg, Maastricht, The Netherlands

INTRODUCTION

The catalytic activity of prothrombin activators from venoms of Australian elapid snakes is greatly enhanced by the presence of accessory components (phospholipids and/or factor V_a) of the prothrombinase complex. In 1966, Jobin and Esnouf reported that phospholipids plus Ca^{2+} and factor V_a stimulated the prothrombin converting activity of the venom of *Notechis scutatus scutatus* (tiger snake) in a multiplicative way. The venom from *Oxyuranus scutellatus* contains a prothrombin activator that readily activates prothrombin, whose activity is greatly enhanced by phospholipids plus Ca^{2+}, but not by factor V_a (Denson, 1969; Pirkle et al., 1972; Owen and Jackson, 1973).

Studies on the precise mode of action of such venom prothrombin activators may help in the elucidation of the mechanism of factor X_a-catalyzed prothrombin activation. Recently, we reported on the purification and the structural and functional characterization of the prothrombin activator from *Notechis scutatus scutatus* (Tans et al., 1985, 1987). This venom activator is a 54,000-dalton protein that shares many properties with factor X_a. It has a high affinity for both negatively charged phospholipids and factor V_a, a property which may explain the stimulatory effects of these accessory components on Notechis-catalyzed prothrombin activation. Much less is known about the prothrombin activator from the venom of *Oxyuranus scutellatus*. This activator has been partially purified by Walker and co-workers (1980), who showed that it is a large protein that by itself already has the ability to rapidly activate prothrombin. Its catalytic action is stimulated several-fold by phospholipids plus Ca^{2+}, and the activator is capable of cleaving both the Arg_{274}-Thr_{275} and Arg_{323}-Ile_{324} bonds of prothrombin, thereby converting it into thrombin. In this chapter, we describe the purification of this venom activator, and present kinetic data that provide further insight in the catalytic properties of this protein.

EXPERIMENTAL PROCEDURES

Proteins

Lyophilized crude venom from *Oxyuranus scutellatus* was obtained from Sigma, St. Louis, Mo. According to the manufacturer, this venom may be from the subspecies *canni* (Papuan taipan) or *scutellatus* (Australian taipan), or a mixture of both. Bovine prothrombin, thrombin, factor X_a, and factor V_a were purified by methods reported in an earlier study (Rosing et al., 1980). The purified coagulation factors were stored at -80°C in 50 mM Tris-HCl (pH 7.9), 175 mM NaCl, and 0.5 mg/ml human serum albumin. The storage buffer of factor V_a also contained 5 mM $CaCl_2$. The purified prothrombin activator from the venom of *Oxyuranus scutellatus* was stored at -80°C in 50 mM Hepes (pH 7.2), 150 mM NaCl, and 10 mM benzamidine.

Protein Concentrations

The molar concentrations of thrombin, factor X_a and prothrombin (after complete activation with *Echis carinatus* venom) were determined by active site titration with *p*-nitrophenyl-*p*'guanidinobenzoate. The factor V_a concentration was determined by kinetic analysis as described by Lindhout and associates (1982).

Phospholipids

Single bilayer phospholipid (PL) vesicles were prepared by sonication of mixtures of 25 mole % 1,2-dioleoyl-*sn*-glycero-3-phosphoserine (PS), and 75 mole % 1,2-dioleoyl-*sn*-glycero-3-phosphocholine (PC) as previously described (Rosing et al., 1980).

Assay System for Measurement of Prothrombin Activation

Prothrombin activation was followed by measuring the generation of amidolytic activity towards H-D-phenylalanyl-L-pipicolyl-L-arginin-p-nitroanilide (S-2238). In a typical experiment, the venom activator was preincubated for 5 minutes at 37°C in 50 mM Tris-HCl (pH 7.9), 175 mM NaCl, and 0.5 mg/ml human serum albumin. When factor V_a, phospholipid, and/or $CaCl_2$ were present, these were included in the preincubation mixture. Prothrombin activation was started with the addition of prewarmed prothrombin in the same buffer. The generation of amidolytic activity (thrombin plus meizothrombin) in the reaction mixture was determined by transferring samples to cuvettes containing 50 mM Tris-HCl (pH 7.9), 175 mM NaCl, 20 mM EDTA, 0.5 mg/ml ovalbumin, and 235 μM S-2238, and measuring the absorbance change on an Aminco DW-2C spectrophotometer ($\Delta A_{405-500nM}$). The amount of activated prothrombin was calculated from a calibration curve of S-2238 conversion by known amounts of active site-titrated thrombin.

RESULTS

Purification

The prothrombin activator from the venom of *Oxyuranus scutellatus* was purified by gel permeation chromatography on Sephadex G-200, followed by ion exchange chromatography on QAE-Sephadex. Details about the purification procedure are given in Table 1. The venom prothrombin activator is a large molecular weight protein that eluted close to the void volume of the Sephadex G-200 column. The partially purified activator adhered to QAE-Sephadex and was eluted from this

Table 1 Purification of the Prothrombin Activator from the Venom of *Oxyuranus scutellatus*

	Total protein (mg)	Specific activity (nmol prothrombin activated/min/μg venom)	Yield (%)	Purification factor
Crude venom	100	0.2	100	1
Sephadex G-200	6.3	2.1	66	10
QAE-Sephadex	1.7	3.5	30	18

Crude venom (100 mg) was dissolved in 4 ml 50 mM Hepes (pH 7.2), 150 mM NaCl and 10 mM benzamidine, and applied to and eluted from a Sephadex G-200 column (2.5 × 90 cm) at a flow rate of 15 ml/hr at 4°c. The fractions containing the prothrombin activator were applied to a QAE-Sephadex A-50 column (1 × 20 cm) and eluted with a linear salt gradient (2 × 60 ml) of 150 to 350 mM NaCl in 50 mM Hepes (pH 7.2) and 10 mM benzamidine. The activator eluted from the QAE-Sephadex A-50 column at about 200 mM. The prothrombin-converting activity of the column eluates was determined by measuring the generation of amidolytic activity towards S-2238 in a prothrombin activation mixture that contained venom in appropriate dilutions, 0.5 μM prothrombin, 50 μM PL (25% PS/75% PC; mole/mole): and 5 mM $CaCl_2$.

material with a linear salt gradient. This procedure resulted in an 18-fold purification and a 30% recovery of prothrombin-converting activity.

Four major protein bands with molecular weights of 220,000, 120,000, 80,000, and 60,000 daltons, respectively, were observed when the purified activator was subjected to SDS gel electrophoresis (Fig. 1). After reduction with β-mercaptoethanol, three bands were shown with molecular weights of 110,000, 80,000 and 30,000 daltons, respectively. The 60,000-dalton subunit presumably contains the active site of the activator since the fluorescent active site-directed chloromethylketone dansyl-Glu-Gly-Arg-CH_2Cl (dansyl-GGACK) almost exclusively incorporated into this subunit (Fig. 1). On the reduced gel, the fluorescence appeared at a considerably lower molecular weight, which indicates that the catalytic 60,000-dalton subunit consists of two disulfide-linked subunits of about 30,000 daltons.

Catalytic Properties of the Purified Activator

The purified venom activator readily activated prothrombin in the absence of accessory components of the prothrombinase complex (Table 2). Venom-catalyzed prothrombin was somewhat stimulated by Ca^{2+} ions and was not affected by factor V_a. Phospholipid plus Ca^{2+} ions caused a 3000-fold increase in the rate of prothrombin activation by the venom activator. This rate enhancement was not further influenced by the presence of factor V_a.

The purified venom activator was also able to cleave a number of commercially available chromogenic substrates for factor X_a (data not shown). The amidolytic activity of the venom activator as measured with S-2337 (N-benzoyl-Ile-Glu-(pip)Gly-Arg-pNA), S-2222 (N-benzoyl-Ile-Glu-Gly-Arg-pNA), or the Pentapharm X_a substrate CH_3OCO-D-CHG-Gly-Arg-pNA was, however, 10 to 40 times lower than that of bovine factor X_a.

Kinetic Parameters of Venom-Catalyzed Prothrombin Activation

To obtain more insight into the mechanism by which the accessory components of the prothrombinase complex affect the venom activator, we decided to determine the kinetic parameters of venom-catalyzed prothrombin activation in reaction mixtures with different compositions. Prothrombin activation, followed by measuring the generation of amidolytic activity toward S-2238, was linear in time and proportional with the amount of venom activator present. The kinetic parameters, which were obtained from Lineweaver-Burk plots, are summarized in Table 3. Prothrombin activation by the purified activator alone had a high K_m

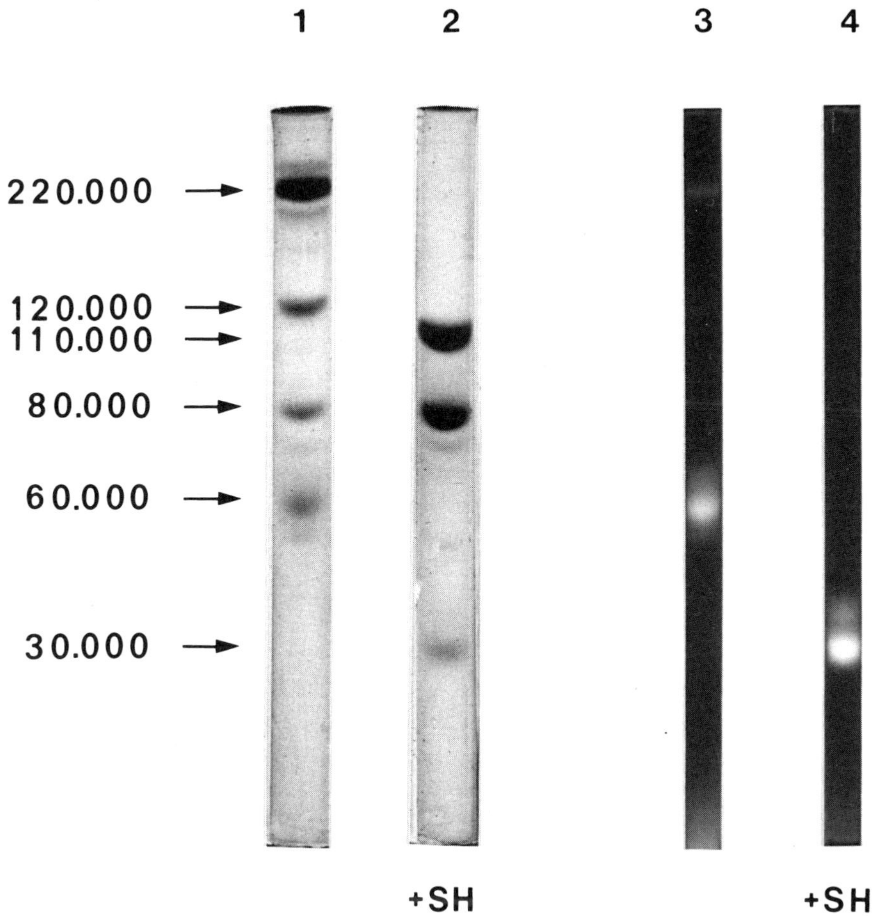

Figure 1 SDS gel electrophoresis of the purified prothrombin activator from the venom of *Oxyuranus scutellatus*. Gel electrophoresis was carried out on 7.5% gels according to Laemmli (1970). The gels contain 33 μg of activator (gels 1 and 2) or active site-labeled activator (gels 3 and 4). Five percent β-mercaptoethanol was present in the reduced samples (gels 2 and 4). Active site-labeled activator was obtained by incubation with 40 μM dansyl-GGACK for 20 minutes at 22°C. Gels containing activator were stained with Coomassie brilliant blue, and the gels containing the active site-labeled activator were photographed under UV light. The molecular weights of the subunits were determined in a separate experiment on 5% gels (subunits with $M_r > 100{,}000$) or 10% gels (subunits with $M_r < 100{,}000$).

Table 2 The Effect of Phospholipid, Ca^{2+}, and Bovine Factor V_a on Prothrombin Activation by the Venom Activator of *Oxyuranus scutellatus*

Activator	Prothrombin activated nmol/min/mg venom	Relative rate
Venom	1.6	1
Venom, Ca^{2+}	13.7	8
Venom, Ca^{2+}, V_a	11.3	7
Venom, Ca^{2+}, PL	5430	3365
Venom, Ca^{2+}, V_a, PL	5620	3487

Prothrombin activation was determined by measuring the generation of amidolytic activity towards the chromogenic substrate S-2238. The reaction mixture contained 50 mM Tris-HCl (pH 7.9), 175 mM NaCl, 0.2 mM EDTA, 0.5 mg/ml human serum albumin, 0.5 μM prothrombin, appropriate dilutions of the purified venom activator, and, when indicated, 5 mM $CaCl_2$, 5 nM factor V_a, and/or 50 μM PL (25% PS/75% PC; mole/mole).

Table 3 Kinetic Parameters of Prothrombin Activation by the Purified Activator from *Oxyuranus scutellatus*

Activator	K_m (μM)	V_{max} (mol prothrombin activated/min/mole venom activator)
Venom, Ca^{2+}	116	750
Venom, Ca^{2+}, V_a	119	550
Venom, Ca^{2+}, PL	0.34	2137
Venom, Ca^{2+}, PL, V_a	0.31	2198

The kinetic parameters of prothrombin activation were determined at 37°C in a reaction mixture containing 50 mM Tris-HCl (pH 7.9), 175 mM NaCl, 5 mM $CaCl_2$, 0.5 mg/ml human serum albumin, appropriate dilutions of the venom activator, varying amounts of prothrombin, and, when indicated, 50 μM PL (25% PS/75% PC; mole/mole) and/or 5 nM bovine factor V_a. Prothrombin activation was measured by determining the generation of amidolytic activity towards S-2238. The molar concentration of the venom activator was calculated from the amount of purified venom activator protein, using a molecular mass of 300,000 daltons.

for prothrombin and a relatively high V_{max}. The kinetic parameters were hardly affected by the presence of factor V_a in the absence or presence of phospholipids. Phospholipids, however, drastically decreased the K_m for prothrombin and had only a small effect on the V_{max}. This effect of phospholipid on the K_m can completely account for the phospholipid-dependent increase in the relative rate of prothrombin activation reported in Table 2.

Determination of γ-Carboxyglutamic Acids in the Venom Activator

The stimulatory effect of phospholipids on *Oxyuranus*-catalyzed prothrombin activation indicates that the venom activator binds to phospholipid and may, like factor X_a or the venom activator from *Notechis scutatus* (Tans et al., 1985, 1987), contain γ-carboxyglutamic acid (Gla) residues that support its Ca^{2+}-dependent binding to the membrane surface. The Gla content of the venom activator was determined by HPLC according to Kuwada and Katayama (1983). As a control, we also determined the Gla in factor X_a. Figure 2A shows a HPLC elution pattern of a basic hydrolysate of factor X_a. The peak in the chromatogram of factor X_a that eluted at 13.88 minutes was identified as Gla since it was not present in an acid hydrolysate in which Gla is decarboxylated and converted into glutamic acid (data not shown). Also, the chromatogram of the basic hydrolysate of the purified venom activator from *Oxyuranus scutellatus* contained a peak (elution time 13.45 minutes) that corresponded to γ-carboxyglutamic acid (Fig. 2B). We were able to calculate a Gla/(Glu + Gln) ratio of 0.042 for the venom activator. Since the absolute Glu-Gln content of the venom activator is not known, it was not possible, however, to calculate the exact number of Gla-residues per venom activator molecule. Considering the large molecular weight of the activator, the venom activator presumably contains sufficient Gla-residues to support Ca^{2+}-dependent binding of this protein to negatively charged phospholipids.

Structural Properties of the Purified Activator

To establish the molecular weight of the prothrombin activator purified from the venom of *Oxyuranus scutellatus*, the venom activator was subjected to gradient polyacrylamide gel electrophoresis under nondenaturating conditions. After electrophoresis, the gradient gel was sliced and eluted, and the gel slices were tested for the presence of amidolytic activity and prothrombin-converting activity (Fig. 3). Two protein bands with molecular masses of 300,000 and 57,000 daltons, respectively, contained amidolytic activity towards S-2337. From these two bands, the 300,000-dalton protein rapidly activated prothrombin

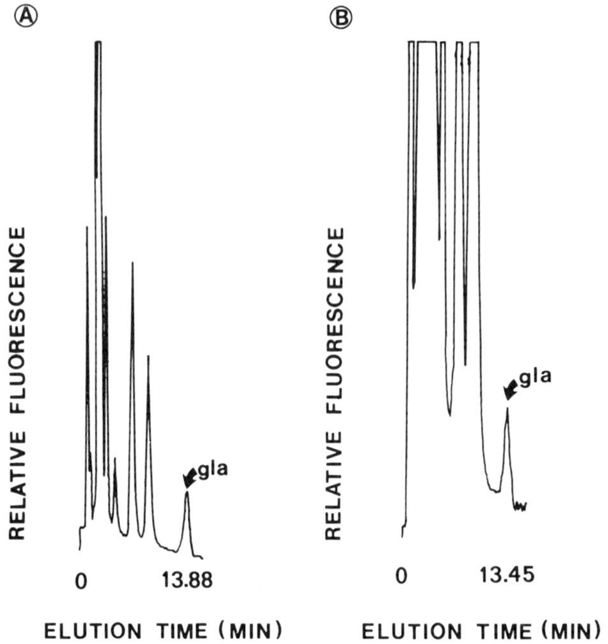

Figure 2 High performance liquid chromatographic determination of γ-carboxyglutamic acid residues in the venom prothrombin activator from *Oxyuranus scutellatus*. The presence of γ-carboxyglutamic acid residues in factor X_a (A) or the venom activator (B) were determined in a basic hydrolysate on a Nucleosil-5SB column according to the method of Kuwada and Katayama (1983). The peaks at the chromatograms eluting after 13.88 minutes (A) or 13.45 minutes (B) represent γ-carboxyglutamic acid.

irrespective of whether factor V_a was present or not. Prothrombin conversion by the 57,000-dalton protein was, however, very slow, but was greatly enhanced by the presence of bovine factor V_a.

Factor V_a was not the only protein that was able to promote prothrombin activation by the 57,000-dalton eluate. Recombinations with gel slice eluates showed that there was a protein band of 220,000 daltons, which had no catalytic activity, but which could restore prothrombin activation by the 57,000-dalton component in a concentration-dependent manner (Fig. 4). The 220,000-dalton protein was, however, not able to stimulate factor X_a-catalyzed prothrombin activation.

Prolonged electrophoresis of the purified venom activator on gradient gels led to the complete disappearance of the 300,000-dalton species of the venom activator and a recovery of its amidolytic activity

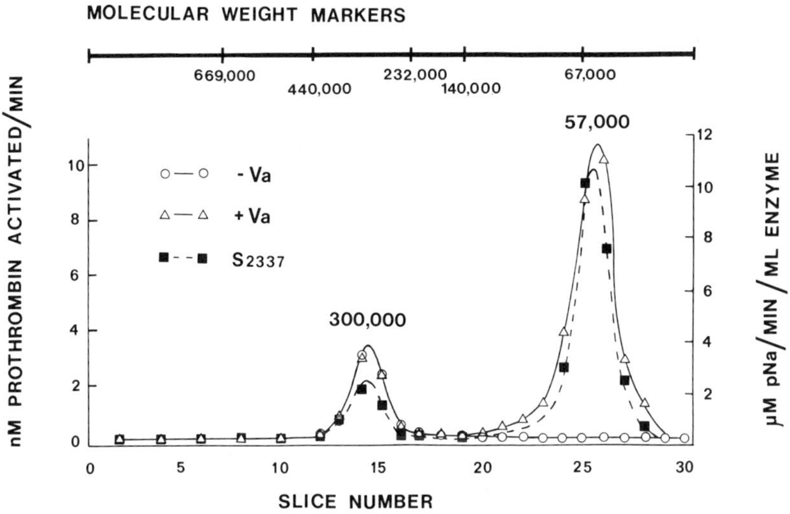

Figure 3 Gradient gel electrophoresis of the purified prothrombin activator from the venom of *Oxyuranus scutellatus*. A total of 62 μg of the purified venom activator was electrophoresed on a Pharmacia PAA 4/30 gradient gel according to the manufacturer's instruction. The migration distance of the molecular weight markers thyroglobulin (M_r 669,000), ferritin (M_r 440,000), catalase (M_r 232,000), lactate dehydrogenase (M_r 140,000), and bovine serum albumin (M_r 67,000) are indicated at the top of the figure. After electrophoresis, the gel was sliced and eluted, and the gel slices were tested for the presence of amidolytic activity towards S-2337 (■--■) and prothrombin-converting activity determined in the absence (○—○) or presence of factor V_a (△—△).

in the 57,000-dalton component (experiment not shown). These data indicate that the venom prothrombin activator form *Oxyuranus scutellatus* is a protein with a molecular mass of about 300,000 daltons, which can dissociate into a catalytic unit of 57,000 daltons with high amidolytic and low prothrombinase activity, and a cofactor part of 220,000 daltons, which is able to stimulate the prothrombin-converting activity of the low molecular weight catalytic unit.

DISCUSSION

In this chapter, we have reported on the purification and the structural and functional characterization of the prothrombin activator from the venom of *Oxyuranus scutellatus* (Taipan snake). The ability of

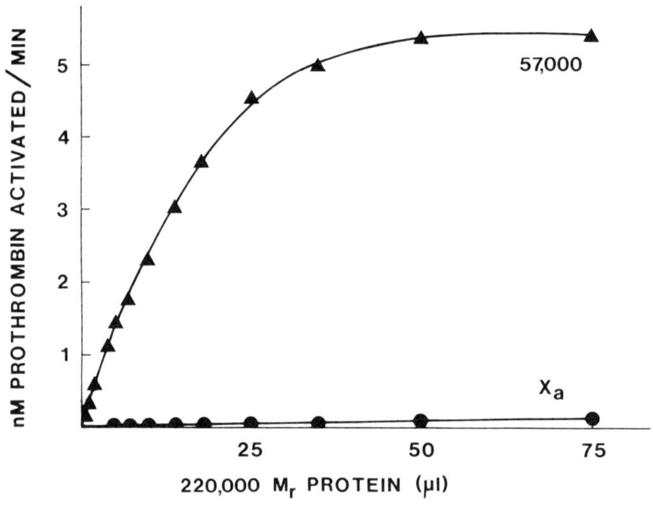

Figure 4 The effect of the 220,000-dalton protein on the prothrombinase activity of the 57,000-dalton subunit of the *Oxyuranus* prothrombin activator. A 100-fold dilution of fraction 26 of the gradient gel (see Fig. 3) containing the 57,000-dalton subunit (▲—▲) or 40 pM factor X_a (●—●) were incubated at 37°C in 50 mM Tris-HCl (pH 7.9), 175 mM NaCl, 5 mM $CaCl_2$, and 0.5 mg/ml human serum albumin with different amounts of gel fraction 17, which contains the 220,000-dalton protein. After 6 minutes, the prothrombinase activity of this incubation mixture was measured in a reaction mixture containing 50 µM PL (25% PS/75% PC; mole/mole), 0.5 µM prothrombin, and 5 mM $CaCl_2$, as described under "Experimental Procedures."

this venom protein to activate prothrombin is greatly enhanced by negatively charged phospholipids plus Ca^{2+}, but not by bovine factor V_a. The venom prothrombin activator appears to be a large protein with a molecular mass of about 300,000 daltons, which is composed of an enzymatic unit of 57,000 daltons and a cofactor part of 220,000 daltons, which by itself has no catalytic activity, but which promotes prothrombin conversion by the 57,000-dalton enzymatic unit.

The native venom activator shares many properties with the factor X_a-V_a complex. Both complexes have a high K_m for prothrombin and catalyze prothrombin activation with a high V_{max}. Using a molecular weight of 300,000 daltons for the venom activator, a k_{cat} of 750 min^{-1} can be calculated for venom-catalyzed prothrombin activation in the presence of Ca^{2+} ions. This is of the same order of magnitude as the k_{cat} of the bovine factor X_a-V_a complex, which was reported to be

373 min^{-1} (Rosing et al., 1980). The venom activator and the factor X_a-V_a complex have a low K_m for prothrombin when negatively charged phospholipids are introduced as part of the prothrombin-activating complex. This effect of phospholipid is presumably caused by the fact that the substrate (prothrombin) and the enzyme (venom activator or factor X_a-V_a complex) simultaneously bind to the phospholipid surface. Such simultaneous binding will promote the formation of the enzyme-substrate complex and result in a decrease in the K_m and an increase in the rate of prothrombin activation. It is likely that the γ-carboxyglutamic acid residues present in the venom activator will support the binding of the activator to the phospholipid surface.

The venom prothrombin activator from *Oxyuranus scutellatus* both functionally and structurally resembles the factor X_a-V_a complex. The enzymatic part of the venom activator has a molecular weight of 60,000 daltons and is composed of two disulfide-linked subunits of about 30,000 daltons. Factor X_a is also a two-chain molecule, which consists of a light chain of 15,000 and a heavy chain of 30,000 daltons, bearing the catalytic site. Two other major protein bands are observed upon SDS gel electrophoresis of the purified venom activator. These subunits, which have apparent molecular weights of 110,000 and 80,000 daltons, respectively, likely comprise the cofactor part of the venom activator and may have the same function as the subunits of factor V_a. Bovine factor V_a (M_r = 180,000) is composed of two polypeptides of about 115,000 and 73,000 daltons (Esmon, 1979). The structural features of the venom prothrombin activator lead to the hypothetical model shown in Figure 5. The data presented in this chapter

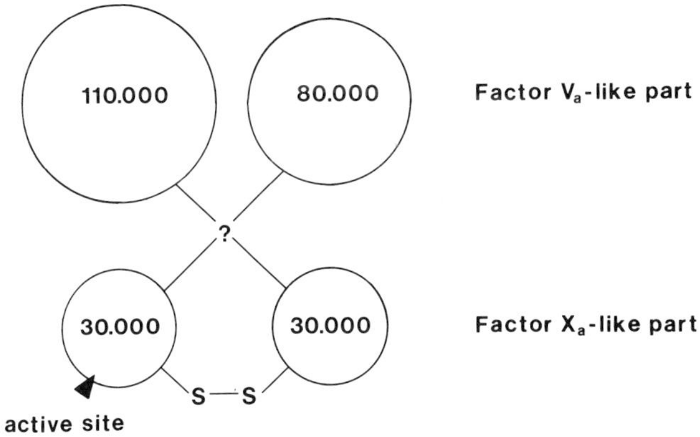

Figure 5 Hypothetical model for the prothrombin activator from the venom of *Oxyuranus scutellatus*.

indicate that the *Oxyuranus* activator consists of a factor X_a-like catalytic unit comprising two subunits of about 30,000 daltons and a factor V_a-like cofactor of approximately 220,000 daltons that consists of two polypeptide chains of about 110,000 and 80,000 daltons, respectively.

It is likely that the presence of such a complex prothrombin activator in a snake venom is not unique for *Oxyuranus scutellatus*. Studies of the coagulant activities of a large number of Australian and Papuan elapid snake venoms (Hermann et al., 1972; Marshall and Herrmann, 1983) indicate that the venoms of *Demansia nuchalis affinis* (dugite), *Parademansia microleptidotus* and *Pseudonaja textilis* (common brown snake) also contain a prothrombin activator that may carry its own factor V_a-like cofactor. This can be inferred from the observation that these venoms have the same coagulant activity in normal plasma and in factor V-deficient plasma. At present, we are investigating whether the prothrombin activators of Australian elapid venoms can indeed be divided into a Notechis-like group that is stimulated by phospholipids and factor V_a and an *Oxyuranus*-like group that has its own cofactor and that is only stimulated by phospholipids plus Ca^{2+}.

REFERENCES

Denson, K. W. E. (1969). Coagulant and anticoagulant actions of snake venoms. Toxicon 7:5-11.

Esmon, C. T. (1979). The subunit structure of thrombin-activated factor V. J. Biol. Chem. 254:964-973.

Herrmann, R. P., Davey, M. G., and Skidmore, P. H. (1972). The coagulation defect after envenomation by the bite of the dugite (*Demansia nuchalis affinis*), a western Australian brown snake. Med. J. Aust. 2:183-186.

Jobin, F. and Esnouf, M. P. (1966). Coagulant activity of tiger snake (*Notechis scutatus scutatus*) venom. Nature 211:873-876.

Kuwada, M. and Katayama, K. (1983). An improved method for the determination of γ-carboxyglutamic acid in proteins, bone and urine. Anal. Biochem. 131:173-179.

Laemmli, U. K. (1970). Cleavage of structural proteins during the assembly of the head of bacteriophage T4. Nature 227:680-685.

Lindhout, T., Govers-Riemslag, J. W. P., van de Waart, P., Hemker, H. C., and Rosing, J. (1982). Factor Va-factor Xa interaction. Effect of phospholipid vesicles of varying composition. Biochemistry 21:5494-5502.

Marshall, L. R. and Herrmann, R. P. (1983). Coagulant and anticoagulant actions of Australian snake venoms. Thromb. Haemostas. 50:707-711.

Owen, W. G. and Jackson, C. M. (1973). Activation of prothrombin with *Oxyuranus scutellatus scutellatus* (Taipan snake) venom. Thromb. Res. 3:705-714.

4

A Prothrombin Activator Found in *Rhabdophis tigrinus tigrinus* (Yamakagashi Snake) Venom

Takashi Morita, Hiroaki Matsumoto, and Sadaaki Iwanaga
Kyushu University, Fukuoka, Japan
Atsushi Sakai
The Japan Snake Institute, Nitta-gun, Japan

INTRODUCTION

Procoagulants and coagulants in snake venoms are useful to study activation mechanisms of blood coagulation factors (1). Several prothrombin activators have been isolated and characterized from the venoms of *Echis carinatus* (2-6), *Oxyuranus scutellatus scutellatus* (7-8), *Notechis scutatus scutatus* (9-10), and *Dispholidus typus* (11). Some of these activators have a different specificity from that of mammalian factor X_a (3-6, 11).

Rhabdophis tigrinus tigrinus (Yamakagashi snake) is one of the most abundant snakes in Japan (Fig. 1). Although it is believed to be a nonpoisonous snake, several cases of serious envenomation by this snake have been reported (12, 13). In 1976, Nahas and coworkers (14) reported that the crude venom displays the ability to activate factor X. The purpose of this chapter is to examine the enzymatic properties of coagulant in the venom extracted from the Duvernoy's gland of *Rhabdophis t. tigrinus*, and identify which mammalian coagulation factors are activated by this venom coagulant. We show here that this venom activates prothrombin but not factor X, and that the activation pathway of prothrombin to α-thrombin by the venom activator is identical to that of the prothrombin activator of *Echis carinatus* venom (4).

Figure 1 Yamakagashi snake (*Rhabdophis tigrinus tigrinus*).

MATERIALS AND METHODS

Sephacryl S-200, Sephadex G-150, QAE Sephadex A-50, and marker proteins for molecular weight determinations were obtained from Pharmacia Fine Chemicals, Uppsala, Sweden. Sodium dodecylsulfate (SDS), Coomassie brilliant blue R-250, and acrylamide were obtained from Nakarai Chemical Co., Kyoto, Japan. *t*-Butoxycarbonyl-Val-Pro-Arg-*p*-nitroanilide (Boc-Val-Pro-Arg-pNA) was a product of the Protein Research Foundation, Minoh, Osaka, Japan. Hydroxylapatite (Bio-Gel HTP) was a product of Bio-Rad Laboratories, Richmond, California. Hirudin (Grade IV, 1000 units/mg protein) was procured from Sigma Chemical Co., St. Louis.

Extraction of the Venom from *Rhabdophis t. tigrinus*

The venom was collected from the Duvernoy's gland of *Rhabdophis t. tigrinus* (Fig. 2). This gland was excised and minced, and the venom was extracted by water. The insoluble material was removed by centrifugation, and the supernatant was then lyophilized.

Proteins. Human prothrombin (15), prothrombin (16), prethrombin 1 (17), factor X (16), factor IX (16), protein C (16), and α-thrombin (18) from bovine plasma were purified by previously published methods. Bovine fibrinogen (type 2, 98% clottable protein) was a product of Daiichi Pure Chemicals Co., Ltd., Tokyo, Japan.

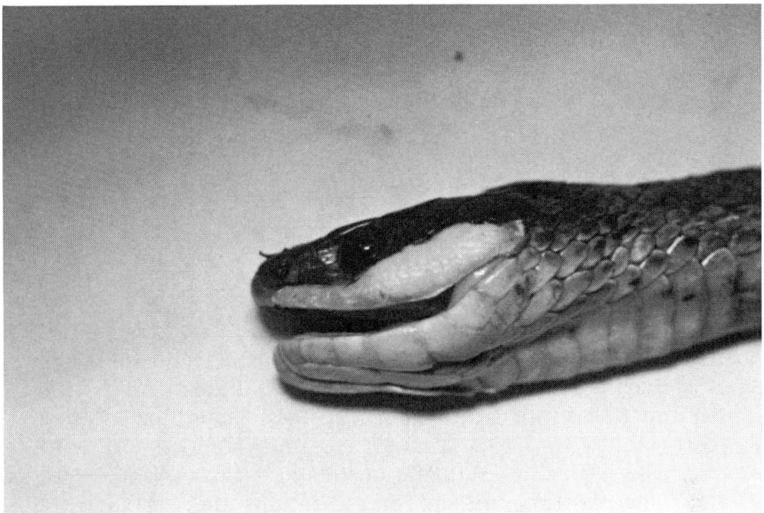

Figure 2 Duvernoy's gland (Dg) of *Rhabdophis t. tigrinus*.

Protein Concentration

For the determination of protein concentration, the following $E_{280}^{1\%}$ values were used: 13.8 for human prothrombin, 15.5 for bovine prothrombin, 16.4 for prethrombin 1, 12.4 for factor X, 12.0 for factor IX, 13.7 for protein C, and 19.5 for α-thrombin.

Electrophoresis

Proteins were analyzed by polyacrylamide gel electrophoresis (PAGE) in the presence of sodium dodecylsulfate (SDS) (19). The gels were stained with Coomassie brilliant blue R-250.

Clotting Assay of α-Thrombin

Clotting assay was performed using 2% fibrinogen as described previously (20).

Assay of Prothrombin Activator

Prothrombin activator was followed by measuring the generation of amidolytic activity towards the chromogenic substrate (Boc-Val-Pro-Arg-*p*-nitroanilide). A reaction mixture containing 10 μl of sample and 100 μl of 1.1 μM human prothrombin (79 μg/ml) was preincubated at 37°C for 10 minutes. Then, 25 μl of 10 mM EDTA, 10 μl of 5 mM chromogenic substrate, and 105 μl of 0.05 M Tris-HCl/0.1 M NaCl (pH 8.0) were added, and the mixture was incubated for a further 5 minutes. Finally, 0.8 ml of 0.6 M acetic acid was added to terminate the reaction, and the absorbance at 405 nm was read in a Hitachi 220A spectrophotometer. One unit of the enzyme activity was defined as 1 μmol of *p*-nitroaniline liberated per minute.

Separation of Prothrombin Activator from the Extracts of Duvernoy's Gland

A 45-mg sample of the dried crude venom was dissolved in 2.0 ml of 0.05 M Tris-HCl/0.1 M NaCl (pH 8.0) and allowed to stand for 30 minutes. After centrifugation at 15,000 rpm for 30 minutes at 4°C, the clear supernatant was applied to a column of Sephacryl S-200 (1.6 × 117 cm) equilibrated with 0.05 M Tris-HCl/0.1 M NaCl (pH 8.0). Elution was performed with the equilibration buffer at a flow rate of 13 ml/hr at 4°C. The prothrombin activator fractions (no. 44-49), indicated by the dashed line in Figure 3, were pooled and concentrated to 7.5 ml by ultrafiltration and applied to a column of Sephadex G-150 (2.1 × 147 cm) equilibrated with 0.02 M Tris-HCl/1.0 M NaCl (pH 8.0), and eluted with the equilibration buffer at a flow rate of 11 ml/hr. The activator fractions (elution volume 174 to 203 ml) were pooled and dialyzed overnight against 0.05 M Tris-HCl/0.1 M NaCl (pH 8.0). The dialyzed solution was chromatographed on a column of QAE-Sephadex A-50 (1.4 × 34 cm) equilibrated with 0.05 M Tris-HCl/0.1 M NaCl (pH 8.0). Prothrombin activator was eluted with a linear concentration gradient of NaCl (0.1 to 0.6 M in 100 ml per chamber) in 0.05 M Tris-HCl (pH 8.0) at a flow rate of 20 ml/hr. The activator fractions (no. 48-54) in Figure 4 were pooled and dialyzed overnight against 0.01 M sodium phosphate buffer (pH 6.8). The dialyzed solution was finally applied to a column of hydroxylapatite (1 × 15 cm). After washing the column with 0.05 M sodium phosphate buffer (pH 6.8), a prothrombin activator was eluted in a stepwise fashion with 0.1 M sodium phosphate buffer (pH 6.8), as shown in Figure 5. A prothrombin activator fraction (no. 18-21) was pooled and concentrated to 0.9 ml by ultrafiltration.

RESULTS

Effects of the Venom on Blood Coagulation Factors

When 10 μg of the crude venom was added to 0.3 ml of bovine citrate plasma, an average clotting time was 50 seconds. However, the clotting time was prolonged to more than 200 seconds in the presence of 1.6 mM EDTA. The venom had no fibrinogen clotting activity and no direct fibrinolytic activity.

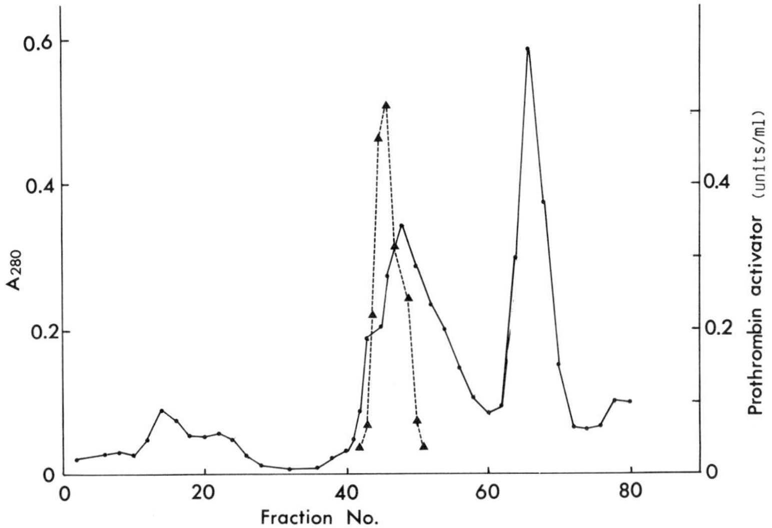

Figure 3 Sephacryl S-200 elution profile of *Rhabdophis t. tigrinus* venom. The conditions are described in the text.

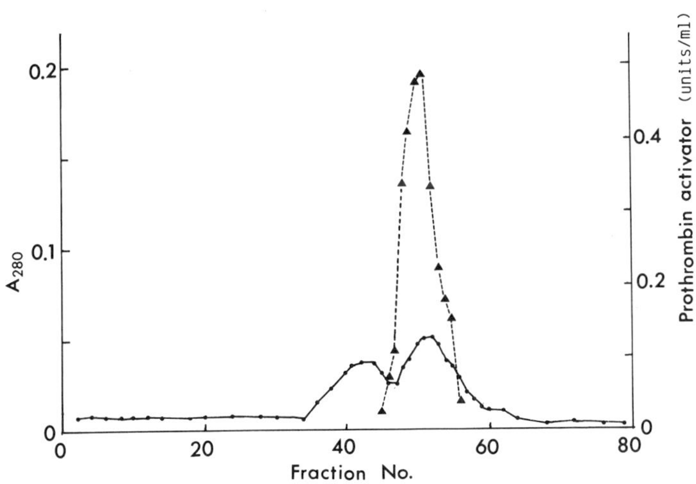

Figure 4 Chromatography of *Rhabdophis t. tigrinus* venom activator on QAE-Sephadex A-50. The conditions are described in the text.

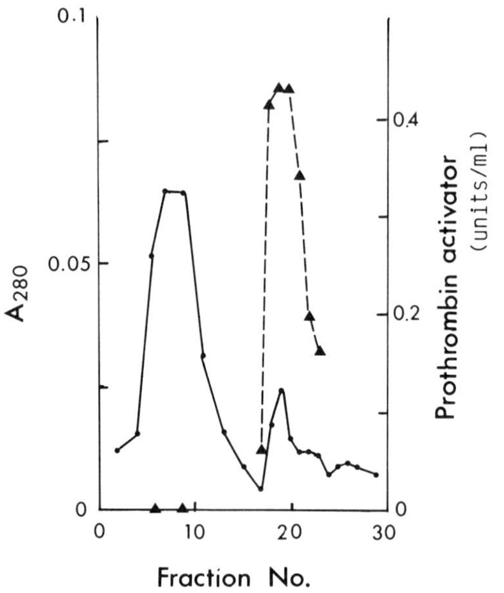

Figure 5 Chromatography of *Rhabdophis t. tigrinus* venom activator on hydroxyapatite. The conditions are described in the text.

We also examined its ability to activate human and bovine prothrombins and factor X, factor IX, and protein C. Among these factors, only prothrombin was converted to α-thrombin, and no molecular change of factor X, factor IX, and protein C was observed on the SDS-PAGE analysis when each purified coagulation factor was treated with the crude venom (1/60, w/w) for 30 minutes at 37°C.

Activation of Prothrombin

Figure 6A shows the activation of human prothrombin by the crude venom of *Rhabdophis t. tigrinus* in a weight ratio of 1:60. The conversion of human prothrombin to α-thrombin was very rapid, even in the absence of Ca^{2+}, phospholipids, and factor V. The maximum activity was observed after 30 minutes when human prothrombin was used. However, the thrombin clotting activity generated from bovine prothrombin was less than 10% of that of human prothrombin. We followed the activation process of human prothrombin by SDS-PAGE at

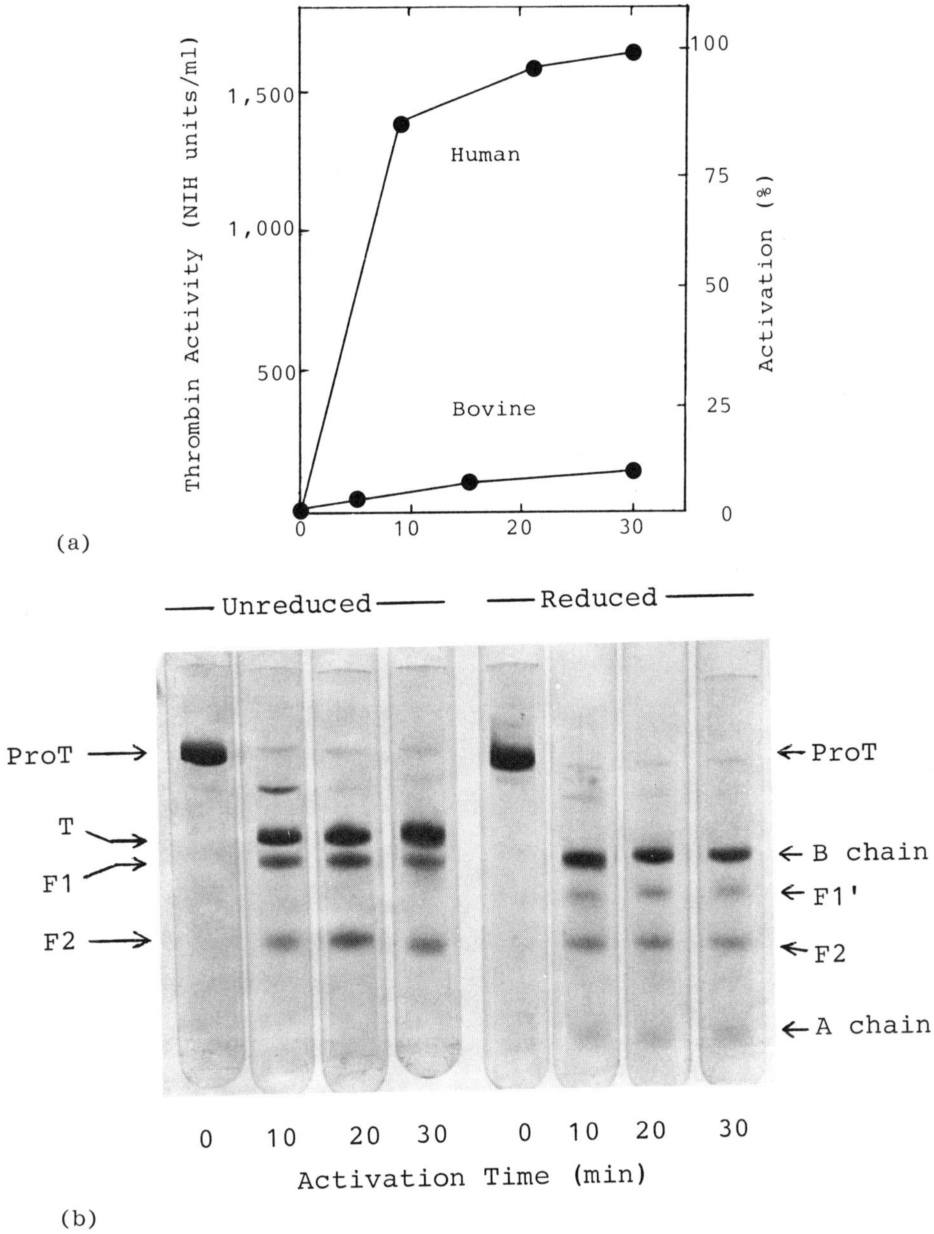

Figure 6 Time course of prothrombin activation by the crude venom of *Rhabdophis t. tigrinus*. (A) A mixture of 200 µl of prothrombin (1.3 mg/ml in 0.05 M Tris-HCl/0.1 M NaCl, pH 7.5) and 20 µl of crude venom (0.22 µg/µl) was incubated at 37°C. At the intervals shown, aliquots of the mixture were taken for assaying the fibrin clot formation. (B) At the intervals shown, aliquots of the mixture were subjected to SDS-PAGE. ProT = prothrombin; T = thrombin; F1 = prothrombin fragment 1; F2 = prothrombin fragment 2.

the same time to determine the activation mechanism of this venom. The result is shown in Figure 6B. After 20 minutes, human prothrombin was converted to α-thrombin, releasing the NH_2-terminal fragment 1 and fragment 2. The reduced SDS-PAGE patterns showed that α-thrombin B-chain was generated after 10 minutes, and fragment 1 was converted to fragment 1', which is a nicked form produced by α-thrombin (3, 21). Since a very low clotting activity was observed in the activation of bovine prothrombin, we compared the molecular change of bovine prothrombin with that of human prothrombin in the course of the activation using SDS-PAGE, as shown in Figure 7. The gel on the right is the SDS-PAGE of α-thrombin as a reference. Although human prothrombin generated α-thrombin after 15 minutes of incubation, bovine prothrombin was converted to prothrombin derivative with a molecular weight of 52,000 daltons. The reduced SDS-PAGE patterns clearly demonstrated this derivative as meizothrombin (free from the NH_2-terminal fragment 1). The venom activator also converted bovine

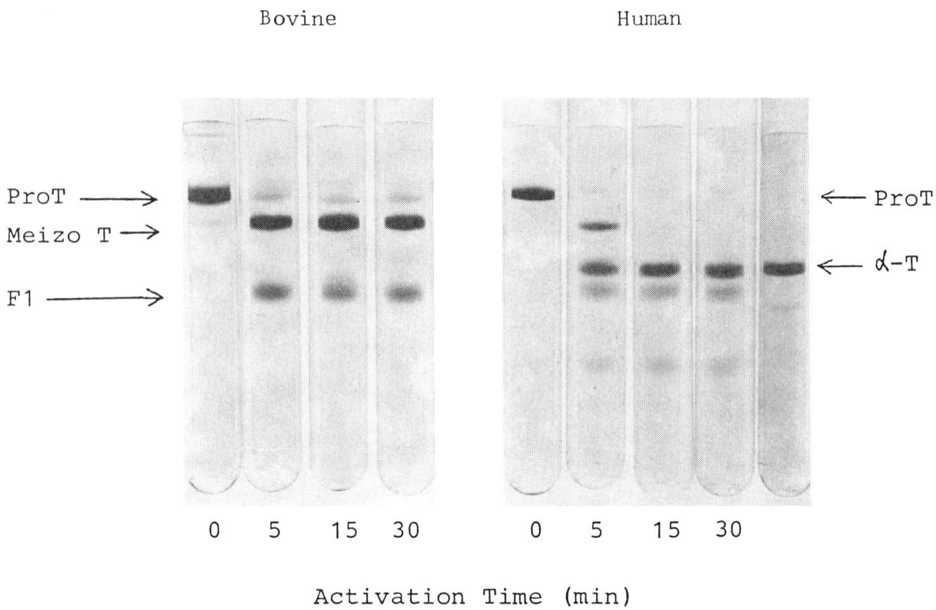

Figure 7 SDS-PAGE patterns of the reaction products derived from bovine and human prothrombins by the crude venom of *Rhabdophis t. tigrinus*. The same conditions as those in Figure 6 were used, and aliquots of the reaction mixture were subjected to SDS-PAGE without reduction. ProT = prothrombin; Meizo T = meizothrombin (free from fragment 1); F1 = fragment 1; α-T = α-thrombin. The gel on the right is bovine α-thrombin.

prethrombin 1 to meizothrombin after 5 minutes under the same conditions as used for the bovine prothrombin activation (data not shown).

We also studied the activation of bovine prothrombin in the presence of hirudin. Figure 8 demonstrates that the venom activator converted prothrombin to meizothrombin with a molecular weight of 72,000 daltons. This meizothrombin has been identified as an active prothrombin derivative produced by the prothrombin activator of *Echis carinatus* venom (4). The reduced SDS-PAGE pattern shows that the meizothrombin consists of two chains. The higher molecular weight chain consisted of fragment 1, fragment 2, and thrombin A chain (designated F1·F2·A chain). The other chain was thrombin B chain (Fig. 5).

Fractionation of a Prothrombin Activator from the Crude Venom

A prothrombin activator was partially purified from the crude venom of *Rhabdophis t. tigrinus* by a combination of several procedures involving gel filtration on Sephacryl S-200 (Fig. 3) and Sephadex G-150, and column chromatographies on QAE Sephadex A-50 (Fig. 4) and hydroxylapatite (Fig. 5). The final preparation contained three proteins with molecular weights of 84,000, 58,000 and 52,000, respectively, estimated on SDS-PAGE analysis. The results of purification procedures are summarized in Table 1.

Table 1 Fractionation of a Prothrombin Activator from *Rhabdophis t. tigrinus* Venom

Step	Total protein[a] (mg)	Total activity[b] (units)	Specific activity (units/mg)	Recovery (%)
1. Crude venom	45.0	43.2	0.96	--
2. Sephacryl S-200	4.59	143.2	13.2	100
3. Sephadex G-150	3.74	89.5	23.9	63
4. QAE-Sephadex	1.22	45.5	37.3	32
5. Hydroxylapatite	0.14	21.8	155.7	15

[a]Protein concentration was determined from the absorbance at 280 nm assuming $E_{280}^{1\%} = 10.0$.

[b]One unit of a prothrombin activator catalyzes the formation of 1 μmol of α-thrombin per minute.

DISCUSSION

Although Nahas and coworkers reported that *Rhabdophis t. tigrinus* venom activates factor X (14), our present results indicate that this venom contains a prothrombin activator, but not factor X activator. As shown in Figures 7 and 8, a prothrombin activator in the venom of *Rhabdophis t. tigrinus* cleaves only at the site of the Arg_{320}-Ile_{321} bond in the prothrombin molecule. The resulting active meizothrombin is rapidly converted to α-thrombin autocatalytically. This activation process of prothrombin is identical to that catalyzed by a prothrombin activator of *Echis carinatus* venom (4, 22). These results suggest that the coagulant found in the venom of *Rhabdophis t. tigrinus* is identified as a coagulant similar to a prothrombin activator present in the venom of *Echis carinatus*.

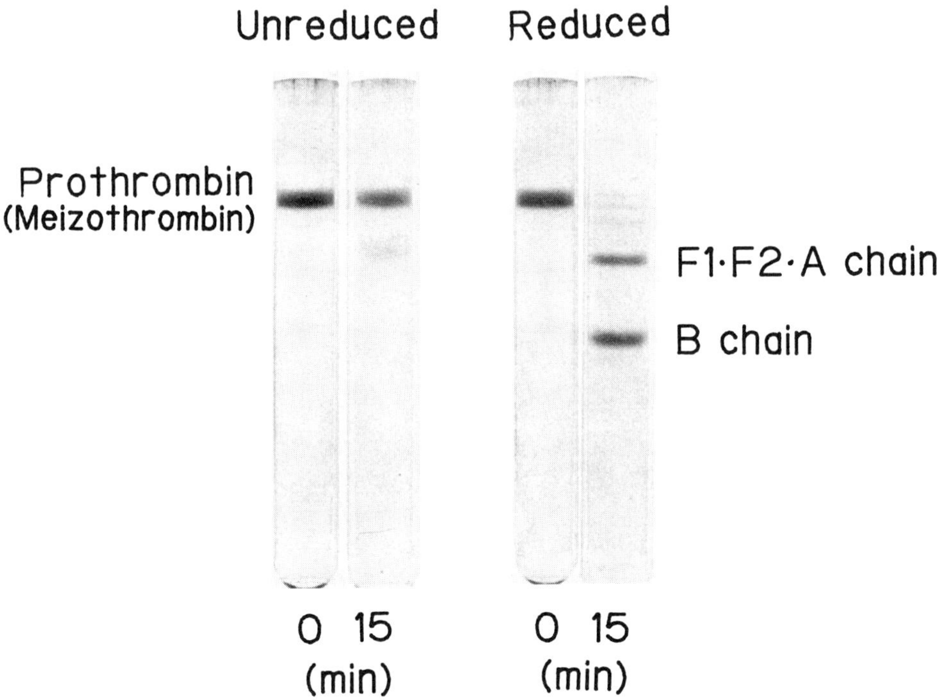

Figure 8 Time course of prothrombin activation by a partially purified venom coagulant in the presence of hirudin. Bovine prothrombin (20.6 μg/10 μl of 0.05 M Tris-HCl/0.1 m NaCl, pH 7.5) was activated with a partially purified coagulant (0.0075 absorbance unit) in the presence of hirudin (50 antithrombin units) in a total volume of 110 μl. After 15 minutes, aliquots were taken and subjected to SDS-PAGE.

The prothrombin activation by *Rhabdophis t. tigrinus* venom is not dependent on Ca^{2+} ions, and the activation rates of prothrombin and prethrombin 1 are essentially identical. This indicates that *Rhabdophis t. tigrinus* venom could be a useful tool in the assay of PIVKA-II (descarboxy prothrombin), which has no functional Ca^{2+} binding site.

We have partially purified a prothrombin activator from the extract of Duvernoy's gland. The final preparation contains at least three components as detected on SDS-PAGE. The molecular weight of the prothrombin activator is estimated to be 170,000 daltons by molecular sieving on a Sephadex G-150 column. Further investigation is required for the biochemical characterization of this coagulant.

The starting material is the extract from parotid-like gland (Duvernoy's gland) of *Rhabdophis t. tigrinus*. Its enlarged posterior maxillary teeth possess no groove but have a sharp posterior cutting edge, and the duct of the Duvernoy's gland opens in the maxillary region immediately anterior to the rearmost teeth. Therefore, the prothrombin activator of the Duvernoy's gland might be one of active components that is responsible for abnormal coagulation due to human envenomation by this snake.

CONCLUSION

A prothrombin activator is found in *Rhabdophis tigrinus tigrinus* (Yamakagashi snake) venom collected from the Duvernoy's gland. The experimental results on the activation of human and bovine prothrombins in the absence and presence of hirudin by the venom indicate that the prothrombin activator of this venom cleaves only at the site of Arg_{320}-Ile_{321} peptide bond linking the A chain and the B chain of α-thrombin. The resulting active meizothrombin is converted to α-thrombin autocatalytically. Thus, this prothrombin activation process is identical to that catalyzed by the prothrombin activator of *Echis carinatus* venom.

REFERENCES

1. Seegers, W. H., and C. Ouyang. In *Handbook of Experimental Pharmacology*, Volume 52 (Snake Venom), C-Y. Lee (Ed.), Springer-Verlag, Berlin, 1979, p. 684.
2. Schieck, A., F. Kornalik, and E. Habermann. Naunyn-Schmiedeberg's Arch. Pharmacol. 272:402 (1972).
3. Granza, B. R., D. L. Arnson, and J. S. Finlayson. J. Biol. Chem. 250:7057 (1975).
4. Morita, T., S. Iwanaga, and T. Suzuki. J. Biochem. (Tokyo) 79:1089 (1976).

5. Morita, T., and S. Iwanaga. J. Biochem. (Tokyo) 83:559 (1979).
6. Phee, M-J., S. Morris, and D. P. Kosow. Biochemistry 21:3437 (1982).
7. Owen, W. T. and C. M. Jackson. Thrombos. Res. 3:705 (1973).
8. Walker, F. J., W. G. Owen, and C. T. Esmon. Biochemistry 19:1020 (1980).
9. Jobin, F. and M. P. Esnouf. Nature 211:873 (1966).
10. Tans, G., J. W. P. Govers-Reimslag, J. L. M. L. van Fijn, and J. Rosing. J. Biol. Chem. 260:9366 (1985).
11. Guillin, M-C., A. Bezeaud, and D. Menache. Biochem. Biophys. Acta 537:160 (1978).
12. Mittelman, M. B. and R. C. Goris. Herpetologica 30:113 (1974).
13. Kono, H. and Y. Sawai. The Snake 7:38 (1975).
14. Nahas, L., A. S. Kamiguti, A. R. Hoge, and R. C. Goris. In *Animal, Plant, and Microbial Toxins*, Volume 1, A. Ohsaka, K. Hayashi, and Y. Sawai (Eds.). Plenum Press, New York, 1976, p. 159.
15. Miletich, J. P., G. J. Broze, and P. W. Majerus, Anal. Biochem. 105:340 (1980).
16. Hashimoto, N., T. Morita, and S. Iwanaga. J. Biochem. (Tokyo) 97:1347 (1985).
17. Morita, T., H. Nishibe, S. Iwanaga, and T. Suzuki, J. Biochem. (Tokyo) 76:1031 (1974).
18. Morita, T., J. Mizuguchi, S. Kawabata, and S. Iwanaga. J. Biochem. (Tokyo) 99:561 (1986).
19. Weber, K. and M. Osbern. J. Biol. Chem. 244:4406 (1969).
20. Magnusson, S. In *Methods in Enzymology*, Volume 19, Perlmann, G. E. and Lorand, L. (Eds.), Academic Press, New York, 1970, p. 157.
21. Aronson, D. L., A. P. Ball, R. B. Franza, T. E. Hugli, and J. W. Fenton, II. Thrombos. Res. 20:239 (1980).
22. Downing, M. B., R. J. Butkowski, M. M. Clark, and K. G. Mann. J. Biol. Chem. 250:8897 (1975).

5

Thrombin-like Snake Venom Enzymes

Kurt F. Stocker and Jürg Meier
Pentapharm AG, Basel, Switzerland

INTRODUCTION

The venoms of snakes belonging to the family of *Viperidae* contain a variety of enzymes, especially hydrolases, that affect lipids, polysaccharides, nucleotides, and proteins in tissues and body fluids of prey animals (1).

A number of proteinases that act on particular proteins of the kallikrein, complement, coagulation, and fibrinolysis system in vertebrates' plasma have been identified. Thus, numerous snake venom proteinases exert a promoting or inhibiting action on blood coagulation. Enzymes that activate or inactivate zymogens, cofactors or inhibitors of the intrinsic prothrombin activation pathway, and proteinases that exert one or several of the multiple functions of thrombin have been detected in snake venoms. Categories of venom enzymes acting on physiologic thrombin substrates are represented by typical examples in Table 1.

SNAKE VENOM ENZYMES ACTING ON PHYSIOLOGIC THROMBIN SUBSTRATES

Actions on Cells

Platelets

Contact of platelets with thrombin induces a contractile response, a release reaction, and aggregation. Similar effects have been observed

Table 1 Thrombin Substrates Affected by Snake Venom Enzymes

Substrate	Product	Thrombin	Thrombocytin (*B. atrox*)	RVV-V	RVV-X	Protac[a] (*A. contortr.*)	Batroxobin (*B. atrox*)	Venzyme (*A. contortr.*)	Gabonase (*B. gabon.*)	Crotalase (*C. adam.*)
Fibrinogen Aα-chain	Fibrin I	+	-	-	-	-	+	(+)	+	+
Fibrinogen Bβ-chain	Fibrin II	+	-	-	-	-	-	+	+	-
V	V_a-V_i	+	+	+	-	-	-			-
VIII	VIII_a-VIII_i	+	+	-	-	-	-			-
II	Frag. 1.2[1]									
	Prethrombin 2	+	-	-	-	-	-	-		+
XIII	XIII_a	+	+	-	-	-	-	-	+	-
C	C_a	+	-	-	+	+	-	-		
Platelets	Aggregation release	+	+	-	-	-	-	+		-
Endothelium	TPA release	+					+			
	PGI_2 release	+	+				-			

[a]Recent findings indicate that Protac might not be an enzyme.
In brackets = weak effect.

with crude preparations of the fibrinogen coagulant proteinase batroxobin from *Bothrops atrox* venom, whereas highly purified batroxobin did not affect platelets. This phenomenon could be explained by the identification of a contaminating venom proteinase, thrombocytin (2, 3). Thrombocytin exerts a thrombin-like action on platelets and thrombin-sensitive plasma clotting factors V, VIII, and XIII, but unlike thrombin, neither converts fibrinogen into fibrin nor activates protein C.

Thrombocytin is a serine proteinase with a molecular weight of 36,000 daltons. It is a single-chain glycopeptide with a carbohydrate content of 5.6%. A comparison of properties and activities of thrombin, thrombocytin and batroxobin is presented in Table 2.

Crotalocytin, isolated from *Crotalus horridus* venom, is a serine proteinase. It has a molecular weight of 55,000 daltons and acts like thrombocytin on platelets (4).

Our observations that in presence of platelet-rich human plasma and the specific thrombin inhibitor hirudin, crude venoms from *Bothrops moojeni, Bothrops neuwiedi, Bothrops jararaca, Crotalus durissus terrificus, Sistrurus miliarius*, and *Echis carinatus* produce a retracting clot, could indicate the presence of venom enzymes with similar

Table 2 Thrombocytin: Comparison of Activities

Activity	Thrombin	Thrombocytin	Batroxobin
Platelet aggregation and release	+	+	-
Clot retraction	+	+	-
Clotting of fibrinogen	+	Weak	+
Activation of			
Factor XIII	+	+	-
Factor VIII	+	Weak	-
Factor X	-	Weak	-
Factor V	+	+	-
Protein C	+	-	-
Cleavage of prothrombin to intermediate I	+	+	-
Stimulation of PGI_2 release	+	+	-
Inhibition by			
DFP	+	+	+
Antithrombin III-heparin	+	+	-
Hirudin	+	-	-
α_1-antitrypsin	-	-	-

properties to those of thrombocytin (5). However, the presence in these venoms of nonenzymatic platelet-activating proteins like convulxin found in *C. durissus cascavella* (6), aggregoserpentin from *Trimeresurus mucrosquamatus* (7), or Botrocetin, which is present in numerous viperidae species (8), should also be considered as an initiator of this contractile response.

Endothelium

Plasminogen activator release. Perfusion of the isolated pig ear with batroxobin-containing Tyrode's solution, according to the procedure of Markwardt and Klöcking (9), causes release of plasminogen activator as measured by means of a fibrin plate technique described by Klöcking and associates (10). The liberated plasminogen activator was inhibited by monoclonal antibodies directed against vascular plasminogen activator and was not inhibited by an antibody preparation directed against urokinase (Markwardt, personal communication, 1985).

Budzynski (11) observed a stimulation of plasminogen activator secretion from bovine aortic cells and human umbilical cord cells by ancrod and batroxobin, and Dosekun and co-workers (12) reported on a correction of the reduced plasma level of vascular plasminogen activator during ancrod treatment in patients with systemic lupus erythematosus.

Prostacyclin release. As demonstrated by Czervionke and coworkers (13), the addition of thrombin to cultured endothelial cells from human umbilical vein causes a strongly increased release of the platelet aggregation inhibitor prostaglandin I_2 (PGI_2), as followed by measuring the stable end product 6-keto PGF, using a radioimmunoassay. If batroxobin, instead of thrombin, was given to the culture, PGI_2 release did not increase, whereas thrombocytin exerted a similar stimulating effect as that of thrombin.

Action on Factor V

The venom of *Vipera russelli*, besides being a highly potent activator of factor X (RVV-X) (14), contains a serine proteinase (RVV-V), which by cleavage of a single peptide bond, converts factor V into factor V_a (15). No other protein substrate for this enzyme has yet been identified.

In contrast to thrombin and thrombocytin (which also activates factor V), RVV-V is not inhibited by antithrombin III-heparin complex.

Factor V, after activation by either thrombin, RVV-V or thrombocytin, loses its activity faster than does the native molecule (3, 15, 16).

Action on Factor VIII

Incubation of bovine factor VIII with thrombin causes a significant increase in factor VIII clotting activity. Thrombocytin from *B. atrox* venom exerts a similar activity (3), whereas the related enzyme crotalocytin from *C. horridus* has not been investigated for this effect.

Action on Prothrombin

The prothrombin molecule contains a thrombin-susceptible peptide bond. Its cleavage leads to the liberation of the N-terminal fragment 1.2, which contains the calcium binding sites of prothrombin, and to the formation of prethrombin 2 (17).

Pirkle and colleagues (18) investigated the proteolytic action of five purified fibrinogen coagulant snake venom enzymes on prothrombin. Crotalase from *C. adamanteus* and the coagulant enzyme from *A. c. contortrix* catalyzed the cleavage of the thrombin-susceptible bond in prothrombin, whereas the enzymes from *A. rhodostoma, B. atrox,* and *B. moojeni* had no proteolytic activity on prothrombin.

Teng and Seegers (19) reported the cleavage of fragment 1.2 from prothrombin by acutin from *A. acutus* venom.

Incubation of prothrombin with thrombocytin from *B. atrox* venom did not generate any coagulant activity. However, as shown by SDS gel electrophoresis, two split products were formed, the molecular weights of which corresponded to fragment 1 and prethrombin 1 (3).

Action on Factor XIII

Treatment of factor XIII with thrombin or thrombocytin causes proteolytic cleavage of the α-chain and release of a peptide with a molecular weight of approximately 4000 daltons. The resulting activated factor XIII causes fibrin cross-linkage. A fibrin clot formed by batroxobin is soluble in 2% acetic acid, but in presence of factor XIII and either thrombocytin or thrombin, the clot is rendered insoluble (3).

Ancrod and batroxobin have been reported to cause a partial activation of factor XIII (20). Fibrin produced in the presence of these enzymes contained γ-γ-dimers, whereas the incorporation of dansylcadaverin into casein could not be catalyzed by these preparations (20). There is some doubt about the purity grade of the enzyme preparations used in this study since batroxobin, free from thrombocytin contamination by affinity chromatography on immobilized heparin, produces clots from prothrombin-free plasma in the presence of factor XIII and calcium, which remain soluble in acetic acid 2%, monochloroacetic acid 1%, or urea 5 M, whereas thrombin or crude *B. atrox* venom produces insoluble clots. Moreover, according to Okada and associates (21), batroxobin-treated factor XIII does not catalyze the covalent binding of fibronectin to α-chains of fibrin gels. Factor XIII is activated by the fibrinogen coagulant enzyme gabonase from *B. gabonica* (22).

Action on Protein C

Thrombin catalyzes a limited proteolysis of protein C, and thereby causes the conversion of the inactive precursor into the proteinase protein C_a. In presence of calcium ions, this process is strongly accelerated by the endothelium protein thrombomodulin, whereas in the absence of thrombomodulin, calcium inhibits the thrombin-mediated activation of protein C (23, 24). A slow protein C activation is also achieved by incubation of the zymogen with the factor X-activating proteinase (RVV-X) from Russell's viper venom (14, 23).

In our experiments, we succeeded in isolating a fast-acting protein C activator from *A. c. contortrix* venom and in detecting similar activities in the venoms of *A. c. pictigaster*, *A. c. mokeson*, *A. p. piscivorus*, *A. p. leucostoma*, *A. bilineatus*, *B. moojeni*, *B. pradoi*, *V. lebetina*, and *Cerastes cerastes*.

Independent from our experiments, Exner and co-workers (25) used a novel immunochromometric method to identify a potent protein C activator in the venoms of *A. contortrix* and *A. piscivorus*.

Some properties of our protein C activator preparation (Protac*) are presented in Table 3. The velocity of protein C activation induced by thrombin, RVV-X, and Protac is compared in Table 4.

Table 3 Protac and Thrombin: Comparison of Activities

Activity	Thrombin	Protac
Activation of protein C		
Ca^{2+} absent	Slow	Fast
Ca^{2+} present	Slower	Fast
Accelerated by thrombomodulin	+	-
Coagulation of fibrinogen	+	-
Fibrinolysis	-	-
Inhibition by		
Antithrombin III	+	-
Hirudin	+	-

*Pentapharm Ltd., CH-4002 Basel, Switzerland.

Table 4 Activation of Human Protein C by α-Thrombin, RVV-X, and Protac

Activator	PC_a potency[a] (U/ml sample)	Required activation time (min)
α-Thrombin (human), 4 U/ml	12.9	65
RVV-X, 25 μg/ml	5.1	240
Protac, 25 μg/ml	12.2	4

[a]1 U of protein C_a = amount that at pH 8.4, $\gamma/2 = 0.3$, and 37°C converts 1 μM of 2AcOH•H-D-Pro-Pro-Arg-pNA per minute.
Human protein C and α-thrombin were prepared by W. Kisiel, Albuquerque, New Mexico, and J. Fenton, Albany, New York, respectively.

Protac is not inhibited by diisopropylfluorophosphate (DFP), iodoacetamide, or EDTA. It is stable during 10 minutes at 70°C and does not lose its activity when stored for 24 hours at pH 2. In view of these properties, some doubt arises about the enzyme nature of Protac.

The use of Protac for the quantitative determination of protein C in human plasma has significant advantages over functional assay procedures using thrombin as a protein C activator. Whereas thrombin itself exerts clotting activity and catalyzes the hydrolysis of chromogenic substrates and, therefore, affects such test systems, Protac neither acts as a coagulant nor causes chromophore release from synthetic substrates. Protac may, therefore, remain in the test mixture after achieved activation (Table 5), whereas thrombin has to be quenched or removed. Moreover, the fast activation of protein C by Protac may be carried out in dilutions where plasma protein C inhibitor (PCI) remains virtually ineffective, while for the slow thrombin activation process, protein C has to be separated from PCI (26, 27).

Action on Fibrinogen

Properties of Fibrinogen Coagulant Snake Venom Proteinases

The term "thrombin-like" is currently attributed to those venom enzymes that exert the most prominent thrombin action: the limited proteolysis of fibrinogen that leads to fibrin formation.

Among the enzymes listed in Table 1, batroxobin and the fibrinogen coagulant proteinases from *Agkistrodon contortrix* and *Bitis gabonica* exemplify three types of fibrinogen coagulant snake venom enzymes that split fibrinopeptide A or B or both A and B, respectively,

Table 5 Principles of Protein C Determination

Clotting method	Chromogenic method
Pipette into test tube Plasma (diluted) Protac APTT reagent	Pipette into cuvette Plasma (diluted) Protac
Add $CaCl_2$	Add chromogenic substrate
Measure clotting time	Record absorbancy
Calculate % of normal	Calculate enzyme units (U)

from fibrinogen and thereby induce fibrin clot formation. It is this kind of enzyme that has found the widest practical use and which, therefore, has become the best characterized.

Table 6 lists those fibrinogen coagulant proteinases which hitherto have been isolated from crude venom and of which physicochemical and biochemical properties have been studied. Whereas all these enzymes can convert fibrinogen into fibrin, a number of species- or subspecies-dependent characteristic differences in physical, chemical, and immunological properties can be identified (5, 28, 29).

In addition, distinct enzymatic properties with regard to substrate specificity may be observed in many enzymes. Thus, the specificity of ancrod and batroxobin for synthetic chromogenic substrates is quite different (Table 7) and the two enzymes, upon prolonged action, split different fragments from the α-chain of fibrin (30).

A different susceptibility of fibrinogen coagulant proteinases to thrombin inhibitors is also known. The enzyme from *A. contortrix* is inhibited by antithrombin III-heparin, whereas ancrod and batroxobin remain unaffected. A further striking example is the enzyme from *T. flavoviridis*, which is inhibited by hirudin (31). No other fibrinogen coagulant venom enzyme is known to be affected by this highly specific thrombin inhibitor.

The majority of these enzymes have been characterized as serine proteinases consisting of a single peptide chain and a carbohydrate moiety. Their molecular weights range between 19,500 (Defibrizym from *C. horridus*) and 65,000 to 70,000 (fibrinogen coagulant proteinase from *T. flavoviridis*) daltons (31, 32).

The interaction of fibrinogen coagulant enzymes, fibrinogen, and serum proteinase inhibitors depends on a striking species-specificity (5, 28, 33). For example, the enzyme from *T. flavoviridis* splits fibrinopeptide A almost exclusively from rabbit fibrinogen, whereas human, guinea pig, monkey, dog, and rat fibrinogen are virtually unaffected

Table 6 Fibrinogen Coagulant Snake Venom Proteinases

Species	Enzyme names	Molecular weight (daltons)	FP release	Inhibition by DFP	Inhibition by AT III + heparin	References
A. acutus	Acutin	33,500		+	-	62, 63
A. blomhoffi		36,000	A		-	64
A. caliginosus		a) 34,000-37,000 b) 42,000-43,000				65 65
A. contortrix	Venzyme	68,000	B (A)	+	+	66-68
A. halys		43,000	B (A)	+		69, 70
A. rhodostoma	Ancrod, Arvin	35,000	A	+	(+)	46, 71
B. asper	Asperase	30,000	A	+	-	72
B. atrox	Batroxobin, Reptilase-R	42,000	A	+	-	45, 73
B. moojeni	Batroxobin, Defibrase	35,800	A	+	-	45, 73
B. gabonica	Gabonase	40,000	A,B		-	22
C. adamanteus	Crotalase	32,700	A	+		74
C. horridus	Defibrizyme	19, 500	A		(+)	32
L. muta	Clotase	36,300	A	+		75
T. flavoviridis		65,000-70,000	A			31
T. gramineus		29,500	A		-	76
T. okinavensis		35,000	A			77

In brackets = weak effect.

Table 7 Hydrolysis of Synthetic Chromogenic Substrates by Ancrod (Arvin) and Batroxobin (Defibrase)

Substrate	mU per AU[a]	mU per BU[a]
2 AcOH·H-D-Ph'Gly-Phe-Arg-pNA	0.5	4.9
2 AcOH·H-D-CHA-Phe-Arg-pNA	0.4	14.4
2 AcOH·H-D-Pro-Pro-Arg-pNA	0.3	--
Bz-Pro-Phe-Arg-pNA·AcOH	0.2	2.2
2 AcOH·H-D-CHG-Ala-Arg-pNA	0.2	4.2
Tos-Gly-Pro-Arg-pNA	0.2	0.4
Tos-Gly-Pro-Lys-pNA	0	0
CH_3OCO-D-CHG-Gly-Arg-pNA·AcOH	0	--
2 AcOH·H-D-Val-CHA-Arg-pNA	0.4	7.7
AU per BU[b]	0.59	
BU per AU[b]	1.75	

[a]AU = ancrod unit; BU = batroxobin unit.
[b]Fibrinogen clotting assay.
Ph'Gly = phenylglycyl; CHA = cyclohexylalanyl-; CHG = cyclohexylglycyl-; pNA = p-nitroanilide.

by this enzyme (31). In turn, batroxobin from *B. atrox* barely coagulates rabbit fibrinogen, while human, dog, rat, cow, and sheep fibrinogen are good substrates for this enzyme (33).

The protease inhibitors contained in rat serum exert a strong inhibitory capacity toward ancrod and toward batroxobin originating from either *B. atrox* or *B. moojeri*. Unlike rat serum, human serum inhibits these enzymes in increasing order: batroxobin "moojeni," ancrod, and batroxobin "atrox" (28).

Crotalase, the fibrinogen coagulant enzyme from *C. adamanteus*, in addition to its proteolytic action on fibrinogen and prothrombin, exerts a kallikrein-like action, forming split products from high molecular kininogens that exert depressor activity similar to an equivalent amount of bradykinin (34). As observed in our experiments, ancrod and batroxobin do not show any kallikrein-like biological activity.

The specificity of fibrinogen coagulant enzymes and their species-dependent interaction with fibrinogen and serum proteinase inhibitors strongly determine their pharmacologic properties and their potential practical applications.

Practical Application of Fibrinogen Coagulant Snake Venom Enzymes

Fibrinogen coagulant venom proteinases from *A. acutus*, *A. contortrix*, *A. rhodostoma*, *B. atrox*, *B. moojeni*, and probably some less well known enzymes have found a practical application either as in vitro diagnostic agents or as pharmaceutical preparations. The batroxobin preparations Reptilase-Reagent from *B. atrox* and Defibrase from *B. moojeni* are examples of such applications.

Batroxobin. The Linnean species *Bothrops atrox* has been subclassified according to morphologic and immunologic criteria into *B. atrox*, *B. asper*, *B. marajoensis*, *B. moojeni*, and *B. pradoi* (35). Whether these are true species or rather subspecies has yet to be determined. Leloup (personal communication, 1985) succeeded in cross-breeding *B. atrox* and *B. moojeni*, as well as *B. atrox* and *B. asper*, and received fertile descendants. The venom of all five "species" does immunologically cross-react despite some quantitative differences in the antigen-antibody interaction. The five varieties exert a comparable coagulant effect on fibrinogen, but the isolated thrombin-like enzymes show distinct differences in their physical, chemical, and biochemical properties.

For practical reasons, the venoms of *B. atrox* (reference locality Maranhao, Brazil)* and *B. moojeni* (reference locality Minas Gerais, Brazil) were used for large-scale batroxobin production. A comparison of properties of batroxobin prepared from the venom of the two species or subspecies is presented in Table 8.

In practical terms, the properties listed in Table 8 show that batroxobin from *B. atrox* has a high yield, good physical stability, but bad qualities as an in vivo defibrinogenating agent, whereas batroxobin from *B. moojeni* shows low yield, fair physical stability, and good qualities as a defibrinogenating drug. In consideration of these properties, batroxobin is prepared from *B. atrox* venom for in vitro diagnostic use, whereas batroxobin is prepared from *B. moojeni* venom for in vivo use as a defibrinogenating drug.

Application of Batroxobin in Vitro. Batroxobin is not inhibited by antithrombin III or heparin. A batroxobin clotting time, performed as a screening test in parallel to the thrombin time, will therefore discriminate between delayed coagulation due to thrombin inhibition and that due to a qualitative or quantitative defect in fibrin formation (36).

B. atrox (Maranhao) was erroneously designated as *B. marajoensis* in the published literature. *B. marajoensis* is a proposed species occurring on the Amazonas island of Marajo exclusively.

Table 8 Properties of Batroxobin Isolated from *B. atrox* and *B. moojeni*

Property	Batroxobin *B. atrox*	Batroxobin *B. moojeni*
Molecular weight (SDS PAGE, reduced and alkylated)	41,600	35,800
Specific activity (BU[a]/mg protein)	1,900	500
Electrophoretic mobility pH 8.6 (bromophenol blue = 1.0)	0.2	0.1
Inactivation by human serum (24 hr, 37°C)	83.3%	56.7%
Inhibition by α2M[b]	5.6%	0.5%
Elution from heparin-agarose (NaCl concentration)	0.1 M	0.25 M
Inactivation by elastase (24 hr, 37°C)	0%	100%
In vivo defibrinogenating effect	Weak, short	Strong, persistent

[a]The batroxobin unit (BU) defines the fibrinogen coagulant potency in comparison with the First British Batroxobin Standard (78).
[b]Inhibition by α2-macroglobulin (α2M) was measured as previously described (79).

Since batroxobin causes fibrin formation through the exclusive release of fibrinopeptide A from fibrinogen, the batroxobin clotting time may provide rapid information on the molecular location of a structural defect in pathologic fibrin (37, 38).

A turbidimetric follow-up of fibrin formation induced by batroxobin is the basic principle of a quantitative fibrinogen determination in plasma developed by Becker and colleagues (39). This method has successfully been adapted to automatic systems.

A further current use of batroxobin is the in vitro defibrinogenation of plasma samples, especially in the presence of heparin, for the determination of FDP or plasma clotting factors (40, 41). Batroxobin is also used for the preparation of soluble fibrin I monomer, which serves as a stimulator of TPA-catalyzed plasminogen activation in the functional TPA assay (42).

Batroxobin and platelet-rich human plasma form a nonretracting clot. Simultaneous or subsequent addition of an aggregating agent causes a retraction, which may be followed by measuring the extruded fluid or by direct recording of the contraction of a hanging clot by means of a kymograph (43). The ability of endothelial cells, fibroblasts, and various types of cancer cells to induce fibrin clot retraction has also been studied using batroxobin (44). Moreover, the Reptilase clot retraction test has been used to screen potential inhibitors of platelet aggregation and release reactions (43).

Application of Batroxobin In Vivo

The clinical application of batroxobin as an anticoagulant, antithrombotic agent is based on extensive toxicological, pharmacokinetic, and pharmacodynamic studies. Detailed information on preclinical and clinical work in therapeutic defibrinogenation with batroxobin and ancrod is provided by several reviews (29, 34, 45-47).

Parenteral application of batroxobin causes a dose-dependent reduction in the plasma fibrinogen level and blood viscosity. Plasminogen and α_2-antiplasmin are, to some extent, consumed during the defibrinogenation process; platelet count and coagulation factors other than fibrinogen remain virtually unaffected during the batroxobin treatment (48, 49), and primary hemostasis is only slightly affected (49, 50).

The mechanism of batroxobin-induced defibrinogenation may to date be understood as follows: Batroxobin catalyzes the cleavage of the Arg_{16}-Gly_{17} bond of the Aα-chain of fibrinogen (38) and stimulates the secretion of vascular plasminogen activator (VPA) (Markwardt, personal communication, 1985). Fibrin I monomer (des-A-monomer), a preferential substrate for thrombin and plasmin, and fibrinopeptide A are formed. Fibrin I associates with fibrinogen and probably other plasma proteins to form soluble, nonclottable complexes (51). In vitro, under prolonged batroxobin action in absence of VPA, or in vivo, upon bolus injection of batroxobin, fibrin I_m polymerizes in a linear order to fibrin I_p (52), a highly permeable type of fibrin gel (53) that is highly susceptible to fibrinolysis (54). Systemically, VPA activates plasminogen into plasmin, a process that is stimulated by fibrin I (42). Fibrin I is a preferred substrate of plasmin, which cleaves the Arg_{42}-Ala_{43} bond of the Bβ-chain of fibrin I to form the nonclottable fragment X and Bβ 1-42 (54, 55). Subsequently, plasmin catalyzes further fibrin degradation and finally binds to α_2-antiplasmin to form an inactive complex (56).

In extravasated blood, at the site of a vascular injury, thrombin is generated via the intrinsic and extrinsic activation pathway. Thrombin cleaves the Arg_{14}-Gly_{15} bond of the Bβ-chain of fibrin I to form fibrin II and fibrinopeptide B. Moreover, thrombin exerts its action on fibrinogen and its multiple activator effects within the hemostatic system. Activated factor XIII causes stabilization of fibrin and its crosslinkage with fibronectin (21). Batroxobin is bound to α_2-macroglobulin to form an inactive complex that is metabolized and excreted as inactive fragments in the urine (57).

Toxicologic Aspects

The toxicity of batroxobin in mice is very low. According to Osaki and co-workers (58), the LD_{50} (intravenously) in mice is 192 to 210 BU/kg and, according to Galli (personal communication, 1977), it is 382 to 638 BU/kg in male mice and 452 to 1154 BU/kg in female mice. An LD_{50} could not be calculated using other routes of application (i.e., no mice died after subcutaneous injection of 1690 BU/kg). No lethality could be observed in mice that received 900 BU/kg intravenously following previous defibrinogenation by intraperitoneal injection of batroxobin (59). These figures indicate that intravenous bolus injection may have caused the death of some animals by intravascular coagulation when the fibrinolytic system was overcharged. Such overcharge was excluded in the case of slow resorption by other routes of administration.

The LD_{50} of crude *B. atrox* venom in mice is about 3.9 mg/kg intravenously, regardless of whether the animals were previously defibrinogenated or not (5, 59). The addition of a sufficient amount of an antibatroxobin IgG preparation to a lethal dose of *B. atrox* venom (5 mg/kg body weight) to completely neutralize its fibrinogen coagulant activity did not exert any protective effect in mice (unpublished data). The lethal effect of the venom was, however, strongly reduced by the concomitant administration of the proteinase inhibitor aprotinin, and the depressor activity of the venom in the anesthetized rat was completely prevented by this trypsin and kallikrein inhibitor (59).

These findings indicate that batroxobin does not significantly contribute to the venom's lethal effect in mice, whereas the kininogen-activating principle seems to be of greater importance. This conclusion from mice experiments might be applicable to other mammal species. It is, however, not necessarily true for prey animals other than mammals (e.g., birds, reptiles, amphibians, or fish). Batroxobin exerts, in fact, a stronger toxic effect in pigeons than in mice. As observed in our experiments, five pigeons survived the intravenous injection of 10 BU/kg body weight without symptoms, one animal of the five died after receiving 20 BU/kg, and 50 BU/kg was lethal for the remaining four animals. This higher acute toxicity of batroxobin in a bird species may be due to the lack of a fibrinolytic system in birds (60). In these animals, fibrin-related material formed by batroxobin action on fibrinogen probably has to be removed by phagocytosis which, for this purpose, seems to be less efficient than fibrinolysis.

Thus, fibrinogen coagulant proteinases in viper venoms may represent a means of immobilizing prey animals that do not possess a fibrinolytic system, such as birds and reptiles (61), whereas other toxic properties (e.g., kininogen activators) contained in the same venom are required to affect other classes of prey animals.

REFERENCES

1. Iwanaga, S. and T. Suzuki. In *Snake Venoms* , C. Y. Lee (Ed.), Springer-Verlag, Berlin, Heidelberg, New York, p. 61, 1979.
2. Kirby, E. P., S. Niewiarowski, K. Stocker, C. Kettner, E. Shaw, and T. M. Brudzynski. Biochemistry 18:3564 (1979).
3. Niewiarowski, S., E. P. Kirby, T. M. Brudzynski, and K. Stocker. Biochemistry 18:3570 (1979).
4. Schmaier, A. H. and R. W. Colman. Blood 56:1020 (1980).
5. Stocker, K., H. Fischer, and J. Meier. Toxicon 20:265 (1982).
6. Vargaftig, B. B. Toxicon 72:279 (1982).
7. Ouyang, C., J. P. Wang, and C. M. Teng. Biochim. Biophys. Acta 630:246 (1980).
8. Brinkhous, K. and M. S. Read. Blood 55:517 (1980).
9. Markwardt, F. and H. P. Klöcking. Thromb. Res. 8:217 (1976).
10. Klöcking, H. P., G. Bock, J. Drawert, K. P. Hinsenbrock, B. Kaiser, and U. Sedlarik. Folia Haemat. 103:404 (1976).
11. Budzynski, A. (Abstract). Thromb. Haemost. 54:314 (1985).
12. Dosekun, A. K., V. E. Pollak, P. Glas-Greenwalt, S. Kent, P. Penovich, A. Lebron-Berges, M. A. Weiss, and J. E. Levinson. Arch. Intern. Med. 144:37 (1984).
13. Czervionke, R. L., J. B. Smith, J. C. Hoak, G. L. Fry, and D. L. Haycraft. Thromb. Res. 14:781 (1979).
14. Furie, B. C. and B. Furie. In *Methods in Enzymol*, Lorand, L. (Ed.), Academic Press, New York, London, 45:191 (1976).
15. Kisiel, W. and W. M. Canfield. In *Methods in Enzymol*, Lorand, L. (Ed.), Academic Press, New York, London, 80:275 (1981).
16. Neshaim, M. E., J. A. Katzmann, P. B. Tracy, and K. G. Mann. In *Methods in Enzymol*, Lorand, L. (Ed.), Academic Press, New York, London, 80:249 (1981).
17. Seegers, W. H., H. I. Hassouna, D. Hewitt-Emmett, D. A. Walz, and T. J. Andary. Sem. Thromb. Haem. 1:211 (1975).
18. Pirkle, H., F. S. Markland, and I. Theodor. Thromb. Res. 8:619 (1976).
19. Teng, C. M. and W. H. Seegers. Thromb. Res. 22:213 (1981).
20. Walter, M., D. Nyman, V. Krajnc, and F. Duckert. Thromb. Haemost. 38:438 (1977).
21. Okada, M., B. Blombäck, M. D. Chang, and B. Horowitz. J. Biol. Chem. 260:1811 (1985).
22. Gaffney, P., N. A. Marsh, and B. C. Whaler. Biochem. Soc. Trans. 1:1208 (1973).
23. Kisiel, W. J. Clin. Invest. 64:761 (1979).
24. Esmon, N. L., W. G. Owen, and C. T. Esmon. J. Biol. Chem. 257:859 (1982).
25. Exner, T., J. Koutts, B. Cotton, and M. Howden. (Abstract). Thromb. Haemost. 54:313 (1985).

26. Suzuki, K., J. Nishioka, and S. Hashimoto. J. Biol. Chem. 258:163 (1983).
27. Francis, R. B. and M. J. Patch. Thromb. Res. 32:605 (1983).
28. Stocker, K. In *Natural Toxins*, Eaker, D. and T. Wadström (Eds.), Pergamon Press, Oxford, New York, p. 111 (1980).
29. Blasko, G. and R. Machovich. In *The Thrombin*, Volume II, R. Machovich (Ed.), CRC Press, Boca Raton, p. 89 (1984).
30. Hessel, B. and M. Blombäck. FEBS Letters 18:318 (1971).
31. Kosugi, T., Y. Ariga, K. Kinjo, and H. Mihara. Ketsueki to Myakkan 15:311 (1984).
32. Bonilla, C. Thromb. Res. 6:151 (1975).
33. Wik, K. O., O. Tangen, and F. N. Mc Kenzie.Br. J. Haemat. 23:37 (1972).
34. Markland, F. S. J. Toxicol. Toxin Rev. 2:119 (1983).
35. Hoge, A. R. Mem. Inst. Butantan 32:109 (1965).
36. Funk, C., J. Gmür, R. Herold, and P. W. Straub. Br. J. Haemat. 21:43 (1971).
37. Matsuda, M., E. Saeki, A. Kasamatsu, C. Nakamikawa, S. Manabe, and Y. Samejima. Thromb. Res. 37:379 (1985).
38. Blombäck, B. In *Struktur und Funktion des Fibrinogens-Blutgerinnung und Mikrozirkulation*, H. Schroer, G. Hauck, and E. Zimmermann (Eds.), Schattauer, Stuttgart, New York, p. 5 (1976).
39. Becker, U., K. Bartl, and A. Wahlefeld. Thromb. Res. 35:475 (1984).
40. Avvisati, G., J. W. ten Cate, E. M. van Wijk, L. H. Kahlé, and G. Mariani. Br. J. Haemat. 45:343 (1980).
41. Donati, M. B. In *Blood Coagulation and Haemostasis*, J. M. Thomson (Ed.), Churchill Livingstone, Edinburgh, London, New York, p. 190 (1980).
42. Ranby, M., B. Nossman, and P. Wallén. Thromb. Res. 27:743 (1982).
43. de Gaetano, G., D. Bottecchia, and J. Vermylen. Thromb. Res. 2:71 (1973).
44. Barbieri, B., G. Balconi, E. Dejana, and M. B. Donati. Proc. Soc. Exp. Biol. Med. 168:204 (1981).
45. Stocker, K. In *Handbook of Exp. Pharmacology*, Volume 46, F. Markwardt (Ed.), Fibrinolytics and Antifibrinolytics, Springer-Verlag, Berlin, Heidelberg, New York, p. 451 (1978).
46. Bell, W. R. In *CRC Handbook in Clinical Laboratory Science, Section I: Haematology*, D. Seligson (Ed.), CRC Press, Boca Raton, p. 301 (1980).
47. Seegers, W. and C. Ouyang. In *Handbook of Experimental Pharmacology, Volume 52*, Snake Venoms, C. Y. Lee (Ed.), Springer-Verlag, Berlin, Heidelberg, New York, p. 684 (1979).

48. Egberg, N. Acta Med. Scand. 124:1 (1973).
49. Fukutake, K., M. Fujimaki, H. Nagasawa, and M. Kato. Acta Haem. Jap. 44:1178 (1981).
50. Latallo, Z. Thromb. Haem. 50:604 (1983).
51. Brosstad, I., P. Kierulf, and H. C. Godal. Thromb. Res. 14: 705 (1979).
52. Blombäck, B., B. Hessel, M. Okada, and N. Egberg. Ann. New York Acad. Sci. 370:536 (1981).
53. Blombäck, B. and M. Okada. Thromb. Res. 25:51 (1982).
54. Nossel, H., J. Wasser, K. L. Kaplan, K. S. La Gamma, I. Yudelmann, and R. E. Canfield. J. Clin. Invest. 64:1371 (1979).
55. Bilezikian, S. B., H. L. Nossel, and V. P. Butler. J. Clin. Invest. 56:438 (1975).
56. Collen, D. and M. Verstraete. Thromb. Res. 11:417 (1977).
57. Egberg, N. Thromb. Res. 4:35 (1974).
58. Osaki, M., T. Hiyama, K. Kunikane, S. Sato, K. Tsushima, and N. Mori. Pharmacometrics 25:339 (1983).
59. Meier, J. and K. Stocker. Folia Haemat. 111:877 (1984).
60. Niewiarowski, S. and Z. Latallo. Thromb. Diath. Haem. 3:404 (1959).
61. Hawkey, C. M. In *The Haemostatic Mechanism in Man and Other Animals*, Academic Press, London, New York, p. 133 (1970).
62. Ouyang, C., I. S. Hong, and C. H. Teng. Thromb. Haem. 26: 224 (1971).
63. Ouyang, C. and J. S. Hong. Toxicon 12:449 (1974).
64. Ugarova, T. P., V. N. Melnik, and O. V. Gornitskaya. (Abstract). Chem. Abst. 103:2657 (1985).
65. Suzuki, T. and H. Takahashi. Toxicon 22:29 (1984).
66. Herzig, R. H., O. D. Ratnoff, and J. R. Shainoff. J. Lab. Clin. Med. 76:451 (1970).
67. Shainoff, J. R. and W. R. Welches. (Abstract). Thromb. Res. 54:313 (1985).
68. Dyr, J. E., B. Blombäck, and F. Kornalik. Thromb. Res. 30: 185 (1983).
69. Guan, L., X. Zhang, and C. Chi. (Abstract). Thromb. Haem. 54:313 (1985).
70. Guan, L. F., C. W. Chi, and M. Yuan. Thromb. Res. 35:301 (1984).
71. Reid, H. A. and K. E. Chan. Lancet 9 (March):485 (1968).
72. Ortiz, F. A. and F. Gubensek. Bull. Inst. Pasteur 74:145 (1976).
73. Blombäck, B. Arkiv Kemi 12:321 (1958).
74. Markland, F. S. and P. S. Damus. J. Biol. Chem. 246:6460 (1971).
75. Magalhaes, A., G. J. de Oliveira, and C. R. Diniz. Toxicon 19: 279 (1981).

76. Ouyang, C. and F. Y. Yang. Biochim. Biophys. Acta 351:354 (1974).
77. Andersson, L. Haemostasis 1:31 (1972).
78. Gaffney, P. G. Thromb. Haem. 38:562 (1977).
79. Stocker, K. and H. Yeh. Thromb. Res. 6:189 (1975).

6

Studies of the Preferential Release of Fibrinopeptide B by Copperhead Procoagulant Enzyme

John R. Shainoff and William R. Welches
The Cleveland Clinic Foundation, Cleveland, Ohio

INTRODUCTION

Previously, we had shown that venom from *Ancistrodon contortrix contortrix* contained a procoagulant enzyme that released fibrinopeptide B faster than fibrinopeptide A in the course of coagulating fibrinogen, and that the release of fibrinopeptide A could be suppressed almost entirely by low temperatures. The release of fibrinopeptide B unmasks an aggregation site that is located immediately behind fibrinopeptide B at the amino terminus of the β-chain. The inhibition of fibrinopeptide A release is attributable to a steric hindrance imposed by tightened aggregation of β-fibrin at low temperatures. This inhibited release of fibrinopeptide A that results from aggregation of the β-fibrin contrasts sharply with the converse acceleration of fibrinopeptide B release that accompanies aggregation of α-fibrin (lacking fibrinopeptide A) in the course of coagulation by thrombin. The converse effects of (a) blockage of release of fibrinopeptide A after aggregation through release of fibrinopeptide B by the enzyme, and (b) acceleration of release of fibrinopeptide B after aggregation through release of fibrinopeptide A by thrombin may be viewed as an indication that both enzymes access both the 16-17 Arg-Gly residues of the Aα-chain and the 14-15 Arg-Gly of Bβ-chain from a fibrinopeptide B to A directional vector.

Shortly after venom of the Malayan pit viper *Ancistrodon rhodostoma* had been found to contain a fibrinogen-coagulating enzyme with novel specificities (1), a search was initiated (2) for similar enzymes in a related viper, the Southern copperhead *Ancistrodon contortrix contortrix*. Although the crude venom from the copperhead did not elicit perceptible clotting, a thrombin-like enzyme was found among proteins separated by DEAE chromatogrpahy. Expression of the procoagulant activity in the crude venom was obscured by the much more abundant fibrinolytic enzymes. Analysis of the chemical alterations elicited by the enzyme in the course of its coagulation of fibrinogen indicated that it acted on the same segments as thrombin to release fibrinopeptides A and B, but unlike thrombin-like enzymes described previously (3,4), it released the peptides in a converse order. In the course of coagulating fibrinogen, known thrombin-like enzymes cleaved off fibrinopeptide A much faster than fibrinopeptide B. The copperhead procoagulant venzyme cleaved off fibrinopeptide B faster.

Prior studies had demonstrated that the release of fibrinopeptide A sufficed to transform fibrinogen into a self-aggregating, clot-forming substance (3,5), and the contribution of fibrinopeptide B was secondary. It had been thought that release of fibrinopeptide A functioned in linear aggregation, and that the release of fibrinopeptide B added to clot formation by contributing to an independent, side-by-side mode of aggregation (6,7). However, the precise role of fibrinopeptide B could not be addressed directly previously, because of the relatively rapid release of fibrinopeptide A in the course of coagulation by thrombin. The preferential release of fibrinopeptide B by the copperhead procoagulant venzyme provided an opportunity to address the significance of fibrinopeptide B release directly (8). The results obtained indicated that release of fibrinopeptide B did unmask an independent aggregation site, but that this site functioned to reinforce the linear aggregation that follows release of fibrinopeptide A, rather than induce a separate side-by-side mode of aggregation.

LIMITING THE RELEASE OF FIBRINOPEPTIDE A FOR PRODUCTION OF FIBRIN LACKING FIBRINOPEPTIDE B ALONE

In our initial studies (2), carried out at 37°C, the release of fibrinopeptide B proceeded at an initial rate eight times faster than that for fibrinopeptide A, but the releaseof fibrinopeptide A gradually accelerated, and the reaction ultimately resulted in complete release of both peptides. While searching for ways to restrict the action of the enzyme to release fibrinopeptide B without the concomitant release of fibrinopeptide A, we observed that the overall release of fibrinopeptide A became limited at lower temperatures (8). Release of fibrinopeptide A reached only 25% of completion with reactions at 25°C. When the temperature was lowered further to 2°C, release of fibrinopeptide A reached only 8% of completion. The same highly limited release of fibrinopeptide A occurred at 14°C, at which solubility of fibrinogen remained quite high. Thus, we were able to obtain a full release of fibrinopeptide B with only an 8% loss of fibrinopeptide A from concentrated solutions of fibrinogen (10 to 20 mg/ml) by carrying out reactions at 14°C. The fibrinogen coagulated fully at 14°C, even though little release of fibrinopeptide A occurred. With rabbit fibrinogen, as little as 2% of the fibrinopeptide A was released, and with bovine fibrinogen, it was less than 1%. However, when the solutions were warmed to 37°C, the clots dissolved, and at the elevated temperature, the release of fibrinopeptide A that had been limited by tight aggregation in the cold began to resume, and the fibrinogen would clot again, this time in association with the release of fibrinopeptide A.

The release of fibrinopeptide A that occurred on warming could be blocked by adding phenylmethylsulfonyl fluoride (PMSF) to inactivate the venom enzyme before dissolving the fibrin. With the enzyme activity out of the way, the clots could be dissolved to yield stable solutions of fibrin molecules lacking fibrinopeptide B. We called this derivative β-fibrin to signify that it aggregates through the B-dependent aggregation site located contiguous to B on the β-chain. The fibrin lacking both fibrinopeptides was called αβ-fibrin to designate that its aggregation involves sites on both the α- and the β-chains.

ENERGETICS OF AGGREGATION

Although unexpected at first, the full dissolution of β-fibrin at 37°C was viewed as somewhat analagous to results that we had obtained with regular fibrin at 41°C. One can treat fibrinogen, or EDTA plasma, with thrombin at 41°C to release all the fibrinopeptide without forming a clot. The molecules formed only soluble aggregates at the higher temperature. We felt that the only difference with β-fibrin was that we did not have to elevate the temperature as high as with α-fibrin or αβ-fibrin to obtain full dissolution.

Ultracentrifugation showed that the β-fibrin dissociated into 11S soluble aggregates and 8S monomers at 37°C. Observation that monomer concentrations remained constant while concentrations of the aggregates rose with increasing fibrin concentrations led to a clarification of the equilibrium involved in the aggregation (9). Further, the change in monomer concentrations as temperature was varied enabled us to determine the enthalpy of the aggregation as -15 kcal/mole (8,9); a substantial fraction of full enthalpy involved clot formation by fully aggregated fibrin (10). The 11S aggregates were characterized to correspond to trimers on the basis of their diffusion constant (9). The formation of trimers without substantial buildup of dimers indicated that the individual (D.E) interactions between D and E domains (11,12) involved in the association of monomers into dimers were less stable than in the trimeric assembly (D.E.D). A slight D.D interaction, one with an association constant as slight as $K_A = 10^1$ (liters/mole), was suggested to lead to such a stabilization. The overall association constant for the monomer-trimer transition had an association constant of $K_A = 1.6 \times 10^6$ liters/mole; thus, the component D.E would only be on the order of 0.5×10^6 liters/mole. The association constant for aggregation of α-fibrin monomer was subsequently evaluated to be only about six times stronger. Yet, αβ-fibrin lacking both fibrinopeptides formed very tightly bound aggregates.

Comparison of the chromatographic characteristics of β-fibrin, α-fibrin, and αβ-fibrin indicated that the aggregation accompanying the release of fibrinopeptide B and that accompanying the release of fibrinopeptide A act in concert to reinforce interactions, rather than produce separate modes of interaction. This was confirmed in studies on fibrin-fibrinogen interaction (13), which demonstrated that its αβ-fibrin monomer did not bind more fibrinogen than α-fibrin; it just bound the fibrinogen with higher avidity, with $K_A > 10^{10}$ liters/mole (9).

IDENTIFICATION OF THE β-CHAIN AGGREGATION SITE

It has long been known that plasmin rapidly cleaves amino-terminal Bβ-fragments from fibrinogen without eliciting aggregation, and that the loss of these fragments containing fibrinopeptide B and contiguous segments through β-42 does not alter the aggregation associated with the release of fibrinopeptide A (14,15). When it was observed that release of fibrinopeptide B itself by the copperhead enzyme elicited aggregation, the absence of aggregation accompanying release of the Bβ-fragment indicated that the aggregation site that is unmasked by release of fibrinopeptide B exists immediately behind B in the β-segment 15-42.

Since plasmin concurrently releases a C-terminal segment form the α-chain along with the Bβ-fragments (15,16), the question arose as to whether the α-chain segment rather than the β-15-42 segment was critical for the aggregation that accompanies release of fibrinopeptide B itself. We were able to address this question with a preparation of copperhead enzyme that contained fibrinogenolytic contaminant that rapidly degraded the C-terminal half of the Aα-chain of fibrinogen. The contaminant could be inactivated with tosyl-lysine chloromethyl ketone (TLCK) while the TLCK had no effect on the procoagulant enzyme. Both preparations yielded clots with fibrinopeptide A intact; the preparation with the active contaminant yielded clots with degraded Aα-chains, while the TLCK-treated enzyme yielded clots with intact Aα-chains.

β-FIBRIN AND THE KINETICS OF FIBRIN CROSS-LINKING

Ability of dilute β-fibrin to dissociate fully into monomers at 37°C at concentrations below 0.15 mg/ml and form clots at lower temperatures enabled us to gain insight into the kinetic mechanism of clot stabilization. Dardik and Shainoff (17) observed that the dilute, monomeric β-fibrin underwent very slow cross-linking at the same rate as dilute fibrinogen, at rates that were much slower than α-fibrin or αβ-fibrin, which underwent full aggregation and cross-linking at the same concentration. However, at low temperatures or more concentrated solutions where β-fibrin formed aggregates, it underwent cross-linking at rapid rates essentially equal to that of the other fibrins and much faster than fibrinogen. From these observations, it was concluded that the rapid cross-linking of fibrin was due to its preassembly into proper alignment for cross-linking, rather than to an enhanced association of factor XIIIa with fibrin relative to its association with fibrinogen. Thus, prior observations (18) on an inhibitory effect of fibrinogen on cross-linking of fibrin and the direct cross-linking of fibrinogen by factor XIII were clarified.

CALCIUM AND AGGREGATION

Calcium was found to accelerate clot formation by β-fibrin at room temperature. As in acceleration produced with other types of fibrin, the precise role of the calcium was not clear (19). However, an ability to measure directly and accurately by ultracentrifugation the equilibrium distributions of monomer and soluble trimer in the reversible aggregation of β-fibrin enabled us to determine whether calcium was influencing the monomer/trimer equilibrium (9). Surprisingly, addition of either calcium (2.5 mM) or EDTA (0.1 mM) had no measurable effect on the concentrations of monomer and oligomer in a solution of dissolved β-fibrin. Thus, it could be concluded that the acceleration of clot formation arose from an effect on oligomer/oligomer other than monomer/monomer association. It is our suspicion that calcium promotes a slight conformational change that has no effect on monomer association, but enables oligomers to coalesce more easily into coarse strands.

SIDEDNESS IN THE PROTEOLYSIS OF FIBRINOGEN BY THE VENOM ENZYME AND BY THROMBIN

The steric inhibition that blocks release of fibrinopeptide A in association with the tightened aggregation that accompanies release of fibrinopeptide B by the venom enzyme in colder temperatures may be indicative of a sidedness in association of the enzyme with the peptide bond. Release of fibrinopeptide A by thrombin also becomes inhibited by aggregation through the B-dependent epitope (8). By contrast, aggregation of α-fibrin, lacking fibrinopeptide A in the initial stages of coagulation by thrombin, does not inhibit the release of fibrinopeptide B by the copperhead enzyme. Instead, the aggregation causes an acceleration of

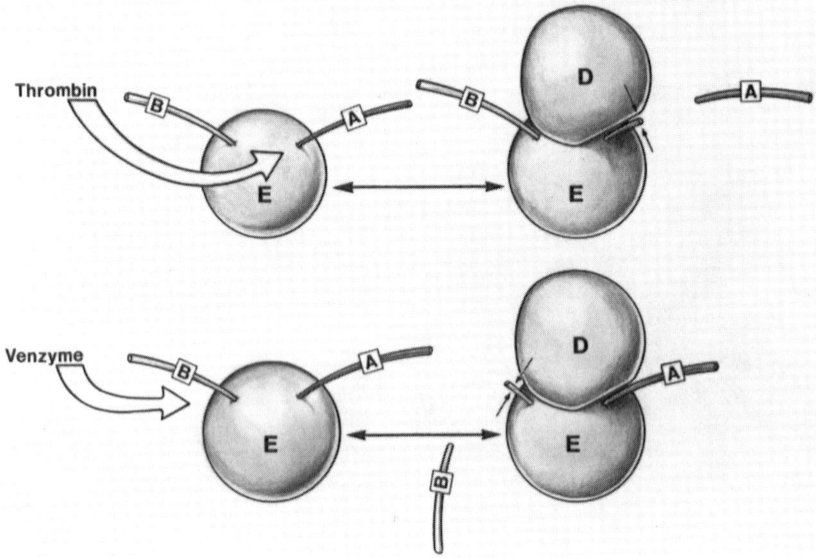

Figure 1 Sidedness in proteolysis of fibrinopeptides. Both thrombin and copperhead enzyme are portrayed as accessing the fibrinopeptide in a B to A directional vector. Cleavage of fibrinopeptide A by thrombin, followed by aggregation and apposition of a D domain, leaves peptide B unprotected. Conversely, preferential cleavage of fibrinopeptide B by the copperhead venzyme, followed by aggregation and apposition of a D domain, blocks access to peptide A.

the release of fibrinopeptide B by thrombin (20,21). As illustrated in Figure 1, the converse effects of release of one peptide on the subsequent release of the other peptide under conditions favoring tight aggregation may be viewed as evidence that the two enzymes access both the A and the B peptides in a B to A direction. When aggregation occurs first through the B-dependent aggregation site, access to the A-peptide becomes blocked, but when aggregation occurs first through the A-dependent site access to the B-peptide remains open, and release of B proceeds uninhibited, or accelerated in the case of thrombin. At high temperatures, the release of both peptides proceeds freely, because the reversible dissociation of oligomers enables the enzyme to gain access to the peptide unhindered.

The viper *Ancistrodon halys bloomhoffi* studied by Guan and coworkers (22) contains an enzyme that appears to be very similar to the procoagulant enzyme of the Southern copperhead. It specifically releases fibrinopeptide B. Release of fibrinopeptide A by this enzyme has not as yet been described; however, the effects of temperature on the reaction have not been examined.

FUTURE STUDIES

The provocative differences between the mechanisms of action of venom enzymes and thrombins in causing the aggregation of fibrinogen may be the result of some rather fundamental structural differences. Crotalase, the thrombin-like proteinase from the venom of *Crotalus adamanteus*, aggregates fibrinogen by releasing fibrinopeptide A, but not fibrinopeptide B (23). Pirkle and associates (24) found that the 37% sequenced portion of the molecule shows greatest homology to kallikrein, which is not particularly closely related to thrombin. From physical properties established thus far (2), the copperhead enzyme is also only distantly related to thrombin.

REFERENCES

1. Ewart, M. R., M. W. C. Hatton, J. M. Basford, and K. S. Dodgson. Biochem. J. 115:17 (1969).
2. Herzig, R. H., O. D. Ratnoff, and J. R. Shainoff. J. Lab. Clin. Med. 76:451-465 (1970).
3. Blombäck, B., M. Blombäck, and I. I. M. Nelsson. Thromb. Diath. Haemorrh. 1:3-13 (1957).
4. Ewart, M. R., M. W. C. Hatton, J. M. Bashford, and K. S. Dodgson. Biochem. J. 118:603-609 (1970).
5. Shainoff, J. R. and I. H. Page. J. Exp. Med. 116:687-707 (1962).
6. Laurent, T. C. and B. Blombäck. Acta Chem. Scand. 12:1875-1877 (1958).
7. Blombäck, B., B. Hessel, D. Hogg, and L. Therkildsen. Nature 275:501-505 (1978).
8. Shainoff, J. R. and B. N. Dardik. Science 204:200-202 (1978).
9. Shainoff, J. R. and B. N. Dardik. Ann. NY Acad. Sci. 408:254-268 (1983).
10. Sturtevant, J. M., M. Laskowski, T. H. Donnelly, and H. A. Scheraga. J. Am. Chem. Soc. 77:6168-6172 (1955).
11. Marder, V. J. and A. Z. Budzynski. Thromb. Diath. Haemorrh. 33:199-207 (1975).
12. Kudryk, B. J., D. Collin, K. R. Woods, and B. Blombäck. J. Biol. Chem. 249:3322-3325 (1974).

13. Shainoff, J. R. *Proc. Workshop in Fibrinogen, Stockholm, Sweden* A. Henschen (Ed.), Walter de Gruyter, Inc., New York. In press, 1985.
14. Shainoff, J. R., B. Lahiri, and F. M. Bumpus. Thromb. Diath. Haemorrh. Suppl. 39:203 (1970).
15. Budzynski, A. Z., V. C. Marder, and J. R. Shainoff. J. Biol. Chem. 249:2294-2302 (1974).
16. Mosesson, M. W., J. S. Finlayson, and D. K. Galanakis. J. Biol. Chem. 248:7913-7929 (1973).
17. Dardik, B. N. and J. R. Shainoff. Thromb. Haemo. 3:864-872 (1979).
18. Kanaide, H. and J. R. Shainoff. J. Lab. Clin. Med. 85:574-597 (1975).
19. Boyer, M. H., J. R. Shainoff, and O. D. Ratnoff. Blood 39:382-387 (1972).
20. Hurlet-Jensen, A., H. Z. Cummins, H. L. Nossel, and C. Y. Liu. Thromb. Res. 27:419-427 (1982).
21. Lewis, S. D., P. P. Shields, and J. A. Shafer. J. Biol. Chem. 260:10192-10199 (1985).
22. Guan, L., C. Chi, and M. Yuan. Thromb. Res. 35:301-310 (1984).
23. Markland, F. S. and H. Pirkle. Thromb. Res. 10:487-494 (1977).
24. Pirkle, H., F. S. Markland, I. Theodor, R. Baumgartner, S. S. Bajwa, and H. Kirakossian. Biochem. Biophys. Res. Comm. 99:715-721 (1981).

7

Differences in Fibrin Polymerization and Fibrinopeptide A and B Release Induced by Human Thrombin and by the Thrombin-like Enzyme from the Venom of *Agkistrodon halys pallas*

Li-feng Guan,* Xia Zhang,† and Chen-wu Chi
Chinese Academy of Sciences, Shanghai, Republic of China

INTRODUCTION

Snake venoms from the family of Crotalidae contain serine proteinases acting on the Arg-X bond of specific proteins. These proteinases are involved in the toxicologic action of the venoms, such as disturbance of coagulation or release of bradykinin. Among these serine proteinases of venoms, the thrombin-like enzymes which interact with fibrinogen, are the most interesting from the standpoint of their potential for therapeutic application. These enzymes clot fibrinogen in vitro, but in vivo act as anticoagulants as a result of the benign state of defibrinogenation that is induced by their action (1).

From the snake of *Agkistrodon halys pallas* (Pallas' viper), a Chinese pit viper distributed in southeast part of China (Zhe Jiang Province), we isolated three proteinases possessing arginine esterase activities. They were identified as bradykinin-releasing, thrombin-like, and fibrinogenolytic enzymes (2). Among them, the thrombin-like enzyme (TLE) has been studied in detail because of its distinct mechanism of action on fibrinogen and its potential use as an anticoagulant agent.

Thrombin-like enzyme from Pallas' viper was purified by ion exchange chromatography and gel filtration (2). The purified enzyme was homogeneous on acrylamide gel electrophoresis. It is a glycopro-

---

Present affiliations:
*University of Southern California, Los Angeles, California
†VA Hospital, University of Miami, Miami, Florida

tein with a molecular weight of about 43,000 daltons. The analysis of amino acid composition of this enzyme showed a low content of aromatic amino acid residues, which was consistent with the low absorbance of the enzyme at 280 nm. A study of substrate specificity of this enzyme toward synthetic arginine ester or amide derivatives indicated that the substrate specificity of this enzyme was similar to trypsin, but with a higher specificity than that of trypsin. The inhibition of the enzyme by DFP (diisopropyl fluorophosphate) and PMSF (phenylmethylsulfonyl fluoride) showed that it was a serine proteinase, although the rate of inhibition was much slower than that of trypsin.

The similarity between the venom enzyme and human thrombin was shown by using the fluorogenic peptide substrate BOC-Val-Pro-Arg-MCA (tert-butoxycarbonyl Val-Pro-Arg-4-methyl-coumarin amide). However, the low clotting activity suggested a different mode of action than thrombin for the venom enzyme. Analysis of fibrinopeptides released by this enzyme from fibrinogen indicated that the specificity of TLE from Pallas' viper was different than that of human thrombin (2,3). It released fibrinopeptide B (FPB) first; then after a lag phase, fibrinopeptide A was released. The TLE from Pallas' viper is different from most of the thrombin-like enzyme from snake venoms that preferentially release fibrinopeptide A (FPA). Another exception is the procoagulant principle from Southern copperhead venom, which has a mode of action similar to our enzyme (4).

Shinoff and Dardik (5) indicated that the prior release of a certain amount of FPA was necessary for the formation of a fibrin clot caused by a procoagulant principle from Southern copperhead venom. Recently Dyr and co-workers (6) reported the effects of four isolated proteolytic fractions from Southern copperhead venom on the polymerization of fibrinogen. On the other hand, as is well known, human thrombin releases FPA first from fibrinogen; then after a lag phase, the release of FPB is accelerated. Blombäck and associates (7) concluded that the FPB release was greatly enhanced by polymerization of fibrin monomer, and that the prior release of FPA was prerequisite. However, Hurlet-Jensen and co-workers (8) reported that fibrin polymerization was not essential to FPB release.

To elucidate the mechanisms of fibrinopeptide A and B release and fibrin polymerization, it is worthwhile to compare the different actions of human thrombin and TLE on fibrinogen and to study the difference in the structure of fibrin formed from the action of these enzymes. This could provide a theoretical basis for the clinical use of TLE. In this chapter, the release kinetics of fibrinopeptide A and B and fibrin polymerization induced either by human thrombin or by Pallas' viper TLE are studied and compared. Moreover, for the first time, differences in the structure of fibrin clots induced by both enzymes are also investigated by electron microscopy.

MATERIALS AND METHODS

Human thrombin was obtained from Shanghai Second College of Medicine. Its specific activity was 250 NIH U/mg. Human FPA and FPB was generously provided by Professor B. Blombäck of Karolinska Institute, Sweden. Acetonitrile was chromatography grade for high pressure liquid chromatography (HPLC). Other reagents for HPLC were of guaranteed grade, and all solutions were prepared with twice glass-distilled water. Fibrinogen was purified by using polyethylene glycol 1000, according to the method of Masri and colleagues (9). The clottability was 95%.

The TLE was prepared with some modifications of the previous procedure (2,10). Crude venom was chromatographed on a DEAE-Sephadex A-25 column; TLE was eluted by 0.01 M phosphate buffer, pH 7.8, with a linear gradient of NaCl from 0 to 0.1 M. The active fractions were pooled and dialyzed against 0.01 sodium acetate buffer, pH 5.0, then chromatographed on a SP-Sephadex C-25 column and eluted with a linear gradient of NaCl from 0 to 0.15 M in the same buffer. The pooled active fractions were then chromatographed on a Sephadex G-75 column. The purified preparations were lyophilized.

The FPA and FPB were analyzed by HPLC on a Hitachi liquid chromatograph (Model 638-50) with a 635 M multiwavelength ultraviolet monitor. A reversed phase column (4 × 250 mm) packed with Licrosorb RP 18 (particle size 5 μm) was used for separation. The FPA and FPB were eluted from the column with a linear gradient of acetonitrile in the mobile phase from 3 to 60%, containing 0.025% phosphoric acid, pH 4.5, over 80 minutes at a flow rate of 1.0 ml/min. The peptides eluted were monitored at 210 nm. The amount of peptide released was represented in terms of peak area eluted from the column. Referred to the standard FPA and FPB, 1 sq. cm of peak area was equal to 6.3 μg of FPA or 8.3 μg of FPB.

The preparation of samples for FPA and FPB measurements was as follows: Fibrinogen was dissolved with 0.05 M Tris-HCl, pH 7.4, to give a final concentration of 5 mg/ml. Then human thrombin or Pallas' viper TLE was added, and the mixtures were incubated for various intervals. The reaction was stopped by heating in a boiling water bath for 2 minutes. After removal of precipitated protein by centrifugation at 10,000 rpm/min, the supernatant liquid was injected for HPLC analysis.

Fibrin polymerization was monitored at 350 nm with a Unicam ultraviolet spectrophotometer (Model SP-1800). The structure of the fibrin clots was studied by electron microscopy according to the method of Kay and Buddigan (11), with slight modification. Some 0.1 ml of human thrombin (40 NIH u/ml) or TLE from Pallas' viper (150 μg/ml) was mixed with 0.1 ml of fibrinogen solution (5 mg/ml in 0.05 M Tris-HCl buffer, pH 7.4) in a small tube, and a drop of mixed solution was pipetted onto a grid coated with carbon film. After standing at room temperature (or 37°C) for various intervals, the grids were washed with a few drops of saline solution and then stained with three drops of 1.0% uranyl acetate solution. The grids were then rinsed with distilled water and dried. The grids were examined by electron microscopy with a magnification of 5000 X.

RESULTS

The Time Courses of FPA and FPB Release Induced by TLE from Pallas' Viper and Human Thrombin

The FPA and FPB released from fibrinogen by thrombin could be separated and detected by HPLC. Figure 1 shows the separation of FPA and FPB on HPLC as peaks 2 and 4, respectively, with respective retention times of 24.8 and 26.8 minutes. Peaks 1 and 3 were identified as AP (phosphorylated at position 3 in FPA) and des-Arg FPB (12,13).

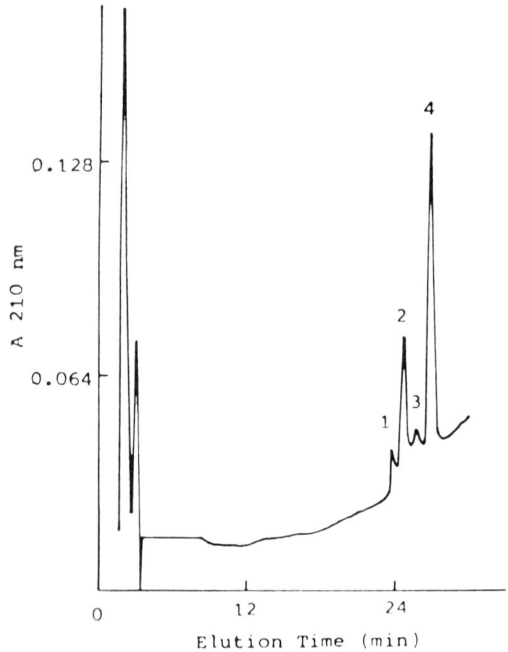

Figure 1 Fibrinogen (5 mg/ml) was incubated with TLE from Pallas' viper (150 μg/ml) at 37°C for 30 minutes. For HPLC analysis, 100 μl of the supernatant was injected.

In contrast to human thrombin, when TLE of Pallas' viper acted on human fibrinogen at a low enzyme concentration (50 μg/ml), FPB was released rapidly and reached the maximum even though no visible clot could be observed, whereas FPA was released slowly after a lag phase (Fig. 2). The clot could be observed only after standing overnight at 37°C.

Figure 3 shows the relationship between the release of FPA and FPB and fibrin polymerization at a high concentration of Pallas' viper TLE (150 μg/ml). When human thrombin acted on fibrinogen, fibrin polymerization occurred as soon as a small amount of FPA was released. Subsequently, FPB release accelerated, and finally the amounts of both fibrinopeptides approached nearly the same level. These results were compatible with those reported by Blombäck and co-workers (7). In the case of Pallas' viper TLE, FPB (not FPA), was released completely before the aggregation of fibrinogen reached the maximum. However, the FPA release remained at a low level and the ratio of FPB to FPA was nearly unchanged after a visible clot was observed.

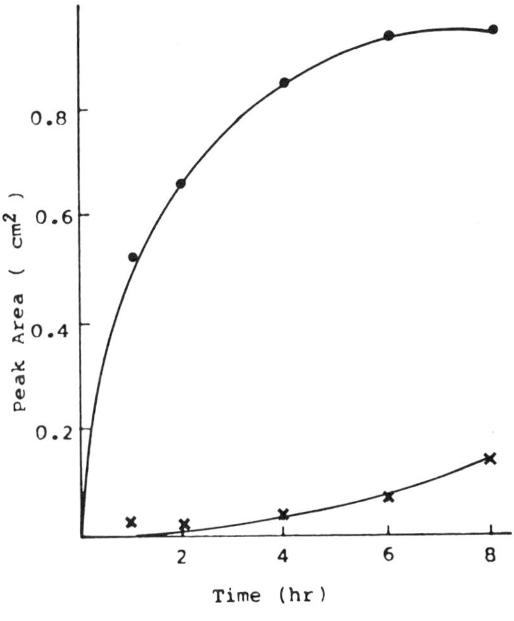

Figure 2 Release of FPA (x——x) and FPB (•——•) from human fibrinogen by TLE from Pallas' viper. The fibrinogen solution (5 mg/ml) was incubated with Pallas' viper TLE (50 μg/ml) at 37°C.

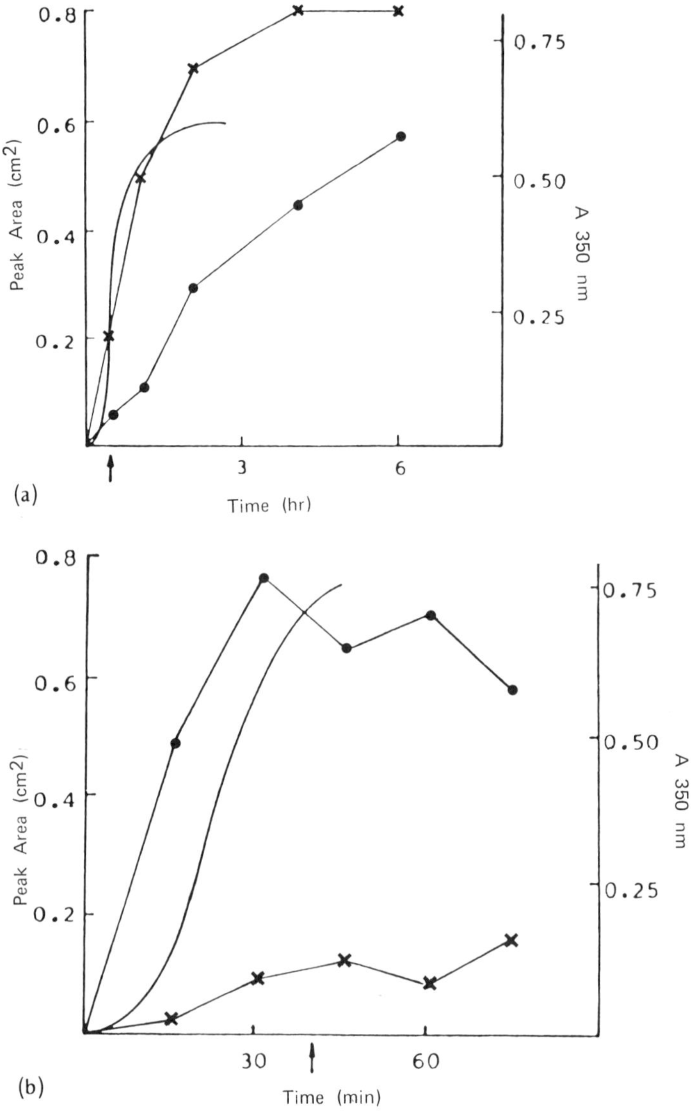

Figure 3 The relationship between fibrinopeptide release and fibrin polymerization. Human fibrinogen was dissolved in 0.05 M phosphate buffer, pH 7.4 (5 mg/ml), containing 100 mM epsilon-amino caproic acid (EACA). Human thrombin (final concentration 0.01 NIH u/ml) or TLE from Pallas' viper (final concentration 150 μg/ml) was added and incubated at 37°C for various intervals. The reaction was stopped by immersion in a boiling water bath. After centrifugation, 100 μl of the supernatant was injected for HPLC analyses of FPA and FPB. a: Human thrombin; b: Pallas' viper TLC. x——x = FPA; ●——● = FPB, —— absorbance at 350 nm. Arrows indicate the time of clot formation.

The Effects of Calcium Ion and Urea on Fibrin Polymerication

Calcium ions shortened the clotting time of fibrinogen caused either by human thrombin or Pallas' viper TLE. With human thrombin, when calcium ions (0.02 M) were added, the release of FPA did not change noticeably; however, the release of FPB was apparently enhanced and approached the maximum level faster (Fig. 4). In contrast, with Pallas' viper TLE as an activator, whether or not in the presence of calcium ions, the release of FPA still remained at the same level, but the release of FPB decreased by nearly 33%. This showed that further FPB release in this case was not necessary for fibrin polymerization after a certain amount of FPB had appeared.

In contrast to the results of Blombäck and co-workers (7), 2 M of urea did not inhibit fibrin aggregation. When fibrinogen was mixed either with human thrombin or with Pallas' viper TLE in the presence of 2 M urea, no visible clot could be observed, though the absorbance at 350 nm still increased. However, the absorbance curves slightly shifted to the right and the absorbance maxima were greater than those observed in the absence of 2 M urea (Fig. 5). When incubated either with Pallas' viper TLE or with human thrombin in the presence of 2 M urea, the absorbance increase of fibrinogen at 350 nm was essentially identical, the only difference being that it required a relatively long time for Pallas' viper TLE to produce the same absorbance increase. Since proteins in the reaction mixture could not be precipitated by heating or adding acetone after aggregation in these conditions, it was not possible to detect FPA and FPB released by HPLC.

Study on the Structure of the Fibrin Clot with Electron Microscopy

The fibrin clots caused by human thrombin or TLE of Pallas' viper were observed by electron microscopy (Fig. 6). The samples examined were prepared directly on grids by mixing fibrinogen separately with each of the enzymes. Since Pallas' viper TLE exhibited a weak clotting ability, it was necessary to incubate the reaction mixture at 37°C for about 2 hours in order for a better network of fibrin clot to form; in contrast, human thrombin aggregated fibrinogen at room temperature within 20 minutes. The clot caused by human thrombin was dense, and less transparent, and the fiber was compact, long, and fine with a diameter of 60 to 150 nm. The clot caused by Pallas' viper TLE was thin and more transparent, the fiber being loose, shorter, and thicker with a diameter of 150 to 250 nm. This indicated that the fibrin clots formed by these enzymes were apparently different.

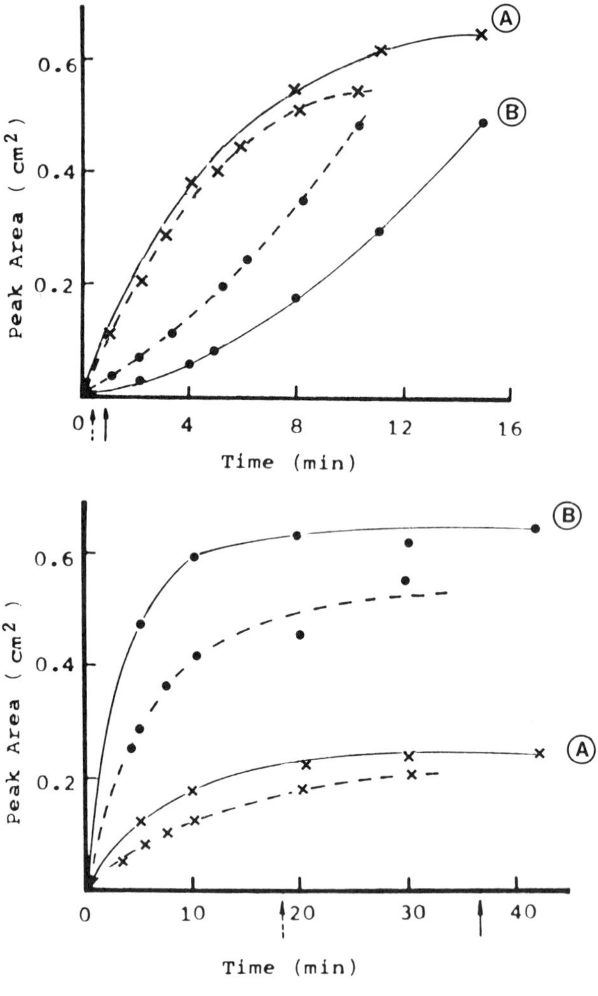

Figure 4 Release of FPA and FPB from human fibrinogen by human thrombin and TLE from Pallas' viper in the presence of 0.02 M $CaCl_2$. Human fibrinogen was dissolved in 0.05 M Tris-HCl buffer (final concentration 5 mg/ml); then human thrombin or Pallas' viper TLE was added and incubated at 37° C for various intervals. x——x = FPA; ●——● = FPB. The solid line shows the experimental values in the absence of 0.02 m $CaCl_2$; the dotted line shows experimental values in the presence of 0.02 M $CaCl_2$. Arrows indicate the time of formation of a visible clot. Top: human thrombin (0.4 NIH u/ml); bottom: Pallas' viper TLE (150 μg/ml).

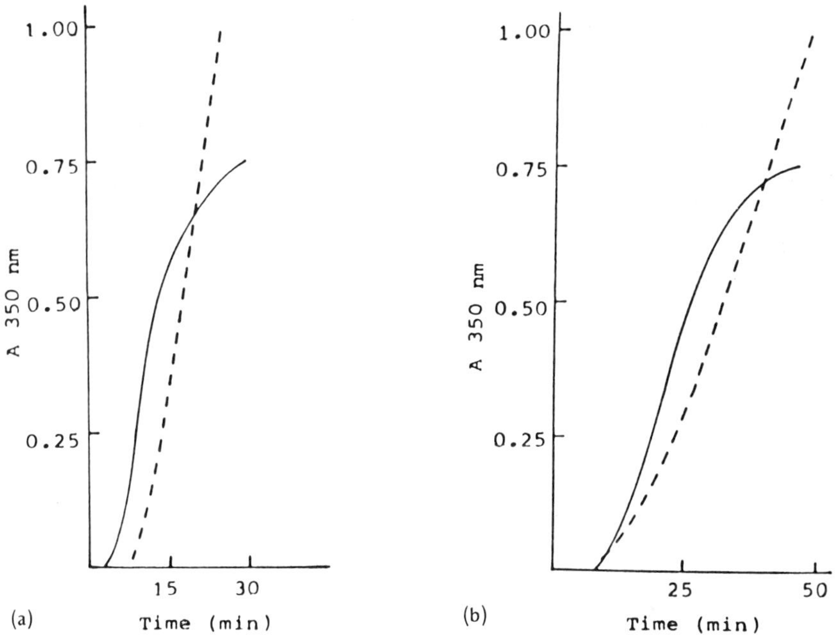

Figure 5 The effect of 2M urea on fibrin polymerization. —— = in the absence of 2 M urea; ---- = in the presence of 2 M urea. Left: polymerization with human thrombin (0.04 NIH u/ml); right: polymerization with Pallas' viper TLE (150 μg/ml).

DISCUSSION

Recently, HPLC has been introduced to separate FPA and FPB (12-14). It was more sensitive and specific for fibrinopeptides than were previous methods of detection. Although it has the drawback of being less sensitive than radioimmunoassay of fibrinopeptides (15), it has the great advantage of identifying FPA and FPB simultaneously, and is without the disadvantage caused by degraded fragments of fibrin with the same antigenic determinants for the antiserum against FPA and FPB. High pressure liquid chromatography has, therefore, become a helpful tool for studies of the release course of fibrinopeptides.

With reference to the mechanism of fibrin polymerization, several proposals have been suggested. Blombäck and co-workers (7) indicated that there were two sets of polymerization sites. With fibrinogen activated by human thrombin, one polymerization site (A, A') was exposed and was found to trigger the "end to end" aggregation of fibrin I $(\alpha, B\beta, \gamma)_2$. It led to a conformational change of fibrin I,

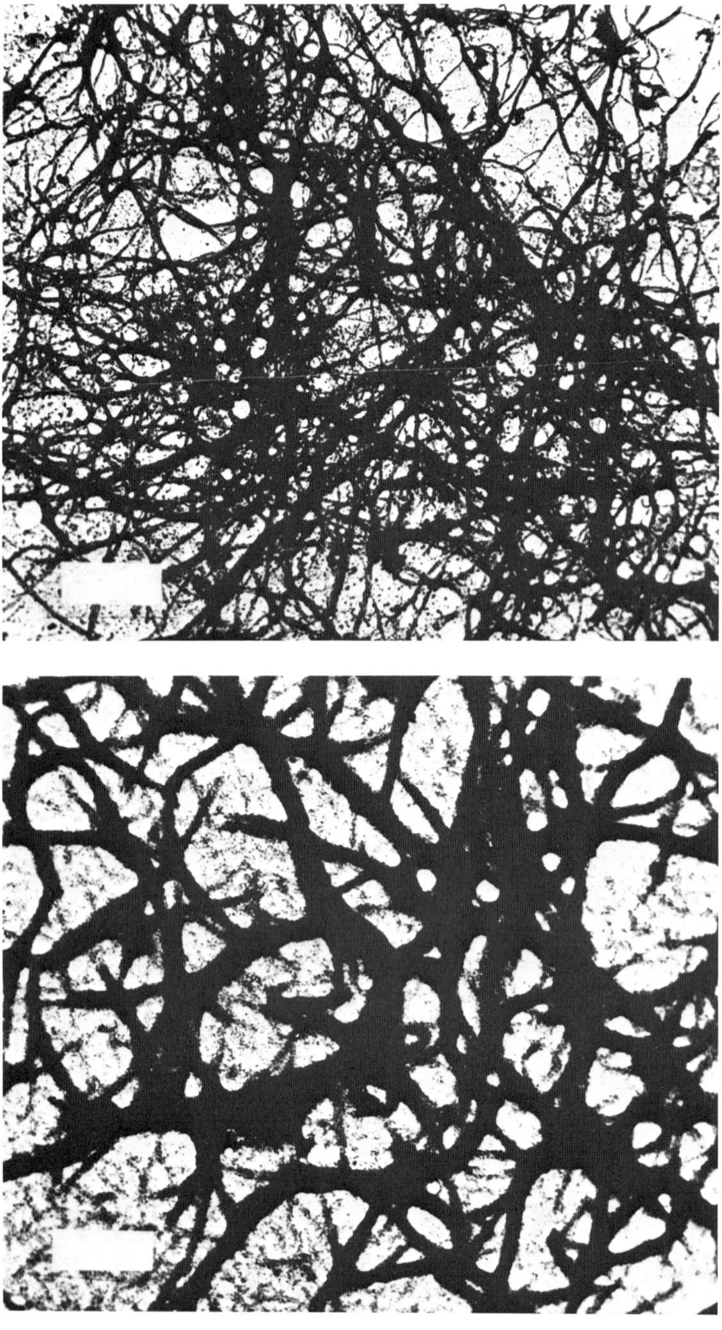

Figure 6 Fibrin clot as viewed by transmission electron microscopy. The bar is equal to 1 μm. Top: clot induced by human thrombin; bottom: clot induced by TLE from Pallas' viper. Original magnification: 5000 X.

making the B-chain accessible to thrombin for cleavage of FPB from fibrin I and consequently to form a clot of fibrin II $(\alpha, \beta, \gamma)_2$. Therefore, the release of FPB had a lag phase and was dramatically accelerated in the course of polymerization of fibrin monomer (16). Recently, Hurlet-Jensen and associates (8) studied the relationship between fibrin polymerization and FPB release caused by human thrombin, using Gly-Pro-Arg-Pro as an inhibitor of polymerization. They indicated that there were three distinct phases in the FPB release reaction. The initial slow but steady rate phase was not inhibited by the tetrapeptide, whereas the second stage of rapid release of FPB from fibrinogen depended on fibrin polymerization (8). Our data showed that when fibrinogen was incubated with Pallas' viper TLE at a low enzyme concentration (50 μg/ml), no visible clot formed over a rather long period of time, whereas the release of FPB rapidly reached the maximum. Hence, it could be concluded that, at least with Pallas' viper TLE, the stage of rapid release of FPB was also independent of fibrin polymerization.

With human thrombin, a visible clot could be observed after about 25% of FPA was released from fibrinogen. This was consistent with a previous report (5). With the increase in opacity, the absorbancy at 350 nm also increased rapidly and reached the maximum before the release of fibrinopeptides was complete. However, with Pallas' viper TLE, the absorbancy at 350 nm gradually increased only after the release of FPB had reached the end point. It has been proposed that the FPA release caused "end to end" fibrin polymerization, while FPB release caused aggregation of fibrin laterally (8,16). With this concept in mind, it is easy to understand that with Pallas' viper TLE as trigger, a visible clot could be observed only after a certain amount of FPA was released though FPB release had already reached the maximum. Moreover, because there was no further release of FPA, the clot caused by Pallas' viper TLE had a different structure than that formed by human thrombin.

Another important difference between human thrombin and Pallas' viper TLE is demonstrated by the fact that in the former case, the amounts of FPA and FPB released during clot formation were nearly the same, while in the latter case the ratio of FPB to FPA still remained at 3:1 after FPB release has reached the maximum. Similar results were reported with TLE from copperhead venom (5). Hence, it is most likely that the FPB release was accelerated when "end to end" fibrin polymerization occurred, and was suppressed during "side to side" polymerization. Shainoff and Dardik (5) indicated that the aggregation induced by FPB release might hinder the release of FPA because of a conformational change in fibrinogen. This was contrary to the case of human thrombin, in which the release of FPA would accelerate FPB release. It is worth pointing out that the clot formed after removing FPB by the copperhead venom enzyme at 14°C dissolved when warmed to

37°C, and FPA release approached 80% of the maximum during the second clot formation. However, FPA release by TLE from Pallas' viper still remained at a level of 30% of the maximum, even though the clot was formed at 37°C.

During the reaction of human thrombin with fibrinogen in the presence of 0.02 M calcium ions, FPA release remained at the same level as in absence of calcium ions, while FPB release was enhanced with a reduction in the clotting time. This provided another piece of evidence showing that fibrin polymerization would accelerate FPB release in the second phase. Recently Dyr and co-workers (6) reported that calcium ions shortened the clotting time of fibrinogen with the copperhead enzyme without effect on FPA release, but caused an apparent decrease in FPB release. In our experiment in the presence of calcium ions (0.02 M), FPA cleaved by TLE from Pallas' viper did not increase, whereas FPB decreased by 33%.

To the best of our knowledge, there has been no report about the structure of fibrin induced by TLE. We compared the structure of fibrin caused by TLE from Pallas' viper and human thrombin by electron microscopy. The results showed that the clot induced by Pallas' viper TLE was more transparent and loose with short and thick fibers, whereas the clot induced by human thrombin was compact with longer and thinner fibers. This might be due to the release of a considerably higher ratio of FPB by Pallas' viper TLE and, as a result, the fibrin molecules became aggregated laterally, leading to the formation of shorter and thicker fibers. Such structure in fibrin would be more susceptible to the native fibrinolytic system. The TLE from *Agkistrodon rhodostoma* has been used for anticoagulant therapy. The mechanism of action of this enzyme is to decrease the level of fibrinogen in plasma. It has been suggested that the microclot caused by the enzyme is especially susceptible to the fibrinolytic system and, therefore, instead of forming thrombi in vivo, it depletes the circulating fibrinogen. Recently, Kornalik (17) reported that ecarin, an enzyme from *Echis carinatus* that converts prothrombin directly to thrombin, could also be used for thromboembolic therapy. Our preliminary experiments showed that TLE from Pallas' viper exhibited an anticoagulant action in vivo in the rabbit (18). The results presented here further suggest that TLE from Pallas' viper might also be potentially useful in clinical therapy.

SUMMARY

Human fibrinopeptides A (FPA) and B (FPB) released either by human thrombin or by thrombin-like enzyme from the venom of *Agkistrodon halys pallas* were quantitatively estimated by HPLC. In contrast to human thrombin, the thrombin-like enzyme predominantly released fibrinopeptide B, which rapidly reached the maximum even before complete aggregation of fibrin when monitored at 350 nm. The ratio of FPA:FPB (1:3) remained unchanged after a visible clot was observed. The effects of calcium ions on the release of FPA and FPB were investigated. The release of FPB was apparently enhanced by calcium with human thrombin and suppressed with the thrombin-like enzyme. In contrast, there was no noticeable change in fibrinopeptide A release. In the presence of 2 M urea, no visible clot could be found though the absorbance at 350 nm was still observed to increase. The structure of fibrin fibers caused by human thrombin or the thrombin-like enzyme was studied by electron microscopy. The fibrin clot formed by human thrombin was more compact, dense, and stretched, with a fiber diameter of 60 to 150 nm, whereas the network of fibers formed by the thrombin-like enzyme was less compact, more transparent, and had shorter fibers with a diameter of 150 to 250 nm. In the latter case, much more lateral aggregation occurred because of the preferential release of FPB.

ACKNOWLEDGMENT

We are deeply grateful to Professor B. Blombäck of Karolinska Institute, Sweden, for gifts of FPA and FPB. We are greatly indebted to Professor Zhao-wen Sheng of our institute and Professor F. S. Markland of University of Southern California for helpful discussion. We also thank Mr. Li Qian and Mr. Hong-jiun Zhang for their excellent technical assistance.

REFERENCES

1. Markland, F. S., and H. Pirkle. In *Chemistry and Biology of Thrombin*, R. L. Lundblad, J. W. Fenton, and K. G. Mann (Eds.), Ann Arbor Science, Ann Arbor, Michigan, p. 71, 1977.
2. Guan, L. F., C. W. Chi, and M. Yuan. Thromb. Res. 35:301 (1984).
3. Guan, L. F. and C. W. Chi. Acta Biochim. Biophys. Sinica. 14:421 (1982).
4. Herzig, R. H., O. D. Ratnoff, and J. R. Shainoff. J. Lab. Clin. Med. 76:451 (1970).
5. Shainoff, J. R. and B. N. Dardik. Science 204:200 (1979).
6. Dyr, J. E., B. Blombäck, and F. Kornalik. Thromb. Res. 30:185 (1983).
7. Blombäck, B., A. Hessel, D. Hogg, and L. Therkildsen. Nature 275:501 (1978).
8. Hurlet-Jensen, A., H. Z. Cummins, H. L. Nossel, and C. Y. Liu. Thromb. Res. 27:419 (1982).

9. Masri, M. A., S. A. Marsri, and N. D. Boyd. Thromb. Haemostas. 49:116 (1983).
10. Guan, L. F. Manuscript in preparation.
11. Kay, D. and B. J. Buddigan. Br. J. Haemat. 13:341 (1967).
12. Kehl, M. and A. Henschen. In *High Performance Liquid Chromatography in Protein and Peptide Chemistry*, F. L. Lottspeich, A. Henschen, and K. P. Hupe (Eds.), Walter de Gruyter, Berlin, p. 339, 1981.
13. Lewis, S. D. and J. A. Shafer. Thromb. Res. 35:111 (1984).
14. Martinelli, R. A. and H. A. Scheraga. Anal. Biochem. 91:246 (1979).
15. Butler, V. P. and H. L. Nossel. In *Methods in Enzymology*, Volume 84, L. Lorand (Ed.), Academic Press, New York, London, 1982, p. 102.
16. Blombäck, B. and L. A. Henson. In *Plasma Proteins*, John Wiley and Sons, Chichester, 1979.
17. Kornalik, F. Folia Haematol. 104:833 (1977).
18. Jiang, Z. P., L. F. Guan, K. X. Jiang, X. Zhang, L. M. Zhi, and H. L. Wang. Zool. Res. 6:361 (1985).

8

Partial Separation and Characterization of a Thrombin-like Enzyme from the Venom of the Peruvian Bushmaster Snake, *Lachesis muta muta*

S. Campos, E. Escobar, F. Lazo, and A. Yarlequé
Universidad Nacional Mayor de San Marcos, Lima, Peru
Neville A. Marsh, Paul M. Peyser, and B. C. Whaler
Queen Elizabeth College, London, England
Lesley J. Creighton and P. J. Gaffney
National Institute for Biological Standards and Control, London, England

INTRODUCTION

The bushmaster, *Lachesis muta*, is the longest if not the largest American pit viper. It is widely distributed across northern South America, Panama, and parts of Costa Rica. The subspecies, *L. muta muta*, is restricted to Peru. In common with several other members of the Crotalidae subfamily of snakes, the bushmaster venom contains a thrombin-like enzyme. These enzymes act as proteinases, cleaving the Aα- and Bβ-chains of fibrinogen in a manner similar to thrombin. The majority of thrombin-like enzymes from snake venoms, in fact, cleave only fibrinopeptide A (FPA), and two enzymes have been extensively studies: ancrod from *Agkistrodon rhodostoma* (1) and reptilase (batroxobin) from *Bothrops atrox* (2). However, a few enzymes cleave fibrinopeptide B (FPB) more readily than FPA, such as *Agkistrodon contortrix contortrix* (3), and we have reported on one snake venom enzyme that cleaves both FPA and FPB with equal facility (4). This chapter reports on the partial separation and characterization of a thrombin-like enzyme from venom of the Peruvian bushmaster, *Lachesis muta muta*, which like that from *Bitis gabonica*, appears to remove both FPA and FPB.

MATERIALS AND METHODS

Crude venom from *L. muta muta* was obtained from snakes originating from the Alto Marañon region of northern Peru and kept in captivity at the Serpentarium of the Instituto Nacaional de Salud, Lima, Peru.

Sephadex G-100, DEAE-cellulose, tosyl-L-arginine methyl ester (TAME), α-N-benzoyl-DL-arginine-p-nitroanilide (BAPNA), α-N-benzoyl-L-arginine ethyl ester (BAEE), and bovine fibrinogen were obtained from Sigma Chemical Company, St. Louis, Missouri, USA; chromozym TH from Boehringer Mannheim GmbH, West Germany; human fibrinogen Kabi Diagnostica from Flow Laboratories, Irvine, Scotland; bovine thrombin USP, Thrombostat, from Parke Davis Company, Morris Plains, New Jersey, USA; bovine heparin from Evans Medical Ltd., Beaconsfield, England; and standard preparations of human fibrinopeptide A and human fibrinopeptide B from Sigma Chemical Co.

Coagulant activity was measured by a parallel line assay using either bovine fibrinogen (2 mg/ml in 0.02 M phosphate buffered saline, pH 7.2) or human fibrinogen (2 mg/ml) as substrate. Clotting times were recorded with dilutions of venom and with dilutions of bovine thrombin as standards. Results are expressed as NIH thrombin units, and specific coagulant activity as the number of NIH units per mg of protein. Protein concentration was determined using the method of Lowry and co-workers (5) with bovine serum albumin as a standard. Esterolytic activity was measured with BAEE (0.33 mM) and TAME (2 mM) in 0.05 M Tris-HCl buffer, pH 7.4. The changes in absorbance was measured over periods of up to 8 minutes using a Pye Unicam SP8-100 UV/VIS spectrophotometer, 1-cm path length, recording at 253 nm and 247 nm, respectively. Amidase activity was measured using BAPNA (0.9 mM in 0.05 Tris-HCl buffer, pH 8.1). After 20 minutes of incubation, the reaction was stopped with 60% acetic acid, and absorbance was read at 430 nm.

Fibrinopeptide release was determined by clotting human fibrinogen for differing periods with thrombin-like enzyme, removing insoluble protein with 1% trichloroacetic acid and then subjecting the supernatant to high pressure liquid chromatography on a Vydac C18 column, 25 × 4.6 mm internal diameter, reversed phase, using a gradient of 0.025 M ammonium acetate and acetonitrile. Identification of FPA and FPB was carried out by passing purified peptides through the column and recording the elution time. Disc polyacrylamide gel electrophoresis was performed on 7.5% gels using the method of Davis (6) and sodium dodecyl sulfate gel electrophoresis by the method of Weber and Osborn (7).

Defibrination experiments were carried out on New Zealand White rabbits (weight range 2.6 to 3.2 kg) anesthetized with sodium pentobarbitone (Sagatal, May and Baker, Dagenham, England), and in Sprague-Dawley rats (weight range 280 to 320 g) anesthetized with ethyl carbamate (urethane, Sigma Chemical Co., St. Louis, Missouri, USA). Experiments were also carried out in conscious rats implanted with an exteriorized cannula inserted into the jugular vein. Plasma fibrinogen levels were assessed by the titer method, using doubling dilutions of blood and 0.9% sodium choride (8).

RESULTS

Purification

A total of 150 mg of crude *Lachesis muta* venom was dissolved in 3 ml of 0.05 M Tris-HCl buffer, pH 8.0, and applied to a Sephadex G-100 column (2.2 × 42 cm). The gel was eluted with the same buffer at a flow rate of 5.0 ml/hr, and 2.5 ml fractions were collected. Active fractions were identified by means of the bovine fibrinogen coagulant assay. The thrombin-like enzyme was eluted as a single peak close to the void volume of the column. The active fractions were pooled, dialyzed against distilled water, and lyophilized. The pooled material from three separations was then dissolved in 0.05 M tris-HCl buffer, pH 8.0, and a total of 54 mg was applied to a 1.5 × 40 cm DEAE-cellulose column. The column had been previously equilibrated with the same buffer and was eluted with 0.05 M Tris-HCl buffer (flow rate 5.0 ml/hr), followed by a linear ionic strength gradient from 0 to 0.3 M NaCl starting after 400 ml of Tris-HCl had passed through the column. The active material was eluted as a single peak immediately upon the commencement of the NaCl gradient. The active fractions were treated as before.

Recovery and Measures of Purity

A total of 12.4 mg of lyophilized thrombin-like enzyme was recovered with a specific activity of 168 NIH units/mg. The specific activity of the crude venom was 16 u/mg and, thus, the separation procedure achieved a 10-fold purification, with a yield of 28%.

Parallel line assays of the thrombin-like enzyme in crude venom and as purified fraction generated activity-dilution curves approximating that for thrombin (Fig. 1) suggested a close similarity between the snake enzyme and purified thrombin.

Purification of the amidase and esterase activities was assessed by means of four synthetic substrates. The results are given in Table 1. The results with synthetic substrates suggest a somewhat higher purification (mean 13.5-fold) than with the functional coagulant assay. The purified thrombin-like fraction was also subjected to SDS polyacrylamide gel electrophoresis and appeared as a single band. Although this is suggestive of a high degree of purity, heavier loading of the gel may well have demonstrated small quantities of impurities.

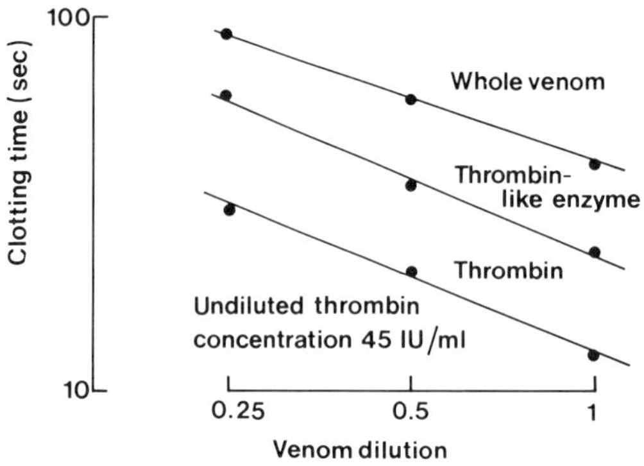

Figure 1 Parallel line assay of *Lachesis muta* thrombin-like activity.

Biochemical and Kinetic Measurements

Human fibrinogen was reacted with the thrombin-like enzyme for varying periods, and the fibrinopeptide release examined by high pressure liquid chromatography. The results (Fig. 2) show immediate cleavage of FPA, but a slow cleavage of FPB, which only became detectable after some 60 minutes of clotting. The rate of FPB release was slower than that following thrombin treatment of fibrinogen, but faster than that after ancrod treatments.

Samples of human fibrinogen were clotted for 24 hours with purified enzyme and then subjected to SDS polyacrylamide gel electrophoresis. There was no evidence of Aα-chain degradation (as seen with ancrod treatment), nor was there any α-polymer or γ-γ dimer formation (as seen with thrombin) suggesting that factor XIII- induced cross-linking was not initiated.

As with other snake venom thrombin-like enzymes, heparin did not inhibit the coagulant activity of the purified material. This effect was noted up to relatively high concentrations (Fig. 3).

Kinetic parameters for the esterolysis of TAME were calculated from a Lineweaver-Burk plot (Fig. 4). The values for V_m (135 μmol/min/mg) and for K_m (2.5×10^{-4} M) are of the same order of magnitude as those recorded for *L. muta* venom by other workers (9). The pH optimum for TAME esterolysis was 8.0 (Fig. 5).

Biological Effects

Purified enzyme was injected intravenously into anesthetized rats and rabbits. Doses as low as 50 μg/kg brought about complete defibrinogenation of rats after 60 minutes, but in the rabbits a somewhat greater dose was required (between 100 and 200 μg/kg), suggesting

Table 1 Amidase and Esterase Activity in *Lachesis muta* Crude Venom and in Purified Thrombin-Like Enzyme

Substrate	Activity in crude venom	Activity in purified fraction	Purification
BAPNA[a]	0.15	3.04	20
BAEE[†]	3.95	52.13	13
TAME[†]	7.16	96.15	13
Chromozym TH[a]	3.13	25.45	8

[a]Results with BAPNA and Chromozym TH are expressed as μmol of p-nitroanilide released/min/mg.
[†]Results with BAEE and TAME are expressed as μmol of substrate hydrolyzed/min/mg.

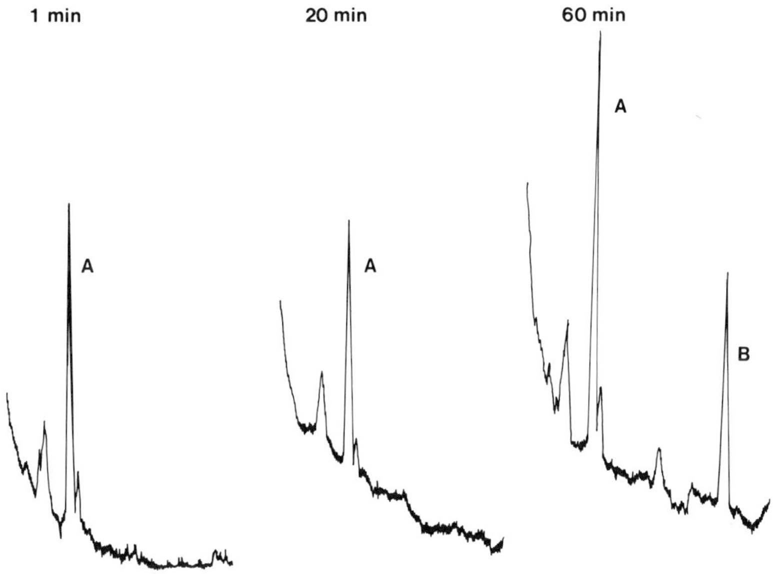

Figure 2 Recordings from high pressure liquid chromatography showing the release of FPA and FPB by *Lachesis muta* thrombin-like enzyme. Times indicate the period of fibrinogen clotting with enzyme prior to measurement of fibrinopeptide. The peaks before and immediately after FPA on all three traces represent fibrinopeptides AP and AY.

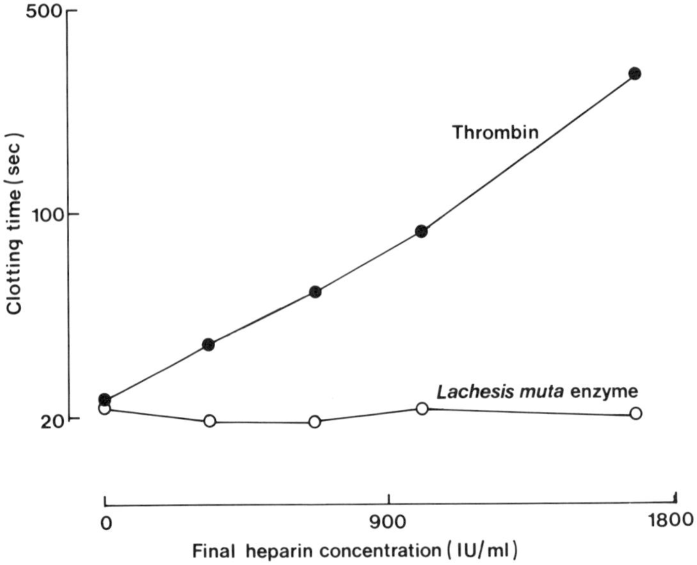

Figure 3 The effect of heparin on the coagulant activity of thrombin and *Lachesis muta* thrombin-like enzyme.

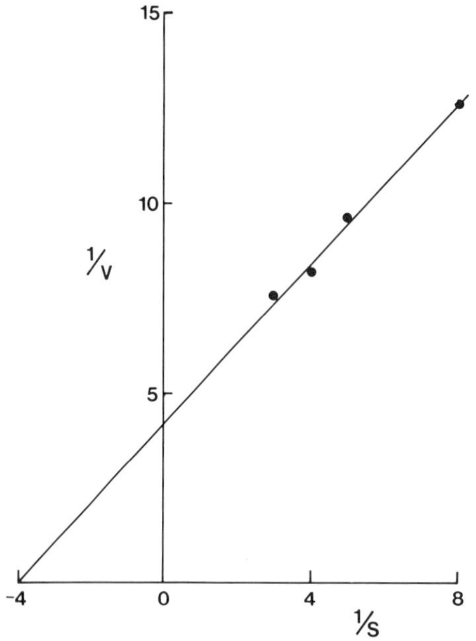

Figure 4 Lineweaver-Burk plot for *Lachesis muta*-catalyzed TAME hydrolysis. Ordinate = reciprocal of rate in min·mg/μmol × 10; abscissa = reciprocal of molar concentration of substrate × 10^{-3}.

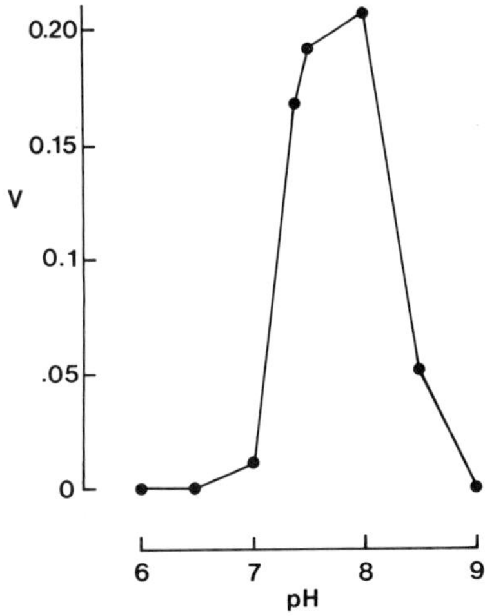

Figure 5 The pH dependence of TAME hydrolysis by *Lachesis muta* enzyme. Ordinate = rate expressed as μmol/min; abscissa = pH at which the enzyme was held for 3 minutes.

a small difference in species specificity. The animals were allowed to survive for 2 to 3 hours postdefibrinogenation, during which time no hypotensive effects were observed. A postmortem examination was carried out subsequently and in no animal were any signs of hemorrhage noted in the abdominal and thoracic viscera.

Enzyme at doses of 60 to 80 μg/kg was also administered to conscious rats by means of an indwelling venous cannula. Blood fibrinogen levels decreased to 25% of resting values after 60 minutes and to zero after 5 hours, indicating that conscious animals were less sensitive to the effects of the venom enzyme. The rats showed no long-term effects, and plasma fibrinogen levels had returned to normal after 3 days.

DISCUSSION

The results of this study indicate that the thrombin-like enzyme from *L. muta muta* venom shares many properties with similar enzymes from other crotalid and viperid snake venoms and, in particular, resembles the thrombin-like enzyme found in the venom of *L. muta noctivaga* (9,10). This subspecies of *L. muta* is found principally in Brazil.

The thrombin-like enzyme separated from *L. muta muta* venom showed a reasonable degree of homogeneity, appearing as a single band on polyacrylamide gel electrophoresis and as a single peak when rechromatographed by gel filtration. However, the specific activity of 168 NIH units/mg was low compared with the value of 1200 NIH units/mg described for ancrod (11) and about one-tenth of that recorded for *L. muta noctivaga* (9).

High pressure liquid chromatography of fibrinopeptides released from human fibrinogen by the thrombin-like enzyme revealed a rapid loss of FPA, followed by a slower release of FPB. This pattern of fibrinopeptide release more closely approximates to that seen with thrombin itself and that after treatment with purified enzyme from *Bitis gabonica*. Factor XIII-induced cross-linking of fibrin was not initiated by the *L. muta* enzyme, in common with a large number of snake venom thrombin-like enzymes. A failure to activate factor XIII undoubtedly contributed to the ability of the enzyme to defibrinogenate conscious and anesthetized rats. The limited proteolysis of the fibrinogen molecule together with a lack of factor XIII-induced cross-linking brings about the formation of partially polymerized, soluble fibrin monomers. These can be readily removed from the bloodstream by the action of the reticuloendothelial system and fibrinolysis (12). The lowest dose of enzyme bringing about defibrinogenation (50 μg) was more than 10 times higher than that used by Magalhães and associates (9), further confirming the differential in specific activity between the two preparations.

The thrombin-like enzyme from *Lachesis muta muta* venom represents yet another member of this interesting class of enzymes that are widely spread throughout the Crotalidae subfamily of snakes (13). Although no biological function has been ascribed to these enzymes, their common occurrence suggests that some advantage has been gained from their presence in venom, presumably involved in the immobilization and dispatch of prey.

REFERENCES

1. Nolan, C., L. S., Hall, and G. H. Barlow. In *Methods in Enzymology*, Volume 45, L. Lorand (Ed.), Academic Press, New York, pp. 205-213 (1976).
2. Stocker, K. and G. H. Barlow. In *Methods in Enzymology*, Volume 45, L. Lorand (Ed.), Academic Press, New York, pp. 214-222 (1976).
3. Herzig, R. H., O. D. Ratnoff, and J. R. Shainoff. J. Lab. Clin. Med. 76:451 (1970).
4. Gaffney, P. J., N. A. Marsh, and B. C. Whaler. Biochem. Soc. Trans. 1:1208 (1974).
5. Lowry, O. H., N. J. Rosebrough, A. L. Farr, and R. J. Randall. J. Biol. Chem. 193:265 (1951).
6. Davis, B. J. Ann. NY Acad. Sci. 121:404 (1964).

7. Weber, K. and M. Osborn. J. Biol. Chem. 244:4406 (1969).
8. Sharp, A. A., B. Howie, R. Biggs, and D. T. Methuen. Lancet ii:1309 (1958).
9. Magalhães, A., G. J. de Oliveira, and C. R. Diniz. Toxicon. 19:279 (1981).
10. Aragon Ortiz, F. and F. Gubenšek. Toxicon 25:364 (1987).
11. Esnouf, M. P. and G. W. Tunnah. Br. J. Haemat. 13:581 (1967).
12. Marsh, N. A. *Fibrinolysis*, John Wiley and Sons, Chichester, 1981.
13. Seegers, W. H. and C. Ouyang. In *Snake Venoms*, C-Y. Lee (Ed.), Springer-Verlag, Berlin, pp. 684-750 (1979).

9

Gabonase

Hubert Pirkle
University of California, Irvine, California

INTRODUCTION

The venom of the Gaboon viper, *Bitis gabonica*, contains a uniquely thrombin-like enzyme which Marsh and Whaler called gabonase (1). As was first noted by Gaffney et al. (2), gabonase not only releases both fibrinopeptides A and B (FPA and FPB) but also activates factor XIII. This stands in contrast to other thrombin-like enzymes which clot fibrinogen through either the exclusive release of FPA (3-5) or the preferential release of FPB (6,7) and which are unable to activate factor XIII (8).

Extending the earlier studies of Marsh and Whaler (9) and of Viljoen et al. (10), Pirkle et al. (11) isolated gabonase and found it to exhibit the properties summarized here.

PROPERTIES

Physical

Gabonase has a single polypeptide chain and an apparent molecular weight of about 30,600 as determined by sodium dodecyl sulfate gel electrophoresis with and without reduction of disulfide bonds. Its pI is 5.3. Its extinction coefficient ($E_{1cm}^{1\%}$) at 280 nm is 9.6.

Chemical

Gabonase, a serine proteinase which is inactivated by phenylmethanesulfonyl fluoride and by the chloromethyl ketone of tosyl-L-lysine, is a glycoprotein whose large carbohydrate component (20.6%) is comprised of sialic acid (5.2%), mannose (4.5%), galactose (3.3%), and glucosamine (7.6%). The molecule has five intrachain disulfide bonds, no free sulfhydryl groups, and the NH_2-terminal amino acid sequence, Val-Val-Gly-Gly-Ala-Glu-Cys-Lys-Ile-Asp-Gly-His-Arg-Cys-Leu-Ala-Leu-Tyr-, 58% homologous to crotalase but only 16% homologous to human thrombin. The activity of the enzyme is stabilized by calcium ion.

Functional

In clotting fibrinogen gabonase exhibits a specific activity of 45 NIH thrombin-equivalent units and is inhibited by fibrinogen concentrations of 3 mg/ml or greater. At gabonase concentrations that release FPA at essentially the same rate as thrombin, the rate of FPB release is about 30% that of bovine thrombin and about 10% that of human thrombin. However, the time course of fibrin formation, monitored turbidimetrically, resembles that induced by thrombin-like venom enzymes that release only FPA, namely, the lag phase relative to the assembly phase is longer than is seen with thrombin (12). This presumably reflects the fact that with gabonase the assembly of fibrin precedes the retarded release of FPB. Gabonase is indistinguishable from thrombin in the activation of factor XIII.

Gabonase is not inactivated by hirudin or by heparin.

Tosyl-arginine methyl ester is a good substrate for gabonase but tripeptide nitroanilide derivatives are not.

SIGNIFICANCE

Thrombin-like venom enzymes, differing in coagulant specificity, offer opportunities for illuminating structure-function comparisons among themselves and with thrombin. In particular, the finding that gabonase, whose single polypeptide chain is homologous to the B chain of thrombin, can release both fibrinopeptides as well as activate factor XIII, suggests that the thrombin B chain alone might be capable of these actions. And, indeed, we have found this to be the case (13).

ACKNOWLEDGMENTS

This work was supported in part by grants from the National Institutes of Health, HL-31267, and the American Heart Association, California Affiliate.

REFERENCES

1. Marsh, N. A. and B. C. Whaler, Toxicon 22:669 (1984).
2. Gaffney, P. J., N. A. Marsh, and B. C. Whaler, Biochem. Soc. Trans., 1:1208 (1973).
3. Blombäck, B. Ark. Kemi 12:321 (1958).
4. Holleman, W. H. and L. J. Coen, Biochim. Biophys. Acta 200:587 (1970).
5. K. Stocker, this volume.
6. Herzig, R. H., O. D. Ratnoff, and J. R. Shainoff, J. Lab. Clin. Med. 76:451 (1970).
7. Guan, L.-F., C.-W. Chi, and M. Yuan, Thrombos. Res. 35:301 (1984).
8. Markland, F. S. and H. Pirkle, In *Chemistry and Biology of Thrombin*, R. L. Lundblad, J. W. Fenton, II, and K. G. Mann (Eds.), Ann Arbor Science Publishers Inc., Ann Arbor, 1977.
9. Marsh, N. A. and B. C. Whaler, Br. J. Haematol. 26:295 (1974).
10. Viljoen, C. C., C. M. Meehan, and D. P. Botes, Toxicon 17:145 (1979).
11. Pirkle, H., I. Theodor, D. Miyada, and G. Simmons, J. Biol. Chem. 261:8830 (1986).
12. Pirkle, H., P. Vukasin, I. Theodor, and D. Miyada, In *Fibrinogen, Fibrin Formation and Fibrinolysis*, D. Lane, A. Henschen, and K. Jasani (Eds.), Walter de Gruyter, Berlin, 1986.
13. Pirkle, H., M. Christofferson, I. Theodor, D. Miyada, and G. Simmons, Fed. Proc. 45:1638 (1986) Abstr.

10

Thrombin-like Enzymes in the Study of Fibrin Formation

Hubert Pirkle and Ida Theodor
University of California, Irvine, California

INTRODUCTION

Thrombin possesses many actions of biological interest (1), but its most familiar or "thrombin-like" property is its capacity to induce directly the clotting of fibrinogen. In 1937 Eagle summarized his own and previous evidence that the venoms of several snake species have clot-promoting activity that is thrombin-like in this sense (2). But the chemical mode of action of these venoms was not clear until after the discovery that the thrombin-induced clotting of fibrinogen entails the proteolytic removal of two different peptides, fibrinopeptides A and B (FPA and FPB), from the amino terminus of the molecule (3,4). This discovery inevitably raised questions regarding the individual functions of the release of each of these peptides. Since then a number of enzymes isolated from snake venoms have been found to cleave off one or the other fibrinopeptide preferentially or exclusively, depending on conditions, thus providing probes to study the effects of removing each fibrinopeptide separately.

A second thrombin-like action that has been attributed to some venom enzymes is the ability to activate factor XIII and thus to affect the ultimate structure of the fibrin polymer by catalyzing covalent cross-linking between its monomeric units (e.g., Ref. 5).

Finally, many thrombin-like enzymes possess secondary proteolytic actions of sufficiently narrow specificity that they can be used to investigate a wide variety of structure-function relationships of fibrinogen and fibrin (e.g., Refs. 6,64).

This review will summarize the characteristics of these thrombin-like venom enzymes as well as their use in exploring the formation and properties of fibrin.

GENERAL CHARACTERISTICS OF THROMBIN-LIKE ENZYMES

Sources

Most of the snakes whose venoms contain thrombin-like enzymes belong to the family Crotalidae (pit vipers) (8), including the genera *Crotalus* (9-11), *Agkistrodon* (12-17), *Bothrops* (18-21), *Trimeresurus* (22-24), and *Lachesis* (25,26). In addition, thrombin-like enzymes have been found in the venoms of *Bitis gabonica* (family Viperidae or true vipers) (5,27-29) and *Disphlolidus typus* or boomslang (family Colubridae) (30).

Molecular Properties

Insofar as has been determined, the thrombin-like enzymes are all single-chain proteins (8,10) with carbohydrate content varying from 5.4% (19) to 36% (31).

While most of their molecular weights lie between 29,100 and 35,400 (8), values outside this range have been reported for the enzyme from *A. c. contortrix* (∿100,000, ∿68,000) (14,32), for the enzyme from *T. flavoviridis* (65-70,000) (23) in contrast to other *Trimeresurus* species (29,500 and 34,000) (24,33) and for one of the two thrombin-like enzymes from *A. caliginosus* (42-43,000) (16). The molecular weight of the thrombin-like enzyme from *Crotalus horridus horridus*, originally reported to be 19,500 (11), was more recently found to be 29,500 (10). There is also a significant discrepancy between the values given for *Bothrops marajoensis*, 41,500-43,000 (20) and 31,400 (18).

With the single exception of the enzyme from *T. okinavensis* (33) all reported so far have N-terminal Val (5,8,34). Partial amino acid sequences have been reported for crotalase (*Crotalus adamanteus*) (34) and gabonase (*Bitis gabonica*) (5) and the complete cDNA sequence for batroxobin (*Bothrops atrox, moojeni*) (35). These sequences are homologous to the B chain of thrombin (29%, 31%, and 26%, respectively), levels that are no greater than to many other serine proteinases. In fact, crotalase showed the most homology with porcine pancreatic kallikrein (34) and indeed was found to have kallikrein-like enzymatic activity (36). As might be expected, the thrombin-like venom enzymes are highly homologous to one another but, curiously, the greatest homology is seen between crotalase and an esterase from the venom of *Crotalus atrox* that is virtually devoid of thrombin-like activity (37,38). The relationship of the thrombin-like enzymes to the B chain of thrombin fits with reports that the B chain alone possesses thrombin-like activity (39,40).

Amino acid analyses of thrombin-like enzymes have regularly shown half-cystines (8), all apparently disulfide-bonded in the instances examined (5,9,10).

The active sites of the thrombin-like enzymes contain a serine residue and a histidine residue (5,8,10,12,23).

Enzymatic Properties

Substrates

The proteolytic specificities of the thrombin-like enzymes for fibrinogen, summarized in a later section of this review, and for other proteins are very narrow; most proteins lack any hydrolyzable peptide bond (5,8,10,16). Exceptions in the case of the *C. adamanteus* enzyme, crotalase, are the kallikrein-vulnerable bonds of high molecular weight kininogen (36) and the thrombin-vulnerable bond of prothrombin, the latter cleaved also by the venom of *A. c. contortrix* (41).

Basic amino acid esters, especially those of arginine, are hydrolyzed by most thrombin-like enzymes (8,12,16,25,26). But a colubrid enzyme failed to cleave the single arginine ester tested, benzoylarginine methyl ester (30). The arginyl bonds of some tripeptide nitroanilide derivatives are readily split by crotalase with specificity constants (kcat/Km) that correlated only roughly with those of thrombin and kallikrein (42), while all of several such compounds are poor substrates for the thrombin-like enzyme from *T. flavoviridis* (23). On the other hand, amide derivatives of arginine as well as lysine were not hydrolyzed by those thrombin-like enzymes tested (8,12).

Inhibitors

Tripeptide chloromethyl ketones with C-terminal arginine inhibit crotalase; their effectiveness in the following order, as judged by kinetic constants, depends upon the preceding limited amino acid sequence derived from the following peptides and proteins: kinin > factor XIII ≅ prothrombin (43).

Agents that reduce disulfide bonds have yielded results that apparently vary with the enzyme, although reaction conditions also vary among the different reports. Thus, crotalase (9) and the *T. flavoviridis* enzyme (23) were inactivated by 2-mercaptoethanol and the *T. okinavensis* enzyme by thioglycollic acid (22), while these agents, respectively, failed to inactivate the enzymes from *C. h. horridus* (10) and *A. acutus* (44).

Ouyang reported that sulfhydryl-reactive mercury salts inactivated the thrombin-like enzyme from *A. acutus*, but found that these compounds also inactivated thrombin which has no free sulfhydryl group (44).

Heparin alone (5,11,15,20,24,26) and heparin in the presence of an antithrombin III source (5,22,23,45,46) do not inactivate most thrombin-like enzymes. Exceptional in this respect is the enzyme from *A. c. contortrix* which is inhibited by heparin alone (14).

Hirudin has been reported not to inactivate thrombin-like enzymes (5,14,20,22) except for the enzyme from *T. flavoviridis* (23).

The effect of soybean trypsin inhibitor varies with the enzyme. Reptilase is not affected (20), but the enzyme from *A. c. contortrix* is partially inhibited (14) and the enzyme from *T. okinavensis* is inhibited only by high concentrations of soybean trypsin inhibitor (22). Aprotinin has been found inactive (20,22).

Blood serum, due to its content of α_2-macroglobulin and possibly antithrombin III, inactivates ancrod (47), whereas the *T. okinavensis* enzyme was not affected by either plasma or antithrombin III (22). The clotting activity of reptilase is inhibited by α_2-macroglobulin while its amidolytic activity is unaffected (48).

ACTIONS ON FIBRINOGEN

Fibrinopeptide Release

The specificities of thrombin-like venom enzymes for fibrinopeptide release fall into two broad groups, those directed preferentially to FPA release* (5,10,22,23,26,27,49-53) and those directed preferentially to FPB release (12,14,53).

The thrombin-like enzymes that preferentially release FPA form a spectrum with regard to FPB release, ranging from those reported not to release any FPB (10,23,49,51-53) through several that release FPB very slowly (22,26,50) to the *B. gabonica* enzyme which releases FPB at about a third the rate of bovine thrombin (5,27). Equivalent amounts of FPB are released by ancrod in 2 hours, crotalase in 30 hours, and reptilase in 72 hours (54). Thus, there is a clear trend, facilitated by improved analytical methods (55), toward finding that enzymes, previously throught to release FPA only, also slowly release FPB.

A most extraordinarily narrow specificity of preferential FPA release is exhibited by the *T. flavoviridis* enzyme which induces clotting of rabbit fibrinogen or plasma, but not human, bovine, or guinea pig fibrinogen or human, guinea pig, monkey, dog or rat plasma (23).

Clotting induced by preferential as well as virtually exclusive release of FPB by an enzyme from *A. c. contortrix* has been extensively

---

*Common names and sources of enzymes in this group that will be used frequently in this review are reptilase (*Bothrops atrox*), ancrod (*Agkistrodon rhodostoma*), and crotalase (*Crotalus adamanteus*).

investigated (14,32,56,57) and is reviewed in this volume (58). A similar enzyme has been isolated from the venom of *A. h. pallas* (12,59).

Other Peptide Bonds in Fibrinogen

α Chain

Ancrod cleaves the α chain (49,60-65) to yield an N-terminal fragment of molecular weight 39,000 and a 28 kDa C-terminal fragment which is then further digested into a series of pieces ranging from 10 to 27 kDa (60,63). The initial cleavage is at the Arg-Gly bond, Aα 348-349 (66). Ancrod also releases the septapeptide Aα 17-23 from NDSK of human fibrinogen but does not release the tripeptide Aα 17-19 as do reptilase and thrombin (67). The α chain is not attacked by crotalase (49) or by reptilase (62) except after prolonged incubation (68).

β Chain

Crotalase degrades the C-terminal β chain of human fibrinogen, leaving a large N-terminal fragment intermediate in size between the β and α chains (49). Bβ 1-42 is released from NDSK by both reptilase and ancrod (67) and from fibrinogen in 1.6 M urea by reptilase (69).

γ Chain

The γ chain is digested to a limited extent by ancrod (63) but is resistant to proteolysis by crotalase (49).

FIBRIN FORMATION

Fibrin Structure

In 1947 the groundwork for our current view of the physical assembly of the fibrin network was put forward with extraordinary insight by Ferry and Morrison (70). They postulated that linear, end-to-end linkage of enlongated monomeric units, little altered from the parent fibrinogen molecule, produces fine fibrils which then become laterally associated in bundles that, through lateral sharing of fibrils, create a branched network. Further, Ferry's early suggestion that the fine fibril (intermediate polymer of protofibril) consists of two parallel molecular chains with half-staggered junctions (71) has been borne out directly by electron microscopy (72-75).

In addition, electron microscopy has shown that fibrin fibers are indeed built up by lateral aggregation of protofibrils into bundles whose periodic cross-banding (72,76) implies strict side-to-side molecular alignment of protofibrils. Whether this alignment of protofibrils is itself also half-staggered (79) or is in register (74) has not been demonstrated directly, but the results of binding experiments with plasmic fragments (78) favor a staggered arrangement.

Recently Weisel has demonstrated by electron microscopy two levels of lateral aggregation (79). At the first level protofibrils become laterally associated into fibers of essentially the same average thickness (∿85 nm), independent of specific fibrinopeptide release and, within limits, independent of ambient conditions. In contrast, a second level of aggregation entails bundling together of these 85 nm fibers to an extent that is smoothly graded and is, on the other hand, influenced by fibrinopeptide release and solvent conditions.

Interacting Sites

Blombäck has proposed that the two distinct fibrin-forming processes, end-to-end polymerization and lateral aggregation, are activated by the release, respectively, of FPA and FPB and are mediated by two corresponding sets of molecular interaction sites, A:a and B:b (80). As we shall see, this concept arose largely from experiments with thrombin-like enzymes which, in turn, have played an important role in testing it. Efforts to localize these functional sites have taken the form mainly of investigations either of the binding properties of various fragments of the fibrinogen molecule or, more pointedly, of the ability of fragments to inhibit repolymerization of dispersed fibrin monomers. While the inhibition studies have yielded considerable information bearing on end-to-end polymerization, they have not been designed in such a way as to probe directly the process of lateral aggregation. In particular, they have not been accompanied by determination of fiber thickness by electron microscopy or by light scattering or turbidimetric measurements in relationship to quantity of fiber protein.

Nonetheless, the results of experiments using plasmic fragment DD (the distal domains cross-linked by factor XIII) permit the inference that it contains a lateral aggregation site (b-site) which is newly formed by end-to-end apposition of D nodules during protofibril formation. The DD fragment (but not the D fragment) binds to the central nodule (plasmic fragment E) (78) and interferes in some sense with the clotting of fibrinogen by thrombin or reptilase and with repolymerization of fibrin monomers (81). The reason for doubting involvement of a DD site in protofibril-forming polymerization is that the staggered buildup of the protofibril would never present a DD site for the addition of another monomeric unit. Thus, by exclusion, the DD site becomes an attractive candidate for the lateral aggregation b-site. And it is additionally attractive as it comes into being concurrently with the protofibril, the right time and place for a structure whose function is lateral aggregation. According to this interpretation the complementary B-site would be in the central domain of the molecule, but beyond this there is no evidence pointing to a more specific location or to a specific activating mechanism. The reason that the role of specific fibrinopeptide release in the binding of the central domain to fragment DD is not known is that we do not have a testable central fragment. The available fragment E preparations have not been usable as they already lack both fibrinopeptides (82), nor is NDSK suitable because, upon activation, it binds to a different site present in the solitary D fragment (next paragraph), thus precluding any examination of its interaction specifically with the DD site.

Whereas the solitary fragment D does not bind to fragment E (78), fragment D does bind to the central domain in the form of thrombin-activated NDSK, as well as to thrombin- or reptilase-treated, immobilized fibrinogen (83), and inhibits both thrombin- and reptilase-induced clotting of fibrinogen and the repolymerization of thrombin- and reptilase-induced fibrin monomers (81,84). Thus the D domain contains an interacting site which, distinguishable in its binding properties from the DD fragment, is a possible end-to-end polymerization site or a-site. This putative a-site has been more precisely localized to the γ chain segment 374-411 which binds to and inhibits repolymerization of thrombin- and reptilase-induced fibrin monomers and binds to NDSK after activation with thrombin (reptilase not tested) (85). This site has now been more narrowly located within residues 374-396 of the γ chain (86).

Again, as in the case of the B-site, the complementary A-site also appears to be situated in the N-terminal or central domain of the molecule where its expression is facilitated by the release of FPA. Indeed, binding and inhibition experiments with synthetic peptides implicate the amino terminal α chain in the interaction with fragment D (7,84,87), these in agreement with the clotting and fragment D-binding abnormalities that attend the Aα 19 Arg $\rightarrow$ Ser substitution in fibrinogen Detroit (88).

Moreover, the A-site may also include the amino terminal β chain. Thus, although the N-terminal tetrapeptide of the β chain does not competitively inhibit polymerization of fibrin induced by thrombin (7, 87), a larger fragment, Bβ 1-42, does (89). And selective removal of Bβ 1-42 with a proteinase from the venom of *Crotalus atrox* yields a species of fibrinogen which, while releasing FPA normally, is almost incoagulable (90). Also relevant to the role of the N-terminal β chain is the observation that polymerization induced by the *A. c. contortrix* enzyme (preferential FPB release) is inhibited by synthetic N-terminal β chain tripeptide and, interestingly, more strongly by the N-terminal α chain tripeptide (91).

Conditions That Promote Fibrin Formation

Fibrinopeptide Release

Although the clotting of fibrinogen by thrombin entails the release of two fibrinopeptides (FPA and FPB) (92,93) Blombäck made the signal observation that the reptilase-induced release of FPA alone is sufficient to cause clotting (50). On the other hand, Shainoff and Welches, using the thrombin-like enzyme from *A. c. controtrix*, have found that full clotting of bovine fibrinogen can be induced at 14°C by the release of FPB with less than 1% of its FPA released (58). The requirement for lower temperature in this case is not qualitatively different from thrombin-induced clotting which can be blocked by a temperature of 41°C despite complete release of FPA and FPB (58) or, indeed, reversed at 42°C (94).

The concept that the release of fibrinopeptides activates or exposes specific interaction sites has been widely used as a working hypothesis in recent years. On the other hand, according to earlier views, fibrinopeptide release functioned less specifically by removing mutually repelling electrostatic charges from fibrinogen molecules, thereby allowing them to associate (92,94-96). The observed abrogation of FPA- or FPB-mediated clotting by removal of the positively charged Bβ 15-42 with a *C. atrox* proteinase (90) or plasmin (56) is consistent with either mechanism.

Fibrin-like Aggregates

Aggregates formed from native fibrinogen in a variety of ways display the same highly ordered fine structure that is characteristic of fibrin. These include centrifuged pellets of fibrinogen examined by electron microscopy (75) and x-ray diffraction (97) and electron microscopy of fibrinogen precipitated at high ionic strength (98) or in crystalline form (99). Most impressive are the fibrin-like structures that form in the presence of positively charged protamine sulfate. These exhibit not only characteristic fibrin-like packing structure by electron microscopy, but also a typical network of fibers (96).

Interpretation

The occurrence of fibrin-like aggregates forces the conclusion that the native fibrinogen molecule contains the information required for its assembly into ordered arrays that are indistinguishable from fibrin. Thus, fibrinopeptide release does not create information but, rather, allows preexisting information to be expressed. And the highly ordered structure of fibrin implies a high degree of specificity in this information. That fibrin formation may be induced by the release of either fibrinopeptide or with the release of neither fibrinopeptide suggests that fibrinopeptide removal is of a relatively low level of specificity, one of several ways in which the prerequisites for spatially specific aggregation may be met.

PHYSICAL PROPERTIES OF FIBRIN

Electron Microscopy

The fine structure of fibrinogen clots produced by FPA release (reptilase) (72,79) and by predominantly FPB release (*A. c. controtrix* enzyme) (79,100) is the same as that produced with thrombin. The larger pattern of these clots will be discussed in the next section.

Fiber Thickness

Methods

Besides measurement of fibers in electron micrographs (79), fiber thickness can be inferred from light scattering (101) and from turbidimetry (102). Light scattering may be influenced by branching as well as fiber thickness (103), a complication that can be avoided by using the Holtzer plot to estimate branching (104).

Role of FPB Release

From light scattering studies of reptilase- and thrombin-induced fibrins Laurent and Blombäck first concluded that the lateral growth of fibrin fibers is promoted by the release of FPB (105). This idea was supported by some subsequent light scattering investigations with reptilase (72,101,106) as well as light scattering (107) and turbidimetric (108) work with ancrod, but not by other light scattering (104,109) and turbidimetric (6,84) observations using reptilase.

Contribution of the C-terminal β Chain

Pirkle et al. found a correlation between fiber thinning, estimated turbidimetrically, and ancrod-induced loss of a 28 kDa fragment from the C-terminal α chain (64). This effect was attributed to thinning of the constituent monomeric units (64), the result of loss of lateral molecular mass comprised of the C-terminal portion of the α chain (77,110, 111). But this possibility requires electron microscopic investigation, especially as Weisel and Papsun have reported that plasmic removal of a larger portion of the α chain produces abnormal branching by electron microscopy (112).

Experimental Problems

Quite aside from fibrinopeptide release and solvent conditions, the thickness of fibrin fibers is strongly inversely influenced by their rate of formation (70,100,113). The key parameter is the maximal rate of assembly of fibers from protofibrils (MRA), determined turbidimetrically as the slope of the tangent to the point of inflection of the sigmoidal curve that describes the increasing optical density during clot formation (6). The final turbidity of the clot is an inverse, rectilinear function of MRA.

A problem in experimental design may arise from attempts to equalize the activities of thrombin and thrombin-like enzymes using clotting time assays. In general the fibrinogen clotting time (largely lag time before onset of assembly) varies inversely with the MRA, but between thrombin and the thrombin-like enzymes the quantity, MRA/lag time, is not the same for any value of MRA: the lag time is relatively long in the absence of FPB release (5,114,115). The result is that, under most conditions, when clotting times produced with thrombin and thrombin-like enzyme are equal, the thrombin-like enzyme will induce a greater MRA and a correspondingly thinner final fiber (6). In other words, a clot produced by a FPA-releasing thrombin-like enzyme will regularly be less turbid than a clot produced by the "clotting time equivalent" of thrombin, on no other basis than the difference in their maximal rates of assembly.

Although it should hardly be necessary to point out that light scattering depends not only upon fiber thickness (mass/length) but also on concentration of fiber protein (101), the surprising fact is that a large proportion of the papers presenting light scattering or turbidimetric results fail to give any indication that quantity of fiber protein was determined.

As noted above thrombin-like enzymes have secondary proteolytic actions on fibrinogen which may affect fiber thickness and, indeed, have been found to do so in ways that could easily be mistaken for fibrinopeptide effects (64). Also many commercial preparations of these enzymes contain impurities, among which may be proteolytic enzymes with misleading specificities.

The presence of Tris in the clotting buffer may affect clot turbidity independent of any effect on pH (116).

Electron Microscopic Measurements

Weisel, through careful measurements on fibrin fibers induced under various conditions by reptilase, the *A. c. contortrix* enzyme, and thrombin, has found evidence for mature fibers of standard size (80-85 nm thick) which become bundled together in varying degree, depending upon fibrinopeptide release, pH and ionic strength (79). The thickness of the standard fiber was unaltered by clotting at NaCl concentrations from 0.05 to 0.40 M, pH 6.0 to 7.4, and the differences in fibrinopeptide release that would be expected from the three enzymes used. However, at pH 8.5 and NaCl 0.40 M the thickness of the basic fiber induced by all three enzymes was reduced by about half. No measurement of bundling was attempted, but it appeared to be enhanced by the release of both fibrinopeptides and by reduction of ionic strength of pH. Thus, the behavior of this bundling conforms to that of the "lateral aggregation" of earlier work and is, in fact, referred to as lateral aggregation by Weisel. However, it differs conceptually from the kind of lateral aggregation that encompasses also the side-to-side interaction of protofibrils, a process that plainly must occur in Weisel's experiments though evidently under different rules. Although the comparisons in this work involving the *A. c. contortrix* enzyme had to be carried out at low temperatures in order to obtain clots, observations on reptilase and thrombin clots were also made at room temperature and 37°C.

Mosesson et al. (100) examined by electron microscopy clots produced with the thrombin-like enzyme from *A. c. contortrix* and found, in contrast to Weisel (79), thicker bundles of fibers than with thrombin. Judging from the clotting units of each enzyme, this does not appear to have been a consequence of different rates of assembly. Likewise, preferential FPB release by the *A. h. pallas* enzyme is reported to result in a coarser clot (59).

Summary

FPB release, while not a prerequisite for lateral association of protofibrils, may influence the side-to-side aggregation of larger fibers. However, this influence is conditional; as Hantgan has stated, fiber size is kinetically determined (101). And we have found that reptilase-induced fibers may range from thinner to thicker than thrombin-induced fibers, depending only on maximal rate of assembly as controlled by enzyme concentration (6). When kinetic differences in fibrinopeptide release are excluded by comparing the repolymerization of preformed desAA and des AABB monomers, the turbidimetrically determined fiber size of the two clots is exactly the same throughout a wide range of conditions (6).

Permeability

Blombäck and Okada used liquid permeation studies to determine the porosity of clots induced with reptilase and with thrombin over a wide range of conditions of pH, ionic strength, enzyme concentration, and fibrinogen concentration (117). Their results correlated exactly with our turbidimetric findings (6), as would be expected if average pore size, reflected in permeability coefficient, varies directly with fiber thickness, reflected in turbidity. With variation of pH and ionic strength, reptilase clots were more porous for the most part with the exception of more rapidly forming clots, which were less porous when induced with reptilase than with thrombin. When fibrinogen concentration was varied, the same clotting times with reptilase and thrombin yielded less porous reptilase clots, most likely the result of greater maximal rate of assembly as explained in the section above on "Experimental Problems." With different calcium concentrations during clot formation, permeability coefficients and turbidity tended to dissociate; with increasing calcium, reptilase clots showed increases in turbidity that were the same or nearly the same as thrombin clots, but showed less increase in permeability than thrombin clots (118). This difference was attributed to an enhancement by calcium of the B:b sites rendered operative by the release of FPB.

Effects of Calcium

In contrast to other clot-accelerating factors, such as pH and high concentrations of enzyme or fibrinogen, all of which decrease fiber thickness, calcium increases fiber thickness (115,118).

Carr and co-workers found that the effect of calcium on the time course of fibrin assembly, monitored turbidimetrically, was qualitatively similar for reptilase- and thrombin-induced clots, but the calcium-induced increase in fiber thickness of reptilase clots was modestly greater than that of thrombin clots (115). Häberli et al. used light scattering to determine the effects of calcium on the polymerization process prior to gelation induced by reptilase and thrombin and obtained essentially the same results (114). Furlan et al. found that calcium produced a considerably greater acceleration of polymerization of reptilase-induced monomers as compared to thrombin-induced monomers (84). In contrast, the calorimetric results of Mihalyi and Donovan showed less effect of calcium on clotting induced by ancrod than on clotting induced by thrombin (119). Calcium also produced differential effects as between polymerization of reptilase- and thrombin-fibrin with respect to the actions of synthetic tripeptides corresponding to the N-terminal α and β chains (84).

Thermal Stability

Calorimetric comparisons of clotting brought about by ancrod and by thrombin indicate that the increase in thermal stability of the D domain that accompanies clotting is largely dependent upon the release of FPB (119).

Solubility of Monomers

Reptilase-induced fibrin monomers are more soluble than thrombin monomers in the presence of fibrinogen and fibrinogen delays the polymerization of preformed reptilase monomers more than thrombin monomers (109).

Mechanical Properties

FPB release may affect the mechanical characteristics of fibrin. Thus, for example, the elasticity of reptilase- and ancrod-induced clots is less than that of thrombin-induced clots and, in contrast to thrombin, is influenced by calcium concentration (107). Ancrod fibrin films differ markedly from thrombin fibrin, exhibiting greater stress relaxation and irrecoverable deformation (138). The development of rigidity of the ancrod clot is much more rapid than for a thrombin clot of comparable coarseness (128). Also the two clots differ strikingly in creep compliance and recovery. The failure of these properties to change after long incubation with ancrod or to be altered in thrombin clots by the addition of ancrod was taken as evidence that the responsible action of ancrod was removal of FPB rather than the C-terminal α chain (128).

FACTOR XIII ACTIVATION

The question of which thrombin-like enzymes activate factor XIII has been confused by the use of preparations of varying purity, by different methods for testing, by possible contamination of test systems with traces of prothrombin, and by different, sometimes uncertain, animal sources of enzymes.

Reptilase

Some reptilase-type preparations (i.e., from various Bothrops species) have been reported to activate factor XIII (18,62,65,122-125) while others have been reported not to do so (18,125). At first such differences were thought to depend on the species of snake form which the venom was obtained. Preparations from *B. asper* and *B. moojeni* activated factor XIII while those from *B. marajoensis* did not (125,126). Most of the efforts to resolve this problem have been directed to the species or subspecies *moojeni* and *marajoensis*. It now appears that the venoms of both *B. moojeni* (18,123,125) and *B. marajoensis* (123) contain a factor XIII-activating enzyme which is separable from the thrombin-like enzyme (18,125). But the mode of action of different preparations differs. Some, such as a mixture from *B. moojeni* and *B. marajoensis* (123) and an unspecified commercial preparation (62), result in the formation of both γ-γ dimers and α polymers, while others from *B. marajoensis* (123) and from a commercial source (122), result in the formation of γ-γ dimers only. The factor XIII activated by the *B. marajoensis* preparation was of considerable interest for, upon further treatment with thrombin, it acquired the capacity to catalyze the formation also of α polymers (123). Okada et al. speculated that the qualitatively limited capacity of these reptilase preparations to activate factor XIII reflected different specificities of thrombin and the venom enzyme, resulting in different molecular forms of activated factor XIII (122).

Thus, the factor XIII-activating properties of reptilase-type thrombin-like enzymes reported so far appear to be due to contaminating enzymes.

Ancrod

Like reptilase, ancrod preparations also are evidently contaminated (61-63,65,123) or not contaminated (65,127) with a factor XIII-activating enzyme. However, unlike some reptilase preparations, this enzyme results only in γ-γ dimer formation and, as in the case of its reptilase counterpart, the resulting partially activated factor XIII can then be fully activated to an α polymer-forming form by subsequent treatment with thrombin (123).

Other Thrombin-like Enzymes

The thrombin-like enzyme from *B. gabonica* appears to possess the inherent capacity to activate factor XIII to a form that catalyzes both γ-γ dimer and α polymer formation (5,27).

The FPB-releasing enzyme from *A. c. contortrix* (58) also activates factor XIII such that α polymers as well as γ-γ dimers are formed (32).

That both of these enzymes release FPB and activate factor XIII raises the question of whether the two substrate specificities are somehow linked.

FIBRIN CROSS-LINKING

Brosstad found that the factor XIII-catalyzed formation of both γ-γ dimers and α polymers by reptilase-induced fibrin monomers proceeded more slowly than cross-linking of thrombin-induced monomers and attributed the difference to the release of FPB (129).

Ancrod has been used to localize the acceptor sites for α chain cross-linking to two glutaminyl residues lying on either side of the primary ancrod cleavage site (60) which is located at the Arg-Gly bond in position 348-349 (66). This and other α chain cleavages by ancrod (60-65) may have a bearing on the fact that the ancrod preparations which do activate factor XIII, promote the formation of γ-γ dimers but the formation of α polymers only to a limited extent or not at all.

FIBRINOLYSIS

While both reptilase and ancrod clots are more susceptible than thrombin clots to digestion by various fibrinolytic agents (65,130), the difference cannot be ascribed wholly to failure to release FPB, as Pizzo et al. found ancrod clots to be significantly more vulnerable to plasmic lysis than reptilase clots, a difference which they attributed to degradation of the α chain by ancrod (65). Using reptilase, Olexa and Budzynski found that retention of FPB on fibrin changed the early plasmin degradation products to fragments DD and E3, in contrast to the early (DD)E complex released from thrombin fibrin which had shed both fibrinopeptides (131).

DYSFIBRINOGENS

Reptilase and ancrod have often been used in the initial empirical characterization and classification of slow-clotting dysfibrinogens (132-136), yielding information that later takes on mechanistic significance as the mutant molecule is investigated further. In fibrinogens Seattle I and Christchurch II, for example, the finding of a prolonged ancrod and reptilase clotting times (120,137) coupled with the demonstration of an amino acid substitution at the thrombin cleavage site of the Bβ chain (69,137) suggests either a conformational change that includes also the N-terminal Aα chain or, alternatively, involvement of the N-terminal β chain in the polymerization A-site. Fibrinogen Seattle I also exhibits deficient lateral aggregation of repolymerized monomers, equally so when formed with reptilase and with thrombin even though the reptilase monomers, unlike the thrombin monomers, are normal with respect to FPB release, thus pointing to an abnormality of a lateral aggregation site that is normally facilitated by FPA release (121).

Thrombin-like enzymes have also proved useful in the structural investigation of dysfibrinogens. Galanakis and co-workers were able to exploit the reptilase-incoagulability of doubly substituted fibrinogen Louisville molecules in demonstrating nonrandom assembly of chains into only wholly normal or wholly abnormal fibrinogen molecules in heterozygous individuals (139). Pirkle et al. used the secondary action of reptilase at β chain position 42-43 to isolate the substituted fragment of fibrinogen Seattle I in order to determine the state of its Cys 14 residues and its mode of assembly (69).

CLOSING COMMENT

While the thrombin-like venom enzymes have already served as valuable probes in the investigation fibrinogen and its conversion to fibrin, their value can be considerably enhanced by the use of enzymes that have been more carefully purified and characterized with respect to secondary actions.

ACKNOWLEDGMENTS

The work of the authors of this review was supported in part by grants from the National Institutes of Health, HL-31267 and the American Heart Association, California Affiliate.

REFERENCES

1. Fenton, J. W. II, Ann. N. Y. Acad. Sci. 485:5 (1986).
2. Eagle, H. J. Exper. Med. 65:613 (1937).
3. Lorand, L. Nature 167:992 (1951).
4. Bettelheim, F. R. and K. Bailey. Biochim. Biophys. Acta 9:578 (1952).
5. Pirkle, H., I. Theodor, D. Miyada, and G. Simmons. J. Biol. Chem. 261:8830 (1986).
6. Pirkle, H., P. Vukasin, I. Theodor, and D. Miyada. In *Fibrinogen, Fibrin Formation and Fibrinolysis*, D. A. Lane, H. Henschen, and M. K. Jasani (Eds.). Walter de Gruyter & Co., Berlin (1986).
7. Laudano, A. P. and R. F. Doolittle. Proc. Natl. Acad. Sci. 75: 3085 (1978).
8. Markland, F. S., Jr., and H. Pirkle. In *Chemistry and Biology of Thrombin*, R. Lundblad, J. W. Fenton, II, and K. G. Mann (Eds.). Ann Arbor Science Publishers, Inc., Ann Arbor (1977).
9. Markland, F. S. and P. S. Damus. J. Biol. Chem. 246:6460 (1971).
10. Shu, Y.-Y., J. B. Moran, and C. R. Geren. Biochim. Biophys. Acta 748:236 (1983).
11. Bonilla, C. Thrombos. Res. 6:151 (1975).
12. Guan, L.-f., C.-w. Chi, and M. Yuan. Thrombos. Res. 35:301 (1984).
13. Hatton, M. W. C. Biochem. J. 131:799 (1973).
14. Herzig, R. H., O. D. Ratnoff, and J. R. Shainoff. J. Lab. Clin. Med. 76:451 (1970).
15. Ouyang, C., J.-S. Hong, and C.-M. Teng. Thrombos. Diath. Haemorrh. 26:224 (1971).
16. Suzuki, T. and H. Takahashi. Toxicon 22:29 (1984).
17. Esnouf, M. P. and G. W. Tunnah. Br. J. Haematol. 13:581 (1967).
18. Holleman, W. H. and L. J. Weiss. J. Biol. Chem. 251:1663 (1976).
19. Ortiz, F. A. and F. Gubensek. Bull. Inst. Pasteur 74:145 (1976).
20. Stocker, K. and G. H. Barlow. Methods Enzymol. 45:223 (1976).
21. Blombäck, B., M. Blombäck, and I. M. Nilsson. Thrombos. Diath. Haemorrh. 1:76 (1957).
22. Andersson, L. Haemostasis 1:79 (1972).
23. Kosugi, T., Y. Ariga, M. Nakamura, and K. Kinjo. Thrombos. Haemost. 55:24 (1986).
24. Ouyang, C. and F.-Y. Yang. Biochim. Biophys. Acta 351:354 (1974).

25. Magalhaes, A., G. J. de Oliveira, and C. R. Diniz. Toxicon 19:279 (1981).
26. Campos, S., E. Escobar, F. Lazo, A. Yarleque, N. A. Marsh, P. M. Peyser, B. C. Whaler, L. J. Creighton, and P. J. Gaffney. *Hemostasis and Animal Venoms*, H. Pirkle and F. S. Markland, Jr. (Eds.). Marcel Dekker, New York (1988), Chap. 8.
27. Gaffney, P. J., N. A. Marsh, and B. C. Whaler. Biochem. Soc. Trans. 1:1208 (1973).
28. Marsh, N. A. and B. C. Whaler. Br. J. Haematol. 26:295 (1974).
29. Viljoen, C. C., C. M. Meehan, and D. P. Botes. Toxicon 17:145 (1979).
30. Hiestand, P. C. and R. R. Hiestand. Toxicon 17:489 (1979).
31. Nolan, C., L. S. Hall, and G. H. Barlow. Methods Enzymol. 45:205 (1976).
32. Dyr, J. E., B. Blombäck, and F. Kornalik. Thrombos. Res. 30:185 (1983).
33. Andersson, L. Haemostasis 1:31 (1972).
34. Pirkle, H., F. S. Markland, I. Theodor, R. Baumgartner, S. S. Bajwa, and H. Kirakossian. Biochem. Biophys. Res. Commun. 99:715 (1981).
35. Itoh, N., N. Tanaka, S. Mihashi, and I. Yamashina. J. Biol. Chem. 262:3132 (1987).
36. Markland, F. S., C. Kettner, S. Schiffman, E. Shaw, S. S. Bajwa, K. N. N. Readdy, H. Kirakossian, G. B. Patkos, I. Theodor, and H. Pirkle. Proc. Natl. Acad. Sci. 79:1688 (1982).
37. Bjarnason, J. B., A. Barish, G. S. Direnzo, R. Campbell, and J. W. Fox. J. Biol. Chem. 258:12566 (1983).
38. Pirkle, H. and I. Theodor. Unpublished observations.
39. Hageman, T. C., G. F. Endres, and H. A. Scheraga. Arch. Biochem. Biophys. 171:327 (1975).
40. Pirkle, H., M. Christofferson, I. Theodor, D. Miyada, and B. Simmons. Fed. Proc. 45:1638 (1986) abstr.
41. Pirkle, H., F. S. Markland, and I. Theodor. Thrombos. Res. 8:619 (1976).
42. Simmons, G., M. Bundalian, I. Theodor, J. Martinoli, and H. Pirkle. Thrombos. Res. 40:555 (1985).
43. Markland, F. S., C. Kettner, E. Shaw, and S. S. Bajwa. Biochem. Biophys. Res. Commun. 102:1302 (1981).
44. Ouyang, C. and J.-S. Hong. Toxicon 12:449 (1974).
45. Markland, F. S., Jr. Methods Enzymol. 45:223 (1976).
46. Aronson, D. L. Thrombos. Haemostas. 36:9 (1976).
47. Pitney, W. R. and E. Regoeczi. Br. J. Haemat. 19:67 (1970).
48. Egberg, N. Thrombos. Res. 4:35 (1974).
49. Markland, F. S. and H. Pirkle. Thrombos. Res. 10:487 (1977).
50. Blombäck, B. Ark. Kemi 12:321 (1958).

51. Ewart, M. R., M. W. C. Hatton, J. M. Basford, and K. S. Dodgson. Biochem. J. 118:603 (1970).
52. Holleman, W. H. and L. J. Coen. Biochim. Biophys Acta 200: 587 (1970).
53. Bilezikian, S. B., H. L. Nossel, V. P. Butler, Jr., and R. E. Canfield. J. Clin. Invest. 56:438 (1975).
54. Pirkle, H. and I. Theodor. Unpublished observations.
55. Kehl, M., F. Lottspeich, and A. Henschen. Hoppe-Seyler's Z. Physiol. Chem. 362:1661 (1981).
56. Shainoff, J. R. and B. N. Dardik. Science 204:200 (1979).
57. Shainoff, J. R. and B. N. Dardik. Ann. N. Y. Acad. Sci. 408: 254 (1983).
58. Shainoff, J. R. and W. R. Welches. *Hemostasis and Animal Venoms*, H. Pirkle and F. S. Markland, Jr. (Eds.). Marcel Dekker, New York (1988).
59. Guan, L.-f., X. Zhang, and C.-w. Chi. *Hemostasis and Animal Venoms*, H. Pirkle and F. S. Markland, Jr. (Eds.). Marcel Dekker, New York (1988).
60. Fretto, L. J., E. W. Ferguson, H. M. Steinman, and P. A. McKee. J. Biol. Chem. 253:2184 (1978).
61. Gaffney, P. J. and M. Brasher. Nature 251:53 (1974).
62. Mattock, P. and M. P. Esnouf. Nature New Biol. 233:277 (1971).
63. Edgar, W. and C. R. M. Prentice. Thrombos. Res. 2:85 (1973).
64. Pirkle, H., P. Vukasin, I. Theodor, and G. Simmons. In *Fibrinogen and its Derivatives*, G. Müller-Berghaus, U. Scheefers-Borchel, E. Selmayr, and A. Henschen (Eds.). Elsevier Science Publishers B. V. (Biomedical Division), Amsterdam (1986).
65. Pizzo, S. V., M. L. Schwartz, R. L. Hill, and P. S. McKee. J. Clin. Invest. 51:2841 (1972).
66. Pirkle, H. and I. Theodor. Unpublished observations.
67. Hessel, B. and M. Blombäck. FEBS Lett. 18:318 (1971).
68. Holm, B. and H. C. Godal. In *Fibrinogen and Its Derivatives*, G. Müller-Berghaus, U. Scheefers-Borchel, E. Selmayr, and A. Henschen (Eds.). Elsevier Science Publishers B. V. (Biomedical Division), Amsterdam (1986).
69. Pirkle, H., H. Kaudewitz, A. Henschen, I. Theodor, and G. Simmons. In *Fibrinogen*, G. D. O. Lowe, C. D. Forbes, J. T. Douglas, and A. Henschen (Eds.). Elsevier Science Publishers B. V. (Biomedical Division), Amsterdam (in press).
70. Ferry, J. D. and P. R. Morrison. J. Am. Chem. Soc. 69:388 (1947).
71. Ferry, J. D. Proc. Natl. Acad. Sci. 38:566 (1952).
72. Hantgan, R., W. Fowler, H. Erickson, and J. Hermans. Thrombos. Haemost. 44:119 (1980).
73. Krakow, W., G. F. Endres, B. M. Siegel, and H. A. Scheraga. J. Mol. Biol. 71:95 (1972).

74. Fowler, W. E., R. R. Hantgan, J. Hermans, and H. P. Erickson. Proc. Natl. Acad. Sci. 78:4872 (1981).
75. Williams, R. C. Proc. Natl. Acad. Sci. 80:1570 (1983).
76. Hall, C. E. and H. S. Slayter. J. Biophys. Biochem. Cytol. 5:11 (1959).
77. Mosesson, M. W., J. Hanfeld, R. H. Haschemeyer, and J. Wall. J. Mol. Biol. 153:695 (1981).
78. Olexa, S. A. and A. Z. Budzynski. J. Biol. Chem. 254:4925 (1979).
79. Weisel, J. W. Biophys. J. 50:1079 (1986).
80. Blombäck, B., B. Hessel, D. Hogg, and L. Therkildsen. Nature 257:501 (1978).
81. Dray-Attali, L. and M. J. Larrieu. Thrombos. Res. 10:575 (1977).
82. Olexa, S. A., A. Z. Budzynski, R. F. Doolittle, B. A. Cottrell, and T. C. Greene. Biochemistry 20:6139 (1981).
83. Kudryk, B. J., D. Collen, K. R. Woods, and B. Blombäck. J. Biol. Chem. 249:3322 (1974).
84. Furlan, M., C. Rupp, and E. A. Beck. Biochim. Biophys. Acta 742:25 (1983).
85. Olexa, S. A. and A. Z. Budzynski. J. Biol. Chem. 256:3544 (1981).
86. Horwitz, B. H., A. Varadi, and H. A. Scheraga. Proc. Natl. Acad. Sci. 81:5980 (1984).
87. Laudano, A. P. and R. F. Doolittle. Biochemistry 19:1013 (1980).
88. Kudryk, B., B. Blombäck, and M. Blombäck. Thrombos. Res. 9:25 (1976).
89. Budzynski, A. Z., S. A. Olexa, and B. V. Pandya. Ann. N. Y. Acad. Sci. 408:301 (1983).
90. Pandya, B. V., C. S. Cierniewski, and A. Z. Budzynski. J. Biol. Chem. 260:2994 (1985).
91. Furlan, M., C. Rupp, E. A. Beck, and L. Svendsen. Thrombos. Haemostas. 47:118 (1982).
92. Bettelheim, F. R. and K. Bailey. Biochim. Biophys. Acta 9:578 (1952).
93. Lorand, L. Nature 167:992 (1951).
94. Laki, K. and J. A. Gladner. Physiol. Rev. 44:127 (1964).
95. Doolittle, R. F. Advan. Protein Chem. 27:1 (1973).
96. Stewart, G. J. and S. Niewiarowski. Biochim. Biophys. Acta 194:462 (1969).
97. Stryer, L., C. Cohen, and R. Langridge. Nature 197:793 (1963).
98. Tooney, N. M. and C. Cohen. Nature 237:23 (1972).
99. Gollwitzer, R., W. Bode, H.-J. Schram, D. Typke, and R. Guckenberger. Ann. N. Y. Acad. Sci. 408:214 (1983).
100. Mosesson, M. W., J. P. DiOria, M. F. Müller, J. R. Shainoff, K. R. Siebenlist, D. L. Amrani, G. A. Homandberg, J. Soria, C. Soria, and M. Samama. Blood 69:1073 (1987).

101. Hantgan, R. R. and J. Hermans. J. Biol. Chem. 254:11272 (1979).
102. Carr, M. and J. Hermans. Macromolecules 11:46 (1978).
103. Müller, M. and W. Burchard. In *Fibrinogen. Recent Biochemical and Medical Aspects*, A. Henschen, H. Graeff, and F. Lottspeich (Eds.). Walter de Gruyter, Berlin (1982).
104. Müller, M., H. Lasarcyk, and Walther Burchard. Int. J. Biol. Macromol. 3:19 (1981).
105. Laurent, T. C. and B. Blombäck. Acta Chem. Scand. 12:1875 (1958).
106. Wiltzius, P., G. Dietler, W. Kanzig, V. Hofman, A. Häberli, and P. W. Straub. Biophys. J. 38:123 (1982).
107. Shen, L., L. J. Hermans, J. McDonagh, and R. P. McDonagh. Am. J. Physiol. 232:H629 (1977).
108. Spellman, G. G., Jr., J. A. Macoviak, and H. R. Gralnick. Blood 50:619 (1977).
109. Brosstad, F., P. Kierulf, and H. C. Godal. Thrombos. Res. 14:705 (1979).
110. Erickson, H. P. and W. E. Fowler. Ann. N. Y. Acad. Sci. 408:146 (1983).
111. Weisel, J. W., C. V. Stauffacher, E. Bullitt, and C. Cohen. Science 230:1388 (1985).
112. Weisel, J. W. and D. M. Papsun. Thrombos. Res. 47:115 (1987).
113. Buchanan, J. M., L. B. Chen, T. Hamazaki, E. Lenk, and D. Waugh. In *Chemistry and Biology of Thrombin*, R. L. Lundblad, J. W. Fenton, II, and K. G. Mann (Eds.). Ann Arbor Science Publishers, Inc., Ann Arbor (1977).
114. Häberli, A., P. W. Straub, G. Dietler, and W. Kanzig. Biopolymers 26:27 (1987).
115. Carr, M. E., Jr., D. A. Gabriel, and J. McDonagh. Biochem. J. 239:513 (1986).
116. Pirkle, H. and I. Theodor. Unpublished Observations.
117. Blombäck, B. and Okada. Thrombos. Res. 25:51 (1982).
118. Okada, M. and B. Blombäck, Thrombos. Res. 29:269 (1983).
119. Mihalyi, E. and J. W. Donovan. Biochemistry 24:3443 (1985).
120. Branson, H. E., G. Schmer, and D. H. Dillard. Am. J. Clin. Path. 67:236 (1977).
121. Branson, H. E., G. Schmer, I. Theodor, and H. Pirkle. Acta Haemat. 70:257 (1983).
122. Okada, M., B. Blombäck, M.-D. Chang, and B. Horowitz. J. Biol. Chem. 260:1811 (1985).
123. Walter, M., D. Nyman, V. Krajnc, and F. Duckert. Thrombos. Haemost. 38:438 (1977).
124. Kopec, M., Z. S. Latallo, M. Stahl, and Z. Wegrzynowicz. Biochim. Biophys. Acta 181:437 (1969).
125. McDonagh, J. and R. P. McDonagh. Br. J. Haemat. 30:465 (1975).

126. Stocker, K. and G. H. Barlow. Methods Enzymol. 45:214 (1976).
127. Barlow, G. H., W. H. Holleman, and L. Lorand. Res. Commun. Chem. Path. Pharm. 1:39 (1970).
128. Bale, M. D., M. Müller, and J. D. Ferry. Biopolymers 24:461 (1985).
129. Brosstad, F. Thrombos. Res. 13:799 (1978).
130. Kwaan, H. C., G. H. Barlow, and N. Suwanwela. Thrombos. Res. 2:123 (1973).
131. Olexa, S. A. and A. Z. Budzynski. Biochemistry 19:647 (1980).
132. Henschen, A., M. Kehl, and C. Southan. In *Variants of Human Fibrinogen,* E. A. Back and M. Furlan (Eds.). Hans Huber Publishers, Berne (1984).
133. Henschen, A. and J. McDonagh. In *Blood Coagulation,* F. Zwaal and H. C. Hemker (Eds.). Elsevier Science Publishers, B.V. (Biomedical Division), Amsterdam (1986).
134. Siebenlist, K. R., J. T. Prchal, and M. W. Mosesson. Thrombos. Haemostas. 58:148 (1987) abstr.
136. Galanakis, D. K. and H. A. Mondhiri. In *Fibrinogen,* G. D. O. Lowe, C. D. Forbes, J. T. Douglas, and A. Henschen (Eds.). Elsevier Science Publishers, B.V. (Biomedical Division), Amsterdam (in press).
136. Pirkle, H., I. Theodor, P. Vukasin, B. Nandi, D. Miyada, J. Baker, and M. Dukes. Thrombos. Haemostas. 59:148 (1987).
137. Kaudewitz, H., A. Henschen, C. Soria, J. Soria, O. Bertrand, and D. Heaton. In *Fibrinogen and Its Derivatives,* G. Müller-Berghaus, U. Scheefers-Borchel, E. Selmayr, and A. Henschen (Eds.). Elsevier Science Publishers, B.V. (Biomedical Division), Amsterdam (1986).
138. Müller, M. and J. S. Ferry. Biopolymers 23:2311 (1984).
139. Galanakis, D. K., A. Henschen, M. Keeling, M. Kehl, R. Dinsmore, and E. I. Peerschke. Ann. N. Y. Acad. Sci. 408:644 (1983).

11

Helodermatine, an Enzyme from the Venom of *Heloderma horridum horridum* (Mexican Beaded Lizard) with Kallikreinlike Properties

Wolf-Dieter Schleuning
Centre Hospitalier Universitaire Vaudois, Lausanne, Switzerland
Alejandrô Alagon*
Universidad Nacional Autonoma de Mexico, Mexico City, Mexico

INTRODUCTION

The Mexican beaded lizard (*Heloderma horridum horridum*) is one of the two members of the Helodermatidae family (suborder Sauria) that inhabit the southwestern United States and the Pacific Coast of Mexico. The Heloderma venom has recently been characterized in some detail. A neurotoxin (Hendon and Tu, 1981), a hyaluronidase (Tu and Hendon, 1983), peptides related to the vasoactive intestinal peptide-secretin family (Parker et al., 1984; Robberecht et al., 1984), and two arginine esterases (Alagon et al., 1982) have been described. Fleisher (1913) described a hypotensive effect of Heloderma venom when injected into rabbits. Mebs (1969) observed a kinin-generating (or kallikrein-like) activity in the Gila monster (*H. suspectum*). These observations motivated us to purify the arginine esterases from the venom. We characterized the 63-Kda form, which we named helodermatine (Alagon et al., 1986).

MATERIAL AND METHODS

All chemicals were of the best grades available from commercial suppliers. Heloderma venom was obtained by the technique originally described by Fleisher (1913). Human high molecular weight kininogen was generously provided by Drs. Brigitte Dittmann and Hans Fritz from the University of Munich, West Germany. Benzoyl-arginine ethyl

**Present affiliation:* Universidad Autonoma de Mexico, Cuernavaca, Mexico ester (BAEE)-hydrolyzing activity was determined as previously described (Alagon et al., 1982).

Chromogenic substrate assays were performed, following the procedure of Latallo and co-workers (1979). Kinin-liberating activity was measured according to the method of Trautschold (1970). All column chromatographic procedures were performed in standard laboratory glassware at ambient temperature. Electrophoresis was performed as described by Laemmli (1970). The amino-terminal sequence of heloermatine was determined by sequential chemical degradation as described by Edman (1956), applying the technical specifications of Hewick and colleagues (1981).

RESULTS AND DISCUSSION

Helodermatine was purified using a two-step procedure consisting of benzamidine-Sepharose affinity chromatography and DEAE-cellulose anion-exchange chromatography. The resulting protein was homogeneous by SDS polyacrylamide gel electrophoresis, and more than 90% pure by the yield of the first amino-acid derivative obtained by Edman degradation. Of the first 21 residues at the amino-terminus, 19 were identified by automated sequencing, and a significant homology was found to porcine pancreatic kallikrein (Tschesche et al., 1979), crotalase from *Crotalus adamanteus* (Pirkle et al., 1981), and kallikrein-like enzymes from *C. atrox* (Bjarnason et al., 1983).

Using the substrates D-Pro-Phe-Arg pNa (S-2302) and D-Glu-Gly-Arg pNa (S-2266), kinetic data were obtained which classified helodermatine as kallikrein-like enzyme. This notion was corroborated by the strong competition of the enzyme by Kunitz-type bovine pancreatic trypsin inhibitor (aprotinine, Trasylol) and soybean trypsin inhibitor (SBTI).

Finally it was shown that helodermatine is a potent kininogenase (i.e., an enzyme that cleaves the vasoactive kinins out of kinonogen). Thus, by structural and enzymologic criteria, helodermatine belongs to the kallikrein family. It is the first enzyme from Sauria with such properties.

REFERENCES

Alagon, A. C., M. E. Maldonado, J. Z. Julia, C. R. Sanchez, and L. D. Possani. Venom from two sub-species of *Heloderma horridum* (Mexican beaded lizard): general characterization and purification of N-benzoyl-L-arginine ethyl ester hydrolase. Toxicon 20:463-474 (1982).

Alagon, A., L. D. Possani, J. Smart, and W. D. Schleuning. Helodermatine, a kallikrein-like, hypotensive enzyme from the venom of *Heloderma horridum horridum* (Mexican beaded lizard). J. Exp. Med. 164:1835-1845 (1986).

Bjarnason, J. B., A. Barish, G. S. Direnzo, R. Campbell, and J. W. Fox. Kallikrein-like enzymes from *Crotalus atrox* venom. J. Biol. Chem. 258:12566-12573 (1983).

Edman, P. On the mechanism of the phenyl isothiocyanate degradation of peptides. Acta Chem. Scand. 10:761-765 (1956).

Fleisher, M. S. Influence of Heloderma venom upon blood-pressure, diuresis, peritoneal transudate and intestinal fluid. In *The Venom of Heloderma.* Loeb, L. (Ed.), The Carnegie Institution of Washington, Washington, D.C., pp. 101-117 (1913).

Hendon, R. A. and A. T. Tu. Biochemical characterization of the lizard toxin gilatoxin. Biochemistry 20:3517-3522 (1981).

Hewick, R. M., M. W. Hunkapiller, L. E. Hood, and W. J. Dreyer. A gas-liquid solid phase peptide and protein sequenator. J. Biol. Chem. 256:7990-7997 (1981).

Laemmli, U. K. Cleavage of structural proteins during the assembly of the head of bacteriophage T4. Nature (London) 227:680-685 (1970).

Latallo, Z. S., E. Teisseyre, and E. Raczka. Activators of plasminogen measurement using chromogenic substrates. In *Chromogenic Peptide Substrates,* M. F. Scully and V. V. Kakar (Ed.), Churchill Livingstone, Edinburgh, pp. 154-164 (1979).

Mebs, D. Isolierung und Eigenschaften eines Kallikreins aus dem Gift der Krustenechse Heloderma suspectum. Hoppe Seyler Z. Physiol. Chem. 350:821-826 (1969).

Parker, D.-S., J. P. Raufman, T. L. O'Donohue, M. Bledsoe, H. Yoshida, and J. J. Pisano. Amino acid sequences of helospectins, new members of the glucagon superfamily, found in Gila monster venom. J. Biol. Chem. 259:11751-11755 (1981).

Pirkle, H., F. S. Markland, I. Theodor, R. Baumgartner, S. S. Bajwa, and H. Kirakossian. The primary structure of crotolase, a thrombin-like venom enzyme, exhibits closer homology to kallikrein than to other serine proteases. Biochem. Biophys. Res. Commun. 99:715-721 (1981).

Robberecht, P., M. Waelbroeck, J. P. Dehaye, J. Winand, A. Vandermeers, M. C. Vandermeers-Piret, and J. Cristophe. Evidence that helodermin, a newly extracted peptide from Gila monster venom, is a member of the secretin/VIP/PHI family of peptides with an original pattern of biological properties. FEBS Lett. 166:277-282 (1984).

Trautschold, I. Assay methods in the kinin system. In *Handbook of Experimental Pharmacology. Volume 25. Bradykinin, Kallidin and Kallikrein.* E. G. Erdös (Ed.). Springer-Verlag, Berlin, pp. 52-81 (1970).

Tschesche, H., G. Mair, and G. Godec. The primary structure of porcine glandular kallikreins. Adv. Exp. Med. Biol. 120:245-260 (1979).
Tu, A. T. and R. R. Hendon. Characterization of lizard venom hyaluronidase and evidence for its action as a spreading factor. Comp. Biochem. Physiol. 76(2):377-383 (1983).

II
FIBRINOLYSIS

12

Fibrin(ogen)olytic Enzymes from Snake Venoms

Francis S. Markland, Jr.
University of Southern California School of Medicine, Los Angeles, California

INTRODUCTION

Fibrin(ogen)olytic activity has been described in the venoms of a number of snake species, including members of the Crotalidae, Viperidae, and Elapidae families. For the most part, the snake venom enzymes are direct-acting fibrinolytic proteinases, but there are scattered reports of plasminogen activators or plasminogen activator-releasing activities in snake venoms. There are also numerous reports of fibrinolytic activity in a variety of bloodsucking animals and insects; presumably, the purpose of fibrinolytic activity in these creatures is to prevent the host blood from clotting, thereby facilitating feeding and ingestion. Additionally, there is substantial published literature describing proteinases with fibrinolytic activity from different bacterial (1-3) and fungal species (4,5), including brinase, the fibrinolytic enzyme from the mold *Aspergillus oryzae* (6-9). This review will focus primarily on snake venom enzymes with fibrin(ogen)olytic activity, although a brief description will be given of fibrinolytic enzymes in bloodsucking animals and insects.

Why are fibrinolytic activities present in snake venoms? It is possible that these enzymes act to prevent clotting of blood in an animal that has been envenomated, thereby enabling other components of the venom to spread rapidly throughout the blood and quickly and effectively immobilize the hapless creature. In addition to an interest in the protein chemistry of these enzymes and in their structural and evolutionary relationships with other proteolytic enzymes, investigators are interested in using these enzymes to help elucidate structural features involved in fibrin polymerization, and to remove fibrin(ogen) from blood under in vitro and in vivo conditions. From the clinical perspective, these enzymes may prove useful for the dissolution of blood clots formed, for example, during myocardial infarction, deep venous thrombosis, stroke, and pulmonary embolism.

DIRECT-ACTING FIBRIN(OGEN)OLYTIC ENZYMES IN SNAKE VENOMS

General

This topic has been reviewed previously by several investigators (10-14). In the 1700s, Fontana (15) was probably the first to observe that blood remained fluid in animals that died from viper bite. We now know that this effect of venom is most probably due to the combined action of fibrin(ogen)olytic and defibrinogenating (thrombin-like) enzymes. In the late 1800s Mitchell and Reichert (16,17), in a broad series of investigations, studied the effects of rattlesnake venoms on animal organs including blood. They observed that when an animal dies soon after envenomation, the blood is coagulable, but when death is delayed, the blood becomes incoagulable. They further observed from in vitro experiments that fibrinogen in animal blood, after prolonged contact with venom, loses its power to coagulate. Mitchell (16) also reported that when animal blood was rapidly mixed with rattlesnake venom and then set aside at room temperature after clot formation, the clots in many cases completely dissolved by 24 hours. This suggested the presence of fibrinolytic activity in rattlesnake venom. Mitchell and Reichart (17) extended these studies to snakes of the genus *Agkistrodon*, with similar findings. They further showed that this activity was attributable to proteins in the venoms. However, cobra venom had no such marked effect on the coagulability of animal blood.

The presence of fibrinolytic activity in snake venoms was further studied by Eagle (18). He tested 17 snake venoms and found that, in addition to containing enzymes that caused fibrinogen to clot, several venoms contained enzymes that destroyed fibrinogen. Included in the latter group were venoms from *Agkistrodon piscivorus* (cottonmouth moccasin), *Crotalus atrox* (Western diamondback rattlesnake), *C. horridus* (timber rattlesnake), *Bitis arietans* (puff adder), and *Vipera ammodytes* (sand viper). Eagle attributed the action on fibrinogen to proteolytic enzymes in the venoms.

Didisheim and Lewis (19), in an attempt to find an ideal agent for the intravascular dissolution of clinically important blood clots, screened 16 snake venoms for fibrinolytic activity. Fibrinolysis (in which the fibrin clot was formed in the presence of the venom), fibrin clot lysis (in which venom was added after fibrin clot formation), and whole blood clot lysis times were assessed to obtain an indication of the lytic action of the different venoms. Venoms of the *Crotalus, Bothrops,* and *Agkistrodon* groups were fibrinolytic, whereas those of the cobras were not. Interestingly, cobra venoms exhibited fibrinogenolytic activity, but not fibrinolytic action. These authors were the first to suggest that since these enzymes would be less susceptible to human serum proteinase inhibitors, there would be a theoretical advantage for using purified venom enzymes to lyse human blood clots in clinical situations (19).

Kornalik studied the fibrinolytic activity in 24 different snake venoms and found the strongest activity in venoms from *Agkistrodon piscivorus* (cottonmouth), *A. contortrix* (copperhead), and *Vipera lebetina* (Levantine viper) (20,21). All of these venoms contained potent fibrinolytic activity. The venoms were shown to be directly fibrinolytic because they did not require plasminogen for fibrin plate lysis. The venoms also destroyed fibrinogen by their proteolytic action, and this action was not inhibited by epsilon-aminocaproic acid, soybean trypsin inhibitor, or trasylol. Interestingly, serum from humans and other species inactivated the fibrinogenolytic activity of the snake venoms (20,21).

Mohamed and El-Damarawy (22) studied fibrinolytic activity in five Egyption snake venoms (the vipers *Cerastes cerastes, C. vipera,* and *Echis carinatus*, and the cobras *Naja nigricolis* and *N. haje*). Whereas all five venoms produced fibrinogenolysis, only the three viper venoms lysed fibrin plates. Interestingly, plasminogen-free fibrin plates were much less susceptible to the lytic action of the viper venoms than were plasminogen-containing plates. These findings suggest that there may be both direct-acting fibrinolytic activity as well as plasminogen activator activity in the viper venoms studied, whereas the cobra venoms contain only fibrinogenolytic activity.

From a series of studies by many investigators (19,22-27) with crude venoms over the past 30 years, it is generally believed that members of the Crotalidae family contain strong fibrinolytic activity in their venoms, whereas members of the Elapidae family have little or no lytic activity. Venom from some members of the Viperidae family appear to have fibrin(ogen)olytic activity (28-33).

Pit Viper Fibrin(ogen)olytic Activity

Rattlesnakes (Crotalus)

Since the early work of Eagle in 1937 (18), it has been known that rattlesnake venoms contain fibrin(ogen)olytic activity. These studies were confirmed and extended by Didisheim and Lewis (19). Markland and co-workers (34,35) screened a series of venoms and reported that fibrinolytic activity was present in all rattlesnake venoms tested. All of the studies indicated that fibrinolytic activity was present in *Crotalus atrox* (Western diamondback rattlesnake) venom. Denson and associates (23) reported that adult *C. atrox* venom contained fibrinolytic activity, but lacked the fibrinogen clotting enzyme that typifies rattlesnake venom. Interestingly, Posch and co-workers (36) reported that small *C. atrox* specimens contained the fibrinogen clotting enzyme but lacked fibrinolytic activity, whereas the reverse was true for large snakes of this species. Intermediate-sized snakes apparently had both activities. Reid and Theakston (37) reported that as *C. atrox* snakes aged, they demonstrated significant changes in the coagulant activity of their venoms. Foremost was the loss of the fibrinogen clotting activity. Thus, in adult specimens of this species, fibrinolytic activity predominates, whereas in juvenile specimens, thrombin-like activity is the predominant component. Recently, three groups of investigators have studied fibrinolytic activity in *C. atrox* venom (38-42,45,47-49). Laboratory studies were undertaken to purify and characterize fibrinolytic activity in this venom (38,39). Molecular sieve chromatography of crude venom on Sephadex G-100 revealed two peaks of activity, one of high and the other of lower molecular weights. Plasminogen-free fibrin plates were used to assess fibrinolytic activity (38). Purification of the fibrinolytic enzymes from crude venom was carried out by chromatography on Sephadex G-100. The two peaks of fibrinolytic activity were pooled separately, and the high molecular weight fibrinolytic enzyme was additionally purified by chromatography on hydroxylapatite cellulose, benzamidine Sepharose, QAE-Sephadex, and DEAE-cellulose. The final preparation exhibited a single band on SDS acrylamide gel electrophoresis. However, the procedure yielded a very low recovery of fibrinolytic activity. The low molecular weight fibrinolytic enzyme was purified by a similar procedure; the yield was also quite low. The molecular weights of these enzymes were determined by SDS acrylamide gel electrophoresis and found to be 60,000 and 21,500 for the high and low molecular weight fibrinolytic enzymes, respectively. Neither enzyme activates plasminogen, as observed by direct incubation with plasminogen and analysis of plasmin activity generated using either fibrin plate lysis assay or hydrolysis of the plasmin-specific chromogenic substrate S-2251 (D-Val-Leu-Lys-p-nitroanilide). The high molecular weight enzyme was rapidly inhibited by diisopropylfluorophosphate, but this agent had no inhibitory effect on the low molecular weight enzyme. Although the high molecular weight fibrinolytic enzyme exhibited hemorrhagic activity by the rat skin test (39), the low molecular weight enzyme was virtually devoid of this activity at a comparable dose.

Sapru and co-workers (40) isolated a fibrinogenase from Western diamondback rattlesnake venom with a molecular weight of 22,900 and an isoelectric point at pH 4.65. The enzyme is inactivated by phenylmethylsulfonyl fluoride, but not by EDTA. Activity of the enzyme is directed preferentially toward the B β-chain of fibrinogen, but on prolonged incubation with fibrinogen, there was partial digestion of the A α-chain. The enzyme had no activity on casein or fibrin (fibrin plate lysis), and appears to be similar to a B β-chain fibrinogenase isolated from this venom and called protease II (41,42). The enzyme also appears similar in many of its properties to B β-chain fibrinogenases purified by Ouyang and co-workers from venoms of *Trimeresurus mucrosquamatus* (43) and *Trimeresurus gramineus* (44). Nikai and co-workers (45) separately reported a fibrinogenase from *C. atrox* venom that is similar in many respects to crotalase, the fibrinogen-clotting enzyme from *C. adamanteus* venom (46). However, the *C. atrox* enzyme did not form a clot, but it cleaved the A α-chains of fibrinogen preferentially; the B β-chain was cleaved later. The enzyme is an acidic proteinase with a pI of 4.6 and a molecular weight of 31,000. It is inhibited by diisopropylfluorophosphate. Additionally, Bailey and associates (47) report on another fibrin(ogen)olytic enzyme from *C. atrox* venom elsewhere in this book. Although this enzyme has a molecular weight of 93,000, under reducing SDS polyacrylamide gel electrophoresis, a molecular weight of 48,000 is observed. The enzyme is quite active in digesting fibrin clots, and activity is not affected by treatment with EDTA.

In a separate series of studies, Budzynski and colleagues (41,42, 48,49) characterized several proteinases in *C. atrox* venom that degrade fibrinogen. These investigators characterized the major degradation products of fibrinogen following incubation with crude venom (41). They also made the interesting observations in a patient envenomated by *C. atrox* that fibrinogen was absent, and that plasmin degradation products X, Y, D, and E were present (48). Subsequent in vitro studies led these workers to conclude that *C. atrox* venom stimulated the release of plasminogen activator from cultured endothelial and smooth muscle cells from human blood vessels (48). Thus, the hemostatic defects in the patient following *C. atrox* envenomation resulted both from direct fibrinogenolysis by the snake venom, and from a lytic state that was caused by activation of plasminogen by vascular plasminogen activator. The release of plasminogen activator from endothelial cells and smooth muscle cells was most likely triggered by the snake venom. A state of afibrinogenemia developed in the patient as a result of these two events. Because another chapter by Budzynski and co-workers (50) in this text covers venom stimulation of plasminogen activator release, no more mention of this interesting activity of *Crotalidae* venoms will be made here. However, the direct fibrinogenolytic action of *C. atrox* venom has been studied in some detail by Budzynski and co-workers (41,42,49). Early studies by these investigators revealed fibrinogenolytic activity in *C. atrox* venom that caused a rapid loss of thrombin coagulability of both purified human fibrinogen and fibrinogen in human plasma (41). Using crude venom, differences were noted between the degradation products formed from purified fibrinogen and from fibrinogen in plasma. In plasma, the venom degraded fibrinogen from a molecular weight of 340,000 to 325,000 with impaired thrombin coagulability and intact Aα- and γ-chains. The Bβ-chain was degraded from a molecular weight of about 58,000 to a smaller species with a molecular weight of 52,000. With purified fibrinogen, however, it was found that venom degraded the Aα-chains first and then the Bβ-chains. This apparent inconsistency was resolved by the isolation of four immunologically distinct proteases from the venom (42). All four proteases rendered purified fibrinogen incoagulable. Proteases I and IV contained single polypeptide chains with molecular weights of 20,000 and 46,000, respectively. These proteases do not attack fibrinogen in plasma, but with purified fibrinogen, cleave the Aα-chains first and then the Bβ- and γ-chains. Both enzymes are zinc metalloproteinases with pH optima of 7.0. They are inhibited by EDTA and α_2-macroglobulin. By contrast, proteases II and III degrade both purified fibrinogen and fibrinogen in plasma. Protease II contains a single polypeptide chain with a molecular weight of 31,000. Protease III contains two polypeptide chains, each of molecular weight 13,000; the intact protein has a molecular weight of 26,500. These proteases act only on the Bβ-chain of fibrinogen. Both enzymes are alkaline serine proteinases; they cleave at the COOH-terminal end of arginine residues (and lysine residues for proteinase II) in chromogenic peptide substrates. Both enzymes are inhibited by diisopropylfluorophosphate. Neither enzyme is inhibited by human plasma proteinase inhibitors and, for this reason, it is felt that these enzymes produce the anticoagulant effects in patients bitten by the Western diamondback rattlesnake (42). Pandya and co-workers (49) studied digestion of human fibrinogen by protease III, and found that a single cleavage occurs 42 residues from the NH_2-terminal end of the Bβ-chain. A product of 325,000 daltons is produced with impaired thrombin-clottability. The rate of fibrinopeptide A release and the rate and extent of Aα- and γ-chain cross-linking by factor XIII_a were indistinguishable when comparing intact fibrinogen and protease III-cleaved fibrinogen. Pandya and colleagues (49) used the protease III-cleaved fibrinogen derivative to aid in studying the topography of fibrinogen and the process of fibrin assembly. These studies represent an interesting application for venom proteases.

There are numerous reports of other proteinases isolated from *C. atrox* venom. Most notably, five hemorrhagic zinc metalloproteinases were obtained by Bjarnason and Tu (51). It is possible that a 24,000-dalton molecular weight hemorrhagic toxin, toxin b (52), is identical to protease I described by Pandya and Budzynski (42). Toxin b has fibrinogenolytic activity hydrolyzing the Aα-chain first, followed by cleavage of the Bβ-chain. It does not possess fibrinolytic activity.

Additionally, hemorrhagic toxin f isolated by Nikai and associates (53) is a metalloproteinase with a molecular weight of 64,000 daltons. This proteinase hydrolyzes the γ-chains of fibrinogen, but not the Aα- or Bβ-chains. Interestingly, a hemorrhagic metalloproteinase isolated from timber rattlesnake venom (54,55) exhibits hydrolytic activity on the Aα- and Bβ-chains of fibrinogen. It remains to be determined whether fibrinogenolytic activity of the hemorrhagic proteinases represents an inherent activity due to lack of specificity of these proteinases or is due to a contaminant. Conversely, it is most important to determine whether fibrin(ogen)olytic venom enzymes possess inherent hemorrhagic activity. The presence of this activity would be an obvious deterent to the clinical application of these enzymes in the treatment of thrombotic conditions.

American Pit Vipers (Agkistrodon)

It has been known since the early work of Eagle (18), Didisheim and Lewis (19), and Kornalik and Styblova (20,21) that American pit vipers of the genus *Agkistrodon* contain fibrin(ogen)olytic activity. More recently, Moran and Geren (56) described fibrinogenolytic activity in fractions obtained following CM-cellulose chromatography of five subspecies of copperhead (*Agkistrodon contortrix*) venom. Markland and co-workers (57) described fibrinolytic (fibrin plate lysis) activity in Eastern cottonmouth moccasin (*A. p. piscivorus*) and Southern copperhead (*A. c. contortrix*) venoms. After fractionation of crude venoms on Sephadex G-100, the molecular weights of the fibrinolytic enzymes were estimated as 34,000 and 25,000, respectively (57). Dyr and associates (58) described the partial purification of two fibrinogenolytic (and two fibrinogen clotting) proteinases from Southern copperhead venom. Both fibrinogenolytic proteinases degraded the Aα- and Bβ-chains of fibrinogen. One fibrinogenase had a molecular weight of 32,000 and exhibited activity on a COOH-terminal lysine chromogenic peptide substrate for plasmin. The other enzyme had a molecular weight of 25,000 and hydrolyzed a COOH-terminal arginine chromogenic peptide substrate for thrombin. Hydrolysis of the chromogenic substrates would suggest that these enzymes are serine proteinases with trypsin-like specificity. Moran and Geren (59) reported the purification of a fibrinogenolytic proteinase from *A. contortrix mokasen* (Northern copperhead) venom they had described earlier (56). The enzyme is a metalloproteinase that is reversibly inhibited by EDTA and has a molecular weight of 22,900. Thrombin-induced clotting was prevented after incubation of fibrinogen with the enzyme. Closer analysis revealed that the enzyme degrades the Aα-chain of fibrinogen. In addition to fibrinogenase activity, the enzyme hydrolyzes hide powder azure. Importantly, the enzyme possesses no hemorrhagic activity. This enzyme appears quite similar to the enzyme purified from Southern copperhead venom by Markland (10). A detailed description of the Southern copperhead venom enzyme is provided by Markland and colleagues elsewhere in this book (60), and it will not be discussed further here. It should be noted, however, that the fibrin(ogen)olytic metalloproteinases purified from copperhead venom (59,60) are apparently quite different from the fibrinogenolytic enzymes described by Dyr and associates (58), which possess chromogenic substrate activity and appear to be of the trypsin class of serine proteinases.

Asian Pit Vipers (Agkistrodon)

Ouyang and co-workers have characterized fibrin(ogen)olytic activity in the venom of two Asian pit vipers: *Agkistrodon acutus* (hundred pace snake) (61,62) and *A. rhodostoma* (Malayan pit viper) (63). The enzyme from *A. acutus* has been isolated and shown to have a molecular weight of 24,100 and an isoelectric point at pH 3.8 (61). The enzyme is an Aα-chain fibrinogenase and does not cleave the Bβ- or γ-chains. The enzyme degrades fibrin (fibrin plate lysis) and exhibits caseinolytic and hemorrhagic activities. All of these activities are inhibited by EDTA (62). Ouyang and co-workers (63) also isolated a fibrinogenase from *A. rhodostoma* venom with a molecular weight of 25,400. This enzyme is also a metalloproteinase and is inhibited by EDTA. Although similar in many respects to the enzyme from *A. acutus* venom, this enzyme has an isoelectric point at greater than pH 10. This enzyme cleaves specifically the Aα-chain of fibrinogen and does not cleave the Bβ- or γ-chains; it also possesses fibrinolytic and caseinolytic activities. Thus, from the two species of Asian pit vipers studied, Aα-chain fibrinogenases have been isolated and characterized. Although the enzyme from *A. acutus* possessed hemorrhagic activity that was inhibited by EDTA, it is not clear if this activity is inherent to the fibrinolytic enzyme or is due to a contaminant. In separate studies with the Asian lancehead *Trimeresurus gramineus*, Huang and colleagues (64) characterized a hemorrhagic proteinase with a molecular weight of 23,500 that was inhibited by EDTA. Interestingly, this enzyme possesses proteolytic activity toward the Aα-chain of fibrinogen, and was originally described as an Aα-chain fibrinogenase (44). Since these two activities form *T. gramineus* venom have identical amino acid compositions (44,64), the evidence strongly points to the Aα-chain fibrinogenases from Asian lanceheads and pit vipers as being metalloproteinases with inherent hemorrhagic activity.

Asian Lanceheads (Trimeresurus)

Ouyang and colleagues (43,65) described the properties of two fibrinogenolytic enzymes isolated from the venom of *Trimeresurus mucrosquamatus*. The study of the fibrinogenolytic enzymes from this venom revealed that one fibrinogenase degraded the Aα-chain of fibrinogen (Aα-fibrinogenase), while the other enzyme degraded primarily the Bβ-chain (Bβ-fibrinogenase). On prolonged incubation with fibrinogen, the latter enzyme also degraded the Aα-chain. Bβ-fibrinogenase exhibited caseinolytic and arginine esterase activity, but it did not possess coagulant activity on fibrinogen as do the venom fibrinogen clotting enzymes. The Aα-fibrinogenase possessed caseinolytic activity, but not arginine esterase activity. The two enzymes exhibited distinct amino acid compositions, pH activity profiles, isoelectric points, and heat stability. Both enzymes were glycoproteins. Their molecular weights were also different: the Aα-fibrinogenase had a molecular weight of 22,400 while that of the Bβ-fibrinogenase was 26,000. In addition to having distinct physical and chemical properties, the inhibition profile of the two enzymes was also unique. The Aα-fibrinogenase was inhibited by EDTA and cysteine, whereas these agents had little effect on the activity of the Bβ-fibrinogenase. These findings suggest that the Aα-fibrinogenase is a metalloproteinase, and that disulfide bridges may be important to the structural integrity of the enzyme. In contrast, the Bβ-fibrinogenase, which had arginine esterase activity, was almost completely inhibited by phenylmethylsulfonyl fluoride, suggesting that this enzyme is a serine esterase. The enzyme was also weakly inhibited by the chloromethyl ketone of tosyl-lysine, indicating that it may have a reactive histidine residue. However, neither this enzyme or the Aα-fibrinogenase was inhibited by soybean trypsin inhibitor.

Ouyang and colleagues (66) investigated the fibrinogen degradation products produced by the Aα- and Bβ-fibrinogenase of *T. mucrosquamatus* venom. The degradation products produced in both cases were large molecules since they were precipitated by trichloroacetic acid. Both enzymes were weak anticoagulants, acting to prolong the blood coagulation time mainly by the production of fibrinogen degradation products, and not by the depression of fibrinogen levels. Further, the species specificity of fibrinogen degradation by the Aα- and Bβ-fibrinogenases was investigated by Teng and colleagues (67). It was found that Aα-fibrinogenase digested bovine fibrinogen more rapidly than human fibrinogen, whereas the reverse was true for the Bβ-fibrinogenase. Both enzymes digested human fibrin. In plasma, the Aα-fibrinogenase digested human, dog, guinea pig, and rabbit fibrinogen to the same degree. However, the Bβ-fibrinogenase acted on the plasma fibrinogens in the following order: human > dog > guinea pig > rabbit. Fibrinogenolytic activity of the Aα-fibrinogenase on human fibrinogen was strongly inhibited by sera from all four animal species tested. In contrast, activity of the Bβ-fibrinogenase was inhibited by sera in the following order: rabbit > guinea pig > dog > human. Thus, proteinase inhibitors in blood of the different species are probably responsible for the species specificity exhibited by the Aα- and Bβ-fibrinogenases (67). Teng and colleagues (67) have suggested a possible application for the Bβ-fibrinogenase as an antithrombotic agent due to its high lytic action on fibrinogen, its low level of inhibition by human blood protease inhibitors, and its lack of hemorrhagic activity. The Aα-fibrinogenase, which had marked hemorrhagic activity and was significantly inhibited by plasma protease inhibitors, would be a much less likely candidate.

Ouyang and Huang (44) characterized Aα- and Bβ-fibrinogenases from venom of the green tree viper (*Trimeresurus gramineus*). The Aα- and Bβ-fibrinogenases are both single-chain proteinases with molecular weights of 23,500 and 25,000, respectively. They were both shown to have a direct fibrinogenolytic action, and neither activated plasminogen (44). Lack of plasminogen activation was also observed for the fibrinogenolytic enzymes from *T. mucrosquamatus* and *A. acutus* (44). The two enzymes from *T. gramineus* venom are similar in many respects to the two enzymes just described from *T. mucrosquamatus* venom (43, 65), including physicochemical and enzymatic properties, fibrinogen chain cleavage specificity, and inhibition profiles. However, their amino acid compositions and isoelectric points are slightly different, suggesting that they have amino acid sequence differences (44). As already noted (64), the Aα-fibrinogenase from *T. gramineus* venom possess potent hemorrhagic activity.

Viper Fibrin(ogen)olytic Activity

Although there are numerous reports of fibrin(ogen)olytic activities in venom from snakes of the Viperidae family, there are few reports of purification and characterization of the responsible enzyme(s). It has been shown that *Vipera russelli* venom has fibrinolytic activity, as determined by lysis of heated and unheated fibrin plates (33). *Vipera aspis* venom also possesses fibrinolytic activity (31). Chromatographic fractionation of the venom on DEAE-cellulose, followed by molecular sieve chromatography on Sephadex, revealed a fibrinolytic (and fibrinogenolytic) activity that appeared to be homogeneous by polyacrylamide gel electrophoresis. The enzyme had neither thrombin-like nor amino acid esterase activity. However, it did hydrolyze casein as well as bovine and human fibrinogen and fibrin. The enzyme was inhibited by EDTA, but not by diisopropylfluorophosphate, suggesting that it is a metalloenzyme.

The venom of *Bitis gabonica* has marked anticoagulant action in vitro. This effect appeared to be related to a direct proteolytic attack on fibrinogen (28). The venom did not activate plasminogen. Viljoen and co-workers (68) described the fractionation of *B. gabonica* venom into three similar esterolytic fractions, one of which possessed fibrinolytic activity. However, Markland and associates (57) were unable to detect fibrinolytic activity (fibrin plate lysis) in this venom. Recently and Aα-chain fibrinogenolytic metalloproteinase of molecular weight

22,500 has been purified form Egyptian viper (*Cerastes cerastes*) venom. The enzyme has little or no hemorrhagic activity. Since a detailed report of this enzyme appears in another chapter in this book (69), it will not be mentioned further here.

Elapid Fibrin(ogen)Activity

Venoms from snakes of the Elapidae family are generally believed to have very little proteolytic activity (70). Studies on fibrin(ogen)olytic activity in cobra venom have yielded somewhat conflicting results. The earlier studies of Eagle (18) and Didisheim and Lewis (19) showed that cobra venom did not possess fibrin(ogen)olytic activity. However, other investigators (27,70) showed that several cobra venoms (*Naja nigricollis, N. haje, N. melanoleuca,* and *Ophiophagus hannah*) contain fibrinogenolytic activity. Unfortunately, until recently, no studies were carried out on the isolation and characterization of the responsible enzymes. Because these venoms also possess caseinolytic (proteolytic) activity, it is possible that the fibrinogen degradation was not a specific action, but rather a general proteolytic effect of the venoms. Recently, however, Evans and co-workers (70,71) described an Aα fibrinogenase from the venom of the spitting cobra (*Naja nigricollis*). The enzyme contains a single polypeptide chain of molecular weight 58,000. The Aα-chain of fibrinogen and the α-chain of highly cross-linked fibrin are attacked directly by the enzyme, and it does not activate plasminogen. The enzyme is inhibited by EDTA, and this effect can be reversed by zinc, indicating that the enzyme is a zinc metalloproteinase. The enzyme exhibited no caseinolytic activity, but it did cleave α_2-macroglobulin, apparently in the bait region. A more detailed description is provided by Evans and co-workers (72) in a separate chapter in this book.

PLASMINOGEN-ACTIVATING ACTIVITY IN SNAKE VENOMS

Purified fibrinogen clotting enzymes from snake venoms are known to produce benign defibrinogenation when infused into experimental animals or humans (73,74). Early clinical studies of defibrinogenation with ancrod (the fibrinogen clotting enzyme from *Agkistrodon rhodostoma* venom) showed that there was a minimal and inconstant increase in the concentration of plasminogen activator in plasma of defibrinogenated patients (75). It was further suggested that the initial degradation of fibrin microclots produced in the circulation following ancrod infusion is mediated by the fibrinolytic enzyme system (76). Separate studies had already demonstrated that the defibrinogenating venom enzymes do not directly activate plasminogen (77). These studies prompted the suggestion that the combined use of venom defibrinogenating agents with thrombolytic agents might be an effective way to reduce the dosages and increase the effectiveness of urokinase and streptokinase in fibrinolytic therapy (74,78).

Recently, incisive studies by Budzynski and co-workers (48,50, 79,80) have shown that Crotalidae venoms contain factors that stimulate the release of plasminogen activators from cultured human and bovine endothelial and smooth muscle cells. A component of *Crotalus atrox* (Western diamondback rattlesnake) venom was shown to stimulate plasminogen activator release (48). Further, Budzynski and colleagues (79) found that the purified defibrinogenating enzymes ancrod and batroxobin stimulated the release of plasminogen activators from cultured endothelial cells. Thus, it would appear that the mechanism of formation of the fibrinogen-fibrin degradation products produced during clinical use of venom defibrinogenating enzymes may involve indirect activation of the fibrinolytic system by venom enzyme stimulation of vascular endothelial cells to increase their secretion of plasminogen activators. These studies are reviewed by Budzynski and colleagues (50) in another chapter in this book.

OTHER ANIMAL AND INSECT FIBRIN(OGEN)OLYTIC ENZYMES

As indicated in the introduction to this review, bloodsucking animals and insects have been shown to contain fibrinolytic activity. In this section, a brief account of these interesting activities is given.

The salivary gland of the giant leech *Haementeria ghilianii* has been reported to contain a fibrinolytic metalloproteinase (hementin) of molecular weight 122,000 (81-83). This enzyme is unaffected by plasma proteinase inhibitors and apparently does not hydrolyze other blood coagulation factors. Budzynski and associates (83) characterized proteolytic degradation products of fibrinogen produced by hementin, which cleaves peptide bonds in the Aα-, Bβ-, and γ-chains. Hementin cleaves noncross-linked and cross-linked fibrin clots, but at a much slower rate than it attacks fibrinogen. Interestingly, the salivary gland extract from the Brazilian blood sucking leach *Haemeneria lutzi* (Pinto 1920) does not exhibit direct proteolytic activity on casein, fibrinogen, or fibrin. Instead, this leech contains a plasminogen activator (84). Properties of other interesting leech anticoagulants are reviewed by Sawyer in another chapter of this book (85).

The human hookworm *Ancylostoma duodenale*, which attaches itself to villi in the small intestine and extracts blood from the human host, can cause iron-deficiency anemia and is of concern in developing nations (86). The dog parasite *A. caninum* (and presumably the human parasite) secretes a proteinase of molecular weight 36,000 that promotes fibrin clot dissolution (86). This proteinase not only degrades fibrinogen, but also converts plasminogen to a miniplasminogen-like molecule that may be important in the mechanism of fibrin clot dissolution by the hookworm.

The bloodsucking hemipteran insect *Rhodnius prolixus* Stal has been reported to contain a fibrinolytic enzyme in its hemolymph (87). The enzyme acts directly on fibrinogen and is active in fibrin plate lysis. Interestingly, the enzyme, named prolixase, was reported to have a molecular weight of only 5000. However, these results should be considered as preliminary in nature (87). Separately, a plasminogen activator has been shown to be extractable from the gut of this insect (88,89) and from the related blood sucking insect *Eutriatoma maculatus* (90).

Another bloodsucking insect, the tsetse fly *Glossina austeni*, was also reported to contain a fibrinolytic enzyme in the gut as evidenced by plasminogen-free fibrin plate lysis (91). This fibrinolytic enzyme was inhibited by soybean trypsin inhibitor. Additionally, the salivary gland from this insect contains a plasminogen activator. Plasminogen activators have also been reported in other bloodsucking creatures, including in extracts from the gut but not the salivary glands of the tick, *Ornithodotus monbata* (92), and in saliva from the vampire bat *Desmodus rotundus* (93).

Finally, the Saturniidae moth caterpillar *Lanomia achelous* contains fibrinolytic activity, and an enzyme that degrades fibrinogen has been partially purified from hemolymph and characterized (94-97). The proteinase(s) from the caterpillar apparently attacks fibrinogen by a different mechanism than that observed for plasmin, trypsin, and brinase (96). However, the complex nature of activities present in the proteinase fraction (96) makes the results somewhat difficult to interpret. A detailed description of the fibrinolytic enzyme and procoagulant activities from the caterpillar is presented by Arocha-Pinango and colleagues (98) in a separate chapter in this book.

SUMMARY

In general, it would appear that venoms from the different families of snakes contain predominantly two different fibrin(ogen)olytic activities, one specific for the $A\alpha$-chain and another for the $B\beta$-chain.

The $A\alpha$-chain-specific fibrinogenases have been isolated from *Crotalus atrox* (42), *Agkistrodon c. contortrix* (60), *A. c. mokasen* (59), *A. acutus* (61), *A. rhodostoma* (63), *Trimeresurus mucrosquamatus* (43,65,66), and *T. gramineus* (44) venoms. An $A\alpha$-chain fibrinogenase with similar properties has also been isolated from the viper *Cerastes cerastes* (69). Additionally, a hemorrhagic toxin, toxin b isolated from *C. atrox* venom (51), has many properties similar to those of protease I isolated from the same venom by Pandya and Budzynski (42). This protein is most likely also identical to the one described earlier by

Markland and co-workers (38,39). The Aα-chain fibrinogenases are zinc-containing metalloendoproteinases, and are completely devoid of arginine esterase activity. They have molecular weights ranging from 20,000 to 26,000. The enzymes appear to have disulfide bond(s) that are critical to their structure or function. They apparently degrade the Aα-chain of fibrinogen first, and then the Bβ-chain (Table 1). These enzymes are found in different genera of the family Crotalidae, including *Agkistrodon, Crotalus*, and *Trimeresurus*. They may, therefore, represent a common constituent of venom from these snakes. The Aα-chain fibrinogenases are inhibited in serum, probably by α_2-macroglobulin (42,60,67), and as such, they may not be important in inducing anticoagulation. Interestingly, hemorrhagic activity has been reported to be associated with the Aα-chain fibrinogenases from rattlesnakes (51,52) and Asian pit vipers and lanceheads (64). However, hemorrhagic activity does not appear to be associated with similar enzymes found in American pit vipers (59,60) or vipers (69). There appears to be a subgroup of metalloendoproteinases with Aα-chain fibrinogenase activity. Enzymes in this group have molecular weights ranging from 46,000 to 58,000, which is significantly higher than the other Aα-chain fibrinogenases. Enzymes in this subgroup have been reported in venoms from *Crotalus atrox* (42) and the elapid *Naja nigricollis* (71,72).

The second group of venom fibrin(ogen)ases are those with specificity for the Bβ-chain of fibrinogen. These proteinases, in contrast to the metalloproteinases that cleave specifically the Aα-chain of fibrinogen, are serine proteinases. They possess arginine esterase activity and are inhibited by diisopropylfluorophosphate or phenylmethanesulfonyl fluoride, but not EDTA. These enzymes have been found in venoms form *C. atrox* (40,42), *T. gramineus* (44), and *T. macrosquamatus* (43,65-67). The Bβ-chain fibrinogenases have molecular weights in the range of 23,000 to 31,000, and appear to be more heat-stable than the Aα-chain fibrinogenases (Table 2) (42-44). Although they are serine proteinases, the Bβ-chain fibrinogenases do not resemble venom fibrinogen-clotting serine proteases, which cleave either fibrinopeptide A or B or both from fibrinogen (73,74).

Interestingly, although plasma proteinases inhibitors have been shown to inhibit the Aα-chain fibrinogenases (42,60,67), this does not appear to be the case with the Bβ-chain fibrinogenases from *C. atrox* (42). The two Bβ-chain fibrinogenases from *C. atrox* venom proteases II and III appear to be resistant to serum proteinase inhibitors, although α_2-macroglobulin partially inhibits protease III. However, the Bβ-fibrinogenase from *T. mucrosquamatus* (67) exhibits a species-specific inhibition by serum proteinase inhibitors, with human serum showing the lowest inhibitor activity.

It is noteworthy that *C. atrox* venom contains two Aα-chain fibrinogenases and two Bβ-chain fibrinogenases (42). By contrast, the Asian

Table 1 Properties of Snake Venom Aα-Chain Fibrinogenases

	Venom source*										
	A. acutus	*A. rhodostoma*	*A. c. contortrix*	*A. c. mokasen*	*C. atrox*	*C. atrox*[†]	*T. gramineus*	*T. mucrosquamatus*	*C. cerastes*[†]	*N. nigricollis*	*C. atrox*
Properties					Protease I						Protease IV
Amino acid residues per molecule	208	226	N.D.	N.D.	176	272	203	203	193	515[‡]	406
Molecular weight	24,108	25,360	25,000	22,900	20,000	33,700	23,500	22,400	22,500	58,000	46,000
pI	3.8	>10	6.8	N.D.	Cationic	4.6	>10	8.1	5.2	>10	Acidic
Carbohydrate content	<1%	<1%	±	N.D.	-	1.5%	4%	2%	6%	N.D.	+
Enzymatic inhibitors	EDTA cysteine	EDTA cysteine	EDTA DTT	EDTA DTT	EDTA	DFP	EDTA cysteine	EDTA cysteine	EDTA	EDTA	EDTA
Inhibition by human plasma proteinase inhibitors	N.D.	N.D.	α_2-M	N.D.	α_2-M	N.D.	N.D.	Strong	N.D.	N.D.	α_2-M
Optimal pH	7.4	N.D.	7.5	7.8	7	6.0	7.4	7.4	N.D.	8-10	7
Optimal temperature	Heat labile	Heat labile	N.D.	43°C	37°C	Heat labile	Heat labile	Heat labile	N.D.	N.D.	37°C
Hemorrhagic activity	+	N.D.	-	-	N.D.	-	+	+	-	N.D.	N.D.
Fibrinolytic activity	+	+	+	+	N.D.	-	+	+	+	+	N.D.
References	61,62	63	10,60	59	42	45	44,64	43,65,66,67	69	70,71	42

*Abbreviations: α_2-M = α_2-macroglobulin; DFP = diisopropylfluorophosphate; DTT = dithiothreitol; + = positive; - = negative; N.D. = not determined.

†Shows Aα- and Bβ-activity with preferential cleavage of Aα-chain.

‡Tryptophan not included.

Table 2 Properties of Snake Venom Bβ-Chain Fibrinogenases

	Venom source				
	C. atrox	*C. atrox*	*C. atrox*	*T. gramineus*	*T. mucrosquamatus*
Properties	Protease II	Protease III			
Amino acid residues per molecule	284	198	N.D.	211	229
Molecular weight	31,000	24,000†	22,900	25,000	26,000
pI	N.D.	N.D.	4.6	4.5	5.7
Carbohydrate content	+	-	N.D.	7%	2%
Enzymatic inhibitors	DFP	DFP	PMSF	PMSF	PMSF
Inhibition by human plasma proteinase inhibitors	α_2-M weak	None	N.D.	N.D.	Weak
Optimal pH	8.5	10.5	N.D.	9.0	8.5
Optimal temperature	50-65° C	50-65°C	N.D.	Heat stable	Heat stable
Hemorrhagic activity	N.D.	N.D.	N.D.	N.D.	-
Fibrinolytic activity	N.D.	N.D.	-	+ (weak)	+ (weak)
References	42	42	40	44	43,65,66,67

Abbreviations: DFP = diisopropylfluorophosphate; PMSF = phenylmethylsulfonyl fluoride; + = positive; - = negative; N.D. = not determined.

†Two chains, each with a molecular weight of 13,000.

Trimeresurus venoms, although they contain both Aα-chain and Bβ-chain fibrinogenases, apparently only have one enzyme from each class (43,44). The reasons for and significance of these interesting variations between different genera of the Crotalidae family are unknown.

Finally, Nikai and co-workers (45) recently described an Aα-chain fibrinogenase from *C. atrox* venom that has arginine esterase activity and is inhibited by diisopropylfluorophosphate, but not by EDTA. The enzyme has a molecular weight of 31,000 and cleaves the Aα-chain preferentially, but also shows Bβ-chain-cleaving activity. This enzyme may represent another class of venom Aα-chain fibrin(ogen)ase, but additional studies are needed to confirm these findings. It is interesting that this enzyme has many similarities to crotalase, the defibrinogenating serine proteinase from Eastern diamondback rattlesnake venom (46,99), including bradykinin-releasing activity and amino acid composition.

ACKNOWLEDGMENTS

The author would like to acknowledge partial support by PHS Grant No. HL 31389 awarded by the National Heart, Lung, and Blood Institute, DHHS, and by Cortech, Inc., Denver, Colorado.

REFERENCES

1. Roland, F. P. Interaction of blood with *Enterobacteriaceae*: hemolysis, hemagglutination, fibrinolysis. Am. J. Clin. Pathol. 67:260 (1976).
2. Pulverer, G., Z. Wegrzynowicz, H. L. Ko, and J. Jeljaszewicz. Influence of *Pseudomonas aeruginosa* proteases on prothrombin, plasminogen and fibrinogen. Zentralbl. Bakteriol. Hyg. A248:99 (1980).
3. Smith, R. A. G., J. Green, and P. H. Kopper. The purification and properties of a fibrinolytic neutral metalloendopeptidase from *Streptococcus faecalis*. Arch. Biochem. Biophys. 202:629 (1980).
4. Teisseyre, E., Z. S. Latallo, and M. Kopee. Studies on the proteolysis of fibrinogen and fibrin by *Aspergillus ochraceus* enzyme as compared to the action of plasmin. Folia Haematol. 101: 99 (1974).
5. Roschlau, W. H. E. and D. A. J. Ives. Review of the biochemistry and coagulation physiology of brinolase (fibrinolytic enzyme from *Aspergillus oryzae*). Folia Haematol. 101:22 (1974).
6. Eie, H., E. Amundsen, and J. S. Ro. Thrombolytic treatment with brinase of dogs with experimental pulmonary embolism. Scand. J. Thorac. Cardiovasc. Surg. 6:164 (1972).

7. Lindvall, S., O. Magnusson, and K. Orth. On the inhibition of a fibrinolytic enzyme from *Aspergillus oryzae* by serum. Acta Chem. Scand. 23:2165 (1969).
8. Roschlau, W. H. E. Thrombolytic properties and side effects of brinase (fibrinolytic enzyme from *Aspergilus oryzae*) in the dog. J. Clin. Pathol. 25:635 (1972).
9. Frisch, E. P. Thrombolytic experience with brinase. Clinical experience with brinase. J. Clin. Pathol. 25:653 (1972).
10. Markland, F. S. Rattlesnake venom enzymes that interact with components of the hemostatic system. J. Toxicol. Toxin Reviews 2:119 (1983).
11. Seegers, W. H. and C. Ouyang. Snake venoms and blood coagulation. Handbook Exp. Pharm. 52:684 (1979).
12. Russell, F. E. In *Snake Venom Poisoning*. J. B. Lippincott Company, Philadelphia, 1980, pp. 139-234.
13. Tu, A. T. In *Venoms: Chemistry and Molecular Biology*, John Wiley and Sons, New York, 1977, pp. 329-360.
14. Hellmann, K. Naturally occurring anticoagulants and fibrinolysis. In *The Scientific Basis of Medicine Annual Reviews*, Oxford University Press, New York, 1968, pp. 254-265.
15. Fontana, F. *Treatise on the Venom of the Viper* (2 Vols.), J. Murray, London, 1787.
16. Mitchell, S. W. Researches upon the venom of the rattlesnake. In *Smithsonian Contributions to Knowledge*, Smithsonian Institution, Washington, D.C., 1860, pp. 89-95.
17. Mitchell, S. W. and E. T. Reichert. Researches upon the venoms of poisonous serpents. In *Smithsonian Contributions to Knowledge*, Smithsonian Institution, Washington, D.C., 1886, pp. 135-152.
18. Eagle, H. The coagulation of blood by snake venoms and its physiologic significance. J. Exp. Med. 65:613 (1937).
19. Didisheim, P. and J. H. Lewis. Fibrinolytic and coagulant activities of certain snake venoms and proteases. Proc. Soc. Exp. Biol. Med. 93:10 (1956).
20. Kornalik, F. The influence of snake venoms of fibrinogen conversion and fibrinolysis. Mem. Inst. Butantan Simp. Int. 33:179 (1966).
21. Kornalik, F. and Z. Styblova. Fibrinolytic proteases in snake venoms. Experientia 23:999 (1967).
22. Mohamed, A. H. and N. A. El-Damarawy. The role of the fibrinolytic enzyme system in the haemostatic defects following snake envenomation. Toxicon 12:467 (1974).
23. Denson, K. W. E., F. E. Russell, D. Almagro, and R. C. Bishop. Characterization of the coagulant activity of some snake venoms. Toxicon 10:557 (1972).

24. Marshall, L. R. and R. P. Herrmann. Coagulant and anticoagulant actions of Australian snake venoms. Thromb. Haem. 50:707 (1983).
25. Mitrakul, C. Effect of five Thai snake venoms on coagulation, fibrinolysis and platelet aggregation. Southeast Asian J. Trop. Med. Pub. Health 10:266 (1979).
26. Hardy, D. L., M. Jeter, and J. J. Corrigan. Envenomation by the Northern blacktail rattlesnake (*Crotalus molossus molossus*): report of two cases and the *in vitro* effects of the venom on fibrinolysis and platelet aggregation. Toxicon 20:487 (1982).
27. Mackay, N., J. C. Ferguson, and G. P. McNicol. Effects of three cobra venoms on blood coagulation, platelet aggregation, and fibrinolysis. J. Clin. Pathol. 22:304 (1969).
28. Forbes, C. D., A. G. G. Turpie, J. C. Ferguson, G. P. McNicol, and A. S. Douglas. Effect of Gaboon viper (*Bitis gabonica*) venom on blood coagulation, platelets, and the fibrinolytic enzyme system. J. Clin. Pathol. 22:312 (1969).
29. Mohamed, A. H., M. El-Serougi, and L. Z. Khaled. Effect of *Cerastes cerastes* venom on blood coagulation mechanisms. Toxicon. 7:181 (1969).
30. Moan, B., C. Moroz, and A. DeVries. Activation of the fibrinolytic system of the guinea pig following inoculation of *Echis colorata* venom. Toxicon 1:109 (1963).
31. Boffa, M. C. and G. A. Boffa. Correlations between the enzymatic activities and the factors active on blood coagulation and platelet aggregation form the venom of *Vipera aspis*. Biochim. Biophys. Acta 354:275 (1974).
32. MacKay, N., J. C. Ferguson, and G. P. McNicol. Effects of the venom of the rhinoceros horned viper (*Bitis nasicornis*) on blood coagulation, platelet aggregation, and fibrinolysis. J. Clin. Pathol. 23:789 (1970).
33. Aung-Khin, M., K. Ma-Ma, and T. Zin. Effects of Russell's viper venom on blood coagulation, platelets, and the fibrinolytic enzyme system. Jpn. J. Med. Sci. Biol. 30:101 (1977).
34. Bajwa, S. S., F. E. Russell, and F. S. Markland. Thrombolytic agents in snake venoms. Proc. West. Pharmacol. Soc. 25:353 (1982).
35. Bajwa, S. S., F. S. Markland, and F. E. Russell. Fibrinolytic and fibrinogen clotting enzymes present in the venoms of Western diamondback rattlesnake, *Crotalus atrox*, Eastern diamondback rattlesnake, *Crotalus adamanteus*, and Southern Pacific rattlesnake, *Crotalus viridis helleri*. Toxicon 19:53 (1981).
36. Posch, J., B. Reeb, C. Thornsvard, and G. Mills. Coagulant activity of *Crotalus atrox* venom: differences in small and large snakes. Blood 58:224a (1981).

37. Reid, H. A. and R. D. G. Theakston. Changes in coagulation effects by venoms of *Crotalus atrox* as snakes age. Am. J. Trop. Med. Hyg. 27:1053 (1978).
38. Bajwa, S. S., F. S. Markland, and F. E. Russell. Fibrinolytic enzyme(s) in Western diamondback rattlesnake (*Crotalus atrox*) venom. Toxicon 18:285 (1980).
39. Bajwa, S. S., G. B. Patkos, and F. S. Markland. Clinical potential of fibrinolytic enzyme(s) of Western diamondback rattlesnake (*Crotalus atrox*) venom. Proc. West. Pharmacol. Soc. 24:165 (1981).
40. Sapru, Z. Z., A. T. Tu, and G. S. Bailey. Purification and characterization of fibrinogenase from the venom of Western diamondback rattlesnake (*Crotalus atrox*). Biochim. Biophys. Acta 747:225 (1983).
41. Pandya, B. V., R. N. Rubin, S. A. Olexa, and A. Z. Budzynski. Unique degradation of human fibrinogen by proteases from Western diamondback rattlesnake (*Crotalus atrox*) venom. Toxicon 21:515 (1983).
42. Pandya, B. V. and A. Z. Budzynski. Anticoagulant proteases from Western diamondback rattlesnake (*Crotalus atrox*) venom. Biochemistry 23:460 (1984).
43. Ouyang, C., C-M. Teng, and Y-C. Chen. Physiochemical properties of α- and β-fibrinogenases of *Trimeresurus mucrosquamatus* venom. Biochim. Biophys. Acta 481:622 (1977).
44. Ouyang, C. and T-F. Huang. α- and β-Fibrinogenases from *Trimeresurus gramineus* snake venom. Biochim. Biophys. Acta 571:270 (1979).
45. Nikai, T., R. Kito, N. Mori, H. Sugihara, and A. T. Tu. Isolation and characterization of fibrinogenase from Western diamondback rattlesnake venom and its comparison to the thrombin-like enzyme, crotalase. Comp. Biochem. Physiol. 76B:679 (1983).
46. Markland, F. S. and P. S. Damus. Purification and properties of a thrombin-like enzyme from the venom of *Crotalus adamanteus* (Eastern diamondback rattlesnake). J. Biol. Chem. 246:6460 (1971).
47. Bailey, G. S., A. T. Tu, and Z. Z. Sapru. A fibrinolytic enzyme from the venom of the Western diamondback rattlesnake (*Crotalus atrox*). In *Hemostasis and Animal Venoms*, H. Pirkle and F. S. Markland (Eds.), Marcel Dekker, New York, 1988.
48. Budzynski, A. Z., B. V. Pandya, R. N. Rubin, B. S. Brizuela, T. Soszka, and G. J. Stewart. Fibrinogenolytic afibrinogenemia after envenomation by Western diamondback rattlesnake (*Crotalus atrox*). Blood 63:1 (1984).
49. Pandya, B. V., C. S. Cierniewski, and A. Z. Budzynski. Conservation of human fibrinogen conformation after cleavage of Bβ chain NH_2 terminus. J. Biol. Chem. 260:2994 (1985).

50. Budzynski, A. Z., N. E. Kirschbaum, and R. Reczkowski. Secretion of cellular plasminogen activators upon stimulation of *Crotalinae* snake venoms. In *Hemostasis and Animal Venoms*, H. Pirkle and F. S. Markland (Eds.), Marcel Dekker, New York, 1988.
51. Bjarnason, J. B. and A. T. Tu. Hemorrhagic toxins from western diamondback rattlesnake (*Crotalus atrox*) venom: isolation and characterization of five toxins and the role of zinc in hemorrhagic toxin e. Biochemistry 17:3395 (1978).
52. Komori, Y., S. Hagihara, and A. T. Tu. Specificity of hemorrhagic proteinase form *Crotalus atrox* (Western diamondback rattlesnake) venom. Biochim. Biophys. Acta 829:127 (1985).
53. Nikai, T., N. Mori, M. Kishida, H. Sugihara, and A. T. Tu. Isolation and biochemical characterization of hemorrhagic toxin from the venom of *Crotalus atrox* (Western diamondback rattlesnake). Arch. Biochem. Biophys. 231:309 (1984).
54. Civello, D. J., H. L. Duong, and C. R. Geren. Isolation and characterization of a hemorrhagic proteinase from timber rattlesnake venom. Biochemistry 22:749 (1983).
55. Civello, D. J., J. B. Moran, and C. R. Geren. Substrate specificity of a hemorrhagic proteinase from timber rattlesnake venom. Biochemistry 22:755 (1983).
56. Morgan, J. B. and C. R. Geren. Subspecific variations in *Agkistrodon contortrix* venoms-II. Comp. Biochem. Physiol. 65B:739 (1980).
57. Bajwa, S. S., H. Kirakossian, K. N. N. Reddy, and F. S. Markland. Thrombin-like and fibrinolytic enzymes in the venoms from the Gaboon viper (*Bitis gabonica*), Eastern cottonmouth moccasin (*Agkistrodon p. piscivorus*) and Southern copperhead (*Agkistrodon c. contortrix*) snakes. Toxicon 20:427 (1982).
58. Dyr, J. E., B. Blombäck, and F. Kornalik. The fibrinogenolytic and procoagulant activity of Southern copperhead venom enzymes. Thrombos. Res. 30:185 (1983).
59. Morgan, J. B. and C. R. Geren. Characterization of a fibrinogenase from Northern copperhead (*Agkistrodon contortrix mokasen*) venom. Biochim. Biophys. Acta 659:161 (1981).
60. Markland, F. S., K. N. N. Reddy, and A. L. Guan. Purification and characterization of a direct-acting fibrinolytic enzyme from Southern copperhead venom. In *Hemostasis and Animal Venoms*, H. Pirkle and F. S. Markland (Eds.), Marcel Dekker, New York, 1988.
61. Ouyang, C. and T-F. Huang. Purification and characterization of the fibrinolytic principle of *Agkistrodon acutus* venom. Biochim. Biophys. Acta 439:146 (1976).
62. Ouyang, C. and T-F. Huang. The properties of the purified fibrinolytic principle from *Agkistrodon acutus* snake venom.

Toxicon 15:161 (1977).
63. Ouyang, C., L-J. Hwang, and T-F. Huang. α-Fibrinogenase from *Agkistrodon rhodostoma* (Malayan pit viper) snake venom. Toxicon 21:25 (1983).
64. Huang, T-F., J-H. Chang, and C. Ouyang. Characterization of hemorrhagic principles from *Trimeresurus gramineus* snake venom. Toxicon 22:45 (1984).
65. Ouyang, C. and C-M. Teng. Fibrinogenolytic enzymes of *Trimeresurus mucrosquamatus* venom. Biochim. Biophys. Acta 420: 298 (1976).
66. Ouyang, C., C-M. Teng, and Y-C. Chen. Properties of fibrinogen degradation products produced by α- and β-fibrinogenases of *Trimeresurus mucrosquamatus* snake venom. Toxicon 17:121 (1979).
67. Teng, C-M., C. Ouyang, and S-C. Lin. Species difference in the fibrinogenolytic effects of α- and β-fibrinogenases from *Trimeresurus mucrosquamatus* snake venom. Toxicon 23:777 (1985).
68. Viljoen, C. C., C. M. Meehan, and D. P. Botes. Separation of *Bitis gabonica* (Gaboon adder) venom arginine esterases into kinin-releasing, clotting and fibrinolytic factors. Toxicon 17:145 (1979).
69. Daoud, E. W., A. T. Tu, and M. F. El-Asmar. The effect of anticoagulant proteinase (cerastase) from *Cerastes cerastes* venom on blood coagulation system. In *Hemostasis and Animal Venoms*, H. Pirkle and F. S. Markland (Eds.), Marcel Dekker, New York, 1988.
70. Evans, H. J. Cleavage of the Aα-chain of fibrinogen and the α-polymer of fibrin by the venom of spitting cobra (*Naja nigricollis*). Biochim. Biophys. Acta 660:219 (1981).
71. Evans, H. J. Purification and properties of a fibrinogenase from the venom of *Naja nigricollis*. Biochim. Biophys. Acta 802:49 (1984).
72. Evans, J. H. and A. J. Barrett. The action of proteinase F1 from *Naja nigricollis* venom on the Aα-chain of human fibrinogen. In *Hemostasis and Animal Venoms*, H. Pirkle and F. S. Markland (Eds.), Marcel Dekker, New York, 1988.
73. Markland, F. S. and H. Pirkle. Biological activities and biochemical properties of thrombin-like enzymes from snake venoms. In *Chemistry and Biology of Thrombin*, R. L. Lundblad, J. W. Fenton, and K. G. Mann (Eds.), Ann Arbor Science Publishers, Ann Arbor, Michigan, 1977, p. 71.
74. Bell, W. R. Defibrinogenation with Arvin in thrombotic disorders. In *Platelets and Thrombosis*, S. Sherry and A. Scriabine (Eds.), University Park Press, Baltimore, 1974, p. 273.
75. Pitney, W. R., W. R. Bell, and G. Bolton. Blood fibrinolytic activity during Arvin therapy. Br. J. Haemat. 16:165 (1969).

76. Prentice, C. R. M., W. Edgar, and G. P. McNicol. Characterization of fibrin degradation products in patients on ancrod therapy: comparison with fibrinogen products produced by plasmin. Br. J. Haemat. 27:77 (1974).
77. Turpie, A. G. G., C. R. M. Prentice, G. P. McNicol, and A. S. Douglas. *In vitro* studies with ancrod ("Arvin"). Br. J. Haemat. 20:217 (1971).
78. Kwaan, H. C. Use of defibrinating agents ancrod and reptilase in the treatment of thromboembolism. Thromb. Diath. Haemorrh. Suppl. 56:239 (1973).
79. Soszka, T., N. E. Kirschbaum, G. J. Stewart, and A. Z. Budzynski. Direct effect of fibrinogen-clotting enzymes on plasminogen activator secretion from human endothelial cells. Thromb. Haem. 54:164 (1985).
80. Kirschbaum, N. E., T. Soszka, G. J. Stewart, and A. Z. Budzynski. Compartmentation of fibrinolytic response in human venous circulation. Thromb. Haem. 54:106 (1985).
81. Budzynski, A. Z., S. A. Olexa, B. S. Brizuela, R. T. Sawyer, and G. S. Stent. Anticoagulant and fibrinolytic properties of salivary proteins from the leech *Haementeria ghilianii*. Proc. Soc. Exp. Biol. Med. 168:266 (1981).
82. Malinconico, S. M., J. B. Katz, and A. Z. Budzynski. Hementin: anticoagulant protease from the salivary gland of the leech *Haementeria ghilianii*. J. Lab. Clin. Med. 103:44 (1984).
83. Malinconico, S. M., J. B. Katz, and A. Z. Budzynski. Fibrinogen degradation by hementin, a fibrinogenolytic anticoagulant from the salivary glands of the leech *Haementeria ghilianii*. J. Lab. Clin. Med. 104:842 (1984).
84. Kelen, E. M. A. and G. Rosenfield. Fibrinogenolytic substance (Hementerin) of Brazilian blood-sucking leeches (*Haementeria lutzi* Pinto 1920). Haemostasis 4:51 (1975).
85. Sawyer, R. T. Scientific rationale behind the medical use of leeches. In *Hemostasis and Animal Venoms*, H. Pirkle and F. S. Markland (Eds.), Marcel Dekker, New York, 1988.
86. Hotez, P. J. and A. Cerami. Secretion of a proteolytic anticoagulant by *Ancylostoma* hookworms. J. Exp. Med. 157:1594 (1983).
87. Ben Hamouda, M. H., J. Soria, and C. Soria. Description of new fibrinolytic enzyme, isolated from the haemolymph of *Rhodnius prolixus* Stal (*Heteroptera Reduviidae*). Arch. Inst. Pasteur Tunis 59:251 (1982).
88. Hellmann, K. and R. I. Hawkins. Anticoagulant and fibrinolytic activities from *Rhodnius prolixus* Stal. Nature 201:1008 (1964).
89. Hawkins, R. I. and K. Hellmann. Investigations on a plasminogen activator in two blood-suckers, *Rhodnius prolixus* Stal and *Hirudo medicinalis*. Br. J. Haemat. 12:86 (1966).

90. Hellmann, K. and R. I. Hawkins. An antithrombin (Maculatin) and a plasminogen activator extractable from the blood-sucking Hemipteran, *Eutriatoma maculatus*. Br. J. Haemat. 12:376 (1966).
91. Hawkins, R. I. Factors affecting blood clotting from salivary glands and crop of *Glossina austeni*. Nature 212:738 (1966).
92. Hellmann, K. and R. I. Hawkins. The action of tick extracts on blood coagulation and fibrinolysis. Thromb. Diath. Haemorrh. 18:617 (1967).
93. Hawkey, C. Plasminogen activator in saliva of the vampire bat *Desmodus rotundus*. Nature 211:434 (1966).
94. Arocha-Pinango, C. L. and M. Layrisse. Fibrinolysis produced by contact with a caterpillar. Lancet April 19:810 (1969).
95. Arocha-Pinango, C. L., N. A. Marsh, and D. Robinson. A fibrinolytic agent from a Saturnid caterpillar: partial purification and characterization. Thromb. Diath. Haemorrh. 29:135 (1973).
96. Arocha-Pinango, C. L., J. Perales, and Z. Carvajal. Studies on the degradation of fibrinogen by proteolytic enzymes from the larvae of *Lanomia achelous* (cramer). Thromb. Haem. 45:233 (1981).
97. Arocha-Pinango, C. L. and D. S. Pepper. Studies of a fibrinolytic enzyme from the larvae of *Lonomia achelous* (cramer) using chromogenic peptide substrates. Thromb. Haem. 46:710 (1981).
98. Arocha-Pinango, C. L., N. B. deBosch, A. Nouel, A. Torres, I. Perales, M. Alonso, S. Rodriguez, Z. Carvajal, A. Ojeda, M. L. Tasayco, and W. Chitty. Fibrinolytic and procoagulant agents from a Saturniidae moth caterpillar. In *Hemostasis and Animal Venoms*, H. Pirkle and F. S. Markland (Eds.), Marcel Dekker, New York, 1988.
99. Markland, F. S., C. Kettner, S. Schiffman, E. Shaw, S. S. Bajwa, H. Kirakossian, G. Patkos, I. Theodor, and H. Pirkle. Kallikrein-like action of a fibrinogen clotting snake venom enzyme. Proc. Natl. Acad. Sci. USA 79:1688 (1982).

13

Purification and Characterization of a Direct-Acting Fibrinolytic Enzyme from Southern Copperhead Venom

Francis S. Markland, Jr. and K. N. N. Reddy
University of Southern California School of Medicine, Los Angeles, California
Li-feng Guan*
Shanghai Institute of Biochemistry, Chinese Academy of Sciences, Shanghai, Republic of China

INTRODUCTION

Eagle (1) studied coagulation of blood by snake venoms and reported the presence of fibrinogen-degrading activity in an *Agkistrodon* species (*A. p. piscivorus*, the Eastern cottonmouth). Didisheim and Lewis (2), in a search for an ideal agent to lyse intravascular thrombi, reported that fibrinolytic activity was present in a series of snake venoms including specimens from the *Crotalus*, *Bothrops*, and *Agkistrodon* species. These authors were the first to suggest that it should be possible to prepare a purified fibrinolytic enzyme and use it therapeutically as an agent for dissolving intravascular clots. They further suggested a theoretical advantage for these agents, namely, that their fibrinolytic activity should not be readily inhibited by human blood antiproteinases.

Kornalik (3) reported that fibrinolytic activity was present in venom from the *A. p. piscivorus* and *A. c. contortrix* snakes. He showed that proteinases were responsible for the fibrinolytic activity and that the crude venom did not function as a plasminogen activator. Presumably, the venom had a direct fibrinolytic action. Herzig and co-workers (4) described a procoagulant enzyme from Southern copperhead venom. The enzyme acts directly on fibrinogen and is inhibited by phenylmethylsulfonyl fluoride. The authors briefly mentioned a lower molecular weight fibrinolytic activity in the venom. Moran and

*Permanent affiliation: University of Southern California School of Medicine, Los Angeles, California

Geren (5) described a fibrinogenolytic proteinase from Northern copper venom with a molecular weight of 22,900 daltons. This enzyme acts as an α-chain fibrinogenase and is a zinc metalloproteinase.

Dyr and associates (6) reported that Southern copperhead venom contains four proteinases that act on fibrinogen; two are procoagulant and two are fibrinogenolytic. One procoagulant enzyme with a molecular weight of 68,000 is similar to the one reported by Herzig and coworkers (4). The two enzymes with lower molecular weights degraded fibrinogen, and they were further characterized with respect to their action on several chromogenic substrates (6). One enzyme, with a molecular weight of 32,000, has H-D-Val-Leu-Lys-p-nitroanilide (plasmin substrate)-hydrolyzing activity. The other, with a molecular weight of 25,000, has Tos-Gly-Pro-Arg-p-nitroanilide (thrombin substrate)-hydrolyzing activity. Since these enzymes possess activity with chromogenic substrates, they are different from the enzyme described in this chapter.

An earlier publication from our laboratory reported the presence in Southern copperhead venom of both a procoagulant enzyme (of molecular weight about 73,500) acting directly on fibrinogen, and a fibrinolytic enzyme with a molecular weight of about 25,000 (7-9). These activities were separated by gel filtration on Sephadex G-100. In view of the possible clinical application of venom fibrinolytic enzymes as clot-dissolving agents, we report here on our continuing studies with the fibrinolytic enzyme from Southern copperhead venom.

MATERIALS AND METHODS

Materials

A. c. contortrix snake venom was obtained from Biotoxins, Inc., St. Cloud, Florida. DEAE-cellulose (DE-52) and CM-cellulose (CM-52) were purchased from Whatman, Inc., Hillsboro, Oregon. Sephadex G-100 was obtained from Pharmacia, Piscataway, New Jersey. Ion exchange or molecular sieve columns were used according to manufacturers' instructions. p-Aminobenzamidine-agarose and soybean and lima bean trypsin inhibitors were purchased from Sigma Chemical Company, St. Louis, Missouri. Trasylol (aprotinin) was obtained from Mobay Chemical Corporation, FBA Pharmaceuticals, New York, New York. Casein was obtained from ICN Nutritional Biochemicals, Cleveland Ohio. The immobilized ampholytes were obtained from LKB Instruments, Inc., Pleasant Hill, California. Fibrinogen (Kabi, grade L) and the chromogenic substrates S-2288, S-2251, S-2302, and S-2238 were obtained from Helena Laboratories, Beaumont, Texas. The other chromogenic substrates were from Sigma Chemical Company, except N-succinyl-Gly-Gly-Phe-p-nitroanilide, which was obtained from Vega Biochemicals, Tucson, Arizona. Purified α_2-macroglobulin was kindly provided by Dr. Dudley Strickland, American Red Cross, Bethesda, Maryland. All other chemicals were of analytical reagent grade.

Azocasein Hydrolysis

Azocasein was synthesized from casein by diazotization as previously described (10). Hydrolysis of azocasein was measured following incubation of venom fractions with the diazotized proteins for 30 minutes at 37°C (8).

Fibrin Plate Clearance Assay

Fibrin plate lysis by fibrinolytic enzyme fractions was measured as previously described (11). Samples (10 μl) of crude or purified venom enzyme at the appropriate dilution were incubated for 17 hours at 37°C. Lysis was analyzed by measuring (in mm) diameters of the clear zones produced around the wells. Activity is expressed in plasmin equivalent units per ml (a human plasmin standard, second international reference preparation kindly provided by Dr. Patrick J. Gaffney, National Institute for Biological Standards and Control, London, England was used as a control).

Hydrolysis of Chromogenic Substrates

Chromogenic substrates were incubated with venom fractions at 37°C for a minimum of 3 minutes using the following conditions: 100 μl of crude or pure venom sample diluted appropriately, 200 μl of 50 mM Tris-HCl buffer containing 150 mM sodium chloride, pH 7.4 (pH 8.4 for S-2288), and 100 μl of chromogenic substrate at 1 mg/ml. At the appropriate time, the reaction was stopped by the addition of 100 μl of 50% acetic acid, and the absorbance was read at 405 nm.

Fibrinogen and Nondenaturing Polyacrylamide Gel Electrophoresis

We used a modification of the fibrinogen gel method described by Lacks and Springhorn (12) to detect proteolytic activity following polyacrylamide gel electrophoresis. In this procedure, proteinases act in situ on fibrinogen incorporated into the polyacrylamide gel. Fibrinogen with many peptide bonds is a good substrate for proteinases, and it is not likely to be eluted from the gel during washing, nor to be electrophoresed out of the 7.5% gels. The assay for proteinases is based on the hydrolysis of fibrinogen into fragments small enough to diffuse out of the gel. When the gel is stained for protein, the background blue color (due to fibrinogen staining) is absent at bands demonstrating proteinase activity. Duplicate gels (without fibrinogen) are stained with Coomassie blue to visualize proteins, which migrate with similar mobilities regardless of the presence or absence of fibrinogen in the gels. We slightly modified this procedure (13) by: (a) using non-SDS-containing gels, and (b) using 340 μg/ml of fibrinogen instead of 170 μg/ml as recommended originally (12). The fibrinogen gel was made as follows for a 1.5-mm slab gel with 7.5% acrylamide: 4 ml of 0.30 M Tris-HCl, pH 8.9, containing 0.23% (v/v) TEMED, 8 ml 30% acrylamide, 0.9% bisacrylamide, and 10 ml of distilled H_2O were mixed for the resolving gel. The solution was degassed and, prior to casting, 10 ml of 0.1% fibrinogen in 0.001 N NaOH and 240 μl of 10% ammonium persulfate were added. A duplicate (normal) gel prepared identically except without fibrinogen (10 ml of H_2O containing persulfate added in place of the fibrinogen solution) was used for venom protein staining. Purified fibrinolytic enzyme was used for the fibrinogen gel (100 μg) and for the normal gel (50 μg). Electrophoresis was carried out at 4°C. After electrophoresis, the normal gel was stained for protein with Coomassie blue R-250 in 40% methanol-10% acetic acid in water, while the fibrinogen gel was placed in 0.04 M Tris, 0.002 M $MgCl_2$, 0.002 M $CaCl_2$, pH 7.6, and incubated overnight at room temperature. This gel was then stained with Coomassie blue R-250 for at least 2 hours. Both gels were destained in 40% methanol-10% acetic acid. Proteolytic activity on the fibrinogen gels can be observed as a clear area on a blue background.

SDS Polyacrylamide Gel Electrophoresis (SDS-PAGE)

SDS-PAGE of the venom enzyme was by a vertical slab-gel procedure using the conditions described by Laemmli (14). Proteins of known molecular weights (obtained in kit form from Sigma Chemical Company) were used to construct a standard curve from which the molecular weight of the venom enzyme was determined (15). Carbohydrates were detected by staining gels with the periodic acid-Schiff (PAS) reagent (16).

Inhibition Studies

In general, the venom enzyme was preincubated with an appropriate concentration of the inhibitor at room temperature or 37°C, and aliquots were removed at specified times for assay of azocaseinolytic activity. Specifically, the effect of tetraethylenepentamine (TEP) was investigated by incubating 100 μg of the venom enzyme for 10 minutes at room temperature in 500 μl of 5 mM sodium acetate buffer, pH 7.0, containing 10 mM TEP. To study the effect of dithiothreitol (DTT), the venom enzyme (25 μg) was incubated under nitrogen at room temperature for 180 minutes in 55 μl of 0.5 M Tris-HCl buffer, pH 8.1, containing 6 mM DTT. To study the effect of p-chloromercuribenzoate (pCMB), the venom enzyme (35 μg) was incubated at 37° for 30 minutes, 60 minutes, and 24 hours in 0.05 M Tris-HCl buffer, pH 7.6, containing 1 mM pCMB and 0.15 M NaCl. To study the effect of diisopropylfluorophosphate (DFP), the venom enzyme (150 μg) was incubated for 1 hour at room temperature in 300 μl of 0.1% glycine buffer, pH 7.0, containing 10 mM DFP. To study the effect of 1,10 orthophenanthroline, the venom enzyme (35 μg) was incubated at 37°C for 60 minutes in 0.05 M Tris-HCl buffer, pH 7.4, containing 1 mM chelating agent and 0.15 M NaCl. To study the effect of EDTA, the venom enzyme (35 μg) was incubated for 30 minutes at room temperature in 0.05 M Tris-HCl buffer, pH 7.4, containing 20 mM EDTA and 0.15 M NaCl. To study the effect of α_2-macroglobulin, the venom enzyme (3.5 μg) in 0.05 M Tris-HCl buffer, pH 7.4, containing 0.15 M NaCl, was incubated for 20 minutes at room temperature with 7.5-fold molar excess of the proteinase inhibitor (300 μg).

The venom enzyme (30 μg) was also incubated for 5 minutes at room temperature in 0.05 M Tris-HCl buffer, pH 7.4, containing 0.15 M NaCl, with the following proteinase inhibitors: soybean trypsin inhibitor (500 μg), Trasylol (1000 kallikrein inhibitor units), and lima bean trypsin inhibitor (400 μg). Following the incubation, residual activity was assayed.

Plasminogen Preparation

Plasminogen was prepared from human plasma by lysine-Sepharose affinity chromatography as previously described (17). Briefly, human plasma was diluted 1:1 with 0.1 M sodium phosphate buffer, pH 7.4, and Trasylol (pancreatic trypsin inhibitor, 10^4 units/100 ml plasma) was added. The plasma was passed through a lysine-Sepharose column, washed extensively with 0.3 M sodium phosphate, pH 7.4, and the adsorbed plasminogen was specifically eluted with 0.2 M epsilon-aminocapronic acid. The preparation was dialysed against 0.05 M Tris-HCl buffer, pH 9.0, containing 0.02 M lysine and 0.1 M NaCl, and stored frozen.

Activation of Plasminogen

Plasminogen activator activity of the venom enzyme was tested as follows: purified human plasminogen (186 μg) was incubated with venom enzyme (18 μg) at 37°C in a total volume of 0.5 ml of 0.05 M Tris-HCl buffer, pH 7.4, containing 0.15 M NaCl. At various time intervals, 50 μl aliquots were removed and treated with β-mercaptoethanol in sodium dodecyl sulfate (SDS) containing phosphate buffer (0.1% β-mercaptoethanol-0.1% SDS-0.1 M sodium phosphate, pH 7.1). The samples were analyzed by acrylamide disc gel electrophoresis according to the method of Weber and Osborn (15), stained with Coomassie blue, and scanned for the presence of protein bands form the heavy and light chains of plasmin. In separate experiments, plasminogen activation by the venom enzyme was also assessed by measuring the generation of S-2251-hydrolyzing activity. Purified human plasminogen (62 μg) was incubated at 37°C with venom enzyme (3.1 μg) in 1.0 ml of 0.05 M Tris-HCl buffer, pH 7.4, containing 0.15 M NaCl, and aliquots from the incubation mixture were taken at various times to measure plasmin formed. To assay plasmin, the plasmin-specific substrate H-D-Val-Leu-Lys-p-nitroanilide (S-2251) (100 μl from a 3.5 mM stock solution) was warmed to 37°C in 900 μl of the Tris-NaCl buffer, and the reaction was initiated by adding an aliquot from the venom enzyme-plasminogen incubate. The reaction mixture was incubated for 3 minutes at 37°C and, after adding 100 μl of 50% acetic acid, the absorbance of the sample at 405 nm was read in a spectrophotometer.

Fibrinogen Clotting and Hemorrhagic Activity

To help localize the Southern copperhead venom fibrinogen-clotting enzyme in column fractions, aliquots were assayed for activity at 37°C as previously described (18). Briefly, time for clot formation was measured after adding 100 μl of an appropriate dilution of the column fraction to a prewarmed (3 minutes, 37°C) test tube containing 200 μl of a 1% human fibrinogen solution dialyzed against 0.04 M Tris-HCl buffer, pH 7.4, containing 0.14 M sodium chloride. A standard clotting curve was prepared using highly purified human thrombin (3105 units/mg) kindly provided by Dr. John W. Fenton, New York State Department of Health, Albany, New York. Activity of fibrinogen-clotting enzyme was expressed in NIH equivalent thrombin clotting units/ml. Hemorrhagic activity of purified enzyme or chromatographic fractions was analyzed by rat skin testing as previously described (19). In this procedure, anesthesized mice weighing 25 to 30 g were used. The fur was shaved off their backs, and the purified enzyme or chromatographic fractions were injected intradermally (50 to 100 μg of protein in a volume of 0.02 to 0.20 ml). After 16 hours, the mice were sacrificed by an overdose of anesthetic, and the skin was removed and placed between two pieces of glass for visual observation.

Digestion of Human Fibrinogen

Human fibrinogen (500 μg) was incubated at 37°C with 5 μg of the purified venom enzyme, or with 5 μg of the venom enzyme pretreated for 30 minutes at room temperature with 10 mM diisopropylfluorophosphate or 10 mM tetraethylenepentamine. Aliquots containing 25 μg of fibrinogen were removed from the incubation mixture at 0, 0.5, 1, 3, and 20 hours and applied to 7.5% acrylamide slab gels for SDS acrylamide gel electrophoresis according to Laemmli (14). Staining was with Coomassie blue G-250. Control (untreated) fibrinogen was incubated at 37°C for 0 and 20 hours and also applied to the gel.

RESULTS AND DISCUSSION

Enzyme Purification

The fibrinolytic enzyme from Southern copperhead venom has been purified by a three-step procedure including gel filtration on Sephadex G-100, ion exchange chromatography on CM-cellulose, and affinity chromatography on p-aminobenzamidine-agarose to remove serine proteinase contaminants. The addition of a fourth step utilizing DEAE-cellulose did not appreciably improve the purification obtained by the first three steps. The majority of the characterization reported here was performed with the three- or four-step purified material, which was about 80 to 85% pure and contained two detectable minor contaminants. Purification of the enzyme by this procedure has been previously described (9).

More recently, we developed a new three-step procedure, involving ion exchange chromatography on CM-cellulose, gel filtration on Sephadex G-100, and isoelectric focusing on immobilized pH gradients. For this procedure, the venom (2 to 10 g) was dissolved in a minimal volume of 5 mM sodium acetate buffer, pH 6.5, and dialyzed against this buffer overnight. A CM-cellulose column was equilibrated with the dialysis buffer and, after sample application, was washed with this buffer until a large protein peak was washed through. A linear gradient was then initiated from 5 to 500 mM of sodium acetate, pH 6.5. The major fibrinolytic activity eluted in the first protein peak during the gradient (Fig. 1), and this was pooled and concentrated with an Amicon ultrafiltration apparatus prior to gel filtration. The concentrated material was applied to Sephadex G-100 previously equilibrated with 0.04 M Tris-HCl, 0.14 M sodium chloride, pH 7.1, and elution was with this buffer. As shown in Figure 2, azocaseinolytic and fibrinolytic activities coeluted, and were separated from a higher molecular weight peak that contained hydrolytic activity for the chromogenic substrate S-2302 (H-D-Pro-Phe-Arg-p-nitroanilide). The venom enzyme prepared by this two-step procedure is about 80% pure, and is almost identical to the enzyme purified by the three- or four-step method just described (Fig. 3). Comigration of protein (Fig. 3B) and activity (Fig. 3A) suggests that the major protein band represents the fibrinolytic enzyme. When comparing the electrophoretic pattern of the purified enzyme with that of crude venom, it is obvious that the fibrinolytic enzyme represents a major component of the venom. Additional purification of the fibrinolytic enzyme has been achieved by use of narrow pH range, semipreparative isoelectric focusing in immobilized pH gradients (20). This extremely sensitive technique separated the fibrinolytic protein from several contaminants. However, as shown in Figure 4, this procedure further resolved the fibrinolytic enzyme into two components. Since both the components appear to have fibrinolytic activity, they are most likely isoenzymes resulting from minor carbohydrate or amino

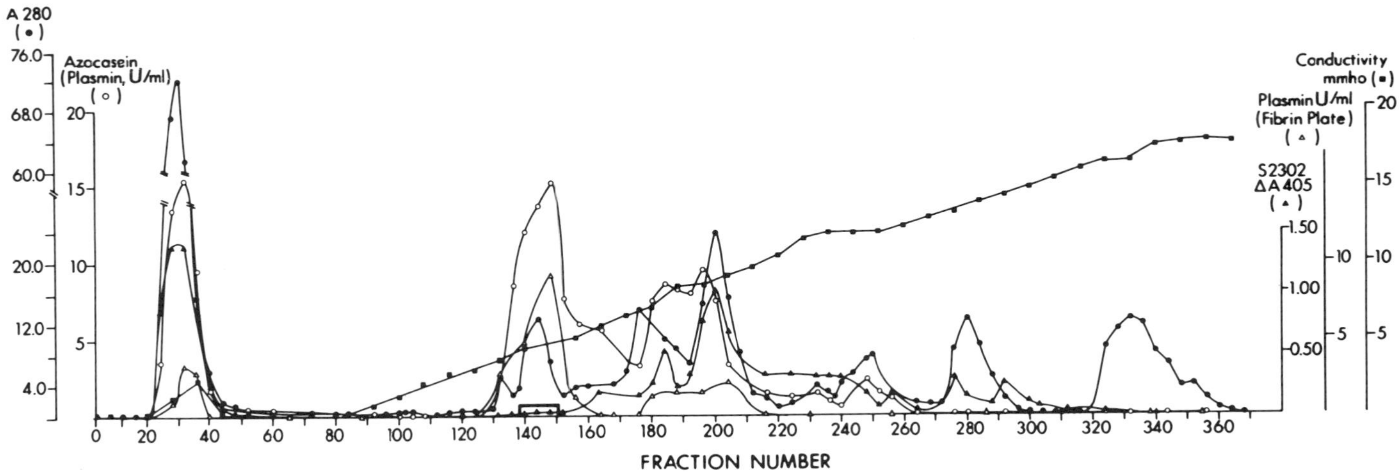

Figure 1 CM-cellulose chromatography of 10 g of crude *A. c. controtrix* venom. The lyophilized venom was dialyzed against 5 mM sodium acetate buffer, pH 6.5, in preparation for chromatography. CM-cellulose (CM-52) was equilibrated with this buffer and packed into a column 2.6 × 43 cm. Elution was initially with this buffer at a flow rate of 20 ml/hour, and fractions of 5 ml/tube were collected. The LKB ultrograd gradient maker was used with limit buffer 500 mM sodium acetate, pH 6.5. Conductivity was recorded as was S-2302 hydrolytic activity, absorbance at 280 mm (A280), azocasein hydrolytic activity, and fibrin plate clearance. Peak fractions pooled are shown by the rectangle (fractions 138 to 152).

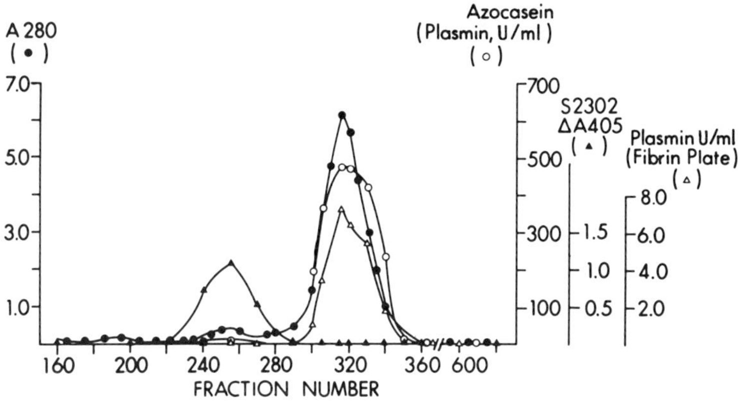

Figure 2 Molecular sieve chromatography on Sephadex G-100. Pooled fractions from the CM-cellulose column were dialyzed against 0.04 M Tris-HCl buffer, pH 7.1, containing 0.14 M sodium chloride, and applied to a Sephadex G-100 column (4.5 × 58 cm) equilibrated in the Tris-NaCl buffer. Elution was with this buffer. Fractions of 2.5 ml were collected at a flow rate of 20 ml/hour. Hydrolytic activity of S-2302 was measured, as was azocasein hydrolytic activity, fibrin plate clearance, and A280. Fractions 300 to 344 were pooled.

acid sequence differences. Preliminary structural analysis of the two partially separated components reveals that they have identical amino-terminal amino acid sequences (data not shown). Preparative isoelectric focusing using a narrow range, immobilized pH gradient is presently being developed for final purification of the enzyme.

Properties of the Enzyme

The enzyme was analyzed by SDS polyacrylamide gel electrophoresis (14,15) and was shown to contain a single Coomassie blue-staining band with a molecular weight calculated to be between 25,000 and 27,000. Periodic acid-Schiff staining (16) of a duplicate gel showed that the enzyme may be a glycoprotein. Isoelectric focusing on immobilized pH gradients (pH 6.5 to 7.5) revealed that the venom enzyme has an isoelectric point of about pH 6.8. Activity of the enzyme can be assayed either by the fibrin plate method (11) or by azocasein hydrolysis (8). The pH optimum for hydrolysis of azocasein is 7.5. The enzyme is not inhibited by diisopropylfluorophosphate, p-chloromercuribenzoate, Trasylol, and soybean or lima bean trypsin inhibitors. However, rapid and virtually complete inhibition of azocasein hydrolysis was obtained by EDTA, tetraethylenepentamine (a specific zinc chelator) (21), and

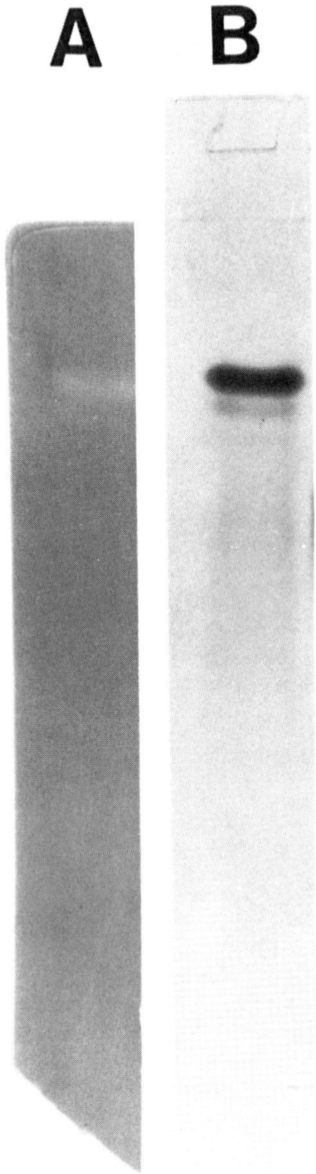

Figure 3 Native polyacrylamide gel electrophoresis of fibrinolytic enzyme from *A. c. contortrix* venom purified by the three-step method (CM-cellulose, Sephadex G-100, and p-aminobenzamidine-agarose). (A) Fibrinogen-containing gel stained with Coomassie blue following electrophoresis of 100 μg of enzyme. (B) Normal gel stained with Coomassie blue following electrophoresis of 50 μg of enzyme. For details, see text.

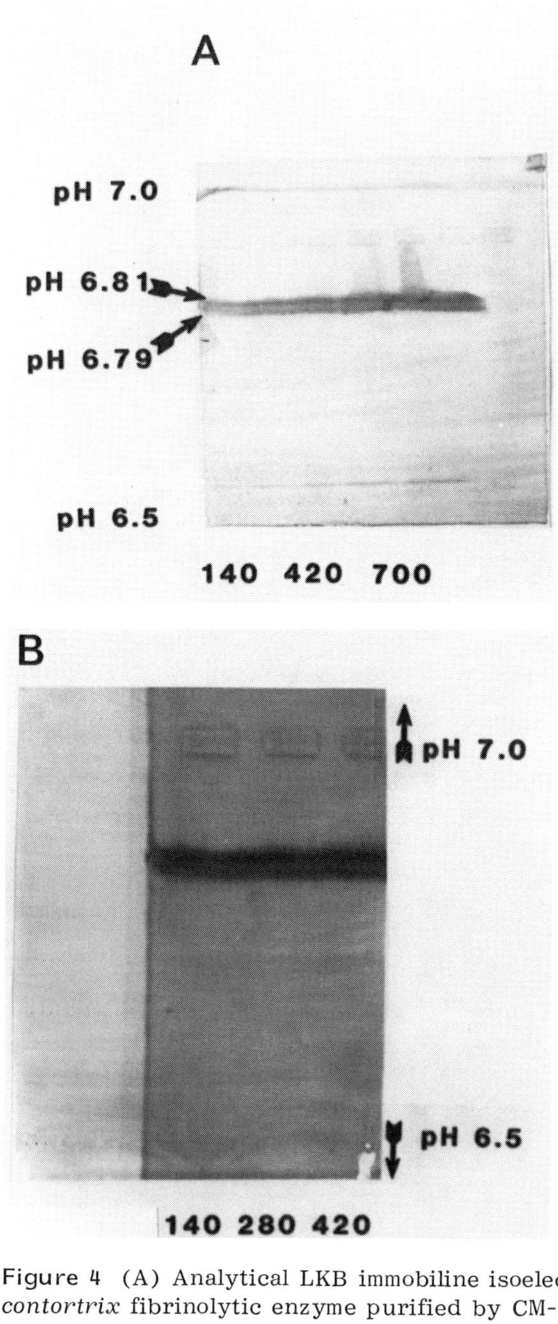

Figure 4 (A) Analytical LKB immobiline isoelectric focusing of *A. c. contortrix* fibrinolytic enzyme purified by CM-cellulose, G-100, and p-aminobenzamidine-agarose. The immobiline acrylamide gel was cast using a pH gradient from 6.5 to 7.0. Electrophoresis was run at 2000 volts for 16 hours followed by 3000 volts for 1 hour and staining with Coomassie blue R-250. A prerun was carried out (2000 volts for 1 hour) before sample application. The numbers under the immobiline runs are protein concentrations (in μg) applied to each well. (B) An enlargement of A, showing the two isozymes more clearly.

o-phenanthroline. These findings suggest that the enzyme is a metalloproteinase and not a serine or cysteine proteinase. α_2-Macroglobulin inhibited azocaseinolytic activity of the venom enzyme following 10 minutes of incubation of the enzyme with a six-fold molar excess of the inhibitor. This was not surprising in that it is known that α_2-macroglobulin is an effective inhibitor of all classes of proteinases (22). It was also found that at least one disulfide bond is critical to the structural or functional integrity of the venom enzyme since incubation with 300-fold molar excess dithiothreitol rapidly inhibited azocaseinolytic activity of the enzyme. These results are presented in Table 1. Activity of the enzyme on a series of p-nitroanilide (pNA)-containing peptide substrates was analyzed by measuring p-nitroaniline release at 405 nm. No activity was detected with the following substrates: N-succinyl-Ala-Ala-Ala-pNA, N-succinyl-Ala-Ala-Pro-Phe-pNA, N-succinyl-Gly-Gly-Phe-pNA, N-carbobenzoxy-Arg-pNA, N-benzyl-Arg-pNA, N-carbobenzoxy-Gly-Gly-Leu-pNA, methoxysuccinyl-Ala-Ala-Pro-Val-pNA, H-D-Val-Leu-Lys-pNA (S-2251), H-D-Ile-Pro-Arg-pNA (S-2288), H-D-Pro-Phe-Arg-pNA (S-2302), and H-D-Phe-pipecolyl-Arg-pNA (S-2238). These findings provide additional verification that the fibrinolytic enzyme is not a serine proteinase since these substrates are hydrolyzed by different subclasses of serine proteinases. Also, the lack of hydrolysis of S-2251 and S-2288 by the venom fibrinolytic enzyme clearly distinguishes it from two fibrinogenolytic proteinases in Southern copperhead venom of similar molecular weight, reported by Dyr and co-workers (6), that hydrolyze these (or similar)

Table 1 Inhibition of Azocaseinolytic Activity of Fibrinolytic Enzyme[a]

Inhibition	%	No inhibition
EDTA (20 mM)	92	Trasylol (1000 KIU)
o-Phenanthroline (1 mM)	76	Lima bean trypsin inhibitor (40 μM)
Tetraethylenepentamine (10 mM)	100	Soybean trypsin inhibitor (50 μM)
Alpha$_2$-macroglobulin (2 μM)	95	Diisopropylfluorophosphate (10 mM)
Dithiothreitol (6 mM)	100	p-Chloromercuribenzoate (1 mM)

[a]Venom enzyme was preincubated separately with each of the inhibitors prior to assay (see text).

substrates. The purified enzyme exhibited no fibrinogen clotting or hemorrhagic (mouse skin test) activity.

In contrast to the action of streptokinase (an agent currently used clinically for thrombolytic therapy by acting as a plasminogen activator), the fibrinolytic enzyme appears to act directly on fibrin and does not activate plasminogen. After incubation of highly purified human plasminogen with the venom enzyme for up to 5 hours, there was no evidence of plasmin formation as demonstrated either by the lack of generation of S-2251-hydrolyzing activity or by the failure to produce the two-chain electrophoretic form as analyzed by SDS polyacrylamide gel electrophoresis. Further, aliquots from the incubation mixture containing plasminogen and the venom enzyme were treated with streptokinase, and S-2251-hydrolyzing activity was generated, indicating that plasmin had been formed. These results showed that the venom enzyme not only did not activate plasminogen to plasmin, but also did not degrade human plasminogen to an inactive form since it was still activatable with streptokinase. The venom enzyme was directly fibrinolytic as evidenced by its ability to form clear zones on plasminogen-free fibrin plates. Similarly, fibrin overlays on native polyacrylamide gels, after electrophoresis of venom chromatographic fractions or purified fibrinolytic enzyme, were rapidly lysed by the enzyme, producing visible zones of clearing against an opaque background immediately above the position of the fibrinolytic enzyme on the acrylamide gel. Additionally, clots formed after the addition of thrombin, either to human plasma or purified human fibrinogen, were rapidly lysed by the venom enzyme.

The purified venom enzyme also exhibits fibrinogenolytic activity as evidenced by SDS polyacrylamide gel electrophoretic analysis of the products resulting from incubation of the enzyme with purified human fibrinogen (Fig. 5). These studies reveal that the enzyme is an α-chain fibrinogenase. Digestion of the α-chain is essentially complete by the first time point at 30 minutes. There is some digestion of the β-chain after 20 hours of incubation, but this is most likely due to a minor contaminant in the enzyme preparation. Digestion of the α-chain is completely blocked by 30 minutes of preincubation of the venom enzyme with 10 mM TEP at room temperature, again confirming the metalloproteinase nature of the venom enzyme. In contrast, diisopropylfluorophosphate has a minimal effect on α-chain digestion by the venom enzyme. Although we have shown that the venom enzyme degrades purified fibrinogen, in plasma the thrombin clottability of fibrinogen is only slightly altered by pretreatment with the venom enzyme (data not shown). These findings suggest that the venom enzyme may be inactivated by plasma proteinase inhibitors, and we have reported that α_2-macroglobulin inhibits azocaseinolytic activity of the venom enzyme. Interestingly, the venom enzyme retains fibrinolytic activity in vivo (as demonstrated in the following section). Presumably this is due either to the relative kinetics of its interactions with fibrin

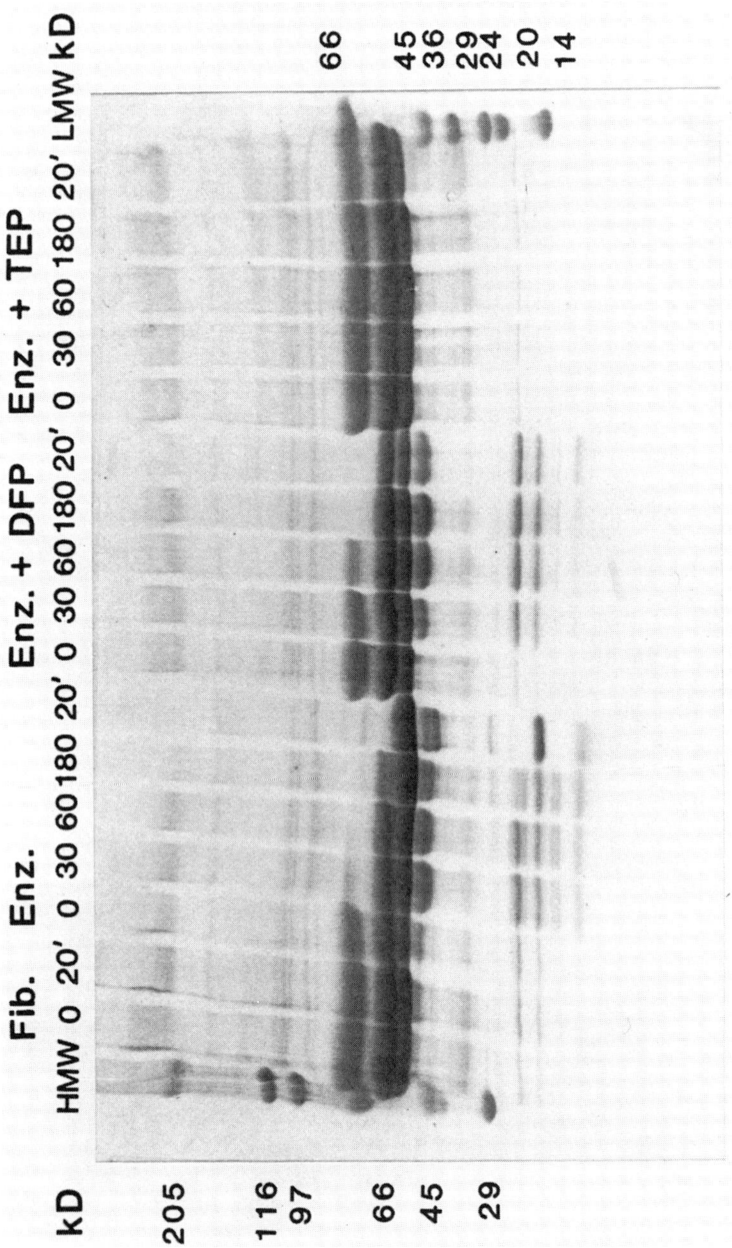
Fib.
Enz.
Enz. + DFP
Enz. + TEP
HMW 0 20' 0 30 60 180 20' 0 30 60 180 20' 0 30 60 180 20' LMW
kD
205
116
97
66
45
29
kD
66
45
36
29
24
20
14

Table 2 Properties of Fibrinolytic Enzyme

Source	*A. c. contortrix* venom
Molecular weight	25,000
pI	6.7-6.8
Enzymatic activity	Metalloproteinase (inhibited by chelating agents)
Degradation of fibrin	Direct-acting (α-chain fibrinogenase)
Plasminogen activation	No

and α_2-macroglobulin, or to overtitration of the available plasma proteinase inhibitor by the venom enzyme. Table 2 summarizes some of the important properties of the venom fibrinolytic enzyme.

In Vivo Studies

In collaboration with J. J. Bookstein, University of California, San Diego, we investigated the in vivo fibrinolytic activity of the purified venom enzyme in a venous thrombosis model system. In the five rabbits studied thus far, excellent results were obtained by infusion of the purified enzyme immediately below the clot. All animals survived

---

Figure 5 Digestion of human fibrinogen by Southern copperhead venom fibrinolytic enzyme purified by the four-step procedure. Fibrinogen (500 μg) was incubated at 37°C with 5 μg of venom enzyme (ENZ) or venom enzyme pretreated for 30 minutes at room temperature with 10 mM DFP (ENZ. + DEP) or 10 mM tetraethylene-pentamine (ENZ. + TEP). Aliquots containing 25 μg of fibrinogen were removed from the venom digest at 0, 0.5, 1, 3, and 20 hours and applied for SDS polyacrylamide gel electrophoresis (14). Staining was with Coomassie blue G-250. Low and high molecular weight markers (LMW and HMW) were applied to the extreme right and left panels, respectively. Marker proteins are: myosin (rabbit muscle), 205,500; β-galactoside (*E. coli*), 116,500; phosphorylase (rabbit muscle), 87,400; albumin (bovine), 66,000; albumin (egg), 45,000; glyceraldehyde-3-phosphate dehydrogenase (rabbit muscle), 36,000; carbonic anhydrase (bovine erythrocyte), 29,000; trypsinogen (bovine pancreas), 28,000; trypsin inhibitor (soybean), 20,100; and α-lactalbumin (bovine milk), 14,200. Control (untreated) fibrinogen (FIB) was incubated at 37°C for 0 and 20 hours, as shown on the left of the gel.

with minimal or no side effects. These studies demonstrate the potential of a venom fibrinolytic enzyme as a clinically useful thrombolytic agent. It appears that additional studies are warranted since the enzyme seems to function independently of the native fibrinolytic system, and may offer promise as a safe, effective, rapid, and specific agent for selective thrombolysis. Detailed results of these studies will be the subject of a separate publication (23).

ACKNOWLEDGMENTS

We would like to acknowledge the expert technical assistance of Mrs. Alicia Perdon throughout these investigations. We also thank Mr. James Stapakis for technical assistance in the α_2-macroglobulin studies and Ms. Lynn Allen for assistance in the tetraethylenepentamine studies. These studies were supported in part by PHS Grand No. HL 31389 awarded by the National Heart, Lung, and Blood Institute, DHHS, and by Cortech, Inc., Denver, Colorado.

REFERENCES

1. Eagle, H. The coagulation of blood by snake venoms and its physiological significance. J. Exp. Med. 65:613 (1937).
2. Didisheim, P. and J. H. Lewis. Fibrinolytic and coagulant activities of certain snake venoms and proteases. Proc. Soc. Exp. Biol. Med. 93:10 (1956).
3. Kornalik, F. The influence of snake venoms of fibrinogen conversion and fibrinolysis. Mem. Inst. Butantan 33:179 (1966).
4. Herzig, R. H., O. D. Ratnoff, and J. R. Shainoff. Studies on a procoagulant fraction of Southern copperhead snake venom: the preferential release of fibrinopeptide B. J. Lab. Clin. Med. 76:451 (1970).
5. Moran, J. B. and C. R. Geren. Characterization of a fibrinogenase from Northern copperhead (*Agkistrodon contortrix mokasen*) venom. Biochim. Biophys. Acta 659:161 (1981).
6. Dyr, J. E., B. Blombäck, and F. Kornalik. The fibrinogenolytic and procoagulant activity of Southern copperhead venom enzymes. Thromb. Res. 30:185 (1983).
7. Bajwa, S. S., F. E. Russell, and F. S. Markland. Thrombolytic agents in snake venoms. Proc. West. Pharmacol. Soc. 25:353 (1982).
8. Bajwa, S. S., H. Kirakossian, K. N. N. Reddy, and F. S. Markland. Thrombin-like and fibrinolytic enzymes in the venom from the Gaboon viper (*Bitis gabonica*), Eastern cottonmouth moccasin (*Agkistrodon p. piscivorus*) and Southern copperhead (*Agkistrodon c. contortrix*) snakes. Toxicon 20:427 (1982).

9. Markland, F. S. Rattlesnake venom enzymes that interact with components of the hemostatic system. Tox. Rev. 2:119 (1983).
10. Charney, J. and R. M. Tomarelli. A colorimetric method for the determination of the proteolytic activity of duodenal juice. J. Biol. Chem. 171:501 (1947).
11. Bajwa, S. S., F. S. Markland, and F. E. Russell. Fibrinolytic enzymes in Western diamondback rattlesnake (*Crotalus atrox*) venom. Toxicon 18:285 (1980).
12. Lacks, S. A. and S. S. Springhorn. Renaturation of enzymes after polyacrylamide gel electrophoresis in the presence of sodium dodecyl sulfate. J. Biol. Chem. 255:7467 (1980).
13. Markland, F. S. and A. Perdon. Comparison of two methods for proteolytic enzyme detection in snake venom. Toxicon 24:385 (1986).
14. Laemmli, U. K. Cleavage of structural proteins during the assembly of the head of bacteriophage T4. Nature 227:680 (1970).
15. Weber, K. and M. Osborn. The reliability of molecular weight determinations by dodecyl sulfate-polyacrylamide gel electrophoreses. J. Biol. Chem. 244:4406 (1969).
16. Gander, J. E. Gel protein stains: glycoproteins. Methods Enzymol. 104:447 (1984).
17. Deutsch, D. and E. T. Mertz. Plasminogen: purification from human plasma by affinity chromatography. Science 170:1095 (1970).
18. Markland, F. S. Crotalase. Methods Enzymol. 45:223 (1976).
19. Ownby, C. L., T. R. Colberg, and G. V. Odell. A new method for quantitating hemorrhage induced by rattlesnake venoms: ability of polyvalent antivenom to neutralize hemorrhagic activity. Toxicon 22:227 (1984).
20. Bjellqvist, B., K. Ek, P. G. Righetti, E. Gianazza, A. Gorg, R. Westermeier, and W. Postel. Isoelectric focusing in immobilized pH gradients. Principle, methodology and some applications. J. Biochem. Biophys. Methods 6:317 (1982).
21. Matsumoto, K., H. Maeda, and R. Okamura. Role of the zinc atom in the activity and conformational stability of *Serratial* 56K protease. A fluorometric study. J. Biochem. 96:1165 (1984).
22. Barrett, A. J. α_2-Macroglobulin. Methods Enzymol. 80:737 (1980).
23. Markland, F. S. and Bookstein, J. J. In preparation.

14

Secretion of Cellular Plasminogen Activators Upon Stimulation by Crotalinae Snake Venoms

Nancy E. Kirschbaum, Robert S. Reczkowski, and Andrei Z. Budzynski
Temple University School of Medicine, Philadelphia, Pennsylvania

INTRODUCTION

Snake venoms and their purified components have been used for many years to elucidate mechanisms of blood coagulation. Enzymes isolated from various snake venoms have a very narrow substrate-specificity in their interactions with blood coagulation factors. For example, reptilase purified from *Bothrops jararaca* venom cleaves an arginine-glycine peptide bond in the Aα-chain of human fibrinogen releasing fibrinopeptide A. However, reptilase does not hydrolyze the arginine-glycine bond in the Bβ-chain and is, therefore, unable to remove fibrinopeptide B (1). Snake venoms also affect fibrinolysis. However, the mechanisms of their interaction and specific sites of action in fibrinolysis are much less understood than are those in blood coagulation. Envenomation of mammals, including humans, by rattlesnakes or vipers belonging to the Crotalidae or Viperidae families causes the blood to become incoagulable. Although fibrinogen-clotting enzymes present in venoms convert circulating fibrinogen to fibrin, the formed clots are degraded by two major pathways. One involves fibrinolysis that degrades circulating soluble fibrin to incoagulable fragments (2). The other requires uptake of undegraded fibrin by macrophages, which has been found to occur in vitro (3-5).

The venom of the adult Western diamondback rattlesnake *Crotalus atrox* does not contain fibrinogen-clotting enzymes (6). Yet, patients bitten by this rattlesnake have no fibrinogen in the circulation, greatly decreased plasminogen concentration, the presence of plasmin-α_2-antiplasmin complex, and a significantly increased amount of fibrinogen degradation products in serum (7). Similar conclusions were derived in a study of patients envenomated by *Crotalus adamanteus* (8). These findings indicated the activation of fibrinolysis without formation of fibrin in the circulating blood. The rattlesnake venoms did not activate plasminogen in vitro. Therefore, one has to postulate involvement of cellular response in the activation of fibrinolysis in vivo. Several studies showed that cultured vascular cells, especially the endothelial and smooth muscle cells, produced and secreted plasminogen activators (9,10). This work examined the effect of several snake venoms on secretion of plasminogen activators from cultured human endothelial and smooth muscle cells. The experiments addressed a hypothesis that the fibrinolytic effect of rattlesnake venoms in vivo may be mediated by stimulation of plasminogen activator secretion from vascular cells.

MATERIALS AND METHODS

All reagents were of the analytical purity grade. The following preparations of dried snake venoms were obtained from Sigma Chemical Company (St. Louis, Missouri): *Agkistrodon acutus*, *Agkistrodon rhodostoma*, *Bothrops atrox*, *Crotalus adamanteus*, *Crotalus atrox*, *Echis carinatus*, *Naja naja*, *Trimeresurus gramineus* and *Vipera musselli*. The venoms were dissolved in phosphate-buffered saline, pH 7.4, to the desired concentration just before use.

Human venous blood, 40 ml, was collected in 10 ml of the ACD anticoagulant. Erythrocytes were prepared by differential centrifugation (11). Polymorphonuclear leukocytes were obtained by sedimentation in 3% (w/v) dextran, hypotonic lysis of erythrocytes, and differential centrifugation (12). Platelets were isolated using low-speed centrifugation and washed (13).

Human venous cells were obtained from umbilical cords collected in Temple University Hospital. Primary cultures of endothelial cells were obtained by perfusion of umbilical cord veins with a mixture of 0.1% (w/v) collagenase and 0.01% (w/v) trypsin. Detached cells were collected by centrifugation and plated on plastic culture dishes of 60-mm diameter. The cells were grown to confluence using Dulbecco's Minimum Essential Medium (DMEM) supplemented with 10% (v/v) fetal bovine serum being kept in an incubator at 37°C in a 95% air 5% carbon dioxide atmosphere (14). Passaged endothelial cells were obtained by detachment of primary cultures with 0.1% (w/v) trypsin in 0.01% (w/v) EDTA, followed by replating. In these experiments, confluent cultures passed between 7 and 20 times that contained about 2.5 million cells per dish were used.

Smooth muscle cells were obtained from vessel wall explants of umbilical veins after removal of the endothelium. The cells were also passaged by detachment from culture dishes using 0.1% (w/v) trypsin in 0.01% (w/v) EDTA, followed by replating (15). In these experiments, cultures in 60-mm dishes that contained between 2.5 to 3.5 million cells per dish were used.

The secretion of plasminogen activators was studied in cultures of vascular cells after replacement of serum-containing DMEM by a serum-free DMEM. Buffer or a snake venom was added and incubated at 37°C in a 95% air and 5% CO_2 atmosphere for 2 hours. Cell-conditioned medium was collected and centrifuged at 8000 × *g* for 3 minutes to remove suspended cell debris. The remaining cells were treated with 0.025% (v/v) Triton X-100 and 0.1% (w/v) EDTA for 30 minutes and centrifuged to obtain cell extract.

The concentration of plasminogen activators in cell-conditioned media was measured using a solid-phase ^{125}I-fibrin lysis assay (16). The results were expressed as either percent of lysed fibrin or converted to milliunits of urokinase activity using a standard curve. Low molecular weight urokinase from Abbott Laboratories (North Chicago, Illinois) was used as the standard.

Polyacrylamide slab gel electrophoresis in the presence of SDS was performed in 9% gels (17). The detection of plasminogen activators in polyacrylamide gels was done using a plasminogen-enriched fibrin-agarose overlay followed by incubation at 25°C for 18 hours (18).

RESULTS AND DISCUSSION

To assess which cell types occurring in the circulation contribute significantly to the total cellular output of plasminogen activators, five cell types were studied: erythrocytes, polymorphonuclear leukocytes, platelets, endothelial cells, and smooth muscle cells. The amount of plasminogen activators secreted by each cell type was determined in the cell-conditioned medium. Erythrocytes did not secrete any activators (Fig. 1). This observation did not support an earlier report describing the isolation of a plasminogen activator from human erythrocytes (19). Leukocyte contamination of those erythrocytes would have explained their lytic activity. Platelets also did not secrete any activators (Fig. 1). This observation was consistent with reports from other investigators, showing that platelets secreted plasminogen-activator inhibitor, rather than the activator itself (20). Polymorphonuclear leukocytes did secrete plasminogen activators, but the amount was small (Fig. 1). Thus, the cells circulating in the blood contained a limited pool of secretable activators that is localized in the leukocytes alone. Leukocyte-derived plasminogen activator appears to be of urokinase type (21). Human fibroblasts of foreskin origin were also tested.

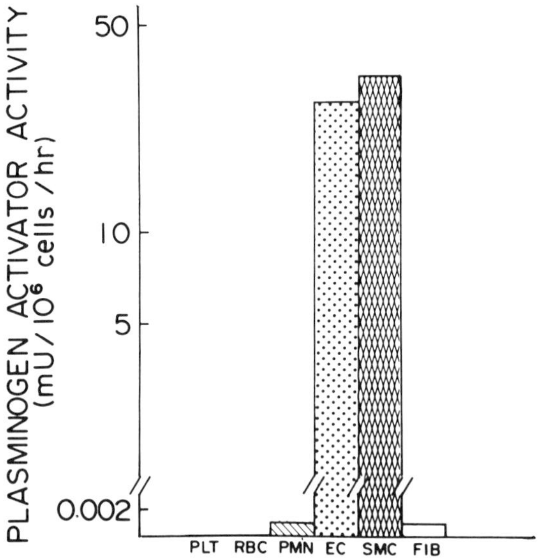

Figure 1 Secretion of plasminogen activators from unstimulated cells. The data show plasminogen activator concentration in cell-free-conditioned media harvested after incubation for 2 hours. The data are mean values from three experiments and are normalized per 10^6 cells. PLT = platelets; RBC = erythrocytes; PMN = polymorphonuclear leukocytes; EC = venous endothelial cells; SMC = venous smooth muscle cells; FIB = foreskin fibroblasts.

Since the basic secretion of plasminogen activators was low (Fig. 1), stimulation was unremarkable, and origin was not vascular, the cells were not investigated further.

Cultured venous smooth muscle cells and endothelial cells secreted about 10,000 times more plasminogen activators per cell than did leukocytes (Fig. 1). This observation supports the concept that the major source of plasminogen activators originates from cells occurring in the blood vessel wall (9,10).

The five cell types were incubated in the presence of various concentrations of *Crotalus atrox* venom, and the activator secretion was measured (Fig. 2). Erythrocytes and platelets did not release any more plasminogen activator activity as compared with that of control cells incubated with serum-free DMEM alone. In contrast, leukocytes, smooth muscle cells, and endothelial cells were stimulated by the venom to secrete more plasminogen activators than were cultures not exposed to the venom. The stimulation was dose-dependent on concentration of the venom: up to 10 μg/ml for polymorphonuclear leukocytes, and up

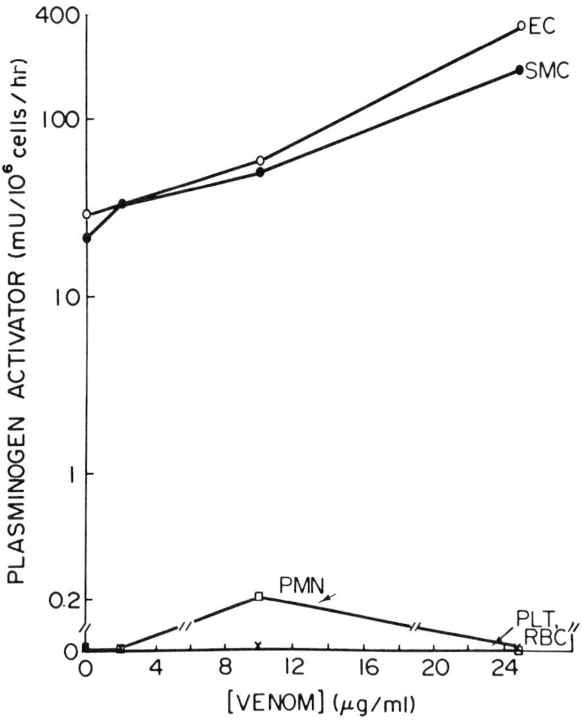

Figure 2 Secretion of plasminogen activators from cells incubated with *Crotalus atrox* venom at different concentrations. The data show plasminogen activator concentration in cell-free-conditioned media collected after 2 hours of incubation, and are normalized per 10^6 cells. Abbreviations as in Figure 1.

to 25 μg/ml for smooth muscle and endothelial cells. However, for smooth muscle and endothelial cells, the highest venom concentration that was tested caused the cells to detach partially from the culture dishes. The detached cells were not significantly lysed by the venom, which was inferred from the intra- and extracellular lactate dehydrogenase activities. Thus, this rattlesnake venom appears to be a good experimental tool to demonstrate the cellular release of plasminogen activators. The results shown in Figures 1 and 2 support our earlier conclusions (22) that only nucleated cells secrete plasminogen activators and that the venom stimulates secretion of plasminogen activators from the nucleated cells only.

In the circulating blood, the number of cells producing plasminogen activators varies in relation to the vessel diameter. Cell number per

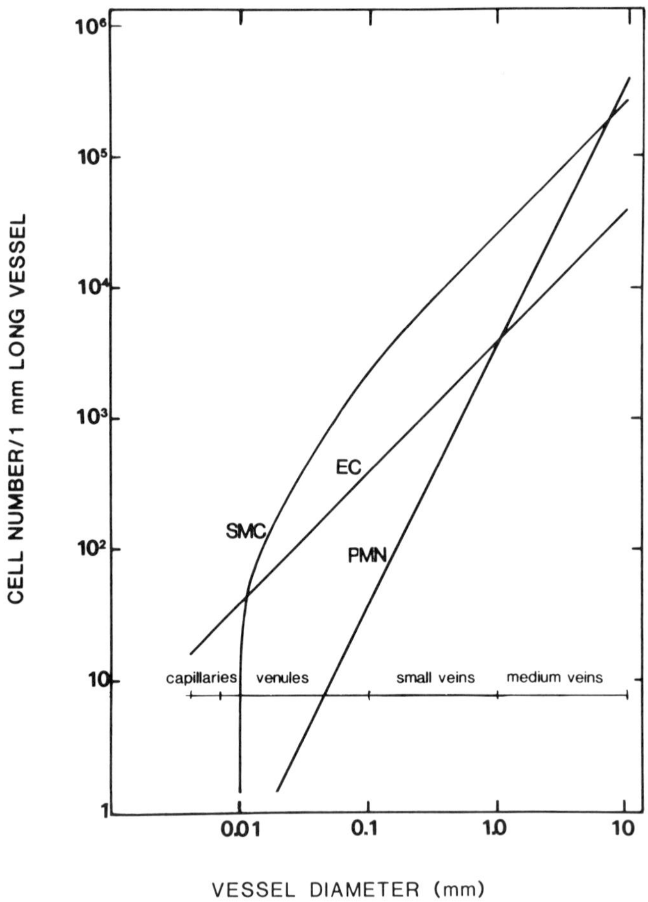

Figure 3 Dependence of cell number on the internal diameter of blood vessels is shown on the abscissa. The ordinate depicts the total cell number contained in a 1-mm long blood vessel. Abbreviations as in Figure 1.

blood vessel length-unit can be calculated. Capillaries are lined with endothelial cells only (Fig. 3). In venules, smooth muscle cells begin to appear, form multicellular layers, and soon their number is larger than that of endothelial cells. Polymorphonuclear leukocytes occur in significant numbers in the blood volume contained in wider venules. In small veins, the number of smooth muscle cells is the largest, endothelial cell number is intermediate, and leukocyte number is the smallest. However, in medium- and large-sized veins, leukocyte number overtakes all other cells. The cell numbers shown in Figure 3 indicate that different cells are the main contributors to the fibrinolytic response in various parts of the circulatory system. This fact is also valid after introduction of a snake venom into the circulating blood.

Several snake venoms were tested for their relative stimulatory effect on secretion of plasminogen activators from cultured endothelial cells (Fig. 4). The magnitude of the stimulation and the dose-response relationship permitted us to categorize the venoms into three groups:

(a) Large stimulation of plasminogen activators secretion and a linear dose-response in the range from 0 to 25 μg/ml of the venom were determined with *Crotalus atrox* and *Crotalus adamanteus* venoms. These two rattlesnake venoms were unparalleled in their effect as compared with all other tested venoms. Interestingly, the two venoms interfere differently with fibrinogen in the blood. *Crotalus atrox* venom contains fibrinogenolytic enzymes that degrade the NH_2-terminus of the Bβ-chain (23), impairing thrombin coagulability (24). *Crotalus adamanteus* venom contains crotalase, a fibrinogen-clotting enzyme cleaving only fibrinopeptide A (25).

(b) Moderate stimulation and a nonlinear dose-response in the range from 0 to 10 μg/ml of the venom was found for *Echis carinatus* and *Agkistrodon rhodostoma* venoms. Although both venoms interfere with blood coagulation, the mechanisms are quite different.

(c) A lack of plasminogen activator stimulation was observed using *Agkistrodon acutus, Bothrops atrox, Naja naja, Trimeresurus gramineus*, and *Vipera russelli* venoms. This negative result demonstrated that the stimulating effect was not characteristic of all snake venoms and confirmed our hypothesis that some venoms indirectly activate plasminogen in vivo.

The results shown in Figure 4 indicate the presence in the species-related diamondback rattlesnake venoms of an agent that stimulates cultured human endothelial cells to secrete significantly increased amounts of plasminogen activators.

The stimulating agent did not seem to be related to either metallo- or serine-proteases present in *Crotalus atrox* venom (23). Inhibition of these enzymes by either EDTA or DFP, respectively, did not alter the stimulation (Table 1). Addition of both inhibitors together to the venom caused even a larger stimulation. This finding indicated that metallo- and serine-proteinases of this venom seemed to interfere with the overall mechanism of stimulation of plasminogen activators secretion.

Cultured human endothelial cells in our laboratory produced both tissue plasminogen activator (TPA) and urokinase (UK). Therefore, these cells provided a good model to answer the question: Which activator type is affected by the venom's agent? To address this point,

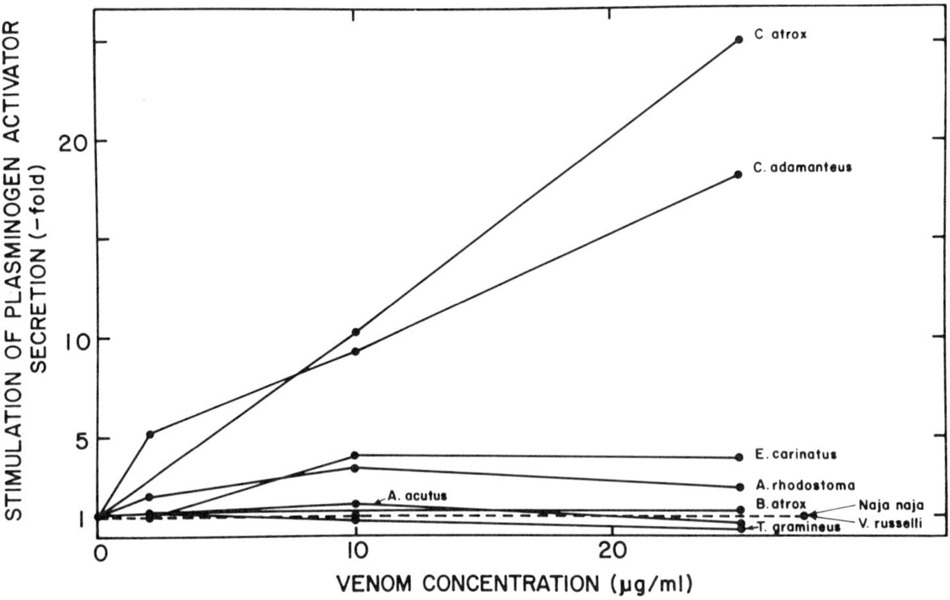

Figure 4 Stimulation of plasminogen activator secretion from human vascular endothelial cells by various snake venoms. The broken horizontal line shows the secretion of activators from untreated cells; the multiples of this value are given on the ordinate. The data are mean values of four determinations.

Table 1 Effect of Protease Inhibitors on *Crotalus atrox* Venom Activity that Stimulates Plasminogen Activator Output from Cultured Human Endothelial Cells

Agent	Fibrin lysed (%)	Stimulation (-fold)
No venom	3.04	0
Venom, 10 µg/ml	8.35	2.7
Venom, 10 µg/ml + EDTA, 10 mM	7.45	2.5
Venom, 10 µg/ml + DFP, 10 mM	7.35	2.4
Venom, 10 µg/ml + EDTA and DFP, 10 mM	14.3	4.7

cultured human endothelial cells were incubated in the presence or absence of 10 μg/ml DFP- and EDTA-treated *Crotalus atrox* venom. Conditioned media containing secreted plasminogen activators or cell extracts containing intracellular plasminogen activators were separated by SDS-PAGE, followed by plasminogen activators detection using fibrin autography (Fig. 5). Extracts of cells not exposed to the venom contained UK (M_r 55,000), TPA (M_r 70,000), UK-inhibitor complex (M_r 95,000), and TPA-inhibitor complex (M_r 110,000). Extracts from venom-treated cells showed increases in free and complexed UK as well as in TPA-inhibitor complex. Cell-conditioned media from unexposed and venom-treated cells differed clearly in one aspect—the latter contained much more UK than did the former. The result is consistent with data presented in Figures 2 and 4 and Table 1; moreover, it showed that the stimulation mainly affected the increased secretion of UK.

The cellular mechanism by which plasminogen activator biosynthesis can be stimulated is, at the present, unclear. The stimulation may occur in several cellular compartments, as shown in Figure 6. First, a receptor for the venom's agent may exist on the endothelial cell surface that transmits a stimulatory signal to the cytosol. Next, the secretion of the existing membrane-bound activators may occur along with transport in receptosomes of receptor-processed venom agent. The latter, after binding to the nuclear genomic DNA, may affect the rate of mRNA synthesis and, in turn, increase the biosynthesis of specific proteins. The secretory mechanism also seems to be affected as UK (but not TPA) appears to be preferentially released from the stimulated endothelial cells. These possibilities have not yet been addressed experimentally.

SUMMARY

The release of plasminogen activators from human erythrocytes, platelets, polymorphonuclear leukocytes, venous endothelial, and smooth muscle cells in cultures was measured in the absence or presence of *C. atrox* venom. Only nucleated cells secreted plasminogen activators; endothelial cells and smooth muscle cells released the activators at concentrations three to four orders of magnitude greater than leukocytes. Both *C. atrox* and *C. adamanteus* venoms showed significant dose-dependent stimulation of plasminogen activator secretion from endothelial cells. The stimulating agent was distinct from metallo and serine proteinases occurring in rattlesnake venoms. The stimulating agent increased intracellular activity of both tissue plasminogen activator and urokinase. However, urokinase was the main activator secreted from endothelial cells into the extracellular pool.

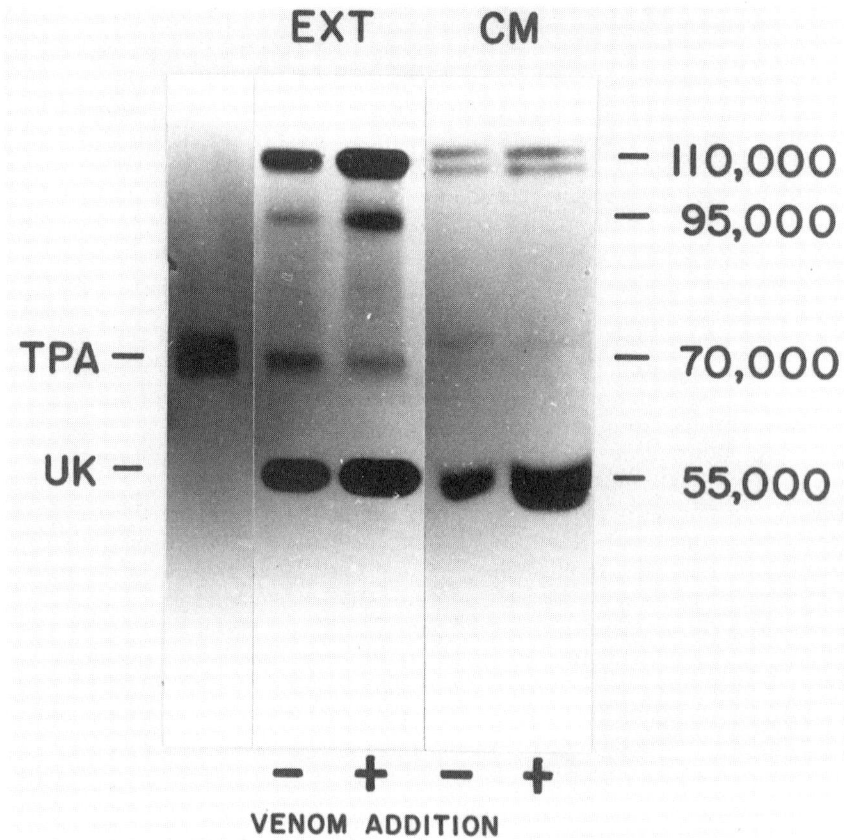

Figure 5 Differentiation between tissue plasminogen activator (TPA) and urokinase (UK). Cell extracts (EXT) and conditioned media (CM) were separated in polyacrylamide gel electrophoresis, overlayed with plasminogen-containing fibrin-agarose and incubated at 37° C for 6 hours. The lysis zones developed in fibrin-agarose overlay were photographed against a dark background. Standard TPA is shown on the left. Apparent molecular weights are shown at the right.

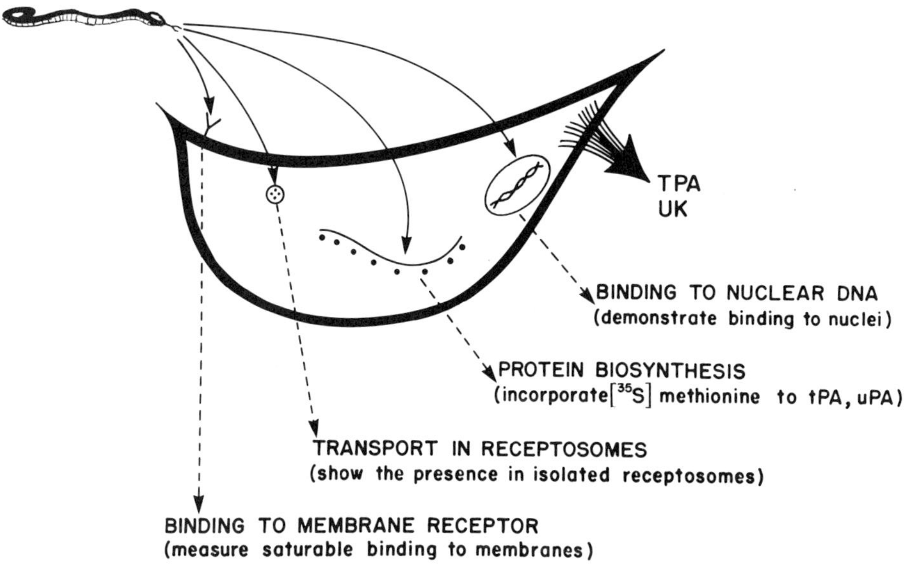

Figure 6 Possible cellular mechanisms for stimulation of plasminogen activator secretion by snake venom agent.

ACKNOWLEDGMENTS

The authors thank Dr. Gwendolyn J. Stewart for providing human venous cells, Dr. Charles Zerweck for help with tissue culture of human smooth muscle cells, and Donald Manuel for the assistance with human endothelial cell cultures.

This work was supported by Grants HL14217 and HL36221 from the National Heart, Lung, and Blood Institute, National Institutes of Health, Bethesda, Maryland.

REFERENCES

1. Blombäck, B., M. Blombäck, and I. M. Nilsson. Thromb. Diath. Haemorrh. 1:76 (1957).
2. McKillop, C. E., C. D. Forbes, and C. R. M. Prentice. Nature 255:638 (1975).
3. Sherman, L. A. and J. Lee. J. Exp. Med. 145:76 (1977).
4. Jilek, F. and H. Hörmann. Hoppe-Seyler's Z. Physiol. Chem. 359:1603 (1978).
5. Gonda, S. R. and J. R. Shainoff. Proc. Natl. Acad. Sci. USA 79:4565 (1982).

6. Reid, H. A. and R. D. G. Theakston. Am. J. Trop. Med. Hyg. 27:1053 (1978).
7. Budzynski, A. Z., B. V. Pandya, R. N. Rubin, B. S. Brizuela, T. Soszka, and G. J. Stewart. Blood 63:1 (1984).
8. Kitchens, C. S. and L. H. Van Mierop. Am. J. Hematol. 14:345 (1983).
9. Loskutoff, D. J. and T. S. Edgington. Proc. Natl. Acad. Sci. USA 74:3903 (1977).
10. Booyse, F. M., J. Scheinbucks, J. Radek, G. Osikowicz, S. Feder, and A. J. Quarfoot. Thromb. Res. 24:495 (1981).
11. Hanahan, D. J. and J. E. Ekholm. Methods Enzymol. 31:169 (1974).
12. Boyüm, A. Scand. J. Clin. Lab. Invest. Suppl. 97:77 (1968).
13. Mustard, J. F., D. W. Perry, N. G. Ardie, and M. A. Packham. Br. J. Haemat. 22:193 (1977).
14. Gimbrone, M. A., E. J. Shefton, Jr., and S. A. Cruise. Tissue Cult. Assoc. Manual 4:813 (1978).
15. Ives, H. E., O. S. Schultz, R. E. Galardy, and J. D. Jamieson. J. Exp. Med. 148:1400 (1978).
16. Moroz, L. and N. J. Gilmore. Blood 46:543 (1975).
17. Laemmli, U. K. Nature 227:680 (1970).
18. Granelli-Piperno, A. and E. Reich. J. Exp. Med. 148:223 (1978).
19. Semar, M., L. Skoza, and A. J. Johnson. J. Clin. Invest. 48:1777 (1969).
20. Erickson, L. A., M. H. Ginsberg, and D. J. Loskutoff. J. Clin. Invest. 74:1465 (1984).
21. Granelli-Piperno, A., J-D. Vassalli, and E. Reich. J. Exp. Med. 146:1693 (1977).
22. Kirschbaum, N. E., A. Z. Budzynski, and G. J. Stewart. Biochemistry 23:3374 (1984).
23. Pandya, B. V. and A. Z. Budzynski. Biochemistry 23:460 (1984).
24. Pandya, B. V., C. S. Cierniewski, and A. Z. Budzynski. J. Biol. Chem. 260:2994 (1985).
25. Markland, F. S. and P. S. Damus. J. Biol. Chem. 246:6460 (1971).

15

A Fibrinolytic Enzyme from the Venom of the Western Diamondback Rattlesnake *(Crotalus atrox)*

Graham S. Bailey
University of Essex, Colchester, England
Anthony T. Tu
Colorado State University, Fort Collins, Colorado
Z. Z. Sapru
University of Essex, Colchester, England

INTRODUCTION

It is well known that the venom of the Western diamondback rattlesnake (*Crotalus atrox*) possesses fibrinolytic activity (1). The potential therapeutic use of such activity for the dissolution of intravascular clots has also been apparent for a considerable period of time (2). More recently, fibrinogenolytic enzymes have been isolated from *Crotalus atrox* venom (3,4), and fibrinolytic activities have also been noted (3,5). This chapter reports on the purification and some properties of one of the fibrinolytic enzymes.

METHODS

Fibrinolytic Assay

The assay was a modification of the procedure of Schumacher and Schill (6). An agar gel plate containing fibrin was produced by addition of thrombin (plasminogen-free, 40 NIH units) to a solution (20 ml) of 0.1% (w/v) fibrinogen (plasminogen-free) and 0.5% (w/v) agar in 0.05 M sodium barbital/HCl buffer, pH 7.8, containing 0.1 M NaCl. A suitable pattern of wells was cut into the gel using a small cork borer. Samples were added (20 µl per well), and each gel was placed in a moist atmosphere at 37°C for 20 hours. Each area of lysis was measured in terms of two perpendicular diameters. For each assay, known amounts of bacterial fibrinolysin were also tested to allow a standard curve to be plotted. The activity of each sample could then be defined in terms of the amount (μg) of bacterial fibrinolysin that produced an area of lysis of the same size under identical conditions.

Fibrinogenolytic Assay

The assay was a modification of the method of Ouyang and Teng (7). Each sample (100 μl) was incubated at 25°C with fibrinogen (100 μl, 2% [w/v]) in 0.05 M Tris-HCl buffer (pH 7.75) for various periods of time. Then 0.05 M imidazole-HCl buffer, pH 7.4, containing 0.09% (w/v) NaCl (30 ml) was added, and the amount of clottable fibrinogen remaining was measured. Thrombin (100 NIH units) was added, and the mixture was kept at room temperature for 30 minutes. The resultant fibrin clot was gathered by centrifugation at 1000 *g* for 20 minutes and was removed with the aid of a small stainless steel loop. After washing with saline solution, the clot was boiled for 30 minutes in 10% (w/v) NaOH (1 ml). Water (5 ml), 20% (w/v) Na_2CO_3 (3 ml) and, Folin-Ciocalteu reagent (1 ml) were added and mixed. The absorbance of the mixture was read at 540 nm after 20 minutes. For each assay, known solutions of fibrinogen (up to 2 mg/ml) were put through the procedure so that a standard curve could be plotted. One unit of activity was defined as the amount (mg) of fibrinogen cleaved per minute.

Arginine-Esterase Assay

A titrimetric assay (8) was employed at 25°C and pH 8.0 using 10 mM α-N-benzoyl-L-arginine ethyl ester as substrate and 10 mM NaOH as titrant. One unit of activity was defined in terms of the amount of enzyme that catalyzed the cleavage of 1 μmol substrate per minute.

Caseinolytic Assay

General proteolytic activity was assessed using casein-yellow (nitrated casein) (9). The sample (100 μl) was incubated for 1 hour at 37°C with 1% (w/v) solution (400 μl) in 0.1 M Tris-HCl buffer (pH 8.0). Twenty percent (w/v) trichloracetic acid (500 μl) was added to stop the reaction, and the mixture was kept at 4°C for 1 hour prior to centrifugation at 10,000 *g* for 20 minutes. After recentrifugation and filtration, 5 M NaOH (100 μl) was added to the resultant solution (500 μl), and the absorbance at 422 nm was measured. One unit of activity was defined as the amount of enzyme that resulted in an increase in absorbance of 0.1 unit/hr.

Analytical Gel Electrophoresis

Samples were analyzed by disc electrophoresis at pH 3.8, pH 7.0, and pH 9.4 (10,11). Electrofocusing was performed using commercially available thin-layer polyacrylamide gels (pH range 3.5 to 9.5), measuring the pH gradient with the aid of internal marker proteins. SDS gel electrophoresis was carried out using a modification of the procedure of Lambin (12), with a 1-mm thick gradient slab gel (from 10 to 18% acrylamide) and the buffer system of Laemmli (13). Samples were prepared for electrophoresis by incubation at 37°C for 2 hours in 0.5 M Tris-HCl buffer (pH 6.2) containing 5% (w/v) SDS and 8 M urea. For reducing conditions, 4% (w/v) β-mercaptoethanol was included.

Concentration of Samples

Concentration and equilibration of samples was achieved by ultrafiltration using appropriate membranes. Protein concentrations were estimated using the method of Warburg and Christian (14).

RESULTS

Gel Filtration of *Crotalus atrox* Venom

The crude venom was fractionated by gel filtration in five batches. Identical conditions were maintained for each run. Crude venom (600 mg) was dissolved in 0.1 M ammonium acetate buffer, pH 4.7 (8 ml). After centrifugation at 10,000 *g* for 20 minutes, the clear supernatant was applied to a column (2.5 × 83 cm) of Sephadex G-100 resin. Components were eluted by the acetate buffer at a flow rate of 16 ml/hr, and fractions of 8 ml were collected. A typical elution profile is recorded in Figure 1. The enzymatic activities of the crude venom and the resultant fractions are recorded in Table 1. None of the enzymatic activities tested for could be detected in fraction 3. In fact, none of the samples showed any fibrinogen clotting activity. In total, 3810 mg of protein in the crude venom had been applied to gel filtration, resulting in 1350 mg of protein in fraction 1, 1320 mg of protein in fraction 2, and 570 mg of protein in fraction 3, representing an 85% yield of protein. Standardization of the column with proteins of known molecular weight resulted in an estimated M_r of 100,000 and 30,000 for fractions 1 and 2, respectively. Because fraction 1 contained the major amount of fibrinolytic activity, it was subjected to further purification.

Ion Exchange Chromatography of Fraction 1

Fraction 1 was thoroughly dialyzed against 5 mM Tris-HCl buffer (pH 7.75) and was processed by anion exchange chromatography in eight identical batches. Fraction 1 (123.5 mg protein in 3 ml) was applied to a column (2.5 × 10 cm) of Whatman DE-52 resin that had been equili-

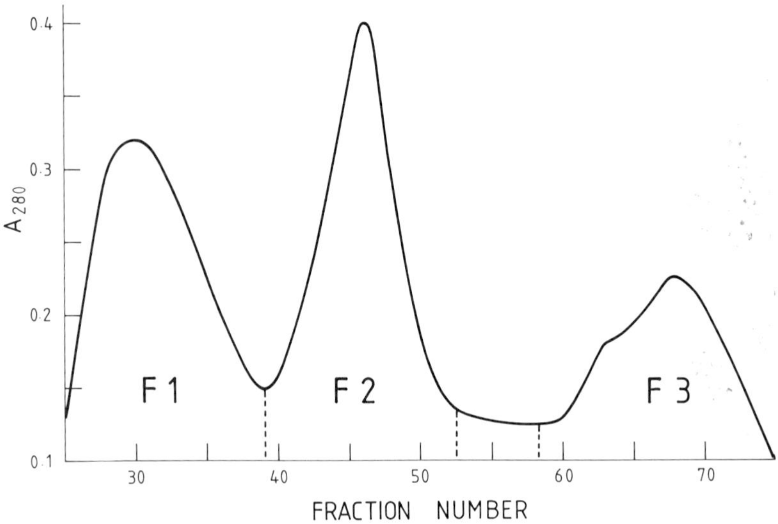

Figure 1 Gel filtration of *Crotalus atrox* venom on Sephadex G-100 resin.

brated with the same Tris-HCl buffer. The column was washed with the buffer (100 ml) at a flow rate of 18 ml/hr for elution of unadsorbed substances. The adsorbed components were collected in 3-ml fractions on elution by application of a linear salt gradient (0.15 to 0.75 M NaCl in Tris-HCl buffer, total volume 300 ml). A representative gradient elution profile is shown in Figure 2. Five postgradient fractions were pooled and concentrated. The amounts of the fractions and enzymatic activities are recorded in Table 2. The total yield of protein from the anion exchange chromatography was 73.5%. The two fractions containing fibrinolytic activity, fraction 1A and fraction 1B, were combined and were dialyzed against 5 mM Tris-HCl buffer (pH 8.75). Considerable precipitation occurred, and a clear supernatant was obtained by centrifugation. The precipitate showed no fibrinolytic activity. The supernatant contained 55.7 mg of protein and fibrinolytic activity of 80 units/mg protein compared with the combined sample of fraction 1A together with fraction 1B prior to dialysis of 170 mg of protein and fibrinolytic activity of 200 units/mg protein. Thus, there had been a considerable loss of activity on dialysis. Nevertheless, the supernatant was subject to anion exchange chromatography at pH 8.75.

Ion Exchange Chromatography at pH 8.75

The supernatant was applied to a column (2.5 × 10 cm) of equilibrated Whatman DE-52 resin. The column was washed with 5 mM Tris-HCl

Table 1 Enzymatic Activities of Crude Venom and Gel Filtration Fractions

Sample	Arginine esterase activity (units/mg)	Fibrinogenolytic activity (units/mg)	Fibrinolytic activity (units/mg)	Caseinolytic activity (units/mg)
Venom	3.0	0.55	40	8.5
Fraction 1	ND	1.50	150	14.2
Fraction 2	8.3	0.63	80	12.5

ND = not detectable.

buffer (pH 8.75, 100 ml) at a flow rate of 18 ml/hr and then with a salt gradient (0.15 to 0.75 M NaCl in the Tris buffer, total volume 300 ml), collecting 3-ml fractions. The postgradient elution profile is shown in Figure 3. Seven fractions were pooled and analyzed. The results are recorded in Table 3. The yield of protein from the column was 67.5%.

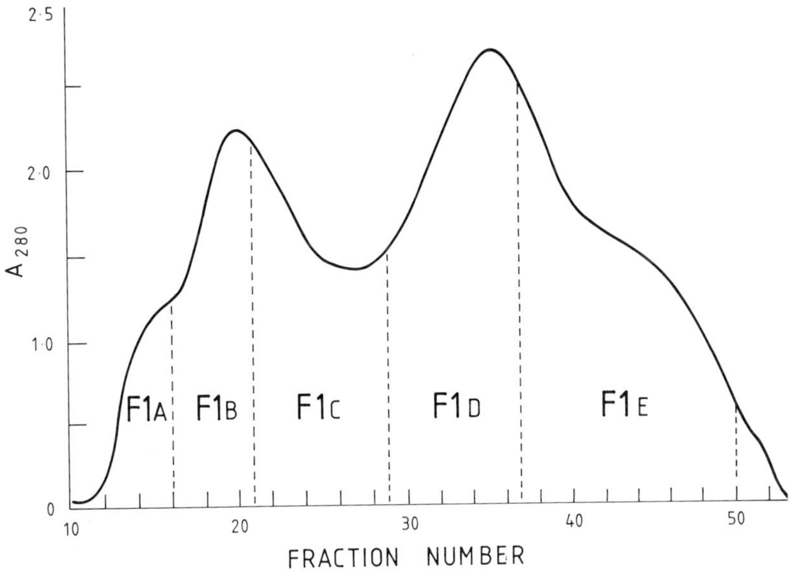

Figure 2 Postgradient elution profile of anion exchange chromatography of fraction 1 on Whatman DE-52 resin at pH 7.75.

Table 2 Enzymatic Activities of Fraction from Ion Exchange Chromatography of Fraction 1

Fraction	Protein (mg)	Fibrinogenolytic activity (units/mg)	Fibrinolytic activity (units/mg)	Caseinolytic activity (units/mg)
1A	51.1	ND	280	10.2
1B	119.7	7.5	150	12.0
1C	65.1	9.2	ND	4.5
1D	274.3	ND	ND	20.3
1E	216.4	ND	ND	50.1

ND = not detectable.

Disc Electrophoresis of Fibrinolytic Fraction

Fraction 1 from the anion exchange chromatography at pH 8.75 was analyzed by disc electrophoresis on both 5% and 10% gels at pH 9.4, using loadings of 25, 50, and 100 μg of protein. After destaining, it was noted that the sample had not moved into the gels at pH 9.4. Similar results were found on electrophoresis at pH 7.0. However, after electrophoresis on 5% gels at pH 3.8, one major and four minor bands were seen.

Analysis of the Fibrinolytic Fraction After Storage at -20°C for 6 Months

After storage of the fibrinolytic fraction at -20°C for 6 months, a precipitate had formed and was removed by centrifugation. The resultant supernatant, containing 2 mg of protein, was found to lack caseinolytic activity and possessed fibrinolytic activity of 500 units/mg protein. Disc electrophoresis on 5% gels at pH 3.8 revealed only one band. Gel isoelectric focusing of the sample revealed three very closely spaced bands near the cathodic strip, indicating isoelectric points of about pH 9.0. The SDS gel electrophoresis of the fibrinolytic sample incubated in the absence of reductant showed a minor component of an estimated M_r of 90,000 and a major component of M_r 93,000. On reduction of the final sample with mercaptoethanol, there was an increase in the mobility of the major component, corresponding to an estimated M_r of 48,000. Preincubation of the fraction for 20 minutes at 20°C with 100 mM EDTA resulted in no loss of fibrinolytic activity.

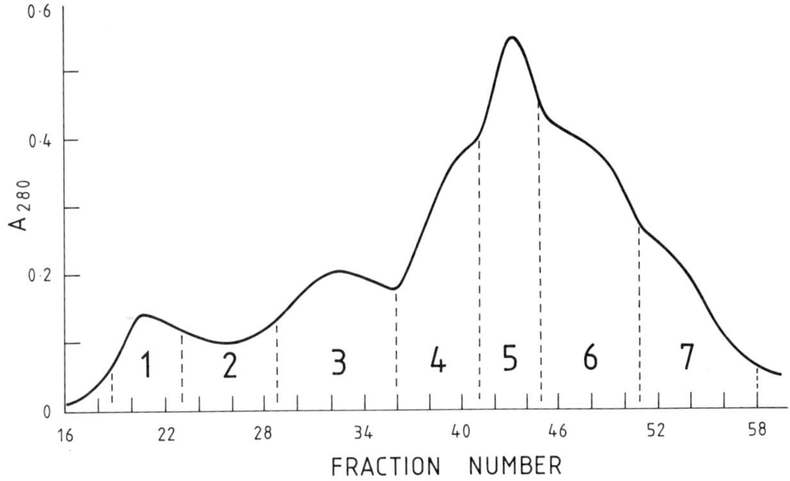

Figure 3 Postgradient elution profile of combined fractions 1A and 1B on Whatman DE-52 resin at pH 8.75.

Table 3 Enzymatic Activities of Fractions from Ion Exchange Chromatography at pH 8.75

Fraction	Protein (mg)	Fibrinogenolytic activity (units/mg)	Fibrinolytic activity (units/mg)	Caseinolytic activity (units/mg)
1	2.9	ND	666	7.4
2	3.0	ND	50	28.9
3	5.0	ND	30	23.4
4	5.7	ND	ND	20.5
5	7.0	ND	ND	22.2
6	8.2	7.5	ND	20.8
7	5.8	10.5	ND	10.2

ND = not detectable.

DISCUSSION

A fibrinolytic enzyme has been highly purified from the venom of *Crotalus atrox*. The method of purification was clearly not ideal. In particular, the change to pH 8.75 resulted in an 80% loss of fibrinolytic activity. Thus, there is a great need to develop a better system of isolation. The fibrinolytic fraction obtained through anion exchange chromatography at pH 8.75 was highly active towards the fibrin clot, but no activity could be detected in the fibrinogenolytic assay. This was a surprising result because fibrinogen and fibrin are very similar molecules (15). The fraction possessed a small amount of general proteolytic activity, as indicated by the activity toward casein, but that was lost on storage at -20°C with a relatively small decrease in fibrinolytic activity. However, that final fraction was not entirely pure as judged by the results of SDS gel electrophoresis. It may well be that one of the fibrinolytic enzymes of *Crotalus atrox* venom consists of three isoenzymes of isoelectric points around pH 9.0 and of an M_r of 93,000, each being composed of two disulfide-linked polypeptide chains of an M_r of 48,000. The confirmation or otherwise of those tentative conclusions must await production of the enzyme by a better purification process. It is interesting to note that other workers have found problems in obtaining reasonable amounts of a purified fibrinolytic enzyme from the venom of *Crotalus atrox* (16).

REFERENCES

1. Eagle, H. J. Exp. Med. 65:613 (1937).
2. Didisheim, P. and J. H. Lewis. Proc. Soc. Exp. Med. Biol. 93: 10 (1956).
3. Sapru, Z. Z., A. T. Tu, and G. S. Bailey. Biochim. Biophys. Acta 747:225 (1983).
4. Pandya, B. V. and A. Z. Budzynski. Biochemistry 23:460 (1984).
5. Bajwa, S. S., F. S. Markland, and F. E. Russell. Toxicon 18: 285 (1980).
6. Schumacher, G. F. B. and W. B. Schill. Anal. Biochem. 48:9 (1972).
7. Ouyang, C. and C. M. Teng. Biochim. Biophys. Acta 420:298 (1976).
8. Proud, D., G. S. Bailey, K. Nustad, and K. M. Gautvik. Biochemical J. 167:835 (1977).
9. Nustad, K., K. M. Gautvik, and J. V. Pierce. In *Chemistry and Biology of the Kallikrein-Kinin System in Health and Disease*, J. J. Pisano and K. F. Austin (Eds.), D.H.E.W. Publication No. 76-791, Washington, D.C., pp. 77-92 (1976).
10. Davis, B. J. Ann. NY Acad. Sci. 121:404 (1964).

11. Williams, D. E. and R. A. Reisfeld. Ann. NY Acad. Sci. 121: 373 (1964).
12. Lambin, P. Anal. Biochem. 85:114 (1978).
13. Laemmli, U. K. Nature 227:680 (1979).
14. Warburg, O. and W. Christian. Biochemistry Z. 310:384 (1942).
15. Marder, V. J., C. W. Francis, and R. F. Doolittle. In *Hemostasis and Thrombosis*, R. W. Colman, H. Hirsh, V. J. Marder, and E. W. Salzman (Eds.), J. B. Lippincott Company, Philadelphia, pp. 145-163 (1982).
16. Bajwa, S. S., G. B. Patkos, and F. S. Markland. Proc. West. Pharmacol. Soc. 24:165 (1981).

16

The Action of Proteinase F1 from *Naja nigricollis* Venom on the Aα-Chain of Human Fibrinogen

Herbert J. Evans
Medical College of Virginia, Virginia Commonwealth University, Richmond, Virginia
Alan J. Barrett
Strangeways Research Laboratory, Cambridge, England

INTRODUCTION

Although it has been recognized for many years that cobra venoms act as anticoagulants, only recently have the reasons for this been clarified. The major anticoagulant effect of the venom of the spitting cobra, *Naja nigricollis*, is due to a basic phospholipase A_2 (Evans et al., 1980), which apparently cleaves or binds the phospholipids needed for the formation of activation complexes in coagulation. The basic phospholipases A_2 from *N. nigricollis* and *N. mossambica mossambica* are much more potent anticoagulants than are the neutral phospholipases from the same venoms or from other sources (Prigent-Dachary et al., 1980).

In addition to the major effect of the basic phospholipase, *N. nigricollis* venom exhibits a time-dependent anticoagulant effect in vitro due to selective proteolytic cleavage of the Aα-chain of fibrinogen. This cleavage, which does not affect the Bβ- and γ-chains of fibrinogen, occurs optimally at neutral to slightly alkaline pH. The proteolytic activity is inhibited by EDTA, and can then be reactivated by zinc ions (Evans, 1981).

N. nigricollis venom also acts on the α-chains of highly cross-linked fibrin (Evans, 1981). The venom does not convert plasminogen to plasmin, which might cause the degradation of fibrinogen and fibrin indirectly.

The major fibrinogenolytic enzyme of *N. nigricollis* venom has recently been purified to homogeneity and named "proteinase F1" (Evans, 1984). The activity of the very selective proteinase was monitored during purification by means of a new assay (Evans and Guthrie, 1984) that takes advantage of the ability of essentially all proteinases to saturate α_2-macroglobulin (Barrett and Starkey, 1973; Starkey and Barrett, 1977). Proteinase F1 has an apparent M_r of 58,000 in SDS polyacrylamide gel electrophoresis. The enzyme is a Zn^{2+}-requiring metalloproteinase. The purified enzyme has been shown to degrade fibronectin, and to cleave the bait region of α_2-macroglobulin, in addition to the Aα-chain of fibrinogen (Evans, 1984). The present study describes the specificity for protein substrates in more detail, and indicates the region of the Aα-chain of fibrinogen cleaved by proteinase F1.

MATERIALS AND METHODS

Venom of *N. nigricollis* (West African) was obtained from the Miami Serpentarium. Proteinase F1 was purified from this venom according to a previously described procedure (Evans, 1984). All other venoms were obtained from Sigma Chemical Company. Human fibrinogen was prepared from outdated plasma obtained from the blood bank of the Medical College of Virginia. The plasma was first passed through a lysine-affinity column to remove plasminogen (Deutsch and Mertz, 1970). Fibrinogen was prepared from this plasminogen-free plasma by precipitation with glycine (Kazal et al., 1963). Electrophoresis reagents were obtained from Fisons Chemicals. All other reagents and proteins were obtained from Sigma Chemical Company.

SDS polyacrylamide gel electrophoresis was with the Ammediol system of Bury (1981) with a 10% polyacrylamide separating gel and a 3.2% stacking gel. Reduced samples were prepared by adding 2-mercaptoethanol to the usual sample buffer to give a final concentration of 1%. Samples were run for 1.5 hours at a constant current of 40 mA/slab.

Protein concentrations were estimated by the Bradford (1976) procedure with concentrated reagent from Bio-Rad and bovine gamma globulin as standard.

RESULTS AND DISCUSSION

Selectivity of Proteinase F1 for Fibrinogen

The effect of proteinase F1 purified from *N. nigricollis* venom on human fibrinogen is seen in Figure 1, which also shows the absence of any visible effects on bovine casein, hemoglobin, or serum albumin. In addition, proteinase F1 showed no significant collagenolytic activity at pH 8.0 when incubated with ^{14}C-labeled rat skin type I collagen that had been thermally reconstituted and was trypsin-resistant (Heath and Evans, unpublished results). Thus, proteinase F1 seems to have a narrow specificity for protein substrates.

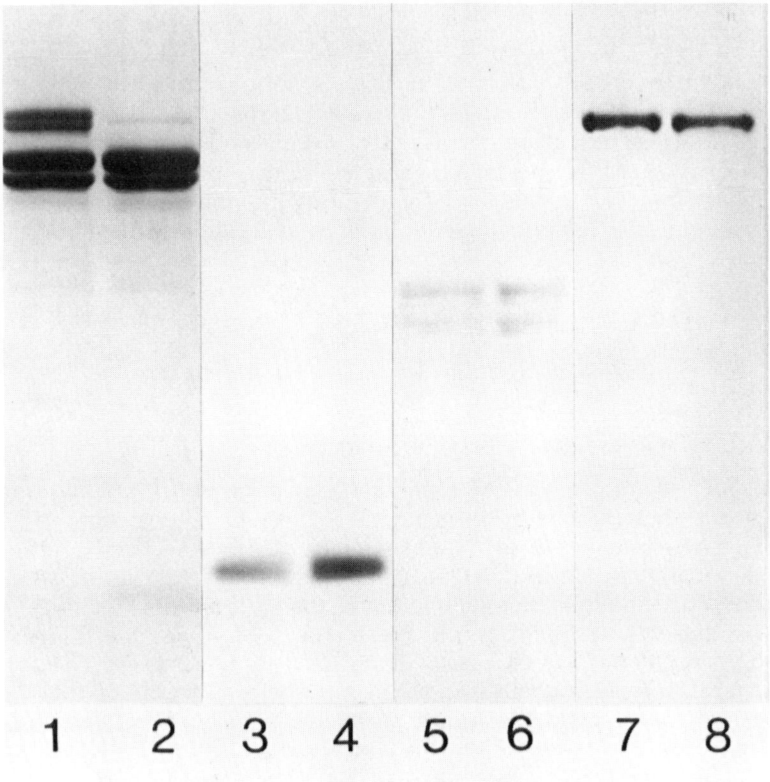

Figure 1 Activity of proteinase F1 on fibrinogen and lack of effect on hemoglobin, casein, and bovine serum albumin. Incubations were for 2 hours at 37°C, and mixtures contained 20 μg of the protein substrate with or without 0.35 μg of proteinase F1, in 0.05 M Tris-HCl (pH 8.0) and 2 mM $ZnCl_2$, with a final volume of 50 μl. Reduced samples containing about 8 μg of protein substrate were applied to the gel. Lanes 1 and 2 = fibrinogen alone and with added proteinase F1; lanes 3 and 4 = hemoglobin alone and with added proteinase F1; lanes 5 and 6 = casein alone and with added proteinase F1; lanes 7 and 8 = bovine serum albumin alone and with added proteinase F1.

The known specificity of proteinase F1 includes only certain proteins found in blood plasma (i.e., fibrinogen, fibronectin, and α_2-macroglobin). A recent report indicates that several Elapid venoms, including that of *N. nigricollis*, are able to inactivate and cleave α_1-antichymotrypsin (Kress and Hufnagel, 1984). Because this venom contains so few proteinases, proteinase F1 seems a likely candidate for this activity as well.

Distribution of the Fibrinogenolytic Activity Among Cobra Species

Elapid venoms have been shown in several previous studies to have little proteolytic activity (Murata et al., 1963; Mebs, 1968; Henriques and Evseeva, 1969; Robertson et al., 1969; Oshima et al., 1969; Kocholaty et al., 1971). We tested the venoms of cobra species from diverse geographical regions, and that of a single non-cobra Elapid, the Australian taipan (*Oxyuranus scutellatus scutellatus*), for the ability to degrade the Aα-chain of fibrinogen. Activity was found to varying degrees in all the African and Asian cobra species tested, but not in the Australian taipan venom (Fig. 2). This suggests that the fibrinogenolytic activity may be limited, within the Elapid family, to the cobras, and it is clearly desirable that all the genera be screened.

Site of Action of Proteinase F1 on Fibrinogen

An aspect of the fibrinogenolytic specificity of proteinase F1 apparent in Figure 2 is that the susceptibility of the Aα-chain to degradation depends on how complete it is. Heterogeneity of the Aα-chain is a common feature of human fibrinogen (Mosesson et al., 1972), and we have observed two bands corresponding to the Aα-chain in SDS polyacrylamide gels of our fibrinogen preparations. Of these, the higher M_r chain predominates, but we have found this to be more rapidly degraded than the smaller chain. In several instances, traces of the smaller chain remain after the larger chain has completely disappeared. This is especially apparent in lanes 3 and 6 of Figure 2, where it is seen that the venoms of *N. nivea* and *N. naja kaouthia* cleaved all of the intact Aα-chain, but left the shorter chain at least partly uncleaved after 2 hours of treatment. This suggests that the enzyme may need the additional peptide of about 2000 M_r at the C-terminus for correct alignment of the active site with the scissile peptide bond, and that the two Aα-chains have slightly different conformations near the cleavage site.

When human fibrinogen is treated with proteinase F1, very few cleavage peptides are generated. The intact Aα-chain has an M_r of 66,125 (Doolittle et al., 1979). The progress of digestion of fibrinogen is shown in Figure 3 as SDS electrophoresis patterns of the reduced products of treatment. It is apparent that as the Aα-chain starts to disappear, the only obvious product is a small peptide with an M_r of about 10,000. The remainder of the Aα-chain, however, appears as a 56,000 M_r peptide which is so nearly the same size as the Bβ-chain (M_r = 57,000) that it appears only as a thickening of the Bβ-chain band after 2 hours of digestion. After a prolonged (20 hours) treatment (Fig. 4), this large (56,000 M_r) peptide is further cleaved to give another with an M_r of about 34,000, and at least one other small

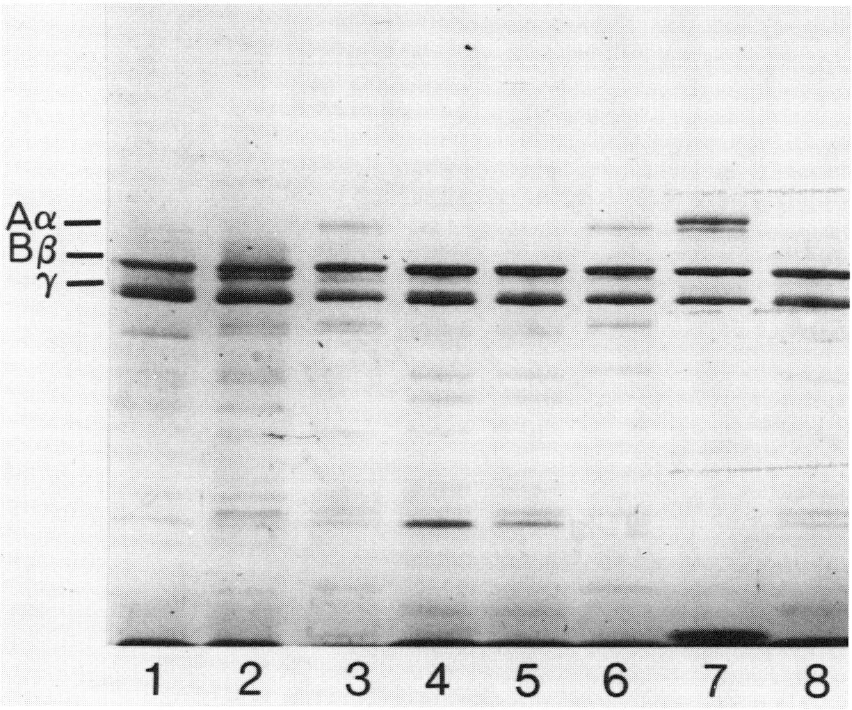

Figure 2 Screening of several cobra venoms and Australian taipan venom for fibrinogenolytic activity. Incubations were for 2 hours at 37° C, and mixtures contained 20 μg of fibrinogen and 10 μg of crude venom in 0.05 M Tris-HCl (pH 8.0) and 2 mM $ZnCl_2$, with a final volume of 100 μl. Reduced samples containing about 10 μg of fibrinogen were applied to the gel. Venoms used were lane 1 = *N. naja atra*; lane 2 = *N. nigricollis* (W. African); lane 3 = *N. nivea*; lane 4 = *N. nigricollis crawshawii*; lane 5 = *N. nigricollis nigricollis*; lane 6 = *N. naja kaouthia*; lane 7 = *O. scutellatus scutellatus*; lane 8 = *N. nigricollis pallida*. Bands appearing below fibrinogen are not necessarily degradation peptides, but are mostly proteins contributed by the crude venom itself (as shown in controls run separately).

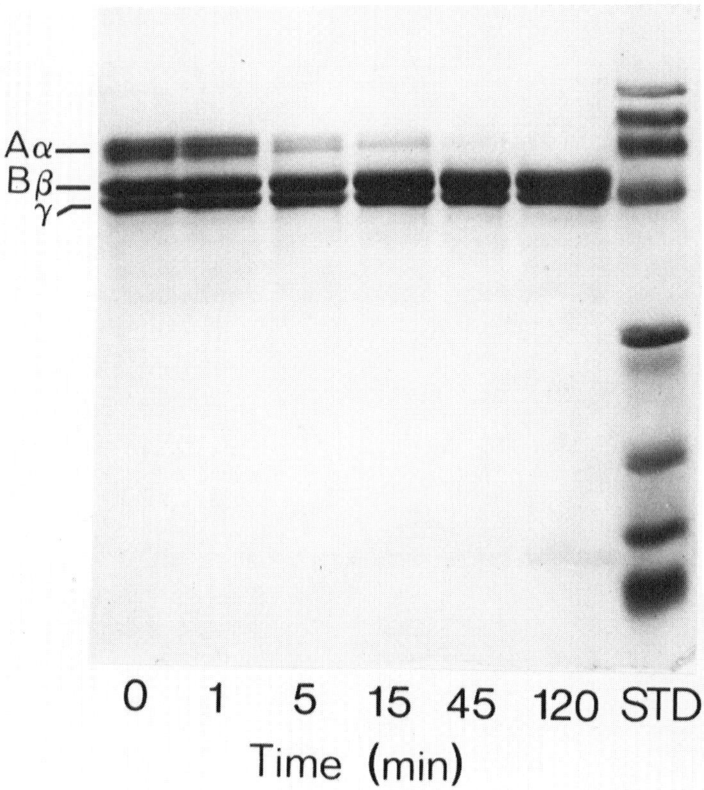

Figure 3 Degradation of fibrinogen by treatment with proteinase F1. The mixtures incubated for up to 2 hours at 37°C contained 320 μg of fibrinogen and 2.8 μg of proteinase F1 in 0.05 M Tris-HCl (pH 8.0) and 2 mM $ZnCl_2$, with a final volume of 400 μl. At the indicated times, 50 μl aliquots were transferred to 50 μl of reducing SDS sample buffer. Reduced samples containing about 16 μg of fibrinogen were applied to the gel. The M_r standard contained about 3 μg each of phosphorylase a (M_r = 97,400), transferrin (M_r = 78,000), bovine serum albumin (M_r = 68,000), IgG heavy and light chains (M_r = 50,000 and 25,000, respectively), carbonic anhydrase (M_r = 29,000), soybean trypsin inhibitor (M_r = 21,500), cytochrome c (M_r = 12,800), and aprotinin (M_r = 6,500).

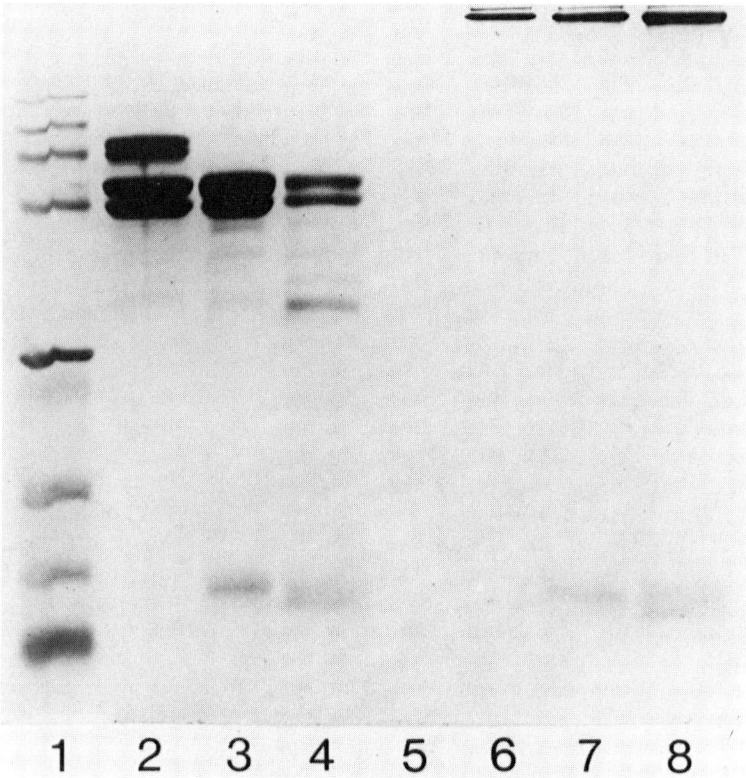

Figure 4 Comparison of reduced and unreduced products of fibrinogen treated with proteinase F1. Incubations set up at 37°C contained 40 μg of fibrinogen and 0.35 μg of proteinase F1, in 0.05 M Tris-HCl (pH 8.0) and 2 mM $ZnCl_2$, with a final volume of 50 μl. At times 0, 10, and 20 hours, 50 μl of reducing (lanes 2 to 4) or nonreducing (lanes 6 to 8) sample buffer was added. Samples containing about 16 μg fibrinogen were loaded on the gel. The M_r standards (lane 1) were as listed in Figure 3.

peptide. The Bβ- and γ-chains are apparently untouched by proteinase F1.

Fibrinogen that has been treated with the proteinase is still clottable by thrombin, although the clotting time after 2 hours of treatment is about 3.5 times longer than normal (Evans, 1981). This suggests that the first peptide cleaved from the fibrinogen molecule by proteinase F1 is from the C-terminal portion of the Aα-chain, since thrombin is still able to act on the N-terminus to initiate clot formation. The Aα-chain consists of 610 amino acid residues (Doolittle et al., 1979), so the cleavage to liberate the C-terminal 10,000 M_r peptide must occur near residue 520.

Further confirmation that the initial peptide is cleaved from the C-terminus of the Aα-chain came from our comparison of the unreduced and the reduced products of proteinase F1 treatment of fibrinogen (Fig. 4). Cysteines at residues 28, 36, 45, and 49 of the Aα-chain form disulfide bonds to link the N-terminus with other sections of the fibrinogen molecule, whereas no cysteine residues are found in the C-terminal 130 residues. This means that any C-terminal peptide cleaved from the molecule should also be separated from the rest of the molecule when run without reduction in SDS electrophoresis, whereas peptides from the N-terminus would remain attached. Figure 4 shows that the peptide generated after 2 hours of treatment with F1 is not attached by disulfide bonds to the rest of the fibrinogen molecule (lanes 3 and 7) and, therefore, comes from the C-terminus.

The other large peptide that is generated after prolonged treatment of fibrinogen with proteinase F1 (Fig. 4, lane 4) has an M_r of about 34,000, and obviously derives from the 56,000 M_r fragment generated by the first cleavage of the Aα-chain, since only a trace of the band is seen after 2 hours of treatment (lane 3). The remaining 22,000 M_r portion of the 56,000 M_r fragment cannot be completely accounted for, but one or more additional bands with an M_r of about 10,000 (and some intermediate sized fragments) were generated (lane 4), suggesting that the 56,000 M_r fragment was cleaved in more than one place during prolonged exposure to proteinase F1. Neither the 34,000 M_r peptide nor any of the intermediate-sized peptides appeared on the SDS gel when the fibrinogen was unreduced (lane 8), indicating that all of the larger fragments derived from the 56,000 M_r initial fragment remained linked to the rest of the molecule by disulfide bridges. It can also be seen in Figure 4 (lanes 2, 3, and 4) that the intensity of the band containing the Bβ-chain is heavier than that of the γ-chain band after 2 hours of treatment with proteinase F1, but returns to the same intensity as the γ-chain after 20 hours of treatment. This is consistent with the appearance of the 56,000 M_r peptide at 2 hours and its disappearance with prolonged incubation.

The results of these experiments with timed digestion of fibrinogen by proteinase F1 purified from *N. nigricollis* venom support the idea that the enzyme specificity is confined to relatively few peptide bonds. The first cleavage of the intact Aα-chain occurs around residue 520 to give 56,000 M_r and 10,000 M_r peptides. The C-terminal 10,000 M_r peptide is not held to the rest of the molecule by disulfide bonds, and thus is revealed as an independent band on nonreduced SDS polyacrylamide gels. Some further limited degradation of the 56,000 M_r peptide occurs after the initial cleavage, but the sequence of generation of these later fragments is not clear. The end result of prolonged treatment of fibrinogen with proteinase F1 appears to be the cleavage of at least two 8000 to 10,000 M_r fragments from the C-terminal end of the Aα-chain, since these fragments do not migrate with the remainder of the molecule on nonreduced SDS polyacrylamide gels.

Solutions of fibrinogen that have been treated with proteinase F1 remain clottable by thrombin, but the clotting time increases with time of treatment. Thus, the fibrinogenolytic proteinases contribute to the anticoagulant effect of the venom, but it is not established whether the effect on clotting is due to lowered clottability of fibrinogen with partially degraded Aα-chain, or to an inhibitory effect of the new peptides generated. We have obtained some preliminary evidence that at least one of the small peptides, isolated by gel filtration of a 20-hour incubation mixture of proteinase F1 with fibrinogen, has an anticoagulant effect in tests of prothrombin time (Evans, unpublished results).

ACKNOWLEDGMENT

This work was supported by a grant from the Virginia Affiliate of the American Heart Association, Inc.

REFERENCES

Barrett, A. J. and P. M. Starkey. Interaction of alpha$_2$-macroglobulin with proteinases: characteristics and specificity of the reaction and a hypothesis concerning its molecular mechanism. Biochem. J. 133:709-724 (1973).

Bradford, M. M. Rapid and sensitive method for quantitation of microgram quantities of protein utilizing principle of protein-dye binding. Anal. Biochem. 72:248-254 (1976).

Bury, A. F. Analysis of protein and peptide mixtures: evaluation of three sodium dodecyl sulfate-polyacrylamide gel electrophoresis buffer systems. J. Chromatogr. 213:491-500 (1981).

Deutsch, D. G. and E. T. Mertz. Plasminogen: purification from human plasma by affinity chromatography. Science 170:1095-1096 (1970).

Doolittle, R. F., K. W. K. Watt, B. A. Cottrell, D. D. Strong, and M. Riley. The amino acid sequence of the α-chain of human fibrinogen. Nature 280:464-468 (1979).

Evans, H. J. Cleavage of the Aα-chain of fibrinogen and the α-polymer of fibrin by the venom of spitting cobra (*Naja nigricollis*). Biochim. Biophys. Acta 660:219-226 (1981).

Evans, H. J. Purification and properties of a fibrinogenase from the venom of *Naja nigricollis*. Biochim. Biophys. Acta 802:49-54 (1984).

Evans, H. J., R. C. Franson, G. D. Qureshi, and W. F. Moo-Penn. Isolation of anticoagulant proteins from cobra venom (*Naja nigri-*

collis): Identity with phospholipases A_2. J. Biol. Chem. 255:3793-3797 (1980).

Evans, H. J. and V. H. Guthrie. Proteolytic activities of cobra venoms based on inactivation of α_2-macroglobulin. Biochim. Biophys. Acta 784:97-101 (1984).

Henriques, O. B. and L. Evseeva. Proteolytic esterase and kinin-releasing activities of some Soviet snake venoms. Toxicon 6:205-209 (1969).

Kazal, L. A., S. Amsel, O. P. Miller, and L. M. Tocantins. Preparation and some properties of fibrinogen precipitated from human plasma by glycine. Proc. Soc. Exp. Biol. Med. 113:989-994 (1963).

Kocholaty, W. F., E. B. Ledford, J. G. Daly, and T. A. Billings. Toxicity and some enzymatic properties and activities in venoms of Crotalidae, Elapidae and Viperidae. Toxicon 9:131-138 (1971).

Kress, L. F. and M. E. Hufnagel. Enzymatic inactivation of human alpha_1-antichymotrypsin by metalloproteinases in snake venoms of the family Elapidae. Comp. Biochem. Physiol. 77B:431-436 (1984).

Mebs, D. Comparative enzyme investigations on snake venoms with special reference to their casein-splitting proteases. Hoppe-Seyler's Z. Physiol. Chem. 349:1115-1125 (1968).

Mosesson, M. W., J. S. Finlayson, R. A. Umfleet, and D. Galanakis. Human fibrinogen heterogeneities. I. Structural and related studies of plasma fibrinogens which are high solubility catabolic intermediates. J. Biol. Chem. 247:5210-5219 (1972).

Murata, Y., M. Satake, and T. Suzuki. Studies on snake venom: XII. Distribution of proteinase activities among Japanese and Formosan snake venoms. J. Biochem. (Tokyo) 53:431-437 (1963).

Oshima, G., T. Sato-Ohmori, and T. Suzuki. Proteinase, arginine ester hydrolase and a kinin-releasing enzyme in snake venoms. Toxicon 7:229-233 (1969).

Prigent-Dachary, J., M. C. Boffa, M. R. Boisseau, and J. Dufourcq. Snake venom phospholipases A_2: a fluorescence study of their binding to phospholipid vesicles correlation with their anticoagulant activities. J. Biol. Chem. 255:7734-7739 (1980).

Robertson, S. S. D., K. Steyn, and G. R. Delpierre. Studies on African snake venoms: III. Caseinase activity of some African Elapidae venoms. Toxicon 6:243-245 (1969).

Starkey, P. and A. J. Barrett. α_2-Macroglobulin, a physiological regulator of proteinase activity. In *Proteinases in Mammalian Cells and Tissues*, Barrett, A. J. (Ed.), Amsterdam, North-Holland Publ. Co., pp. 663-696 (1977).

17

Fibrinolytic and Procoagulant Agents from a Saturnidae Moth Caterpillar

Carmen Luisa Arocha-Piñango
Instituto Venezolano de Investigaciones Cientificas, Caracas, Venezuela
Norma Blumenfeld de Bosch
Banco Municipal de Sangre del Distrito Federal, Caracas, Venezuela
Amara L. Nouel
Hospital Ruiz y Paez, Universidade de Oriente, Ciudad Bolivar, Venezuela
A. Torres
Banco Municipal de Sangre del Distrito Federal, Caracas, Venezuela
J. Perales, Miguel E. Alonso,* S. de Rodriguez, Zoila A. Carvajal, A. Ojeda, M. L. Tasayco, and W. Chitty
Instituto Venezolano de Investigaciones Cientificas, Caracas, Venezuela

It is well known that contact with many species of caterpillar, especially those that are spiny or hairy (for example, Saturnidae), may produce allergic reactions along with urticaria, edema, or fever in humans. The agent or molecule responsible has been identified in many cases. In Venezuela, we have described a bleeding syndrome that appears after contact with the larval form of a moth of the Saturnidae family, namely, *Lonomia achelous* (1-7). In this study, we present the clinical cases and the biochemical tests carried out to determine the causal agent of this syndrome.

MATERIALS AND METHODS

Patients

Fourteen adults, aged between 25 and 60 years (3 females and 11 males), and four children, aged between 7 and 12 years, with a bleeding condition that appeared after contact with caterpillars, were the subject of the present study.

Coagulation Studies

On admission and over the following days of treatment, routine coagu-

**Present affiliation:* Universidad de Los Andes, Merida, Venezuela lation tests including prothrombin time (PT), partial thromboplastin time (PTT), and thrombin time (ThT) were performed using conventional methods, and plasma lysis time was measured as described by Markwardt (8). Euglobulin lysis time (ELT) was measured by the clot lysis method (9) and on fibrin plates (10).

Fibrinogen/fibrin degradation products (FDPs) were quantified by the Latex method using the Thrombo-Wellcotest kit (Wellcome Diagnostics, England) (11). The antigenic activities of plasminogen, α2-antiplasmin, and factor XIII and its subunits were determined by rocket electrophoresis (12) using specific antisera obtained from Diagnostica Stago (France) and Behringwerke AG (Marburg, West Germany). For α2-macroglobulin and antithrombin III, M-Partigen plates (Behringwerke AG) were used. Antithrombin III was also determined by two-dimensional gel electrophoresis (13) using antithrombin III antiserum (Behringwerke AG).

The biological activity of the antithrombin III was measured by the coagulation method using the Ortho Diagnostic System Inc. kit, as well as using chromogenic substrates (14). Plasminogen and α2-antiplasmin were determined using chromogenic substrates, according to the methods described by the manufacturer (Kabi Diagnostica, Stockholm, Sweden). For one patient, thrombin generation (15) was determined over 7 consecutive days. Immunodiffusion (ID) (16) and immunoelectrophoresis (IEF) (17) were performed on the patients' plasma and sera using total antifibrinogen antiserum prepared in our laboratory and specific anti-D and anti-E antisera (homemade and from Diagnostica Stago, France). Reference fragments were also obtained from Diagnostica Stago.

Etiologic Agent

Specimens of the caterpillar with which each patient had come into contact were collected for identification and for the biochemical studies.

Biochemical Studies

Hemolymph was collected by aspiration with a needle and syringe from the body of the caterpillars. A hair extract was prepared according to the method of De Jongh, Kawamoto, and their co-workers (18,19). The hemolymph was purified by molecular exclusion chromatography on Sephadex G-75 (Pharmacia Fine Chemicals, Sweden) using 0.05 M imidazole/0.15 M NaCl buffer, pH 7.4, as eluent, and then by ion exchange on DE-23-cellulose (Whatman Biochemical, Maidstone, England), as previously described (20).

Proteolytic activity was determined by two methods: (a) fibrin plates (10), and (b) chromogenic substrates S-2444, S-2302, S-2222, S-2322, S-2266, S-2227, S-2251, S-2238, S-2160, S-2288, and S-2388 (kindly supplied by Dr. L. Mellstan of Kabivitrum, Stockholm, Sweden)

(21). The activity was expressed as the change in optical density (OD) with time (OD/min). For some substrates, the K_m was also determined. Electrophoretic analysis was performed on all the samples using polyacrylamide gel electrophoresis (PAGE), with (22,23) and without (24) SDS. Isoelectric focusing was also performed (25).

The action of the hemolymph and of the different fractions on purified human fibrinogen obtained from AB Kabi (Stockholm, Sweden) was determined by incubating them with the substrate and stopping the reaction at different times with either urea/SDS or aprotinin. Aliquots taken at each time were characterized by both PAGE/SDS and IEF using total antifibrinogen antiserum. Proteins in the gels were visualized with Coomassie brilliant blue R-250.

RESULTS

Clinical Observations

The patients experienced pain and a burning sensation at the site of contact with the caterpillar, followed by headache and, in children, fever. Between 2 and 72 hours after contact, bleeding started (hematomas, ecchymosis, hematuria, gynecological hemorrhages). Healing wounds opened and bled. One patient showed hemorrhagic blisters. On necropsy of a very severe case, diffuse cerebral hemorrhage and about 1 liter of blood in the peritoneal and renal cavities were found (Fig. 1).

Coagulation and Fibrinolysis Studies

On admission, the coagulation tests in the patients showed normal platelet counts, low values of fibrinogen, with very prolonged PT, PTT, and ThT, which were completely corrected by the addition of an equal volume of normal plasma. The clot that formed in the mixed plasma lysed in less than 30 minutes. Plasma lysis time and ELT (with added fibrinogen) were very short. On fibrin plates, 20 μl of plasma or euglobulin precipitate produced lysis zones with areas greater than 300 mm^2 in only a few hours, compared with normal values of 0 to 50 mm^2 in 18 hours.

The concentrations of the different coagulation and fibrinolytic proteins are shown in Table 1. In two cases, the levels of factor VIII:C II) factor II were but in one of them, the contact with the caterpillars had been extremely severe. The other patient was bitten by a vampire bat some hours after the caterpillar contact. In one patient, FDPs were quantified by both the latex and tanned red cell hemagglutination inhibition immunoassay (TRCHII) methods (26), the latter giving values 10 times higher than the former. In two patients, α2-antiplasmin was determined by chromogenic sub-strates; one appeared normal (70%) and the other was somewhat low (40%). Two-dimensional

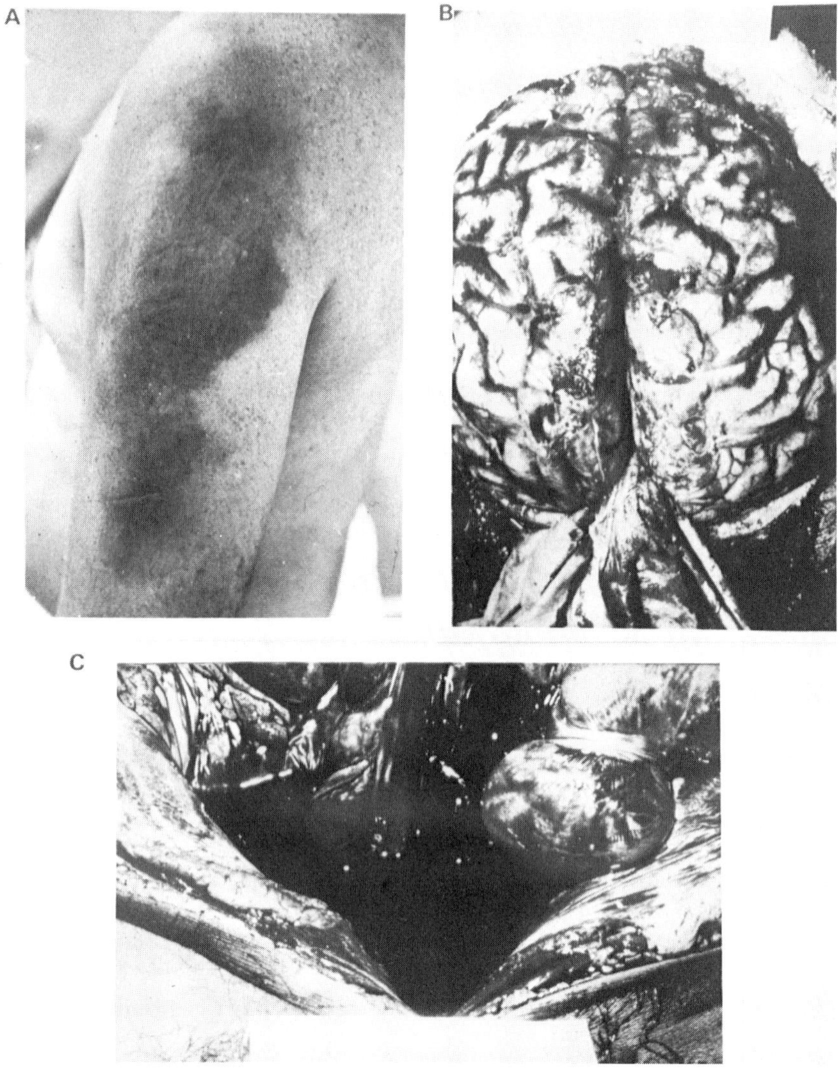

Figure 1 (A) Ecchymotic lesions in one patient. (B) Diffuse cerebral hemorrhage. (C) Liquid blood in the peritoneal cavity found at the necropsy of one patient.

Table 1 Coagulation and Fibrinolytic Factors

	Admission $\bar{X}$ (± SE)	Normal range
Platelets (10^9/liter)	233 (± 48.5)	200-400
Factor XIII (%)		
Functional	20 (± 8.4)	>80
Immunologic		
Subunit A	10.5 (± 1.3)	73-100
Subunit B	35.9 (± 12.9)	70-114
Factor XII (%)	118.3 (± 37.7)	80-200
X (%)	97.1 (± 4.9)	55-260
IX:C(%)	102.0 (± 10.8)	70-260
VIII:C(%)	159.0 (± 9.6)	55-166
VIII:v/v(%)	111.0 (± 29.3)	55-166
VII(%)	123.3 (± 21.3)	60-150
V(%)	37.0 (± 6.1)	55-204
II(%)	54.7 (± 13.3)	60-287
I(mg/dl)	31.3 (± 7.8)	190-450
FDPs (Latex μg/ml)	295.0 (± 84.5)	0-8
Plasminogen		
Chromogenic (%)	52.7 (± 8.2)	82-183
Immunologic (%)	92.5 (± 21.7)	80-120
α_2-Antiplasmin		
Chromogenic (%)	55.0	50-110
Immunologic (%)	82.5 (± 10.3)	80-120
Antithrombin III		
Chromogenic (%)	88.0 (± 4.4)	76-120
Immunologic (mg/dl)	28.6 (± 3.5)	9-25
Coagulant (%)	79.0 (± 8.5)	84-134
α_1-Antitrypsin		
Immunologic (mg/dl)	59.1 (± 7.1)	17-56
α_2-Macroglobulin		
Immunologic (mg/dl)	83.1 (± 19.0)	12-157

FDPs = fibrin(ogen) degradation products.

electrophoretic analysis of antithrombin III, both with and without heparin, carried out on the plasma specimens immediately after admission to the hospital, revealed the formation of thrombin-antithrombin complexes (Fig. 2).

Immunodiffusion of the patients' plasma using total antifibrinogen antiserum showed two precipitation bands, and the immunoelectrophoresis produced two arcs similar to those identified as fragments X and Y. The precipitation arc obtained by immunoelectrophoresis with specific antisera corresponded to that of the reference fragment in the case of anti-D, but not anti-E (Fig. 3). The nonclotting plasma of these patients did clot spontaneously after prolonged storage at -20°C, but the dissociated clot showed protein bands on PAGE/SDS different from those of a normal dissociated clot (Fig. 3).

Clinical Course

As many of the patients were far from a suitably equipped hospital when they came into contact with the caterpillars, only eight of the patients could be followed systematically, and these only within the limitations of the local hospital to which they were admitted. For these reasons, the treatments varied and may be classified as follows:

Group I: received either whole blood, with or without cryoprecipitate or fresh-frozen plasma.
Group II: were treated with aprotinin (Trasylol, Bayer, Germany) and/or EACA (Capramol, Choay, France) in addition to the replacement therapy.
Group III: received only therapeutic human fibrinogen (Immuno AG, Vienna, Austria) and aprotinin.

The results of the laboratory tests performed 72 hours after the beginning of therapy are shown in Table 2. The patients in Group I showed a decrease in platelet, plasminogen, factors II, V, VII, and VIII:C and factor XIII (subunit A) levels, with a slight improvement in fibrinogen. Patients from this group recovered slowly. In Group II, a discrete decrease in platelet levels was observed, without the severe alterations in the coagulation factors observed in Group I. In Group III, hemorrhaging ceased after a few hours of treatment. There was no change in the platelet levels nor a decrease in the concentrations of the different factors. Complete recovery was achieved within 8 days. The thrombin generation test performed in one of the patients in Group III showed an accelerated generation rate, which returned to normal levels only on the 11th day despite a clinical improvement starting on day 2 (Fig. 4).

Etiologic Agent

The caterpillars with which the patients had come into contact were classified as the larval form of the moth *Lonomia achelous* (Cramer),

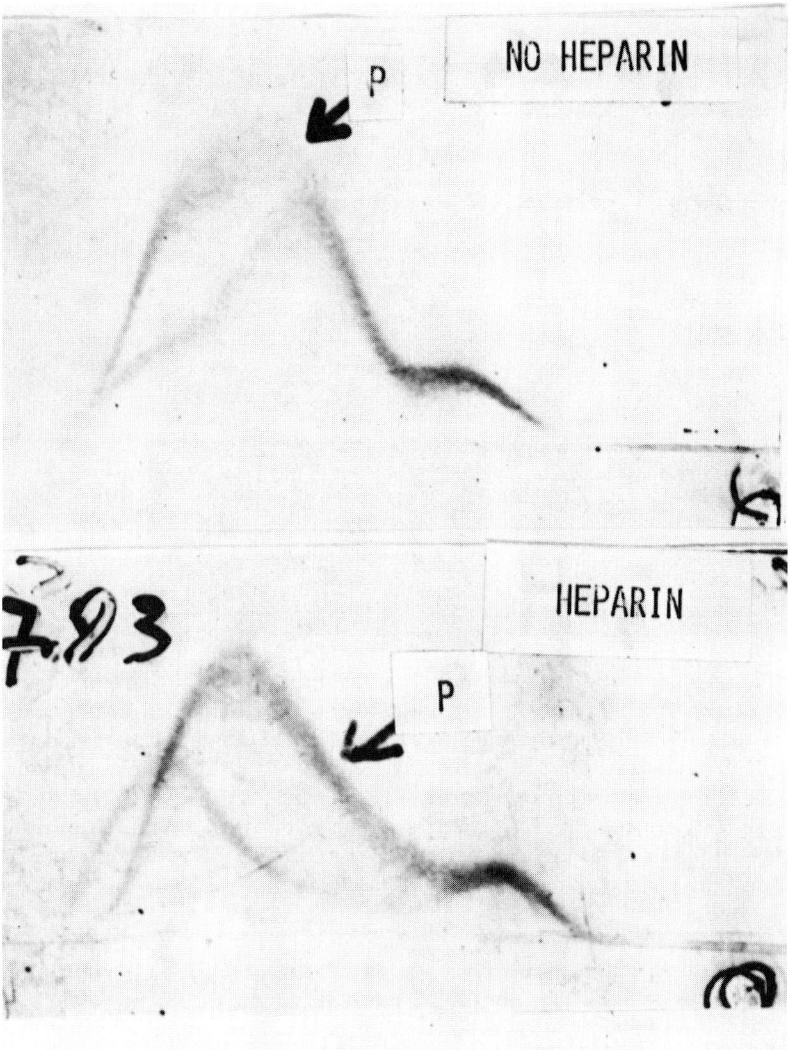

Figure 2 Two-dimensional electrophoresis of antithrombin III of the patients (P) compared with control subjects. Upper section: taken without heparin. Lower section: taken with heparin. Note the differences between the characteristics of the affected and control subjects.

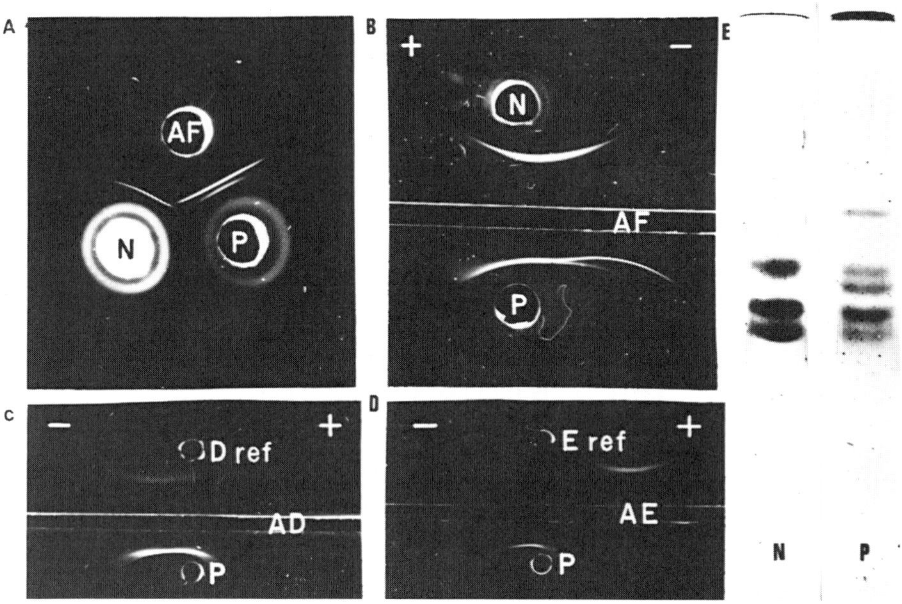

Figure 3 (A) Immunodiffusion and (B) immunoelectrophoresis of plasma from a patient and control plasma using total antifibrinogen antiserum. (C) and (D) Immunoelectrophoresis of a patient's serum compared with reference fragments using specific anti-D and anti-E antisera. Upper wells: reference fragments. Lower wells: patient's serum. Note the precipitation arcs of the patient's serum, identical to the reference fragment D, but not to fragment E. (E) PAGE/SDS of reduced clots from a patient (P) compared to control (N).

which is found in much of South America (Brazil, Guyana, Peru, Ecuador, and Venezuela). The nocturnal moth is yellow-ochre in color, with a brown band along the wings. The larval form or caterpillar is light brown, with slightly darker brown bands. The most striking characteristic is a whitish "V" on the head. Scanning electron microscopy shows the spines to resemble hypodermic needles (Fig. 5). In Venezuela, the larval form of *L. achelous* is found in patches of forest scattered across the plains. Colonies of up to 200 individuals, showing mimetism, live on the host tree *Tapirira guianensis*. The caterpillar appears twice during the year (at the end of the rainy season, and for a brief period in the middle of the dry season).

Table 2 Coagulation and Fibrinolysis Factors after 72 Hours of Treatment

	Group I N = 2	Group II N = 3	Group III N = 3
Platelets	↓↓↓↓	↓↓	↔
Factor XIII			
Functional	↔	↑ ±	↑ ±
Immunologic			
A	↓	↑ ±	↑ ±
B	↑	↑ ±	↑ ±
Factor X	↔	↔	↔
Factor IX	↑	ND	ND
Factor VIII: C	↓↓	ND	↔ ±
Factor VIII: vW	↑	↔	↔
Factor VII	↓ ±	↔	↔
Factor V	↓↓	↑↑	↑↑
Factor II	↓	↓	↔
Factor I	↑	↑	↑
Plasminogen			
Immunologic	↓	↔	↓
Chromogenic	↓	↔?	↔?
α_2-Antiplasmin			
Immunologic	↔	↔	↔
Chromogenic	ND	ND	↔
FDPs	↑↑	↓	↓

↓ = decreased; ↑ = increased; ↔ no changes; ND = not done; FDPs = fibrin(ogen) degradation products.

Biochemical Studies

A fraction with fibrinolytic activity on fibrin plates and with a molecular weight between 16,000 and 18,000 daltons was isolated from the hemolymph by molecular exclusion chromatography, and then purified 41 times on a DE-cellulose column with a recovery of 12 to 22%. On PAGE/SDS, a number of protein bands with molecular weights between 24,500 and 200,000 daltons were detected in the crude hemolymph, but none were observed in the Sephadex G-75 fibrinolytic fraction

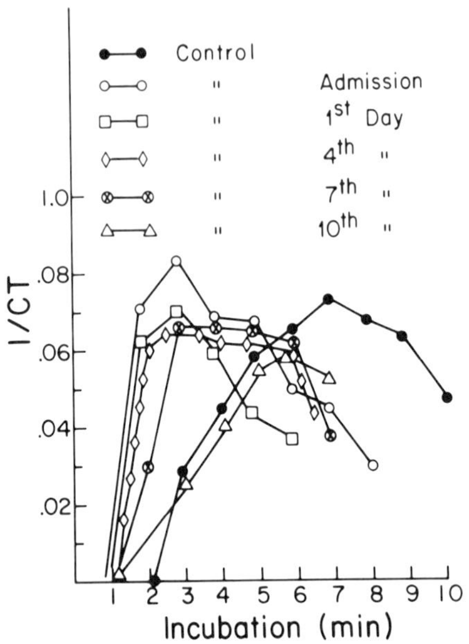

Figure 4 Thrombin generation test on consecutive days in the plasma of a patient in Group III compared with a control plasma assayed on the same day.

(Sephadex FII) or DE-cellulose preparations (results not shown). Carbohydrate was not detected by acid-Schiff staining. Electrophoresis of the DE-cellulose samples on polyacrylamide gel without SDS revealed protein bands with anodic mobility, but the fibrinolytic activity was observed both at the site of application of the sample and towards the cathode where protein could not be detected by the usual methods. These results were corroborated by isoelectric focusing in which the crude hemolymph showed two areas with proteolytic activity on fibrin plates and with the substrate S-2444 (one between pH 3 and 5, and the other at about pH 9). The Sephadex G-75 fibrinolytic peak (Sephadex FII), the hair extracts, and the fraction recovered from the DE-cellulose columns contained only one activity (at pH 9) (unpublished observation).

The fibrinolytic activity in the crude hemolymph was stable to temperature changes between -2°C and -60°C, to pH values between 3 and 9.5, to storage for 16 months at 20°C and to different denaturing agents (1 M Tris-HCl, pH 9.5, 8 M urea, and 1 M potassium thiocyanate). The activity disappeared when the crude hemolymph was heated at 70°C for 1 hour or treated with 6 M guanidine/HCl. The purified

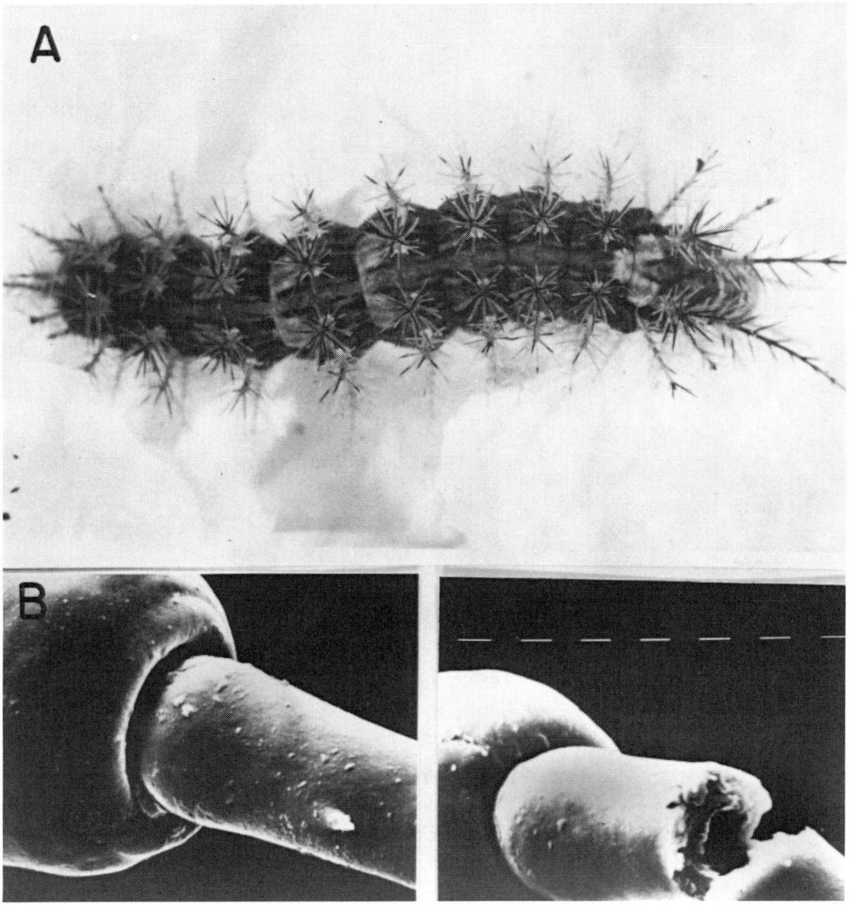

Figure 5 (A) Caterpillar of *Lonomia achelous* (Cramer). (B) Scanning electron micrograph of the hairs of the caterpillar.

hemolymph was found to be stable at 70°C for 1 hour and at 20°C for 7 years.

The degradation of fibrinogen by the Sephadex FII preparation of hemolymph was followed using immunoelectrophoresis and PAGE/SDS and compared with fibrinogen degradation produced by plasmin and trypsin. These preparations gave immunoprecipitation arcs indicating fragments different from those obtained with the latter two enzymes (Fig. 6). When the degradation products were studied on PAGE/SDS, only one fragment (between 190,000 and 290,000 daltons, depending

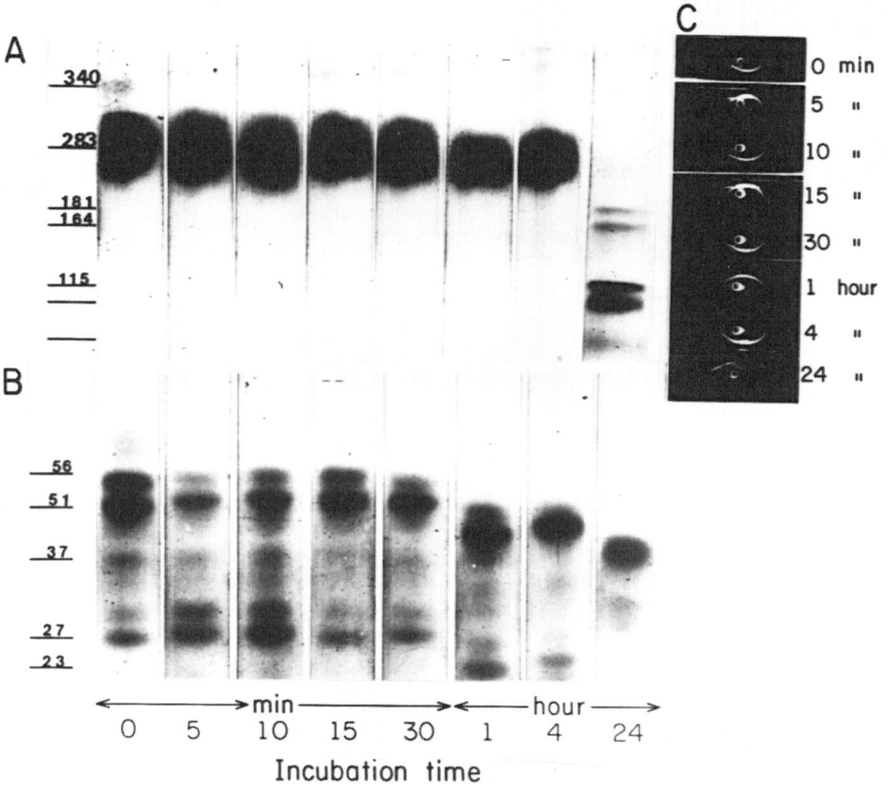

Figure 6 (A) PAGE/SDS of fibrinogen degradation products and (B) the subunits formed after incubation of 1% fibrinogen with Sephadex fraction II and the reaction stopped with aprotinin at the indicated times. (C) Immunoelectrophoresis of the fragments formed.

on the concentration of the fraction used) was observed during the first 4 hours. At 24 hours, fragments were obtained that did not correspond to either the D or E fragments resulting from proteolysis with plasmin or trypsin. In the dissociated samples, it was observed that even at time zero (hemolymph incubated with inhibitors for 30 minutes, and then mixed with the fibrinogen and an aliquot taken immediately), the fibrinogen α-chain was not apparent, and that the other chains were degraded heterogeneously (Fig. 6). This degradation was more evident at higher concentrations of the fibrinolytic fractions.

Of the chromogenic substrates, only S-2444, S-2302, and S-2222 (designed for urokinase, kallikrein, and factor Xa, respectively) and

Table 3 Kinetic Parameters (V_{max} and K_m) of the Amidolytic Activity of Sephadex Fraction II with Different Chromogenic Substrates

Substrate	V_{max} (OD/min)	Substrate	K_m
S-2251	0.21*		
S-2302	0.21	S-2322	$1.2 - 3.2 \times 10^{-4}$
S-2222	0.25	S-2444	$5.8 - 6.9 \times 10^{-5}$
S-2444	0.33	S-2288	$2.1 - 2.6 \times 10^{-6}$

*Preincubated with purified human plasminogen.

S-2322 and S-2288 (for plasminogen activators) were good substrates for the amidolytic activity of the Sephadex FII. None of the substrates for thrombin were hydrolyzed. Optical density changes were not observed with S-2251 when used directly, but when the fraction was preincubated with purified human plasminogen, hydrolysis occurred and caused an OD change of 0.21 per minute. Table 3 shows the OD changes and the K_m of the sensitive substrates.

The hydrolysis of S-2444 showed a pH optimum around 10. The hydrolytic activity increased in the presence of 10 mM cysteine or 10 mM DTT, reaching a maximum in the latter case after about 6 hours of incubation.

The dose-response curve for hemolymph followed a straight line when tested on S-2444, but not when fibrin plates were used. With the latter, an increase in the area of lysis was obtained up to the 1/60th dilution; thereafter, the area of lysis decreased and paralleled the S-2444 curve.

DISCUSSION

The coagulation and fibrinolysis tests performed in this study suggest a double mechanism of action of the larval *L. achelous* venom in the production of the hemorrhagic syndrome, with both intense fibrinolytic and disseminated intravascular coagulation-like characteristics. Fibrinogen values were extremely low, and there was a general tendency towards a decrease in plasminogen and factors XIII, V, and II. The levels of factors VIII:C and VIII:vW were generally found to be on the higher side of the normal range.

In one case, slight thrombocytopenia was present. In two, factor VIII:C and factor II were severely reduced. However, one of the patients was also bitten by a vampire bat and the other suffered the consequences of massive contact with a colony of caterpillars, the latter being the only death among the patients in our study. The necropsy findings of this case are shown in Figure 1.

The values of the FDPs do not correlate with the severity of the bleeding. It is interesting to note the differences in the values of the FDPs obtained when using the latex and TRCHII methods, and also the fact that these FDPs showed no inhibitory effect on coagulation since PT, PTT, and ThT returned to normal levels with the addition of normal plasma. Since the FDPs in the patients' sera and those experimentally produced by the purified fractions were different from those formed by plasmin or trypsin, it may be postulated that during the degradation of the fibrinogen/fibrin molecule by the venom of larval *L. achelous*, damage may occur to sites that are important for the inhibition of the fibrinogen-thrombin reaction. This may explain the poor inhibitory effect of the FDPs in the patients' plasma.

Detection of thrombin-antithrombin complexes in most patients suggests that thrombin generation does occur. One of our patients showed accelerated thrombin formation, even several days after a marked clinical improvement in his bleeding condition. Clot formation in the patients' plasma stored at -20°C seems also to suggest a gradual activation of the coagulation system and degradation of the fibrin subunits by unphysiologic mechanisms. These findings, taken along with the results of the in vitro tests using chromogenic substrates, suggest that the thrombin formation may be due to the factor X_a-like activity detected with the substrate S-2222, since none of the appropriate substrates revealed thrombin-like activity in either the crude hemolymph or the Sephadex FII.

The reduction in platelet and factor II levels and the increase in FDPs observed in those patients receiving replacement therapy only seem to confirm the presence of procoagulant activity. It may be that this is masked by the fibrinolytic activity in untreated cases and only becomes apparent when additional clotting factors are added during the replacement therapy.

In Brazil, Fraiha-Neto and co-workers (27) recently presented three cases of hemorrhagic syndrome from the Amazon Delta which, in laboratory tests, showed prolonged PT, PTT, and ThT, very low fibrinogen levels, and slightly lowered platelet and factor VIII: C concentrations. However, in a recent visit to the area, one of us (C.L.A.P) treated a patient who showed normal platelet levels. According to Fraiha-Neto and Ballarini (28), approximately 60 cases of this syndrome have been reported in the area, with over 30% mortality. However, five cases reported by Pereira (29) in Rio Grande do Sul showed the same laboratory characteristics, but with complete recovery within 6 days of transfusion therapy only. The caterpillars with which the patients in the Amazon Delta had come into contact were identified as the larva of the brown variety of *Lonomia achelous*. Its host tree in that zone is *Hevea brasiliensis*, which lives alongside *Tapirira guianensis*. The enormous distances between these reported cases suggest that the syndrome is as widely distributed as is the caterpillar in South America, but with regional differences in its evolution.

From the biochemical point of view, various proteolytic activities have been identified in the venom of larval *L. achelous*, namely, fibrinolytic, urokinase-like, kallikrein-like, factor X_a-like, and nonurokinase plasminogen activator. Taken together, all these activities found in vitro might explain the clinical syndrome observed in the patients.

The proteolytic activity on the natural substrate (fibrin plate) seems to suggest the presence of an inhibitor since dilutions increase the activity. This phenomenon is not seen when chromogenic peptide substrate S-2444 is used. It is known that the activity of an enzyme on a chromogenic substrate is not always related to the activity on the natural substrate (30,31). In the case of the larval *L. achelous* venom, the dilution effect could be explained assuming that the fibrinolytic activity needs two complementary sites, but the chromogenic substrate needs only one. In the hemolymph, a circulating inhibitor exists that occupies the complementary site different from the site for S-2444. This inhibitor can be diluted out to obtain the maximum fibrinolytic activity.

The fractions with fibrinolytic and urokinase-like activities have a molecular weight of 16,000 to 18,000 daltons, are stable in the presence of different denaturing agents and at room temperature, have a pH optimum at about 10, show two isoelectric points (one at pH 3.5 and the other at about 8.5), and produce a fibrinogen degradation pattern different from that of plasmin, trypsin, brinase, or ochrase (32-35). Only the activity having an isoelectric point of 8.5 is found in purified fractions.

In preliminary experiments, plasminogen activation was followed by PAGE/SDS with the crude hemolymph and the Sephadex FII and DE-cellulose preparations. The protein bands formed during activation were different from those obtained when using urokinase as the activator (in preparation).

The urokinase-like activity was inhibited to various degrees by the following inhibitors: DFP, iodoacetamide, TLCK, and PMSF (which inhibit only at higher concentrations). Aprotinin, soybean trypsin inhibitor, and arginyl chloromethylketone peptides with glycine in position P2 were very potent inhibitors. The inhibition was most effective in those samples preincubated with 10 mM DTT for 6 hours (in preparation).

In other insects of medical importance that produce dermatitis (such as the caterpillar of the *Euproctis* genus), extracts of the spicules have been found mainly to hydrolyze the chromogenic substrate S-2302 (designed for kallikrein), and to a lesser extent S-2444 (a substrate for urokinase) (36). This contrasts with the actions of the venom of larval *L. achelous*, which despite the finding that it hydrolyzes S-2302, is more active with S-2444 and substrates designed for plasminogen activators.

In *Rhodnius prolixus*, a proteinase has been identified that shows a pH optimum between 5.0 and 5.5, is activated by DTT and cysteine, and is inactivated by 20 mM iodoacetamide but not by SBTI or TLCK (37). The larval *L. achelous* urokinase-like proteinase is different in that it shows a pH optimum at 10 and is inhibited by PMSF, DFP, and iodoacetamide but only at high concentrations. Cysteine and DTT are activators, and aprotinin, SBTI, and chloromethylketones are strong inhibitors of this protease.

In summary we may say that:

(1) The poison of larval *L. achelous* produces a hemorrhagic syndrome that cannot be classified as being either totally fibrinolytic or disseminated intravascular coagulation-like.
(2) The body fluids of this insect contain proteolytic activities similar to tissue plasminogen activator, kallikrein, factor X_a, and urokinase.
(3) Urokinase activity is increased in the presence of reducing agents and inhibited to a varying extent by different natural synthetic inhibitors, thus resembling a sulfydryl-dependent enzyme.

ACKNOWLEDGMENTS

Part of this work has been supported by Grants CONICIT S1-0925 and Maraven Petroleos de Venezuela 1327. We are grateful to Drs. L. Mellstam and E. Berggren from AB Kabi Peptide Research, Stockholm, Sweden, for the supply of chromogenic substrates; to Dr. R. Apitz and Dr. P. Taylor, from IVIC, Caracas, Venezuela, for their advice in preparing this paper; to Dr. S. Gorzula, Dr. and Mrs. S. Rodriguez, and Mr. and Mrs. E. Lezama for their help in collecting caterpillars; and to Ms. Elena Suarez, Mrs. A. M. Romero, and Ms. L. Rybak for secretarial work.

REFERENCES

1. Arocha-Piñango, C. L. Acta Cient. Ven. 18:136 (1969).
2. Arocha-Piñango, C. L. and M. Layrisse. Lancet 1:810 (1969).
3. Marsh, N. A. and C. L. Arocha-Piñango. Proc. Royal Entomol. Soc. Lond. 36:2-3 (1971).
4. Arocha-Piñango, C. L., N. A. Marsh, and D. J. Robinson. J. Clin. Pathol. 25:614 (1972).
5. Arocha-Piñango, C. L., J. E. Perales, and S. Gorzula. Acta Cient. Ven. 28:284 (1977).

6. Nouel de Brito, A., E. Santos, L. Caldera, J. Sanchez, V. Chacon, and C. L. Arocha-Piñango. In *X Jornadas de la Sociedad Venezolana de Hematologia*, Maracaibo, Venezuela, Abstract Book, Talleres LaBodiana, Caracas, p. 67 (1981).
7. Torres, A., A. Duarte, R. Pabon, C. L. Arocha-Piñango, A. Rodriguez, A. Argüello, S. Gorzula, N. Pabon, S. Rodriguez, A. Ojeda, and Z. Carvajal. In *XI Jornadas Venezolanas de Hematología y Transfusión*, Valencia, Venezuela, Abstract Book, Talleves LaBodiana, Caracas, p. 80 (1983).
8. Markwardt, F. In *Fibrinolytics and Antifibrinolytics*, Berlin, Heidelberg, New York, Springer-Verlag, p. 487 (1978).
9. Austen, D. E. G. and L. L. Rhymes. In *A Laboratory Manual of Blood Coagulation*, Oxford, Blackwell Scientific, p. 80 (1975).
10. Marsh, N. A. and C. L. Arocha-Piñango. Thromb. Diath. Haemorrh. 28:75 (1972).
11. Arocha-Piñango, C. L. J. Clin. Pathol. 25:754 (1972).
12. Laurell, C. B. Anal. Biochem. 15:45 (1966).
13. Laurell, C. B. Anal. Biochem. 10:358 (1965).
14. Ødegard, O. R., U. Abildgaard, M. Lie, and M. Miller Andersson, Throm. Res. 11:205 (1977).
15. MacFarlane, R. G. and R. Biggs. J. Clin. Path. 6:3 (1953).
16. Outcherlony, O. and L. A. Nilsson. In *Handbook of Experimental Immunology*, D. M. Weir (Ed.), London, Blackwell Scientific, p. 655 (1973).
17. Work, T. S. and E. Work. In *Laboratory Techniques in Biochemistry and Molecular Biology*, Amsterdam, Elsevier North Holland (1969).
18. De Jong, M. C., J. M. Kawamoto, E. Bleumink, A. J. Kloosterhuis, and G. T. Meijer. Toxicon 20:477 (1982).
19. Kawamoto, F. (personal communication).
20. Arocha-Piñango, C. L., N. A. Marsh, and D. Robinson. Thromb. Diath. Haemorrh. 29:135 (1973).
21. Arocha-Piñango, C. L. and D. S. Pepper. Thromb. Haem. 46: 710 (1981).
22. Laemmli, U. K. Nature 277:680 (1970).
23. Weber, K. and M. Osborn. J. Biol. Chem. 244:4406 (1969).
24. Davis, B. Ann. NY Acad. Sci. 121:404 (1964).
25. Svendsen, P. S. and C. Rose. LKB Instrument 17:13 (1970).
26. Merskey, C., P. G. Lalezari, and A. J. Johnson. Proc. Soc. Exp. Biol. Med. 131:871 (1969).
27. Fraiha-Neto, H., I. M. Amaral, A. J. Ballarini, L. B. Dias, D. Costa, Jr., and R. N. Q. Lea. In *XXXI Congresso da Sociedade Brasileira de Medicina Tropical*, Abstract Book, Soc. Brasileva. Med. Tropical Press, Sao Paulo (1985).
28. Fraiha-Neto, H. and A. Ballarini. (personal communication).
29. Pereira, D. V. (personal communication).

30. Gaffney, P. J., M. Miller-Andersson, and T. B. L. Kirkwood. Hemostasis 7:109 (1978).
31. Laemle, B., F. Duckert, Ch. Bondelli, and L. Hanni. Thromb. Haem. 43:112 (1980).
32. Marder, V. J. and A. Z. Budzynski. Thromb. Diath. Hemorrh. 33:199 (1975).
33. Mihalyi, E., R. M. Winberg, D. W. Towne, and M. E. Friedman. Biochemistry 15:5372 (1976).
34. Gaffney, P. J. In *Progress in Fibrinolysis and Thrombolysis*, Vol. 3, J. F. Davidson, R. M. Rowan, M. M. Samama, and P. C. Desnoyers (Eds.), Raven Press, New York, p. 349 (1978).
35. Teisseyre, E., Z. S. Latallo, and M. Kopec. Folia Hematologica (Leipz) 101:99 (1974).
36. Bleumink, E., M. C. J. M. De Jong, F. Kawamoto, G. T. Meyer, A. J. Loosterhuis, and I. J. Slijper-Pal. Toxicon 20:607 (1982).
37. Garcia, E. S., J. A. Guimaraes, and J. L. Prado. Arch. Biochem. Biophys. 188:315 (1978).

18

The Effect of Anticoagulant Proteinase (Cerastase) from *Cerastes cerastes* Venom on the Blood Coagulation System

Emad W. Doaud
Ain Shams University, Cairo, Egypt
Anthony T. Tu
Colorado State University, Fort Collins, Colorado
M. Farid El-Asmar
Ain Shams University, Cairo, Egypt

INTRODUCTION

Snake venoms, especially those from Crotalidae and Viperidae, have a profound effect on blood coagulation. *Cerastes cerastes* is a sand viper commonly found in the Sahara Desert area, including Egypt. The coagulant and anticoagulant nature of its venom has been documented (1,2). We were able to isolate an anticoagulant proteinase from *Cerastes cerastes* venom by a combination of gel filtration, ion exchange chromatography, and high pressure liquid chromatography (HPLC) (3). This anticoagulant enzyme is a single polypeptide chain without subunits, having a molecular weight of 22,500. It has an isoelectric point of 5.2. It contains 6.14% carbohydrate. It is a proteolytic enzyme that hydrolyzes casein, fibrinogen, and fibrin. In this chapter, more detailed studies on fibrinogen degradation and the effect on other components of the blood coagulation system are presented.

MATERIALS AND METHODS

The anticoagulant proteinase (cerastase) was prepared as described by our group (3) from lyophylized *Cerastes cerastes* crude venom obtained by milking mature vipers kept in the serpentarium of the Biochemistry Department, Faculty of Medicine, Ain Shams University. Human fibrinogen (90% clottable), plasmin, S-2238 (H-D-Phe-Pip-Arg-pNA), S-2251 (H-D-Val-Leu-Lys-pNA), and S-2302 (H-D-Pro-Phe-Arg-pNA) were obtained from Parke-Davis. Prothrombin, plasminogen, factor X_a, urokinase, and trypsin were supplied by Sigma Chemical Company.

Study of the Mode of Fibrinogenolytic Action of Cerastase

One ml of human fibrinogen (1% in 5 mM imidazole/10 mM NaCl, pH 7.4) was incubated with 1 ml cerastase (10 or 100 μg/ml in imidazole-saline buffer) at 37°C. Separate experiments were run under the same conditions using plasmin (0.025 or 0.25 U/ml) or trypsin (1 or 10 μg/ml) instead of cerastase. At various time intervals, 0.1 ml of the incubated mixture was withdrawn and pipetted into a small test tube containing 0.1 ml of denaturating solution (10 M urea, 4% SDS, and 4% B-mercaptoethanol in water) and incubated at 37°C over night. Ten to 20 μl of the denatured sample was applied to the top of 7.5% SDS polyacrylamide disc gels, and electrophoresis was performed (4). Another 10 μl was run on an SDS gradient polyacrylamide slab gel (5 to 15% acrylamide concentration) (5).

Study of the Mode of Fibrinolytic Action of Cerastase

Fibrinogen solution (0.1 ml) in imidazole-saline buffer was clotted by the addition of 0.1 ml thrombin (10 U/ml). Next, 0.1 ml of the test solutions (cerastase, plasmin, or trypsin) was added to the clot in different tubes and incubated at 37°C. At different time intervals, the reaction in a tube was stopped by addition of 0.3 ml denaturating solution and incubated overnight. The denatured protein (30 μl) was subjected to electrophoresis by reduced SDS-PAGE (5).

Study of Thrombin-Like Activity, Antithrombin, and Prothrombin Activation

To test for thrombin-like activity of cerastase, 400 μl Tris buffer (50 mmol Tris, 175 mmol NaCl, pH 8.4) was incubated at 37°C for 2 minutes. One hundred μl of the test solutions (cerastase, plasmin, trypsin, or thrombin) was added and incubated for 30 seconds. Three hundred μl of substrate S-2238 was added and incubated for another 30 seconds. The reaction was then stopped by the addition of 300 μl of 50% acetic acid. The absorbance of the mixture was measured at 405 nm. To test for antithrombin activity, cerastase was incubated with a known amount of thrombin and the thrombin activity was tested using S-2238. To test for the effect of cerastase on prothrombin, cerastase was incubated with prothrombin for various time intervals. Aliquots were then tested for any thrombin generation by S-2238 and on gradient SDS-PAGE. As a comparison, prothrombin was incubated with activated factor X.

Study of Plasmin-Like Activity, Antiplasmin, and Plasminogen Activation

The same procedure as above with some modification was followed using S-2251 as a specific substrate for plasmin. As a comparison, plasminogen was incubated with urokinase and plasmin and was tested by S-2251 and SDS-PAGE.

Study of Kallikrein-Like and Kallikrein-Generating Activity

S-2302 was used to check for kallikrein activity in cerastase. Generation of kallikrein from plasma prekallikrein was checked by incubating cerastase with human plasma and testing for kallikrein using S-2302.

RESULTS

Effect on Fibrinogen

The $A\alpha$-chain was readily hydrolyzed by cerastase and disappeared completely after 20 minutes. The $B\beta$-chain was more resistant, but with 120 minutes of incubation, the $B\beta$-band disappeared completely (Fig. 1). The γ-chain was highly resistant, and the band was visible even at 120 minutes of incubation. On slab gels with gradient acrylamide degradation, products with lower molecular weights were viewed (Fig. 2). Cerastase digested the $A\alpha$-chain first, with the production of two main fragments with molecular weights of 42,000 and 20,000 daltons. This was followed by the degradation of the $B\beta$-chain, with the production of more fragments with molecular weights of 40,000, 18,000, 16,000, 13,000, and 10,000 daltons.

Plasmin also acts on fibrinogen, but the fibrinogen degradation products had different molecular weights from those obtained from cerastase (Fig. 3). When trypsin was used, fibrinogen degradation products were different from those of plasmin and cerastase.

Degradation of Fibrin by Cerastase

Degradation of noncross-linked fibrin can be visualized from the degradation patterns of the α- and β-chains (Fig. 4). The α-chain was the first to be degraded; the β-chain was also degraded eventually. The γ-chain was more resistant to hydrolysis. The molecular weight of the initial degradation product was 46,000 daltons. This was followed

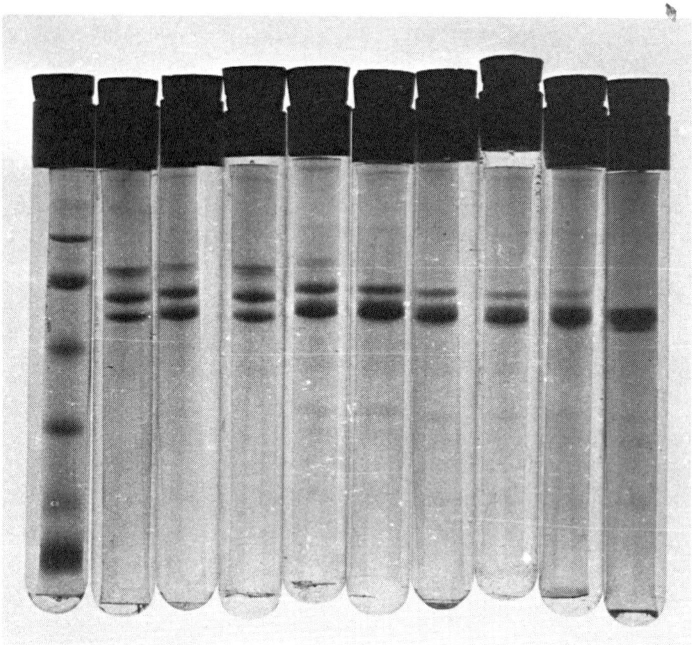

Figure 1 Reduced SDS polyacrylamide disc gel electrophoretic patterns of fibrinogen incubated with cerastase (10 μg/ml) for the time periods indicated (from left to right): 0, 5, 15, 30, 60, and 120 minutes, respectively. The condition of the run is described in the text. The tube at the extreme left contains molecular weight markers.

by the appearance of two fragments of 42,000 and 19,000 daltons and the disappearance of the 46,000-dalton fragment. Finally the γ-chain was digested and disappeared completely after 48 hours of incubation with the appearance of two fragments of molecular weights 44,000 and 13,000 daltons. Plasmin and trypsin produced completely different degradation products.

Thrombin-Like Activity and Effect on Prothrombin

Cerastase did not hydrolyse S-2238, the substrate specific for thrombin (Fig. 5). This excludes the possibility that cerastase has a thrombin-like activity since it is not able to hydrolyze the arginyl bond. The crude venom did hydrolyze S-2238. Using S-2238, it was also concluded that cerastase did not inhibit thrombin, so it is not acting as antithrombin.

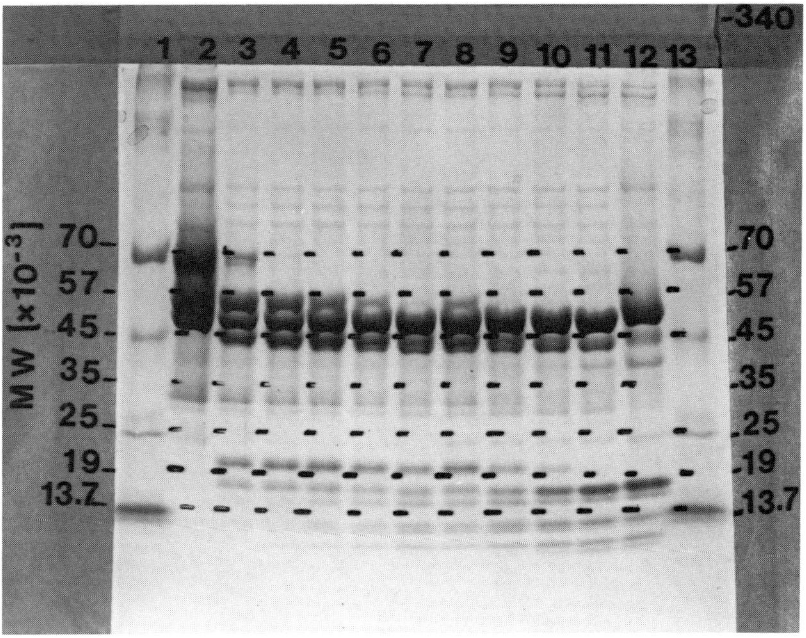

Figure 2 Reduced SDS gradient gel electrophoretic patterns of fibrinogen degradation products produced by cerastase. Lane 2 through 7 = fibrinogen + cerastase (10 μg/ml) incubated for 0, 5, 15, 30, 60, and 120 minutes, respectively. Lane 8 through 12 = fibrinogen + cerastase (100 μg/ml) incubated for 5, 15, 30, 60, and 120 minutes, respectively. Lane 1 and 13 = protein standards with shown molecular weights.

Incubation of cerastase with prothrombin did not generate thrombin, yet prothrombin was degraded to products different from that produced by the action of activated factor X (Fig. 6).

Plasmin-Like Activity and Effect on Plasminogen

Using S-2251, the substrate specific for plasmin, it was shown that cerastase has no plasmin-like activity, whereas crude venom hydrolyzes

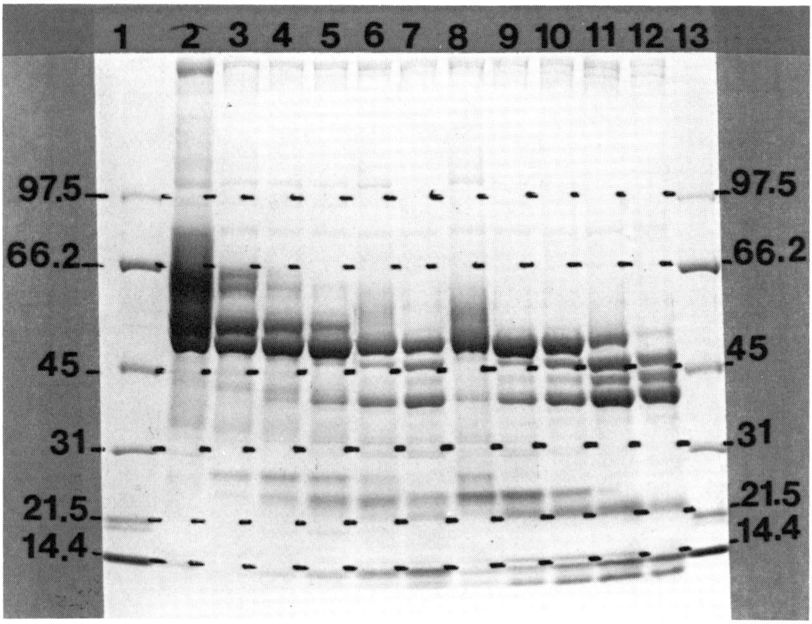

Figure 3 Reduced SDS gradient gel electrophoretic patterns of fibrinogen degradation products produced by plasmin. Lane 2 through 7 = fibrinogen + plasmin (0.025 U/ml), incubated for 0, 5, 15, 30, 60, and 120 minutes, respectively. Lane 8 through 12 = fibrinogen + plasmin (0.25 U/ml), incubated for 5, 15, 30, 60, and 120 minutes, respectively. Lane 1 and 13 = protein standards with shown molecular weights.

this substrate. Cerastase did not activate plasminogen to plasmin when incubated with it, yet it degraded plasminogen very slowly to different products as compared with those produced by urokinase (Fig. 7).

Kallikrein-Like Activity

Cerastase did not show any kallikrein-like activity as tested by hydrolysis of S-2302, a substrate specific for plasma kallikrein. Crude venom showed such activity. When incubated with plasma, cerastase did activate prekallikrein to kallikrein, which could be tested with S-2302 (Fig. 8).

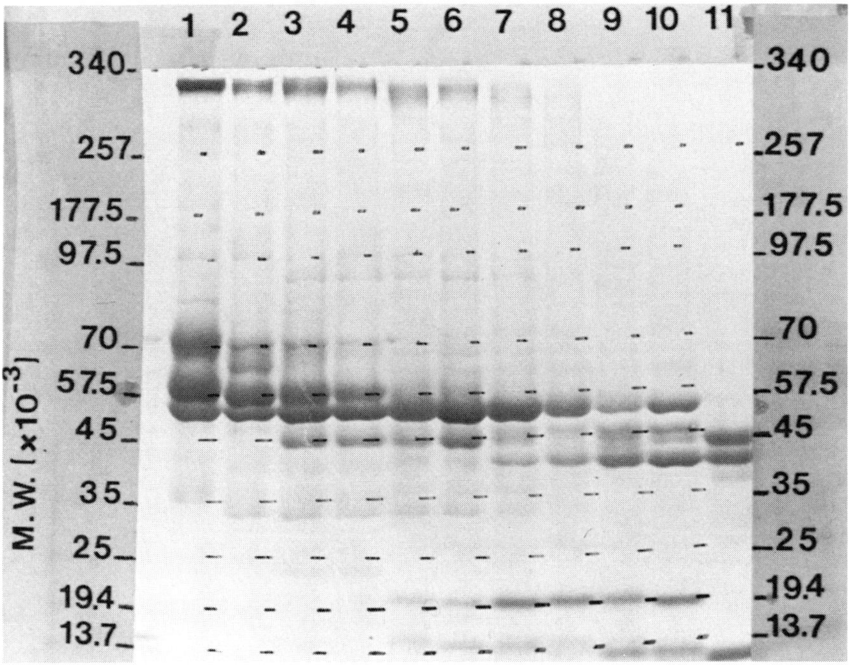

Figure 4 Reduced SDS gradient gel electrophoretic patterns of fibrin degradation products produced by cerastase. Lane 2 through 11 = fibrin + cerastase (10 μg/ml) incubated for 0, 15, 30, 60, 120, 180, 360, 720, 1440, and 2880 minutes, respectively. Lane 1 = fibrinogen. The conditions of the run are described in the text.

DISCUSSION

It is evident form this work that the fibrinogenolytic and fibrinolytic activities of cerastase are probably responsible for its anticoagulant action as demonstrated by in vitro experiments. It is of interest to notice that in in vivo experiments, injecting cerastase (5 μg/gm body weight) into mice caused a decrease in plasma fibrinogen concentration by more than 60% 1 hour after injection. The specific fibrinogenolytic activity of cerastase was calculated to be 15.69 mg fibrinogen hydrolysed/min/mg protein. This is on the same order as that reported by fraction 8 of *Trimeresurus mucrosquamatus* (6).

The pattern of fibrinogen degradation by cerastase seems to be of special interest. It degrades the Aα-chain rapidly, followed by the Bβ-chain. This is in contrast with other purified proteases that

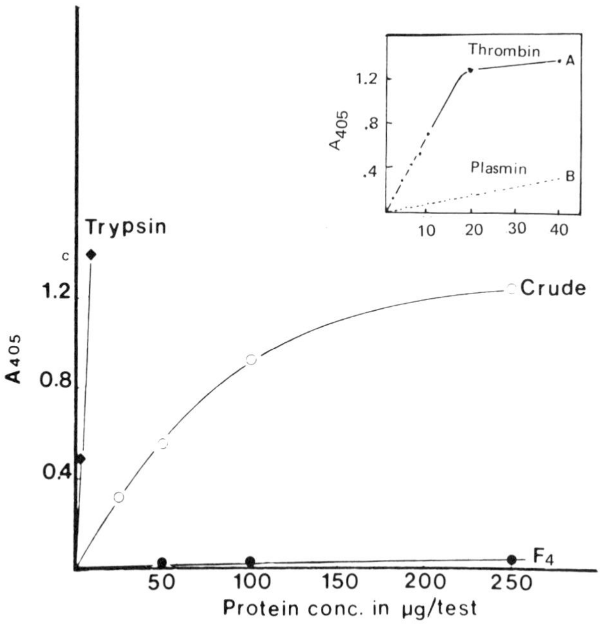

Figure 5 Effect of different concentrations of cerastase (F_4) on S-2238 hydrolysis compared with trypsin and crude venom. A = effect of thrombin on S-2238, B = effect of plasmin on S-2238.

were reported to act specifically on fibrinogen, degrading only the Aα-chain (7) or the Bβ-chain (8). Cerastase seems to be similar in action to the fibrinogenase isolated from the Western diamondback rattlesnake venom (9). The fibrinogenase from *A. rhodostoma* venom (10) or *T. mucrosquamatus* venom (8) digests the Aα-chain with the appearance of two main fragments; they do not affect the Bβ-chain. Another fibrinogenase from *T. mucrosquamatus* (8) digests the Bβ-chain first, with the production of four fragments. This is followed by digestion of the Aα-chain after prolonged incubation, with the production of two more fragments. Cerastase digests the Aα-chain first, with the appearance of two main fragments. This is followed by digestion of the Bβ-chain, with the appearance of many fragments. The molecular weights of these fragments differ from those produced by α-fibrinogenase and β-fibrinogenase, indicating a different site of action. Furthermore, the site and mode of fibrinogenolytic and fibrinolytic actions of cerastase are different from those of plasmin and trypsin.

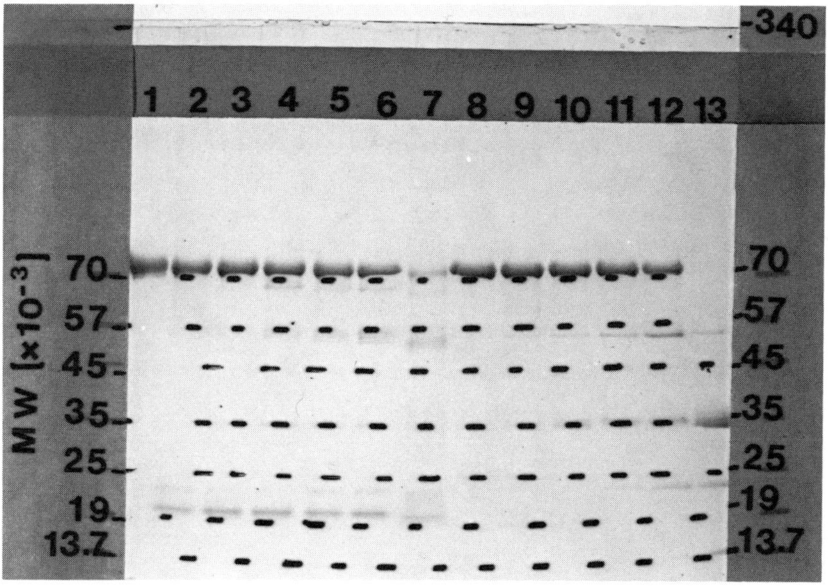

Figure 6 Effect of cerastase on prothrombin compared with activated factor X. Lane 1 = prothrombin + saline incubated for 24 hours. Lane 2 through 7 = prothrombin + cerastase incubated for 5, 15, 30, 60, 120, and 1440 minutes, respectively. Lanes 8 through 13 - prothrombin + factor X_a, incubated for 5, 15, 30, 60, 102, and 1440 minutes, respectively. The conditions of the run are described in the text.

Until now, thrombin-like enzymes from snake venom, such as Ancord and Reptilase, are the ones used experimentally and therapeutically as defibrinating agents. Cerastase appears to have advantage over thrombin-like enzymes as a defibrinogenating agent. Its mode of action on fibrinogen is quite different. No microclots are formed, and fibrinogen is degraded directly to a nonclottable form (3). Cerastase also has a direct fibrinolytic activity that could cause dissolution of any formed clot.

Apart from the fibrinogenolytic and fibrinolytic activities, it was found that cerastase could also activate plasma prekallikrein to kalli-

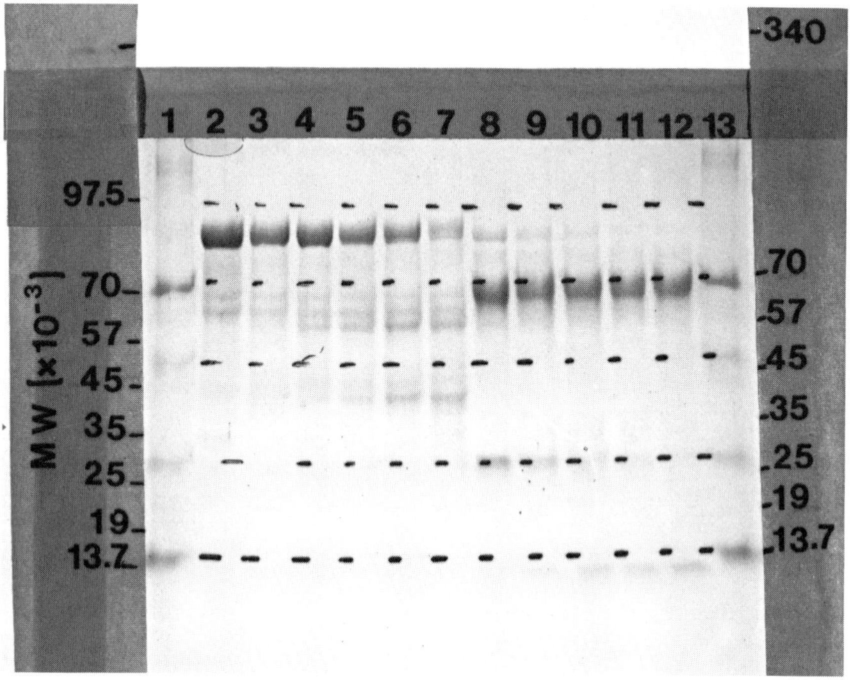

Figure 7 Effect of cerastase on plasminogen compared with urokinase. Lane 2 through 7 = plasminogen + cerastase (10 μg/ml) incubated for 0, 5, 15, 30, 60, and 120 minutes, respectively. Lane 8 through 12 = plasminogen + urokinase (0.29 U/ml) incubated for 5, 15, 30, 60, and 120 minutes, respectively. Lane 1 and 13 = protein standards with shown molecular weights.

krein in in vitro experiments. Whether this behavior has any significance in vivo requires more investigation.

Finally, cerastase has shown a very low toxicity. No gross pathologic changes could be detected after intravenous administration of cerastase in mice (3).

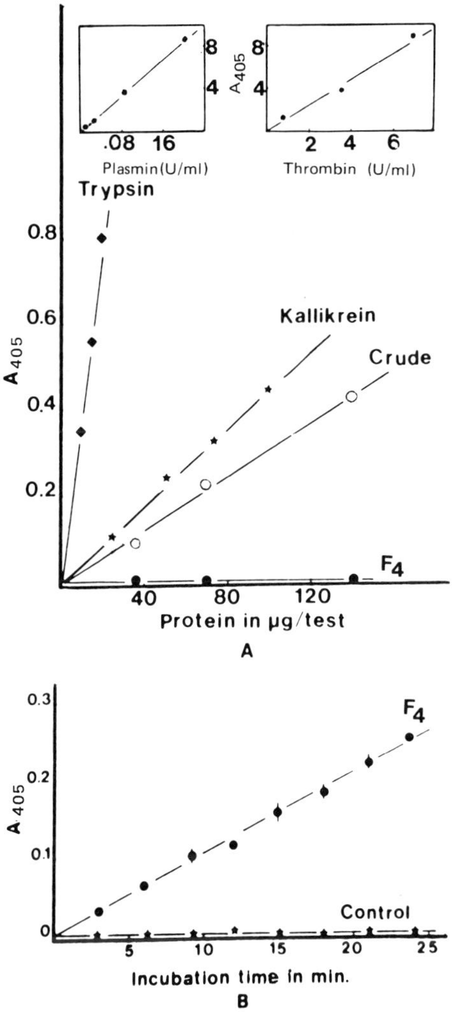

Figure 8 (A) Effect of different concentration of cerastase (F_4) on S-2302 compared with crude venom, kallikrein, and trypsin. The effect of plasmin on S-2302 as well as the effect of thrombin on S-2302 are shown. (B) Effect of cerastase on activation of plasma prekallikrein. The conditions of the experiment are described in the text.

ACKNOWLEDGMENTS

This research was supported by FRCU Grant 840505 (M.F.E.) and NIH Grant GM15591 (A.T.T.).

REFERENCES

1. Mohamed, A. H., M. El-Serougy, and L. Z. Khalid. Toxicon 7:181 (1969).
2. Labib, R. S., M. H. Azab, and N. W. Farag. Toxicon 19:86 (1981).
3. El-Asmar, M. F., E. W. Doaud, and A. T. Tu. Toxicon 23:562 (1985).
4. Weber, K. and M. Osborn. J. Biol. Chem. 244:4406 (1969).
5. Laemmli, U. K. Nature 227:6880 (1970).
6. Ouyang, C. and C. M. Teng. Biochim. Biophys. Acta 278:155 (1972).
7. Ouyang, C. and T. F. Huang. Biochim. Biophys. Acta 439:146 (1976).
8. Ouyang, C. and C. M. Teng. Biochim. Biophys. Acta 420:298 (1976).
9. T. Nikai, R. Kito, N. Mori, H. Sugihara, and A. T. Tu. Comp. Biochem. Physiol. 76:679 (1983).
10. Ouyang, C., L. J. Hwang, and T. F. Huang. Toxicon 21:25 (1983).

III

INHIBITORS

19

Biochemistry and Pharmacology of Hirudin

Fritz Markwardt
Institute of Pharmacology and Toxicology, Medical Academy Erfurt, Erfurt, Federal Republic of Germany

INTRODUCTION

In 1884, John B. Haycraft, working in Schmiedeberg's pharmacological laboratory in Strasbourg, recognized that medicinal leeches (*Hirudo medicinalis*) contain a substance with anticoagulant properties (1). Until the discovery of heparin, it was the only means to prevent the blood from clotting.

In early studies, aqueous extracts and dry preparations from leeches were used. They contained minute amounts of the active substance. Its chemical nature and mode of action were unknown. At that time, it was a mere presumption that the active principal might be a protein-like substance, and the concepts of the anticoagulant mechanism were erroneous (for references, see 2).

Attempts to isolate the anticoagulant agent and to characterize its site of action were successful after protein chemistry had developed and the biochemistry of blood coagulation had been elucidated.

Within the framework of our studies on blood clotting inhibitors from bloodsucking animals (3), we succeeded in isolating the pure anticoagulant substance from medicinal leeches in the late 1950s (4,5). It was characterized as a selective thrombin inhibitor with a polypeptide structure and was named hirudin (6). Due to its pronounced and specific anticoagulant effect, hirudin preparations are employed for diagnostic and scientific uses (7). It is estimated that about 50,000 leeches are required annually to obtain adequate amounts of hirudin for these purposes (8).

Medicinal leeches are available in a limited number only, and they have been recently placed on the list of endangered species. Most recently, the clinical application of hirudin as a highly effective anticoagulant has been brought into the focus of interest again. Advanced methods of peptide isolation and gene technology are expected to provide the opportunity to obtain hirudin in sufficient yield for therapeutic purposes.

BIOCHEMISTRY

Hirudin is produced by the salivary glands of medicinal leeches. During leeching, it is secreted into the blood being sucked up. This substance is of vital importance for the animals, since without the anticoagulant, the blood flow would be arrested by clotting and the leeches would be prevented from sucking.

Isolation

Hirudin is extracted from homogenized whole leeches or their head parts and enriched by various isolation procedures based on precipitation, electrophoresis, ion exchange chromatography, gel filtration, and isoelectric focusing (5,9-13). Recently, we developed a chromatographic procedure for isolation of the anticoagulant agent from crude material, steps of separation used previously being replaced by advanced methods. By affinity chromatography on matrix-bound thrombin, highly purified hirudin was obtained in a yield of 70 to 90% (14, 15). Chromatography on SE-Sephadex or high pressure liquid chromatography of hirudin obtained by these procedures led to separation into several fractions with approximately the same specific activity. The reason why different fractions occurred is not clearly understood. Conceivably, artifacts are formed during preparation. They might be due to formation of dimers or splitting off of the nonessential groups. The occurrence of different hirudins might be also explained by intra- and interindividual variations since the starting material differs with regard to the source, the time of collection, and the age of the animals.

Structure

The analysis of the pure substance showed a one-chain, carbohydrate-free polypeptide stabilized by three intramolecular disulfide bridges. The peptide does not contain arginine, methionine, or tryptophan. Of special interest is the proportion of basic to acidic amino acid residues. Besides only three lysine residues, it contains large amounts of glutamic and aspartic acid that account for the low isoelectric point (5,13, 16,17). Establishment of the primary structure showed that the polypeptide chain contains 65 amino acids from which a molecular weight of about 7000 daltons was calculated (Fig. 1) (18,19).

The results obtained by complete sequencing of highly purified hirudin preparations varied with different investigators. Different N-terminal amino acids of hirudin and slight differences in the amino acid sequence were determined (11,13,20). These variations and the possibility of separating hirudin by sensitive methods into different fractions are suggestive of the existence of iso-inhibitors.

Hirudin is stable in aqueous solution and resistant to heating at 80°C for 15 minutes. It is digested by pepsin, whereas it is not destroyed by trypsin and chymotrypsin. The resistance to these proteolytic enzymes is probably due to the tertiary structure of the peptide. This assumption is supported by the finding that oxidized hirudin is readily split by trypsin (13,17,21).

Mechanism of Action

Analysis of the anticoagulant effect of hirudin showed that it possesses special affinity for the clotting enzyme thrombin. Hirudin forms with thrombin an inactive enzyme-inhibitor complex with a very small dissociation, whereby the active center of the enzyme is blocked (Fig. 2). The K_i is about 50 pmol/l (6,22). To which structural features is the specific and pronounced antithrombin effect due? It is most likely that the secondary and tertiary structure of hirudin is of decisive importance for its antithrombin activity, which is lost during oxidation or reduction of the disulfide bridges and proteolytic degradation of the molecule (17,23).

The knowledge of the hirudin-thrombin reaction requires an understanding of the specific structure and function of the serine proteinase thrombin. The active-site region of thrombin is divided by the catalytic site consisting of the histidine-57, aspartic acid-102, and serine-195 triad into the fibrinopeptide recognition site and the fibrin recognition site. A variety of evidence has indicated that a major noncovalent protein binding site resides on one side of the catalytic site and that an important apolar binding site exists on the other side of the catalytic site (24).

The structures of the fibrinopeptides are characterized by clusters of acidic and hydrophobic amino acids. The phenylalanine residue and the neighboring aspartic acid residue located within a certain position in the fibrinopeptide A moiety are of particular importance for the binding of the substrate.

Hirudin contains a highly acidic C-terminal segment, which shows structural similarities to fibrinopeptides, if one compares the amino acid sequence from the reversed direction (19,23). There are clusters of acidic amino acids and one Tyr-SO_3H within this segment. Moreover, it contains the phenylalanine residue in the immediate vicinity

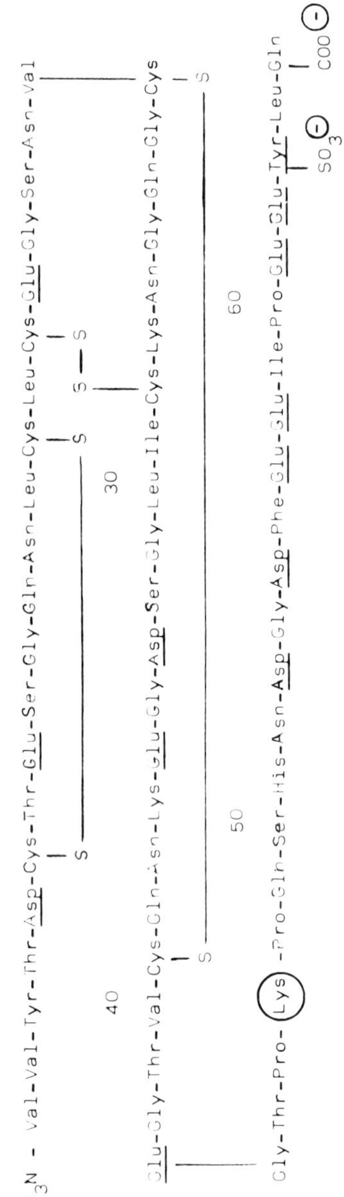

Figure 1 Structure of hirudin (according to references 13,19,20).

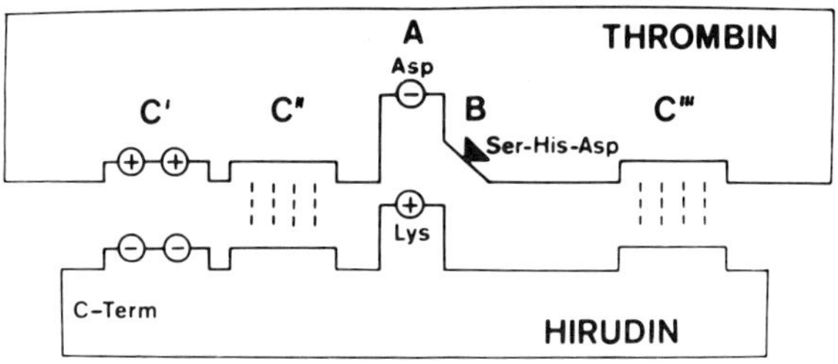

C',C" = FIBRINOPEPTIDE - BINDING SITE
C''' = FIBRIN - BINDING SITE
B = CATALYTIC SITE
A = ARGININE SIDE - CHAIN POCKET

Figure 2 Scheme of hirudin-thrombin interaction.

of an acidic acid important for the recognition by thrombin. This sequence similarity raises the possibility that thrombin may recognize its substrate through those similar acidic regions.

It has been suggested that the C-terminal segment of hirudin binds to the region of the fibrinopeptide interaction site of the thrombin molecule and associates with the apolar binding sites neighboring the arginine side-chain pocket of the active center of thrombin. Therefore, the affinity of the inhibitor is lost during esterification of carboxyl groups or removal of the acidic C-terminal amino acids by chemical or enzymatic methods (6,23). On the other hand, the affinity of hirudin for thrombin decreases by acetylation of basic groups of the enzyme located in the active site (17).

The compact N-terminal portion of hirudin stabilized by disulfide bridges can be attached to the fibrin binding site of the enzyme. Among the three lysine residues, which are possibly the active site of the inhibitor, two are buried in the amino-terminal segment and one is flanked by two prolines. That is why the lysine bonds are not accessible to trypsin digestion.

From binding studies with acylated thrombin, it is concluded that the catalytic serine is not necessary for hirudin-thrombin complexing (6,17). This was corroborated by displacement experiments of proflavine from various modified thrombin preparations (24).

Hirudin also differs from other proteinase inhibitors in that its reactive site is located outside the three disulfide-bridged peptide loops, which would be a unique feature among the numerous proteinase inhibitors described so far (19).

Anticoagulant Activity

The fast and specific reaction of hirudin with thrombin is utilized for quantitative determination of the tight-binding inhibitor. One μg of pure hirudin inhibits about 5 μg of human thrombin. The antithrombin activity corresponds to a 1:1 reaction calculated by means of the molecular weight. Therefore, the activity of hirudin is measured in antithrombin units (AT-U); 1 AT-U is the amount of hirudin that neutralizes 1 unit of thrombin (25). On the other hand, standardized hirudin is used for quantitative determination of thrombin or of prothrombin via its activation to thrombin (26). Furthermore, the selective thrombin inhibitor is employed in biochemical research into blood coagulation (7,27-31) and is particularly suited in clotting assays to remove thrombin specifically (7,32-35).

Of special interest is the pharmacologic control by hirudin of blood coagulation. Due to its high affinity for thrombin, relatively low inhibitor concentrations are necessary to neutralize the clotting enzyme. Hirudin prevents not only fibrinogen clotting, but also further thrombin-catalyzed reactions such as activation of factors V, VIII, and XIII (17,36,37) and the thrombin-induced platelet reaction (38-40). By instantaneous inhibition of the small amount of thrombin generated after activation of the clotting system, the autocatalytic reaction is prevented, which otherwise would lead to accelerated generation of further thrombin.

Depending on the hirudin concentration in blood, coagulation is retarded or completely inhibited. The clotting variables such as thrombin time, partial thromboplastin time, and prothrombin time vary with different hirudin concentrations (Fig. 3) (41,42).

Thrombin loses its activity on endothelial cells and blood platelets by binding to hirudin (21,43,44). Detailed studies on the influence of hirudin on the thrombin-platelet reaction were carried out by determination of thrombin-induced serotonin release (21,39). In contrast to the inhibition of fibrinogen clotting, prevention of the effect of thrombin on platelets was found to require the addition of hirudin in molar excess, since minute amounts of thrombin, as they dissociate from the hirudin-thrombin complex, suffice to trigger the release reaction. Under these conditions, the release of serotonin could be immediately stopped by hirudin at any time.

EXPERIMENTAL PHARMACOLOGY

In early in vivo studies, crude or partially purified hirudin preparations were used. They had a low content of active substance and they contained other proteinase inhibitors besides hirudin. Severe side effects occurred, and repeated injection led to allergic reactions. The results of these studies were inconclusive. After isolation of the pure

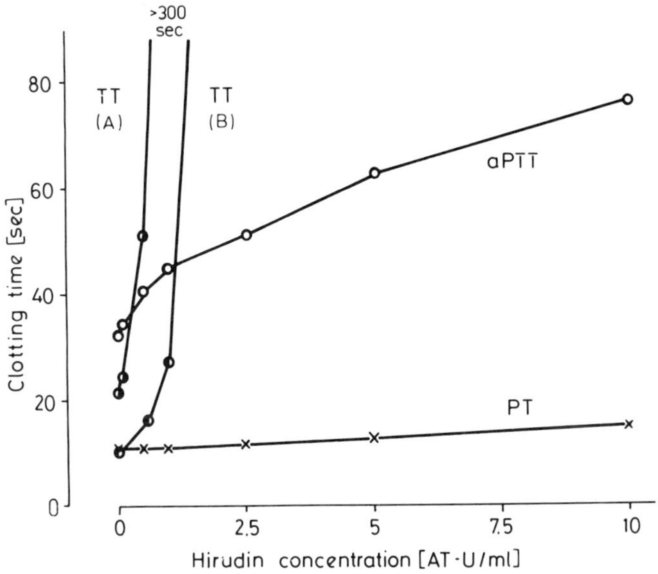

Figure 3 Relationships between hirudin concentration and thrombin time (TT (A) ◑ : induced by 0.2 NIH-U/ml, TT (B) ◐ : 1.0 NIH-U/ml), partial thromboplastin time (aPTT ○), and prothrombin time (PT x) in human citrated plasma.

substance and after elucidation of its mode of action, the in vivo use of hirudin became possible.

Pharmacokinetics

To study the pharmacokinetics of hirudin, methods are required that assess the inhibitor content in body fluids. The assay of hirudin is based on its complexing with thrombin and determination of the residual thrombin activity. Measurement of the clotting activity of thrombin may be used. Recently, we determined the content of hirudin in plasma and urine by a chromogenic substrate method (45). This assay is superior to the clotting assay in that a low detection limits is obtained (about 10 ng/ml). Furthermore, for pharmacokinetic studies, radiolabeled hirudin and immunologic methods were used (46).

Hirudin given intravenously to experimental animals (dog, rabbit, rat) is relatively rapidly eliminated. After the initial distribution phase (t 1/2 ∿ 10 minutes), first-order elimination kinetics followed. The elimination half-life was estimated to be 50 to 60 minutes. Most of the hirudin was excreted in the urine in biologically active form

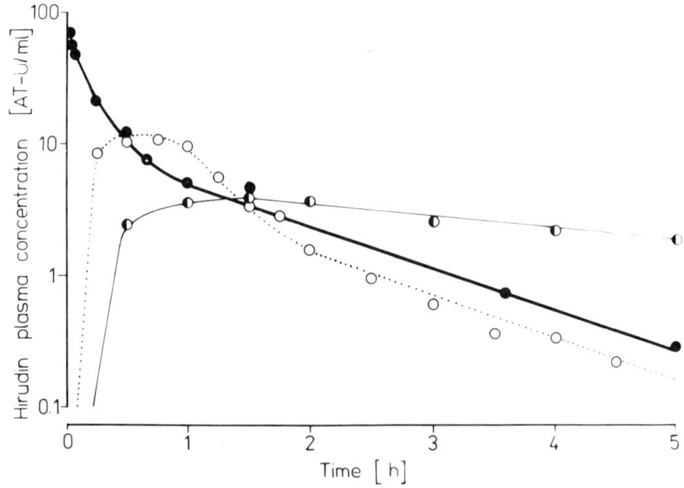

Figure 4 Mean plasma concentration of hirudin (5000 AT-U/kg) following single intravenous (●) and subcutaneous (◐) doses or infusion for 60 minutes (○) in dogs.

(47). After infusion, plateau values were reached within 30 minutes (Fig. 4). After oral administration, no hirudin was detectable in blood.

Pharmacodynamics

Investigation of the pharmacologic properties of hirudin have shown that apart from its anticoagulant effect, it is inert pharmacodynamically and is well tolerated in vivo (47-49). Its acute toxicity is very low. After intravenous administration in mice, the lethal dose was found to be greater than 50 mg/kg. In animal experiments, the intravenous injection of 10 mg/kg was tolerated without changes in heart and respiratory rates; the blood pressure remained constant. Intravenous injection of the anticoagulant did not cause any changes in platelet counts, fibrinogen level, or plasma hemoglobin. With anticoagulant doses, the bleeding time was slightly prolonged. After repeated injection at 2-week invervals, the animals did not show signs of sensitization.

Antithrombotic Action

Although there are various factors inducing formation of intravascular thrombi, the underlying process is coagulation caused by thrombin. Therefore, a thrombin inhibitor would be expected to prevent thrombosis or its progression.

The antithrombotic action of hirudin was studied in animal models of thrombosis that corresponded to the pathogenetic mechanism of venous and arterial thrombosis or disseminated intravascular coagulation (DIC) in man (Table 1). In all conditions, hirudin prevented the development of intravascular thrombi. The antithrombotic effect depended on the dose and the blood level (47, 50-53). For instance, the development of clotting thrombi induced in the jugular vein of rats and rabbits by the stasis model of Wessler was prevented. Likewise, thrombotic occlusion of the carotid artery induced by a lesion of the vessel wall did not occur. Determination of the hirudin plasma level showed that the preventive effect required a concentration of approximately 5 to 10 AT-U/ml.

To assess the antithrombotic activity of hirudin in DIC, microthrombosis induced by thrombin infusion in the lungs of rats pretreated with radiolabeled fibrinogen was followed (52-54). At doses of 200 At-U per kg·hour, blood levels of about 0.5 AT-U/ml were reached 60 minutes after the beginning of the infusion and were maintained throughout the experiment. At this dosage, an accumulation of labeled fibrin deposits in the lungs was prevented.

Furthermore, the influence of hirudin on endotoxin-induced disseminated microthrombosis was studied in weaned pigs (54). After infusion of endotoxin, changes in the clotting system typical of consumption coagulopathy occurred. When hirudin was given simultaneously, the endotoxin-induced changes in the clotting system were less pronounced. After pretreatment with radiolabeled fibrinogen, fibrin deposits appeared during endotoxin infusion, as was obvious from the

Table 1 Effect of Hirudin on Experimental Thrombosis in Rats

Thrombosis model	Hirudin dose	Incidence of thrombosis
Clotting thrombus, jugular vein	—	16/18
(stasis following injection of contact-activated serum)	1.0 mg/kg IV (5 minutes before serum)	1/10
Deposition thrombus, carotid artery	—	10/12
(lesion of the vessel wall by electrical current)	1.0 mg/kg IV (after provocation)	2/10
Disseminated thrombi, lungs	—	5/6
(thrombin infusion 500 NIH U/kg·hr)	0.02 mg/kg·hour	0/6

increase in radioactivity in the organs investigated. Under hirudin treatment, the increase was significantly reduced.

Investigation of cardiovascular functions showed characteristic changes under endotoxin infusion. The right ventricular pressure increased, whereas left ventricular pressure decreased slightly. The respiratory rate and the heart rate increased continuously. These changes in cardiovascular function were reduced by simultaneous infusion of hirudin.

Hirudin also inhibited the development of the localized Sanarelli-Shwartzman phenomenon, whose pathogenesis corresponds to that of the generalized form (55,56).

To assess the antithrombotic activity in experimental extracorporeal circulation, an arteriovenous shunt model was used (47). Depending on the hirudin dose, occlusion of the shunt was prevented or retarded. Heparin at the same concentration was less effective.

CLINICAL PHARMACOLOGY

The use of hirudin has only recently become of therapeutic interest. The results obtained in pharmacotoxicologic investigations allowed study of the effect of this naturally occurring anticoagulant in humans.

Pharmacokinetics

The results of clinicopharmacologic studies in humans corresponded to those obtained in animal experiments (57). After intravenous injection of 1000 AT-U/kg in man, the course of the hirudin plasma level can be described by a rapid initial distribution phase (α-phase), followed by an elimination phase (β-phase). After the initial disappearance, which was interpreted as a distribution phenomenon, first-order elimination kinetics followed. The decline in plasma hirudin concentration was most adequately expressed by an equation describing a two-compartment model (Table 2). The elimination half-life and the volume of distribution indicated distribution into the intravascular space.

Subcutaneous injection of the same dose caused an increase in the hirudin level in blood within 60 minutes, reaching plateau values in plasma of 0.5 AT-U/ml, which then fell below the detection limit after 5 hours. The subcutaneous absorption rate constant indicated quick absorption of the drug. The elimination half-life after subcutaneous dosing was similar to that after intravenous injection.

The cumulative urinary excretion of hirudin showed that much of the administered amount was excreted in the urine in the active form. It must be emphasized that a large percentage of the acidic polypeptide is excreted through the kidneys in unchanged form and is obviously not metabolized or bound in the organism.

Table 2 Pharmacokinetics of a Single Dose of Hirudin 1000 AT-U/kg IV Given to Six Volunteers

Parameter	Mean	±	SD
$t_{1/2\alpha}$ (hr)	0.15		0.06
$t_{1/2\beta}$ (hr)	0.84		0.19
Vd (l)	12.9		2.8
AUC (μg·hr/ml)	0.57		0.17
Cl_{tot} (ml/min)	230.00		98.00
Cl_{ren} (ml/min)	100.00		62.00

The results of the clotting test obtained at different times after administration varied with the hirudin level in plasma. The prolongation of clotting time corresponded to the quantitative reaction of hirudin with thrombin either added to or generated in plasma. Low hirudin concentrations in plasma as they occurred following subcutaneous administration are best detected by prolongation of thrombin time after the addition of small amounts of thrombin.

Side Effects

Administration of hirudin was tolerated without perceptible side effects. It had no influence on fibrinogen level or platelet counts, nor were any changes seen in euglobulin lysis time. Fibrinogen degradation products were detected. Only in the case of the highest hirudin concentration in plasma was the bleeding time prolonged. No changes in blood pressure or heart or respiratory rates were observed.

Hirudin-specific antibodies have not so far been observed in human sera after hirudin treatment. Thus, the polypeptide appears to be a comparatively weak immunogen.

Clinical Investigations

For obvious reasons, pure hirudin is not available in adequate amounts for therapeutic purposes. There are only a few preliminary clinical reports on the antithrombotic action. Successful treatment of superficial thrombophlebitis has been reported following intracutaneous injection, and local administration of hirudin has been found to prevent occlusion of arteriovenous fistulas in dialysis patients.

Based on clinicopharmacologic investigations, we started clinical trials in patients suffering from chronic DIC (Fig. 5). Subcutaneous injection of 0.1 mg/kg at 8-hour intervals resulted in normalization of fibrinogen level and platelet counts and in disappearance of fibrin-monomer complexes and degradation products.

FINAL REMARKS

Due to their different mode of action, it is difficult to compare the antithrombotic effect of hirudin with that of other anticoagulants such as heparin. In contrast to heparin, hirudin has the advantages that its effect does not require the presence of antithrombin and it is not bound or inactivated by platelet factors or other proteins. In contrast to hirudin, heparin reacts with blood platelets at the vascular endothelium in such a way that bleeding might be promoted. Furthermore, hirudin differs from heparin in that it does not cause nonhemorrhagic side effects such as lipolysis. Hirudin is superior to the oral anticoagulants in that it does not interfere with the biosynthesis of clotting factors.

In conclusion, hirudin is an anticoagulant of high quality. Its introduction into clinical medicine will be a decisive contribution to prophylaxis and therapy of thrombosis.

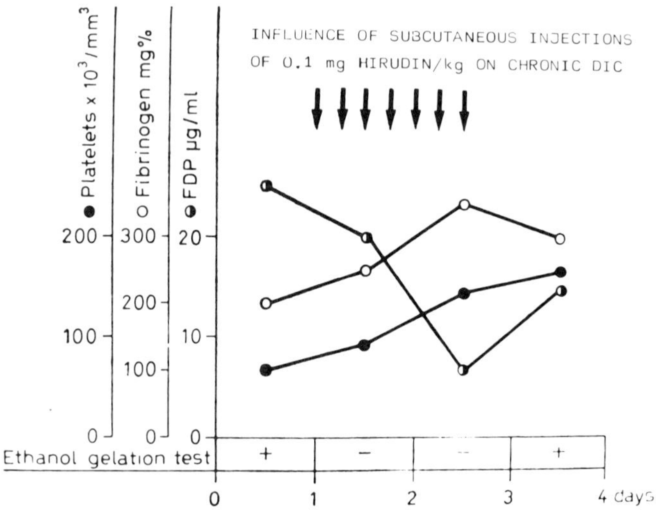

Figure 5 Influence of hirudin (0.1 mg/kg, subcutaneously) on clotting variables in chronic disseminated intravascular coagulation.

REFERENCES

1. Haycraft, J. B. Naunyn-Schmiedeberg's Arch. Exp. Pathol. Pharm. 18:209 (1884).
2. Markwardt, F. *Blutgerinnungshemmende Wirkstoffe aus blutsaugenden Tieren*, Fischer, Jena (1963).
3. Markwardt, F. and H. Landmann. In *Handbuch der experimentellen Pharmakologie, Volume 27*, F. Markwardt (Ed.), Springer-Verlag, Berlin, Heidelberg, New York (1971).
4. Markwardt, F. Naturwissenschaften 42:537 (1955).
5. Markwardt, F. Hoppe-Seyler's Z. Physiol. Chem. 308:147 (1957).
6. Markwardt, F. and P. Walsmann. Hoppe-Seyler's Z. Physiol. Chem. 312:85 (1958).
7. Walsmann, P. and F. Markwardt. Pharmazie 36:653 (1981).
8. Leake, L. P. Endeavour 7:88 (1983).
9. Bagdy, D., E. Barabás, and L. Gráf, L. Thrombos. Res. 8:229 (1973).
10. Baskova, I. P., D. U. Cherkesova, and V. V. Mosolov. Biochimija 41:939 (1976).
11. Baskova, I. P., D. U. Cherkesova, and V. V. Mosolov. Thromb. Res. 30:459 (1983).
12. Markwardt, F., G. Schäfer, H. Töpfer, and P. Walsmann. Pharmazie 22:239 (1967).
13. Markwardt, F. and P. Walsmann. Hoppe-Seyler's Z. Physiol. Chem. 348:1381 (1967).
14. Walsmann, P. Pharmazie 36:860 (1981).
15. Walsmann, P. and F. Markwardt. Thromb. Res. 40:563 (1985).
16. Bagdy, D., E. Barabás, L. Gráf, T. E. Petersen, and S. Magnusson. In *Methods in Enzymology, Volume 45*, L. Lorand (Ed.), Academic Press, New York, San Francisco, London (1976).
17. Markwardt, F. In *Methods in Enzymology, Volume 19*, P. E. Perlmann and L. Lorand (Eds.), Academic Press, New York, London (1970).
18. Dodt, J., H.-P. Müller, and J.-Y. Chang. FEBS Lett. 165:180 (1984).
19. Dodt, J., U. Seemüller, R. Maschler, and H. Fritz. Biol. Chem. Hoppe-Seyler 366:379 (1985).
20. Gráf, L., A. Pathy, E. Barabás, and D. Bagdy. Biochim. Biophys. Acta 310:416 (1973).
21. Hoffmann, A. and F. Markwardt. Haemostasis 14:164 (1984).
22. Landis, B. H., M. P. Zabinski, G. J. Lafleur, D. H. Bind, and J. W. Fenton, II. Fed. Proc. 37:1335 (1978).
23. Chang, J-Y. FEBS Lett. 164:307 (1983).
24. Sonder, S. A. and J. W. Fenton, II. Biochemistry 23:1818 (1984).
25. Markwardt, F. Naunyn-Schmiedeberg's Arch. Exp. Pathol. Pharm. 229:389 (1956).

26. Markwardt, F. Arch. Pharmacol. 290:280 (1967).
27. Perduet, J. F., W. Lubenski, E. Kivity, S. Sonder, and J. W. Fenton, II. J. Biol. Chem. 256:2767 (1981).
28. Pochon, F. and M. Steinbuch. FEBS Lett. 177:109 (1984).
29. Salem, H. H., G. J. Broze, J. P. Miletich, and P. W. Majerus. Proc. Natl. Acad. Sci. USA 80:1584 (1983).
30. Van Obberghen-Schilling, E., R. Pérez-Rodriguez, and J. Pouysségut. Biochem. Biophys. Res. Commun. 106:79 (1982).
31. Wilner, G. D., M. P. Dantiz, M. S. Mudd, K.-H. Hsieh, and J. W. Fenton, II. J. Lab. Clin. Med. 97:403 (1981).
32. Latallo, Z. S. In *Thrombosis and Bleeding Disorders*, N. U. Bang, F. K. Beller, E. Deutsch, and E. F. Mammen (Eds.), Thieme, Stuttgart/Academic Press, New York, London (1971).
33. Seghatchian, M. J. Thromb. Res. 19:757 (1980).
34. Soumela, H., M. Blombäck, and B. Blombäck. Thromb. Res. 10:267 (1977).
35. Van Wijk, E. M., L. H. Kahle, and J. W. Ten Cate. Clin. Chem. 26:885 (1980).
36. Hultin, M. B. J. Clin. Invest. 69:950 (1982).
37. Hultin, M. B. and J. Jesty. Blood 57:476 (1981).
38. Glusa, E. and F. Markwardt. Haemostasis 9:188 (1980).
39. Markwardt, F., W. Barthel, and A. Hoffmann. Naunyn-Schmiedeberg's Arch. Exp. Pathol. Pharm. 251:255 (1965).
40. Reinhold, D. S. and H. Gersham. Thromb. Res. 37:513 (1985).
41. Galdal, K. S., S. A. Evensen, and F. Brosstad. Thromb. Res. 27:575 (1982).
42. Galdal, K. S., S. A. Evensen, and E. Nilsen. Thromb. Res. 37:583 (1985).
43. Detwiler, T. C. Thromb. Haemost. 50:326 (1983).
44. Layes, B. and H. J. Weiss. Thromb. Haemost. 45:173 (1981).
45. Griessbach, U., J. Stürzebecher, and F. Markwardt. Thromb. Res. 37:347 (1985).
46. Bruckbauer, H. *Dissertation der Universität München*, 1984.
47. Markwardt, F., J. Hauptmann, G. Nowak, C. Klessen, and P. Walsmann. Thromb. Haemost. 47:226 (1982).
48. Markwardt, F. Arch. Exp. Pathol. 234:516 (1958).
49. Markwardt, F. Arch. Exp. Pathol. 236:286 (1959).
50. Ishikawa, A., R. Hafter, U. Seemüller, J. M. Gokel, and H. Graeff. Thromb. Res. 19:351 (1980).
51. Markwardt, F., G. Nowak, and J. Hoffmann. Thromb. Res. 11:275 (1977).
52. Markwardt, F., G. Nowak, and J. Hoffmann. Thromb. Haemost. 49:235 (1983).
53. Markwardt, F., G. Nowak, W. Meerbach, and K.-D. Rüdiger. Thromb. Diath. Haemorrh. 34:513 (1975).
54. Nowak, F. and F. Markwardt. Exp. Pathol. 18:438 (1980).

55. Haustein, K.-O. and F. Markwardt. Thromb. Diath. Haemorrh. 13:60 (1965).
56. Haustein, K-O. and F. Markwardt. Folia Haematol. 97:104 (1972).
57. Markwardt, F., G. Nowak, J. Stürzebecher, U. Griessbach, P. Walsmann, and G. Vogel. Thromb. Haemostas. 52:160 (1984).

20

Scientific Rationale Behind the Medical Use of Leeches

Roy T. Sawyer
Biopharm (UK) Ltd., Swansea, Wales

INTRODUCTION

The leech was popular in medicine from prehistoric times and fell into disrepute only comparatively recently in the latter part of the last century. Ironically, it was about the time the leech became unfashionable that the first scientific justification for its use began to emerge. In fact, the world's first anticoagulant (hirudin) was discovered over 100 years ago in 1884 in the salivary gland of the medicinal leech *Hirudo medicinalis*. Since then, scientists have revealed that the leech is a living pharmacopeia that holds considerable promise for medical biotechnology (1).

HISTORICAL BACKGROUND

The earliest documented use of leeches may have been in ancient Egypt. A wall painting was discovered in the tomb of the scribe Userhat of the 18th Dynasty (1567-1308 BC), depicting what has been interpreted by some as an early application of leeches by a barber surgeon. The earliest unequivocal record occurs in ancient India in an impressive Sanskrit system of medicine called *The Sushruta Samhita*, dating back 2000 years or more. The *Sushruta* has a chapter on leeching so detailed that some of the 12 species mentioned can still be recognized. The Greeks and Romans also documented the use of leech therapy, dating back several centuries BC. Leeching persisted through the centuries, and medieval manuscripts are replete with illuminated letters and other illustrations depicting their extensive use in medicine. By the 17th century, the leech was widely used in western Europe, especially in France, Germany, and Britain, but it reached a peak in the first half of the 19th century in France. At this time, France was undoubtedly the biggest user of leeches thanks to the influential physician, Broussais (1772-1838), an army surgeon in Napoleon's campaigns. As a result of his influence, every hospital in France was using enormous numbers of leeches. For example, at Val-de-Grace, Paris, where Broussais was in charge, records show that more than 2 million were used in the 7 years from 1830 to 1826. In 1817, the French government levied an import duty on leeches and, consequently, reliable documentation of their importation is available--over 1,000,000,000 leeches were imported into France alone in the 19th century!

Very few animals have suffered such overcollection for so long (2). Consequently, by the beginning of the last century, the leech populations in western Europe were already depleted, so that the demand was met by importing them from eastern Europe. In fact, the leech trade at this time was sizeable and many people earned a living by collecting and transporting them. By the beginning of the 20th century, the European medicinal leech *Hirudo medicinalis* was in serious decline. In 1910, *Hirudo* was declared extinct in the British Isles. Today, we know it still survives in a few isolated pockets in Britain, but is extinct in Ireland where it was once common. One by one, the same pattern was repeated in other European countries until, in 1983, *Hirudo medicinalis* was listed in the *Red Data Book* of endangered species (3). The reasons for the threatened status of the medicinal leech are problematic. Clearly, overcollection played a major role, but changes in farming practices undoubtedly altered favorable habitats. Furthermore, a concomitant decline in European frog populations may have exacerbated the problem. The hatchling leech has such weak teeth that the first blood meal must be on a frog.

The therapeutic interest in the leech probably had its origins in some accidental discovery in prehistoric times. That the practice has persisted over the centuries to the present day gives credence to a nucleus of truth lying behind the application of leeches. Like so many prescientific remedies, we cannot look to the ancient practitioners for a scientific explanation of their rationale. For many centuries, leeching was often simply a painless alternative form of bloodletting (4). However, by the heyday of leeching from 1820 to 1850, the leech had become a panacea, the aspirin of the day. They were used to treat almost everything from obesity to gout, to insomnia and headaches. It is very difficult for us to understand the "leech mania," but physicians generally believed the leech to be endowed with special abilities to extract bad blood, leaving good blood behind. Some sophisticated schools of thought even argued whether the leech should be applied to the same side as the injury or to the opposite side. Inevitably, it was this indiscriminate and wholesale use of leeches, rather than scientific proof, which resulted in the practice falling into general disrepute by the end of the last century. In recent years, however, there has been a revival in the medical application of leeches, especially by microsurgeons who use them to restore circulation to grafted tissue (5-8).

SCIENTIFIC BACKGROUND

The first scientific revelation that the leech contains a medically important substance occurred in 1884 and is attributable to John B. Haycraft, Professor of Physiology at the Welsh National School of Medicine (Fig. 1) (9). Like most medically oriented scientists at the time, Haycraft was aware of the well known observations that blood within the gut of leeches does not coagulate and that blood continues to flow for abnormally prolonged periods from the wound caused by leeches. He decided to look closer at the phenomenon in an attempt to answer the general medical question, "Why doesn't blood coagulate within the living blood vessels?"

Because of the nature of his experiments, which necessitated withdrawing blood from various mammals, an historic footnote reveals that in order to continue his work on the anticoagulant, Haycraft left Britain for Strasbourg. In his own words, "The difficulties in obtaining a licence to perform the experiments in England being very great, I availed myself of an invitation from my friend Prof. Schiedeberg, of Strasbourg, who kindly placed his laboratory at my service." This explains why, in 1884, Haycraft published essentially the same paper in English (28 March) and in German (24 September). This paper reported on the discovery of the world's first anticoagulant, later called *hirudin*. Heparin from vertebrate mast cells came in 1916, and dicoumarol (warfarin) from moldy clover was produced in the late 1930s. Until the discovery of heparin, hirudin was the only substance which, in relatively small quantities, prevented the coagulation of blood without gross side effects. Even from its inception, hirudin has been used in analyses of the blood clotting process itself. In fact, the specific antithrombin property of hirudin was used extensively in the now accepted Morawitz-Fuld model of blood coagulation (developed in 1904).

In subsequent decades, attempts to isolate pure hirudin met with repeated technical difficulties. Finally, in 1955, following improved techniques in protein chemistry, hirudin was isolated by Professor F. Markwardt, then working in Greifswald, Germany. Markwardt is directly responsible for stimulating a new era of studies on the biochemical nature of hirudin (10,11). We now know that hirudin is a small peptide comprised of 65 amino acids (12) and is an extremely

Figure 1 Photograph of Professor John Berry Haycraft (1857-1922), discoverer in 1884 of hirudin, the world's first anticoagulant isolated from the medicinal leech *Hirudo medicinalis*. Haycraft was the first Professor of Physiology at the University of Wales School of Medicine. (Photograph courtesy of the University College, Cardiff.)

specific inhibitor of thrombin (13). In the treatment of certain blood disorders, hirudin has been demonstrated to have advantages over heparin (14,15). Preliminary in vivo studies indicate no acute side effects with pure hirudin, and no evidence of antigenicity has emerged. Hirudin is discussed in detail in subsequent chapters in this volume so need not be discussed further. However, the leech story does not end here.

DIFFERENT SPECIES

The European medicinal leech *Hirudo medicinalis* is only one of approximately 650 species of leech (1). This species represents one of two major groups of leeches characterized by the presence of jaws (*Arhynchobdellida*). The other major group of leeches (*Rhynchobdellida*) feeds by means of a straw-like organ, the proboscis. This latter group is epitomized by the giant Amazonian leech *Haementeria ghilianii*, known only from two museum specimens collected in the 19th century (Fig. 2). In 1977, an expedition supported by the American Philosophical Society went to Amazonia in search of the giant leech (16). Animals brought back from this expedition were successfully bred in the laboratory of Professor Gunther Stent from the Molecular Biology Department, of the University of California, Berkeley, California, until enough salivary material was available to analyze what kind of blood clotting factor this leech contained.

In 1981, Professor Andrei Budzynski and colleagues at the Thrombosis Research Center at Temple University, Philadelphia, reported a remarkable new anticoagulant, called *hementin*, from the salivary gland of the Amazon leech (17,18). Hementin (Fig. 3) is an enzyme with potent and specific fibrinolytic properties, with unique cleavage sites on the fibrin (and fibrinogen) molecule (19). Furthermore, the *Haementeria* salivary gland itself has a number of useful properties, including representing a model for the experimental study of the relationship between secretory cell activity and subsequent secretion (20).

One significance of this work on *Haementeria* is the demonstration that different leech species have solved the problem of feeding on the blood of mammals in different ways. Hirudin from *Hirudo medicinalis* is not an enzyme, but is a thrombin inhibitor of low molecular weight (7000). On the other hand, hementin from *Haementeria ghilianti* is an enzyme with a molecular weight of 120,000. In other words, the medicinal leech prevents the blood from clotting, whereas the Amazon leech dissolves blood clots after they form by degrading fibrin.

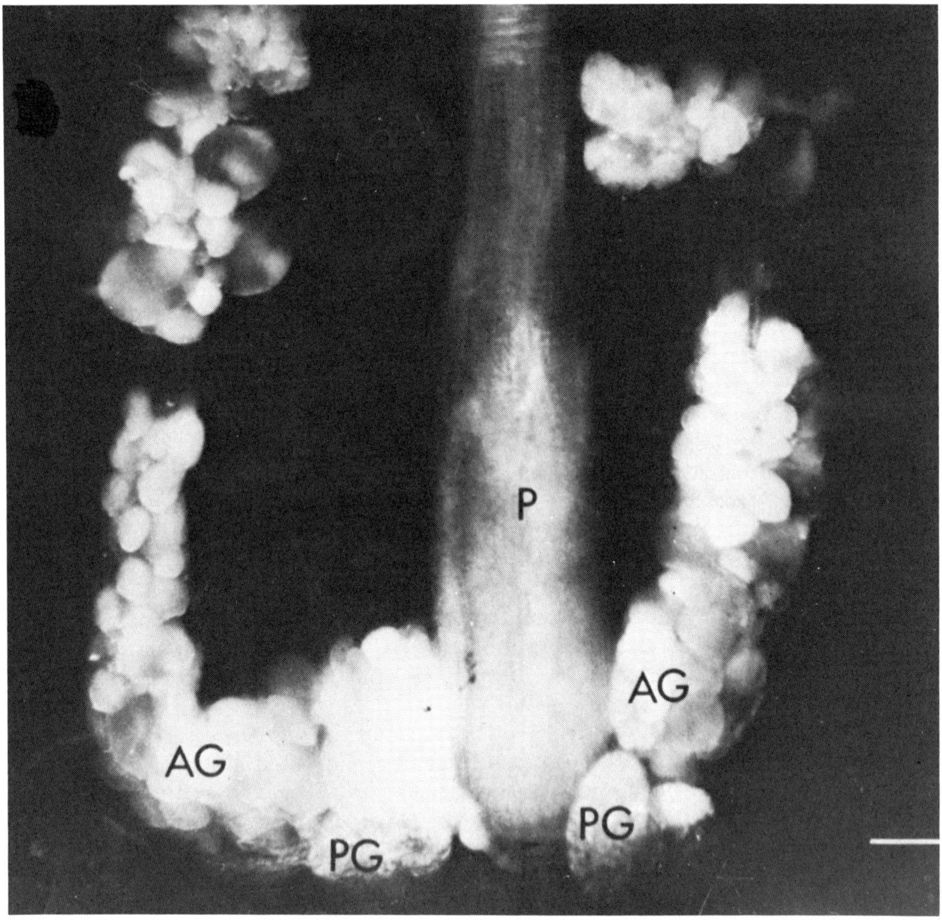

Figure 2 Salivary gland complex of the Amazon leech *Haementeria ghilianii*. AG = anterior gland; P = proboscis; PG = posterior gland. Scale bar = 1 mm.

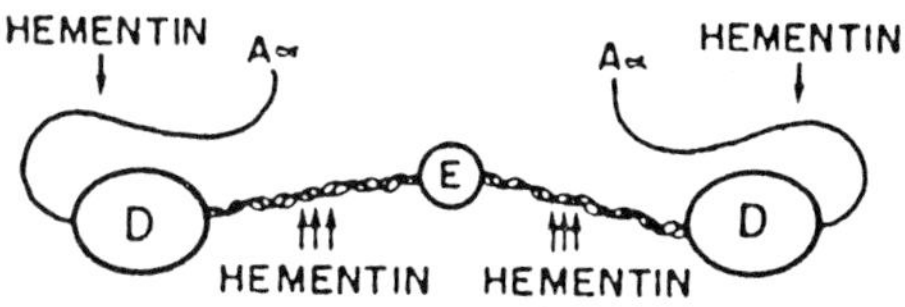

Figure 3 The enzyme hementin from the salivary gland of the Amazon leech *Haementeria ghilianii* specifically degrades fibrin and fibrinogen at one or more sites indicated by arrows. (From reference 10, with permission.)

OTHER ACTIONS

In recent years, a number of other biologically active substances have been found in the salivary glands of leeches, including a local anesthetic, vasodilator, hyaluronidase, antielastase, and antitrypsin. Our current understanding of these substances and their biological significance are reviewed in detail by Sawyer (1), who showed that in the biological context, these secretions are concerned primarily with aspects of the bloodsucking process itself and have little, if anything, to do with digestion. After attaching to an appropriate host, the leech begins the process of sucking blood for which it is well adapted. From the beginning of a bite, diverse salivary secretions play important roles. An anesthetic prepares the area for a bite, while other secretions maximize penetration of the secretions around the bite (spreading factor or hyaluronidase, commercially available under the trade name Orgelase) (Fig. 4) and flow of blood by dilation of blood vessels in the vicinity of the bite (vasodilator). Species-specific anticoagulants prevent or reverse the coagulation of blood in the crop. Antibiotics (from endosymbiotic bacteria) prevent putrefaction of blood prior to digestion in the intestine.

The recent scientific revelation that the leech is a living pharmacopeia containing a diversity of small proteins of low antigenicity has once again focused attention on the leech. Sadly, the medicinal leech, *Hirudo medicinalis* is today endangered, thus threatening full exploitation of the medical potential contained in the leech salivary gland. To overcome this problem of supply and to protect the remaining natural populations of the medicinal leech, Biopharm has set up a leech farm in Wales, from which it makes hirudin, hementin, Orgelase and other pharmacologically active leech proteins available to researchers worldwide.

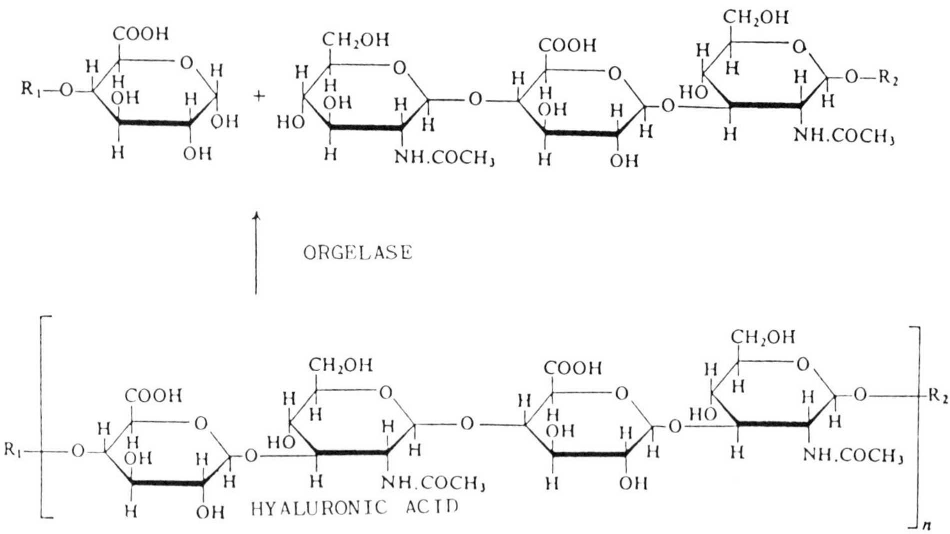

Figure 4 Leech hyaluronidase (Orgelase) is a potent spreading factor or dispersal agent in the skin. This enzyme is specific for hyaluronic acid, an important constituent of connective tissue.

ACKNOWLEDGMENTS

Thanks are expressed to Dr. L. A. Moritz, Vice-Principal, University College, Cardiff, and to Dr. R. F. Mottram, Physiology Department, University College, Cardiff, for their assistance in obtaining information about Professor Haycraft.

REFERENCES

1. Sawyer, R. T. *Leech Biology and Behaviour, Volumes 1-3*, Oxford University Press, Oxford (1986).
2. Sawyer, R. T. Oryx 16:165 (1981).
3. Wells, S. M., R. M. Pyle, and N. M. Collins. In *The IUCN Invertebrate Red Data Book*, International Union for Conservation of Nature and Natural Resources, Monitoring Centre, Cambridge, England (1983).
4. Turk, J. L. and E. Allen. Ann. Roy. Coll. Surg. Eng. 65:128 (1983).
5. Bunker, T. D. Injury: Br. Acc. Surg. 12:430 (1980).
6. Derganc, M. and F. Zdravic. J. Plast. Surg. 13:187 (1960).
7. Henderson, H. P., B. Matti, A. G. Laing, S. Morelli, and L. Sully. Br. J. Plast. Surg. 36:235 (1983).

8. Rao, P., F. B. Bailie, and B. N. Bailey. Practitioner 229:901 (1985).
9. Haycraft, J. B. Proc. R. Soc. Lond. 36:478 (1884).
10. Markwardt, F. Naturwiss. 42:537 (1955).
11. Markwardt, F. Z. Physiol. Chem. 308:147 (1957).
12. Dodt, J., H. P. Müller, U. Seemüller, and J. Y. Chang. FEBS Lett. 165:180 (1984).
13. Markwardt, F. In *Methods in Enzymology, Volume 19*, G. E. Perlmann and L. Lorand (Eds.). Academic Press, New York, London (1970).
14. Ishikawa, A., R. Hafter, U. Seemüller, and H. Gräff. Thromb. Res. 19:351 (1980).
15. Markwardt, F., G. Nowak, and J. Hoffmann. Thromb. Res. 11:275 (1977).
16. Sawyer, R. T. Yearbook Am. Phil. Soc. 212 (1978).
17. Budzynski, A. Z., S. A. Olexa, and R. T. Sawyer. Soc. Exp. Biol. Med. 168:259 (1981).
18. Budzynski, A. Z., S. A. Olexa, B. S. Brizuela, R. T. Sawyer, and G. S. Stent. Proc. Soc. Exp. Biol. Med. 168:266 (1981).
19. Malinconico, S. M., J. B. Katz, and A. Z. Budzynski. J. Lab. Clin. Med. 103:44 (1984).
20. Jones, P. G., R. T. Sawyer, and M. S. Berry. Nature 315:679 (1985).

21

α-Thrombin Specificity Inferred from the Hirudin-Thrombin Interaction

Jui-yoa Chang
Pharmaceuticals Research Laboratories, Ciba-Geigy Ltd., Basel, Switzerland

INTRODUCTION

α-Thrombin is a trypsin-like serine proteinase that catalyzes the final step of blood coagulation (1-3). In this reaction, α-thrombin converts fibrinogen to fibrin by selective cleavage of two Arg-Gly bonds from a total of 181 arginyl and lysyl bonds within fibrinogen (4). In addition, α-thrombin also cleaves a large number of polypeptide substrates with high specificities (Table 1) (1,5). The molecular basis of these proteolytic specificities is still not completely clear. However, mechanisms that govern the selective cleavages of fibrinogen and nonfibrinogen substrates are apparently controlled at two different levels. Selective cleavages of nonfibrinogen substrates require only the intact active site region of α-thrombin, whereas selective cleavages of fibrinopeptides A and B (FPA and FPB) require, in addition to the active site, an independent recognition site (1,3,5).

There are also protein proteinase inhibitors that are specific for α-thrombin (3). One, which has been well characterized, is hirudin (3,7-10), an inhibitor isolated from the bloodsucking leech *Hirudo medicinalis*. Hirudin possesses several unique properties that are useful for probing the molecular mechanism of α-thrombin specificity.

(a) Hirudin is an α-thrombin-specific inhibitor and is the most potent thrombin inhibitor known (7).
(b) Binding of hirudin to α-thrombin also includes a site that is independent of the reactive site-active site region (3,9,10). This independent binding region has exceedingly high binding

Table 1 α-Thrombin Susceptible Arg/Lys-Xaa Bonds

P_5	P_4	P_3	P_2	P_1 ↓	P_1'	P_2'	P_3'	P_4'	P_5'	Substrate[a]
V	R	G	P	R	V	V	E	R	H	Fibrinogen
E	V	I	P	R	S	G	G	S	T	Bovine prothrombin
F	F	N	P	R	T	F	G	S	G	Human prethrombin 2
D	D	A	P	R	A	V	F	P	S	Actin
V	G	R	P	R	A	V	F	P	S	Actin
P	L	N	P	K	A	N	R	E	K	Actin
D	V	T	P	R	A	G	Q	I	L	Ovine growth hormone
D	G	T	P	R	A	G	Q	I	L	Bovine growth hormone
D	G	S	P	R	T	G	Q	L	F	Human growth hormone
P	L	R	P	R	C	R	P	I		Human chorionic gonadotropin
I	V	T	P	R	T	P	P	P	S	Rabbit myelin basic protein
G	F	L	P	R	H	R	D	T	G	Rabbit myelin basic protein
K	H	V	P	R	A	V	F	V	D	Tublin
R	I	R	P	K	L	K	W	D	N	Dynorphin
A	M	A	P	R	E	R	K	A	G	Somatostatin-28
Q	T	Y	P	R	T	N	T	G	S	Salmon calcitonin
Q	T	Y	P	R	T	D	V	G	A	Eel calcitonin
P	L	A	P	R	D	A	G	S		PTH fragment
N	T	L	P	R	T	F	G	G	G	Antibody κ-chain
G	T	A	P	R	L	L	I	Y	S	Antibody λ-chain
G	K	A	P	K	L	M	I	Y	E	Antibody λ-chain
L	G	Q	P	K	A	A	P	S	V	Antibody λ-chain
L	G	Q	P	K	S	S	P	S	V	Antibody λ-chain
P	G	C	P	R	G	V	N	P		Human chorionic gonadotropin
Q	G	V	P	R	G	V	B	L	Z	Human factor XIII
G	L	V	P	R	G	F	B	P	Z	Human factor XIII
K	M	Y	P	R	G	N	H	W	A	Gastron-releasing peptide
G	G	G	V	R	G	P	R	V	V	Fibrinogen

Table 1 (*Continued*)

P_5	P_4	P_3	P_2	P_1 ↓	P_1'	P_2'	P_3'	P_4'	P_5'	Substrate[a]
F	F	S	A	R	G	H	R	P	L	Fibrinogen
A	A	Q	Q	R	G	W	V	T	D	Apolipoprotein CIII
Q	A	W	I	R	G	C	R	L		Lysozyme
A	D	V	G	K	G	Q	P	S	V	Human α-thrombin
Q	G	K	G	R	G	T	V	L	S	Rabbit myelin basic protein
A	I	E	G	R	T	S	E	D	H	Bovine prothrombin
Y	I	E	G	R	I	V	E	G	Q	Bovine prethrombin 2
Y	I	D	G	R	I	V	E	G	S	Human prethrombin 2
V	R	I	G	K	H	S	R	T	R	Bovine α-thrombin
V	I	A	G	R	S	L	N	P	N	Antithrombin III
G	Y	G	G	R	A	A	D	Y	K	Rabbit myelin basic protein
D	S	I	G	R	F	F	S	S	D	Rabbit myelin basic protein
A	P	S	G	R	V	S	M	I	K	Cholecystokinin
Q	N	P	G	K	A	P	K	L	M	Antibody λ-chain
F	D	I	G	R	N	S	V	N	W	Antibody λ-chain
L	S	R	L	R	D	S	A	R	L	Secretin
D	I	L	I	R	L	F	K	S	H	Myoglobin
G	V	Y	A	R	V	T	A	L	V	Chymotrypsinogen A
G	I	Q	V	R	L	G	Q	D	N	Trypsinogen
D	L	L	V	R	I	G	K	H	S	Huamn α-thrombin
A	E	I	F	R	A	S	G	E	H	Troponin
T	V	L	S	R	F	S	W	G	A	Rabbit myelin basic protein
V	S	M	I	K	N	L	Q	S	L	Cholecystokinin
N	F	F	W	K	T	F	T	S	C	Somatostatin-28
S	P	I	V	K	S	F	N	R	N	Antibody κ-chain
G	V	P	A	R	F	S	G	S	L	Antibody λ-chain

[a]References for the listed substrates are documented in reference 5. Arrows indicate the thrombin cleavage sites.

affinity. Experiments have shown that diisopropylphosphoryl-thrombin binds to hirudin almost as tightly as the native α-thrombin does (3,10).

(c) Binding of hirudin to α-thrombin can prevent tryptic conversion of α-thrombin to β- and γ-thrombin (10).

The structure of hirudin is also unique (Fig. 1).

(a) Hirudin is a relatively small protein (65 amino acids) with three disulfide linkages located within the first 39 N-terminal residues (8,11).

(b) Hirudin has a highly acidic C-terminal segment (8). There are seven acidic amino acids within the last 13 C-terminal residues.

(c) Hirudin contains three lysines and no arginine, and one of the lysines is flanked by two prolines (8). These interesting aspects suggest that the highly acidic C-terminal segment may constitute a functional domain, which is exposed on the surface of hirudin and interacts with the protein recognition site of α-thrombin.

This work presented here was undertaken to investigate the functional role of the acidic C-terminal segment of native hirudin, as well as the proteolytic specificity of α-thrombin.

MATERIALS AND METHODS

Human α-thrombin was kindly provided by Dr. J. W. Fenton II (Albany, New York). Hirudin from leeches was kindly provided by Dr. U. Seemüller (Munich, West Germany). The sources of synthetic peptide hormones and antibody light chains have been described elsewhere (5,6). Trypsin (TPCK-treated) and carboxypeptidase Y were purchased from Worthington. Staphylococcal protease V8 was obtained from Miles Laboratories. Chromozym TH (Tos-Gly-Pro-Arg-pNA) was from Boehringer.

Quantitative N-terminal analysis was carried out with the DABITC method (12,13). Amino acid sequence was analyzed by both manual DABITC/PITC method (14) or automatic sequenator (15). Amino acid composition was determined by the DABS-Cl precolumn derivatization method (16). C-terminal sequencing was performed by the combined use of carboxypeptidase Y and DABS-Cl methods (16). The inhibitory activity of hirudin on α-thrombin was analyzed by using chromozym TH as substrate. Carboxyl groups of nonfibrinogen polypeptide substrates were specifically modified with glycine amide (17).

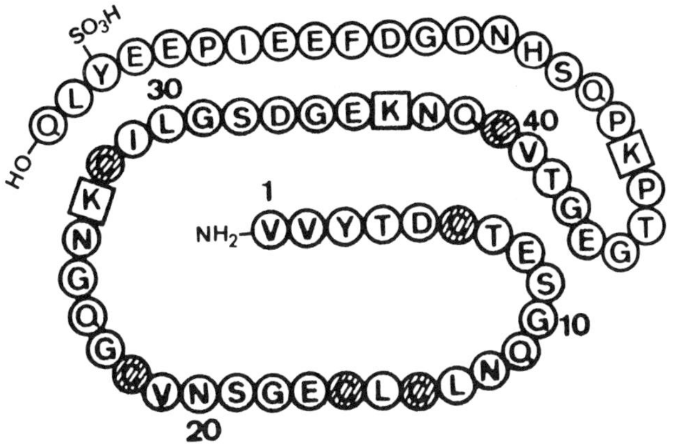

Figure 1 The primary structure (8) and the proposed domain structure (9) of hirudin. Hirudin contains a compact N-terminal domain and an acidic C-terminal domain, which is exposed on the surface of the inhibitor molecule.

RESULTS

Access of Acidic C-Terminal of Native Hirudin Molecule to Endopeptidase and Exopeptidase Digestion

The native hirudin molecule was subjected to limited digestion by aminopeptidase M, carboxypeptidase Y, trypsin, and staphylococcal protease V8.

(a) Native hirudin was not digested by aminopeptidase M after 4 hours of incubation at room temperature with the enzyme:substrate ratio of 1:100. This result suggests that the amino terminus of native hirudin is sterically inaccessible to aminopeptidase M.

(b) Carboxypeptidase Y releases the C-terminal amino acids of both native hirudin and reduced carboxymethylated hirudin with essentially similar efficiencies (Fig. 2), evidence that the C-terminal region of native hirudin is exposed on the surface of the inhibitor molecule.

(c) Reduced carboxymethylated and oxidized hirudins were quantitatively cleaved by trypsin at Lys-Cys (residues 27-28) and Lys-Asn (residues 36-37) bonds. However, native hirudin was not digested by trypsin at all. These results indicate that both Lys-Cys and Lys-Asn bonds are buried within a compact domain in the native hirudin.

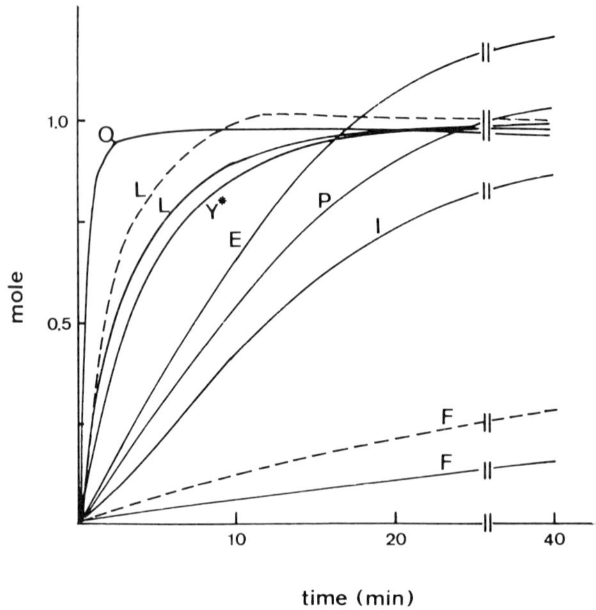

Figure 2 Release rates of the C-terminal amino acids of native hirudin (solid curves) and reduced carboxymethylated hirudin (broken curves, only Leu and Phe are presented) by carboxypeptidase Y. Enzyme digestions were carried out in acetate buffer (pH 5.4) with the weight ratio of enzyme to substrate of 1:10 at room temperature. The digestions were stopped by freeze-drying, and the released amino acids were quantitatively determined by the DABS-Cl method (16). Y* = tyrosine sulfate.

(d) When native hirudin was digested by staphylococcal protease V8, only those Glu-Xaa bonds, which are located within the last 22 C-terminal residues, were found to be cleaved. Amino acid sequence analysis of a 1 hour-digested native hirudin indicates a 45% cleavage at Gly-Gly (residues 43-44), a 16% cleavage at Glu-Glu (residues 57-58), a 17% cleavage at Glu-Ile (residues 58-59), and a 31% cleavage at Glu-Glu (residues 61-62).

These results together support the model that hirudin contains a compact amino-terminal domain held together by three disulfide linkages and an acidic C-terminal domain exposed on the surface of the inhibitor molecule (Figs. 1 and 3).

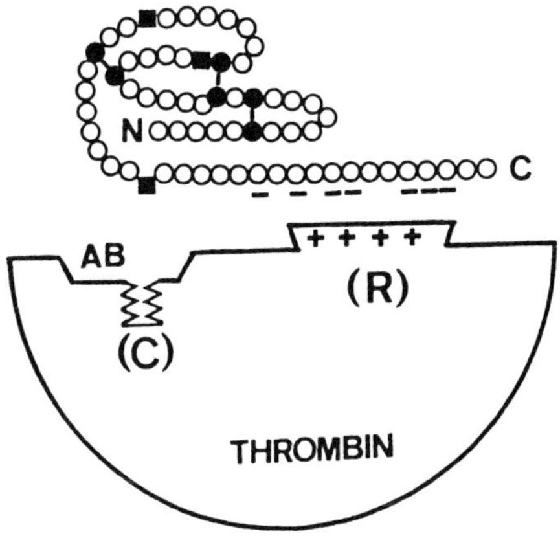

Figure 3 The proposed binding model of hirudin-thrombin complex. α-Thrombin contains a constrained catalytic site (C) (5), a hydrophobic binding pocket (AB), and a protein recognition site (R), which is comprised of clusters of basic amino acids including those within residues 62 to 73 of the B-chain. Binding of the acidic C-terminal domain of hirudin to the basic recognition site of α-thrombin resulted in a stable inhibitor-enzyme complex.

Intactness of the Acidic C-Terminal Segment of Native Hirudin and Its Inhibitory Activity

The function of the acidic C-terminal segment of native hirudin was investigated by chemical and enzymatic modifications.

(a) hirudin with residue 63 (tyrosine sulfate) desulfated,
(b) hirudin with the last seven C-terminal amino acids removed by carboxypeptidase Y, and
(c) hirudin with the last 22 C-terminal amino acids removed by staphylococcal protease V8.

The structures and remaining inhibition activities of those modified hirudins are summarized in Table 2. Interpretations of those data are as follows.

(a) The acidic tyrosine sulfate at residue 63 is required for hirudin activity. However, it probably accounts for no more than 30% of the total activity.
(b) Removal of the last seven C-terminal amino acids (including three acidic amino acids) resulted in a loss of at least 90% of hirudin's inhibitory activity.
(c) Removal of the last 22 C-terminal amino acids resulted in a nearly quantitative loss of hirudin inhibition activity.

Table 2 The Structure and Surviving Activity of Modified Hirudin

Treatment of hirudin[a]	Structure[a]	Surviving activity[b]
Native hirudin	12% desulfated	100%
30% TFA, 50°C, 25 min	46% desulfated	89%
30% TFA, 60°C, 15 min	60% desultated	87%
30% TFA, 70°C, 15 min	93% desulfated	73%
30% TFA, 80°C, 15 min	100% desulfated	45%
Carboxypeptidase Y, 20 min	At least 95% of hirudin with the last five C-terminal amino acids removed (see Fig. 2)	8%
Carboxypeptidase Y, 40 min	See Figure 2	7%
Staphylococcal protease, 20 min	Not determined	42%
Staphylococcal protease, 1 hour	45% of the C-terminal 22 amino acids deleted	27%
Staphylococcal protease 2 hours	74% of the C-terminal 22 amino acids deleted	16%
Staphylococcal protease 4 hours	Not determined	3%

[a]Three types of modified hirudin were preapred. (a) Four samples of 1.5 nmol of native hirudin were incubated with 20 μl of 30% aqueous trifluoroacetic acid under varied conditions. The samples were freeze-dried and redissolved in 30 μl of water. One-third of the sample (0.5 nmol) was removed for C-terminal analysis (see Fig. 3) to determine the extent of desulfation. The remaining samples were used for bioassay. (b) Two samples of native hirudin (1.5 nmol each) were digested with 1.5 μg of carboxypeptidase Y in 30 μl of acetate buffer, pH 5.4, at room temperature for 20 and 40 minutes. Ten μl (0.5 nmol) of each sample was removed for the analysis of the released C-terminal amino acids, and the remaining samples were freeze-dried and used for bioassay. A blank sample (only with 1.5 μg of carboxypeptidase Y) was carried out as a control experiment. (c) Four samples (2 nmol each) of native hirudin were incubated with 2 μg of staphylococcal protease in 40 μl of 0.1 M ammonium bicarbonate, pH 8.0, at 37°C for 20 minutes, 1 hour, 2 hours, and 4 hours. Fifty percent of each sample was removed for four-step N-terminal sequencing to determine the cleavage sites, and the remaining samples were used for bioassay. A blank sample (only contains 2 μg of staphylococcal protease) was carried out as a control experiment.

[b]The inhibitory activity of native hirudin (12% desulfated) was taken as 100%. The data are taken from reference 9.

These experiments demonstrate that the intactness of the C-terminal domain is essential for hirudin inhibition activity.

α-Thrombin-Specific Cleavages of Nonfibrinogen Polypeptide Substrates (5,6)

Approximately 40 nonfibrinogen polypeptides and their derivatives were digested by α-thrombin. The specific cleavage sites and their cleavage rates were determined by quantitative N-terminal analysis and amino acid sequencing. Together with the documented α-thrombin-susceptible peptide bonds, these nonfibrinogen substrates can be grouped according to the structures adjacent to their cleavage sites (Table 1 and Discussion section). Results of two further experiments are particularly interesting.

(a) All nonfibrinogen polypeptide substrates studied in this laboratory were cleaved by α-thrombin and autolyzed α-thrombin (mixtures of α-, β-, and γ-thrombins) with essentially similar efficiencies and specificities (unpublished results).

(b) An acidic amino acid at P1' or P2' positions is shown to hinder critically the α-thrombin susceptibility of those with optimal structures at the P1 to P4 positions. This retarded susceptibility can be drastically recovered when these acidic residues are chemically modified with glycine amide. An interesting example is shown in Figure 4.

DISCUSSION

Delineation of the structural requirement for the efficient release of FPA was extensively investigated with the fragments of the fibrinogen α-chain (18,19) and synthetic peptides that contain varied portions of FPA (20,21). These studies have shown that residues 1-23 of the Aα chain contain the fundamental structural elements required for recognition and FPA release by α-thrombin. In particular, a hydrophobic residue at the P9 position (Phe) (19) and acidic residues at P10 (Asp) (20) and P14 (phosphoserine) (21) were shown to be crucial for efficient cleavage of the Arg-Gly bond in the Aα-chain.

A large number of nonfibrinogen polypeptide substrates are also selectively cleaved by α-thrombin (Table 1) (1,5,6). These nonfibrinogen substrates provide a vast source of information for unraveling the mechanism of α-thrombin specificity. However, one major difficulty in drawing conclusions from Table 1 is the lack of quantitative data. Only approximately 25% of the selective cleavages listed in Table 1 were demonstrated quantitatively. Nonetheless, the structural re-

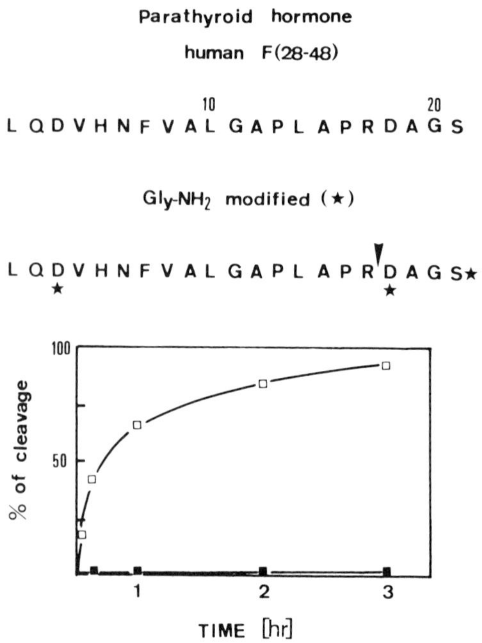

Figure 4 Chemical modification (with glycine amide) at the acidic P1' residue enhances the efficiency of α-thrombin cleavage of human parathyroid hormone fragment. Cleavage rate of unmodified peptide (-■-) and cleavage rate of modified peptide (-□-) at Arg-Asp bond are shown.

quirements for an α-thrombin-specific peptide bond can be tentatively summarized as follows.

(a) α-Thrombin has an Arg preference over Lys.
(b) Those with Pro at the P2 position and hydrophobic residues at the P3 and P4 positions are preferential α-thrombin substrates.
(c) Those with Gly at the P2 or P1' positions are also preferential α-thrombin substrates.
(d) Those with acidic residues at the P1' or P2' positions, despite their optimal structures at the P1 to P4 positions, are poor α-thrombin substrates. In addition, a Ser or Thr residue at the P1' position may also enhance the α-thrombin susceptibility for those arignyl and lysyl bonds that have the above-mentioned structures.

These rules are, of course, only qualitative. However, they have been successfully applied to the selective cleavages of numerous antibody light chains and denatured human antithrombin III (unpublished results).

One interesting question relating to both fibrinogen and nonfibrinogen substrates is whether their selective cleavages by α-thrombin share the same mechanism. Available evidence suggests that they are governed at two different levels and that selective cleavages of nonfibrinogen substrates (Table 1) are apparently controlled only by the active site region of α-thrombin. This evidence may be summarized as follows.

(a) α-Thrombin cleaves fibrinogen much more efficiently than any other nonfibrinogen substrate (10,18,20). Even the best documented nonfibrinogen substrate (gastrin-releasing peptide) was cleaved by α-thrombin with the kcat/Km value only approximately 5 to 10% of that of fibrinogen α-chain (5).

(b) There are derivatives of α-thrombin that have no clotting activity, but that retain full activity toward nonfibrinogen substrates (1,10,22-24). One notable example is β-thrombin. β-Thrombin is derived from α-thrombin through limited tryptic digestion or autolysis at Arg/Lys-Xaa bonds located within residues 62 to 73 of the B-chain. In both bovine and human α-thrombin, at least five different Arg/Lys-Xaa bonds within residues 62 to 73 of the B-chain have been shown to be susceptible to autolysis during α- to β-conversion (22-24). β-Thrombin is thus comprised of fragments that are associated through noncovalent binding. β-Thrombin is stable only in the alkaline buffer (pH 8). In the acidic solution, β-thrombin loses its proteolytic activity, and fragments can be separated by gel filtration or high pressure liquid chromatography.

The specific interaction between hirudin and α-thrombin suggests a basis for the clotting activity of α-thrombin (3,10). Hirudin binds to α-thrombin more tightly than to β- or γ-thrombin (10). Hirudin binding to α-thrombin also prevents tryptic conversion of α- to β- and γ-thrombin (10). These observations suggest that the region of residues 62 to 73 in the B-chain of α-thrombin is somehow protected by hirudin in the hirudin-α-thrombin complex. Since the size of hirudin is only approximately 20% of that of α-thrombin (7,8), this protection most likely involves a direct binding. In addition, since the protected site in α-thrombin is basic, the corresponding binding site in hirudin is most likely an acidic segment. Hirudin does contain a highly acidic segment at its C-terminal region (9). There are seven acidic amino acids including one posttranslationally modified tyrosine sulfate within the last 13 C-terminal amino acids. Evidence (Table 2 and Fig.

FPA (dog)						T	N	S	K	-	**E**	G	**E**	F	I	A	**E**	G	G	G	V	R	G
FPA (sheep)			A	**D**	**D**	S	**D**	P	V	-	G	G	**E**	F	L	A	**E**	G	G	G	V	R	G
FPA (cat)						G	**D**	V	Q	-	**E**	G	**E**	F	I	A	**E**	G	G	G	V	R	G
FPA (horse)								T	**E**	-	**E**	G	**E**	F	L	H	**E**	G	G	G	V	R	G
FPA (pig)						A	**E**	V	Q	**D**	K	G	**E**	F	L	A	**E**	G	G	G	V	R	G
FPA (bovine)			**E**	**D**	G	S	**D**	P	A	-	S	G	**D**	F	L	T	**E**	G	G	G	V	R	G
FPA (monkey)						A	**D**	T	G	-	**E**	G	**D**	F	L	A	**E**	G	G	G	V	R	G
HIRUDIN				Q	L	**Y***	**E**	**E**	P	I	**E**	**E**	F	**D**	G	**D**	N	H	S	Q	P	K	P
FPB (dog)				H	Y	**Y***	**D**	**D**	T	**D**	**E**	**E**	**E**	R	I	V	S	T	V	**D**	A	R	G
FPB (sheep)		G	Y	L	**D**	**Y***	**D**	-	**E**	V	**D**	**D**	N	R	A	K	L	P	L	**D**	A	R	G
FPB (cat)		I	I	**D**	**Y***	**Y***	**D**	-	**E**	G	**E**	**E**	**D**	R	**D**	V	G	V	V	**D**	A	R	G
FPB (horse)				L	**D**	**Y***	**D**	H	**E**	**E**	**E**	**D**	G	R	T	K	V	T	F	**D**	A	R	G
FPB (pig)			A	I	**D**	**Y***	**D**	-	**E**	**D**	**E**	**D**	G	R	P	K	V	H	V	**D**	A	R	G
FPB (bovine)	Z	F	P	T	**D**	**Y***	**D**	-	**E**	G	Q	**D**	**D**	R	P	K	V	G	L	G	A	R	G

Figure 5 Structure similarity between the acidic C-terminal segment (20 C-terminal residues) of hirudin and FPA and FPB. Notice that the hirudin is aligned from the C-terminal toward the N-terminal end. The acidic amino acids Asp (D), Glu (E), and tyrosine sulfate (Y*) are boxed.

2) also shows that this acidic segment is located on the surface of the hirudin molecule (9) and that intactness of this acidic C-terminal segment is essential for the inhibitory function of hirudin (9).

The most interesting feature of this acidic C-terminal segment is its intriguing structural similarity to both FPA and FPB (Fig. 5) (9). If one aligns the last 20 C-terminal amino acids (from the C-terminal toward the N-terminal end) with FPA and FPB (from the N-terminal toward the C-terminal end), one observes an interesting sequence similarity between the two regions (Fig. 5). However, aligned with the selective α-thrombin cleavage sites of fibrinogen α- and β-chains is a possible reactive site of hirudin composed of a Lys flanked with two prolines, hardly an optimal structure, not only because α-thrombin has an Arg preference, but also because a Pro at the P1' position is uncommon for a protein inhibitor of serine proteinase. On the other hand, evidence has shown that the major binding site between hirudin and α-thrombin is independent of the reactive site-active site region (i.e., the DIP-thrombin binds to hirudin as tight as the native α-thrombin does) (3,10). The similarity between hirudin and the fibrinopeptides lies in the acidic clusters approximately 6 to 15 amino acids distant from the α-thrombin cleavage sites. In addition to Asp and Glu, there is a posttranslationally modified tyrosine sulfate in FPA and FPB as well as in hirudin.

This intriguing structural similarity raises the possibility that α-thrombin may recognize both hirudin and fibrinogen through those acidic regions (Fig. 3). This hypothesis is further sustained by the findings that Asp at the P10 position (20) and phosphoserine at the P14 position (21) of human fibrinogen α-chain are crucial for the binding affinity of fibrinopeptide derivatives to the α-thrombin. The recognition site of α-thrombin is comprised of basic amino acids, including those located at residues 62 to 73 of the B-chain. Disruption of the recognition site in α-thrombin is accompanied by the loss of clotting activity and binding affinity to hirudin (Fig. 3).

REFERENCES

1. Fenton, J. W., II. Ann. N.Y. Acad. Sci. 267:469 (1981).
2. Lundblad, R. L., H. S. Kingdon, and K. G. Mann. Methods Enzymol. 45:156 (1976).
3. Magnusson, S. In *The Proteins*, Edition 3, P. D. Boyer (Ed.), 3:277 (1972).
4. Blombäck, B., M. Blombäck, B. Hessel, and S. Iwanaga. Nature (London) 215:1445 (1967).
5. Chang, J.-Y. Eur. J. Biochem. 151:217 (1985).
6. Chang, J.-Y., S. S. Alkan, N. Hilschmann, and D. G. Braun. Eur. J. Biochem. 151:226 (1985).

7. Markwardt, F. Methods Enzymol. 19:924 (1971).
8. Dodt, J., H.-P. Müller, U. Seemüller, and J.-Y. Chang. FEBS Lett. 165:180 (1984).
9. Chang, J.-Y. FEBS Lett. 164:307 (1983).
10. Fenton, J. W., II, B. H. Landis, D. A. Walz, D. H. Bing, R. D. Feinmann, M. P. Zabinski, S. A. Sonder, L. J. Berliner, and J. S. Finlayson. In *The Chemistry and Physiology of Human Plasma Protein*, Bind, D. H. (Ed.), Pergamon Press, Oxford, pp. 151-183.
11. Dodt, J., U. Seemüller, R. Maschler, and H. Fritz. Biol. Chem. Hoppe-Seyler 366:397 (1985).
12. Chang, J.-Y. Anal. Biochem. 102:384 (1980).
13. Chang, J.-Y. Methods Enzymol. 91:79 (1983).
14. Chang, J.-Y. Methods Enzymol. 91:455 (1983).
15. Edman, P. and G. Begg. Eur. J. Biochem. 1:80 (1967).
16. Chang, J.-Y., R. Knecht, and D. G. Braun. Methods Enzymol. 91:41 (1983).
17. Carraway, K. L. and D. E. Koshland, Jr. Methods Enzymol. 25:616 (1972).
18. Hogg, D. H. and B. Blombäck. Thromb. Res. 12:953 (1978).
19. Blombäck, B., B. Hessel, D. H. Hogg, and G. Claesson. In *Chemistry and Biology of Thrombin* Lundblad, R. L., J. W. Fenton, II, and K. G. Mann (Eds.), Ann Arbor Scientific Publications, Ann Arbor, pp. 305-310 (1977).
20. Marsh, H. C., Y. C. Meinwald, T. W. Tannhauser, and H. A. Scheraga. Biochemistry 22:4170 (1983).
21. Hanna, L. S., H. A. Scheraga, C. W. Francis, and V. J. Marder. Biochemistry 23:4681 (1984).
22. Lundblad, R. L., C. M. Noyes, K. G. Mann, and H. S. Kingdon. J. Biol. Chem. 254:8524 (1979).
23. Fenton, J. W., II, B. H. Landis, D. A. Walz, and J. S. Finlayson. In *Chemistry and Biology of Thrombin*, Lundblad, R. L., J. W. Fenton II, and K. G. Mann (Eds.), Ann Arbor Scientific Publications, Ann Arbor, pp. 43-70 (1977).
24. Boissel, J.-P., B. L. Bonniec, M.-J. Rabiet, D. Labie, and J. Elion. J. Biol. Chem. 259:5691 (1984).

22

Studies on the Expression of a Synthetic Hirudin Gene in *Escherichia coli*

Michael Rieger,* Elke Fortkamp, Silvia Schweitzer, Gudrun Heisterberg-Moutses, and Reinhold Sommer
Gen-Bio-Tec, Heidelberg, Federal Republic of Germany

INTRODUCTION

For thousands of years, the leech *Hirudo medicinalis* has been used in medical therapy. Several polypeptides, including proteinase inhibitors, are expressed in the jugular glands of the animal and are secreted into the wound upon leeching. Hirudin, one of these polypeptides, prevents blood coagulation by inhibiting the enzymatic activity of α-thrombin, which plays a prominent role in the process of blood clotting. Hirudin is of considerable medical importance in antithrombotic therapy; it can be applied to inhibit blood coagulation (e.g., during hemodialysis and in the treatment of blood clotting disorders). Because of its physiologic properties, including highly specific binding to α-thrombin, rapid clearance, and very low antigenicity, hirudin is of great interest (for reviews see the preceding chapters by Markwardt and Chang).

So far, hirudin has not been used clinically because it could not be purified in sufficient amounts from its restricted source, the leech *Hirudo medicinalis*. To provide a constant source of the polypeptide, we synthesized chemically an artificial hirudin gene and expressed it in *Escherichia coli*.

CHEMICAL SYNTHESIS OF THE HIRUDIN GENE

Gene Design

The amino acid sequence of hirudin has been determined recently by Dodt and co-workers (1984). This sequence was translated into an

*Present affiliation: German Cancer Research Center, Heidelberg, Federal Republic of Germany

artificial DNA sequence. Concerning the use of codons, the bacterium *E. coli* discriminates between strongly and weakly expressed proteins (Grosjeans and Fiers, 1982). We chose codons typical for highly expressed polypeptides. Since the mature and active hirudin polypeptide starts with valine, an additional ATG/methionine start site was added at the amino terminus. This allowed the independent expression of the polypeptide or, alternatively, the construction of a hybrid protein and subsequent cleavage at the fusion point using cyanogen bromide. To avoid any translational readthrough products, two adjacent stop codons were added to the end of the gene. Two different restriction endonuclease sites at the ends of the gene allowed it to be inserted into appropriate plasmid vectors in a predetermined direction.

Strategy of Gene Synthesis

A new strategy for rapid DNA synthesis was developed (Fortkamp et al., 1986). This method greatly enhances both the speed of gene assembly and the yield of correct DNA fragments. In brief, this method involves assembly of the gene from the double-stranded DNA subfragments produced by chemical synthesis of single-stranded oligodeoxyribonucleotides and pairwise annealing. The dotted lines in Figure 1 demarcate the ends of these subfragments. The preformed fragments were easily joined by ligation at their short, overlapping, complementary sequences without further purification or subcloning.

Chemical Synthesis and Purification of the Oligonucleotides

Chemical synthesis of the oligodeoxyribonucleotides was performed by the phosphite method (Adams et al., 1983; McBride and Caruthers, 1983) using an automatic DNA synthesizer. The reaction products were thoroughly purified by ion exchange high-pressure liquid chromatography and polyacrylamide gel electrophoresis. Purification according to two different separation principles reduced all contaminants to a minimum, especially oligonucleotides less than full length, which may seriously affect the subsequent ligation and cloning steps.

Assembly of Fragments and Cloning of the Gene

The oligonucleotides were mixed pairwise and thermally annealed to yield double-stranded DNA subfragments. The whole gene was built up by ligating a mixture of equimolar amounts of the preformed subfragments. The plasmid vector pEMBL8+ (Dente et al., 1983) was chosen for the first steps of cloning and analysis, because it is favorable for selection, expression, and sequencing of inserted DNA. Under appropriate conditions, transcription and, hence, expression of inserted DNA can be derepressed by inducers such as isopropyl-β-D-

```
     asnsermetValValTyrThrAspCysThrGluSerGlyGlnAsnLeuCysLeuCysGluGlySerAsnValCys

1    AATTCTATGGTTGTTTACACTGACTGCACCGAATCTGGTCAGAACCTGTGCCTGTGCGAAGGTTCTAACGTTTGC
         GATACCAACAAATGTGACTGACGTGGCTTAGACCAGTCTTGGACACGGACACGCTTCCAAGATTGCAAACG

     GlyGlnGlyAsnLysCysIleLeuGlySerAspGlyGluLysAsnGlnCysValThrGlyGluGlyThrProLys

76   GGTCAGGGAAACAAATGCATCCTGGGTTCTGACGGTGAAAAGAACCAGTGCGTTACCGGTGAAGGTACTCCGAAA
     CCAGTCCCTTTGTTTACGTAGGACCCAAGACTGCCACTTTTCTTGGTCACGCAATGGCCACTTCCATGAGGCTTT

     ProGlnSerHisAsnAspGlyAspPheGluGluIleProGluGluTyrLeuGln******

151  CCGCAGTCTCACAACGACGGTGACTTCGAAGAAATCCCGGAAGAATACCTGCAATAGTAAGTGAGCGTCG
     GGCGTCAGAGTGTTGCTGCCACTGAAGCTTCTTTAGGGCCTTCTTATGGACGTTATCATTCACTCGCAGCCTAG
```

Figure 1 Synthesis strategy of the hirudin gene. The DNA double strand is shown together with the corresponding amino acid sequence (three-letter code). Dashes between the DNA strands symbolize base pairing during the annealing of complementary oligonucleotides. Dotted lines represent the borders of the subfragments that were ligated in a second step. The amino acid sequence of hirudin is represented by initial capital letters; preceding amino acids are initiated by lower case letters. Asterisks denote stop codons.

thiogalactoside (IPTG). Plasmid DNA was prepared by cleavage with the restriction endonucleases EcoRl and BamHl. The gene was ligated to the vector DNA in a forced direction using these two different restriction sites. The reading frame of the synthetic gene had been designed so that its insertion into the plasmid resulted in a fusion protein composed of the amino-terminal part of β-galactosidase with hirudin. Thus, we took advantage of the preexisting signal sequences encoded by the vector plasmid (i.e., the moderate *lac* UV5 promoter and its adjacent ribosomal binding site). Figure 2 shows the structure of the hybrid plasmid, which was named pRudi1. Appropriate *E. coli* cells were transformed (Cohen et al., 1972) by the ligation mixture, and subsequently screened for inserts in the vector plasmid on selective agar plates.

EXPRESSION OF THE SYNTHETIC GENE

E. coli Strains

The synthetic gene construct was expressed in *E. coli* W3110 (Brent and Ptashne, 1981), in OM214, which is a derivative of RR1 (Bolivar et al., 1977), and in several strains of the JM series (Yanisch-Perron et al., 1985). In all cases, we found no striking differences concerning the expression level of hirudin; therefore, no reference is made to the strain used in a particular experiment.

ELISA Screening

Bacterial colonies propagating plasmids with synthetic DNA inserts were isolated and investigated further for hirudin expression. The clones were grown in culture, induced, lysed, and subjected to an enzyme-linked immunosorbent assay (ELISA) using the method of Voller and associates (1976). In brief, sheep antihirudin antibody was adsorbed to microtiter plate wells, the supernatants of lysed *E. coli* cells were allowed to bind, and bound hirudin was detected using antihirudin IgG conjugated to alkaline phosphatase. More than 50% of the bacterial clones expressed hirudin antigenic determinants.

Thrombin Inhibition Assays

Encouraged by this result, we investigated the biological properties of the fusion polypeptide by examination of its inhibitory effects on α-thrombin activity. Extracts from *E. coli* cells harboring the recombinant plasmid were compared with authentic hirudin originating from the leech. In a first step, we scrutinized the inhibition of the esterolytic activity of α-thrombin. The chromogenic substrate Tos-Gly-Pro-Arg-4-nitroanilide acetate, releasing 4-nitroaniline upon

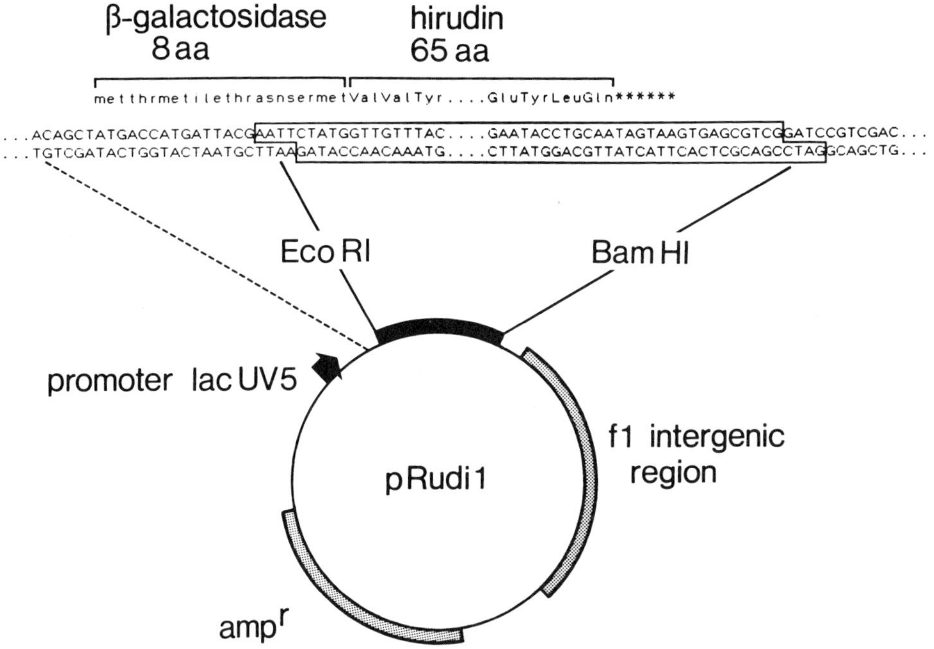

Figure 2 Structure of pRudi1. The synthetic hirudin gene (black bar) inserted into the polylinker site of pEMBL8+ (Dente et al., 1983) via its EcoRI and BamHI cloning sites is shown with part of the vector sequence as a magnification. The synthetic hirudin DNA double-strand (boxed) is abbreviated. The arrow indicates the direction of transcription. The corresponding polypeptide, given above the DNA sequence, is a fusion with eight amino acids of the β-galactosidase amino terminus preceding hirudin. Asterisks denote stop codons; amp^r denotes β-lactamase gene.

thrombin action, was used (Chang, 1983). The results are shown in Figure 3. Since control extracts from bacterial cells not propagating the chimeric plasmid were negative, the assay revealed an inhibitory effect of the biosynthetic polypeptide against α-thrombin. The next step was to check the inhibition of the biological activity of thrombin (i.e., the proteolytic cleavage of fibrinogen). The inhibitory activity of hirudin is measured by titrating a thrombin solution of standardized activity. Using this "clotting test," proteolysis of fibrinogen can be

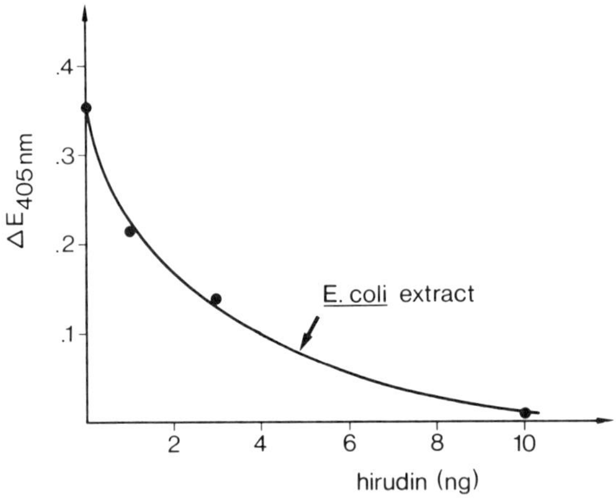

Figure 3 Inhibition of the α-thrombin esterolytic acitivity by an extract from *E. coli* cells harboring pRud1. Different amounts of hirudin were allowed to bind to 0.1 IU of thrombin in a total volume of 960 μl triethanolamine buffer (0.1 ml/l, pH 8.4) at 25°C. After 5 minutes, 125 μl 1.5 mM Tos-Gly-Pro-Arg-4-nitroanilide acetate were added. The release of 4-nitroaniline was monitored at 405 nm after 5 minutes (ΔE_{405nm}). Black dots represent the test calibration using a standard of authentic hirudin. The value of the *E. coli* extract (arrow) corresponded to 50 μg/l of bacterial culture. No effects were observed using cells not harboring pRudi1.

recorded by the thrombin-catalyzed formation of a fibrin gel by inverting the tube after a given reaction time (Markwardt, 1970). The results reported in Figure 4 show, that an extract prepared from bacterial cells harboring the synthetic gene could also inhibit the fibrinogen-clotting activity of α-thrombin, whereas an *E. coli* control extract displayed no inhibition. Interestingly, the few amino acids of β-galactosidase preceding the hirudin sequence in the fusion polypeptide encoded by pRudi1 did not prevent thrombin inhibition. However, the expression of hirudin in pRudi1 only corresponded to 50 μg/l of bacterial culture. Possible explanations for this phenomenon are: low efficiency of transcription (depending on the control region of the gene), low efficiency of translation (depending on the ribosome binding site), or rapid degradation of the protein in the cell.

Expression Under the Control of the *tac* Promoter

Since it was apparent that the biological activity of hirudin can be expressed in bacterial cells, experiments were performed to enhance

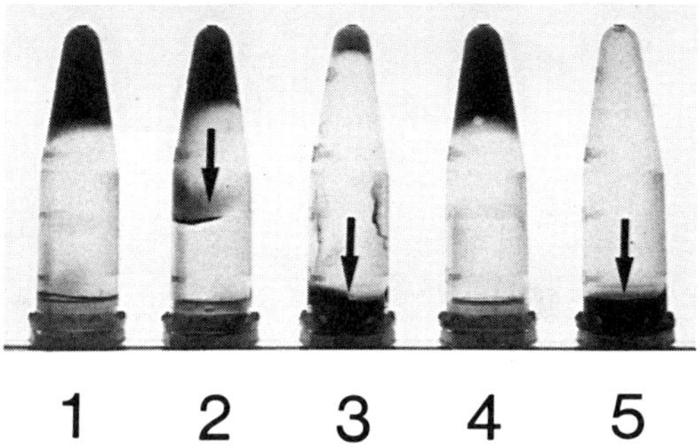

Figure 4 Inhibition of the fibrinogen-clotting activity of α-thrombin by recombinant hirudin. Fibrinogen (20 mg/ml final concentration) was mixed with either buffer (tube 1), 3 ng or 12 ng authentic hirudin (tubes 2 and 3, respectively), or extracts from *E. coli* cells without (tube 4) or with (tube 5) the plasmid pRudi1. Traces of bromophenol blue allowed better reading of the results. After addition of thrombin (0.15 IU), the mixture was incubated at room temperature. After 1 minute, the tubes were inverted. The formation of a stable fibrin clot (tubes 1 and 4) showed unaffected α-thrombin activity, whereas thrombin inhibition could be monitored when the solution flowed down after tube inversion (tubes 2, 3, and 5, arrows).

its production using stronger signal sequences and different amino-terminal fusion moieties. To achieve this, we made several constructs, which are summarized in Figure 5. The uppermost part of Figure 5 schematically shows the fusion polypeptide of pRudi1, already described.

The hybrid *tac* promoter is known to be approximately 10 times stronger when compared with the *lac* UV5 promoter (Amann et al., 1983; DeBoer et al., 1983). It is a hybrid of the -35 region of the *trp* promoter and the -10 region, operator, and ribosomal binding site of the *lac* UV5 promoter. This IPTG-inducible control region is contained in the plasmid vector pKK223-3. The synthetic hirudin gene was inserted directly behind the signal sequences so that only the initial methionine preceded the synthetic hirudin sequence. Lysates from *E. coli* cells harboring the recombinant plasmid were tested against α-thrombin, as described above. However, despite the increased promoter strength, the yield of hirudin activity per liter of bacterial culture revealed essentially no increase. Additional experiments demonstrated that variations of the distance between the ribosomal binding site and the translational initiation codon, by deleting or inserting nucleotides, had no significant effect on the expression level (data not shown).

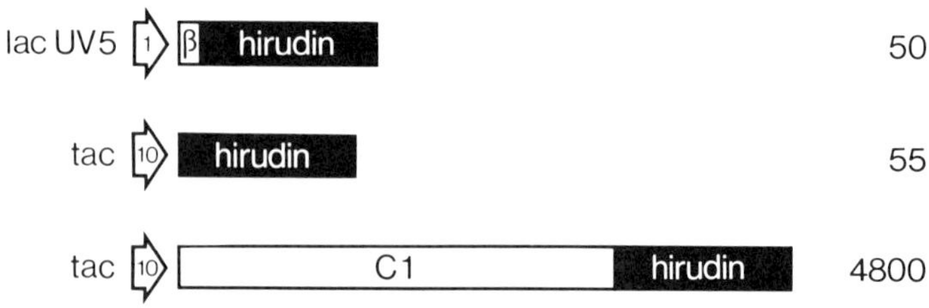

Figure 5 Constructs for hirudin expression. Three genes and their preceding control regions are drawn to scale schematically. The hirudin moiety is shown in black, and fused amino-terminal parts of bacterial genes are represented as white boxes. β = 24 base pairs of β-galactosidase; C1 = 480 base pairs of bacteriophage λ C1 repressor protein. The relative strengths of adjacent upstream promoter regions are indicated by numbers in the open arrows. For specification of the different promoters, see text. Numbers at the right represent the yield of active hirudin (μg/l of *E. coli* culture).

The conclusion drawn from these experiments was that the reported hirudin sequence cannot be sufficiently expressed in *E. coli* without further manipulation. Low expression rates may be due to instability of the corresponding mRNA or rapid degradation of the foreign protein in the cell. This is a well known phenomenon (Hershko and Ciechanover, 1982; Buell et al., 1985), and *E. coli* cells contain a protein degradation system that is even induced by the expression of foreign proteins (Goff and Goldberg, 1985). Extreme stability and high-level expression of synthetic genes encoding eucaryotic proteins in bacteria have, however, been reported (Ikehara et al., 1984; Kempe et al., 1985).

High Level Expression of Hirudin

Problems of instability can be circumvented by creating a fusion polypeptide with a considerable part consisting of an *E. coli* protein. The degradation of such hybrid proteins may be reduced because it is recognized by the cell as self. Following this strategy, we constructed a hybrid gene coding for a truncated coliphage λ C1 repressor protein (Amann et al., 1984) fused to hirudin in the same translational frame (Fig. 5). Again, the gene was expressed under the control of the *tac* promoter described above. *E. coli* cells propagating the corresponding plasmid were grown in liquid culture, induced with IPTG, harvested by centrifugation, and lysed. The fusion protein represented approximately 34% of the total cellular protein, as judged by a densitometer analysis of a $NaDodSO_4$-polyacrylamide gel stained with Coomassie brilliant blue (Fig. 6). Promoter strength also depends on

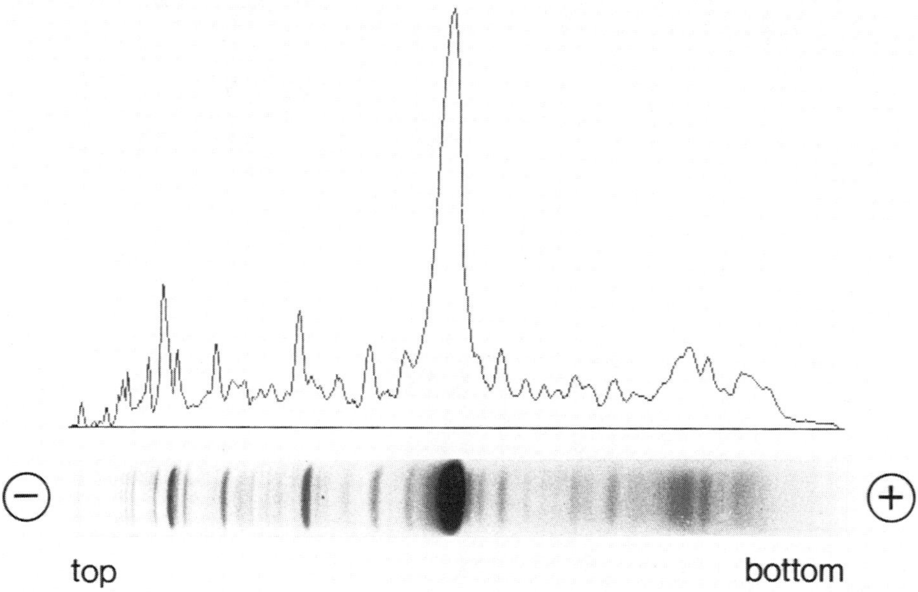

Figure 6 Densitometer scan of a $NaDodSO_4$-polyacrylamide gel of total proteins from *E. coli* cells harboring the plasmid-encoded C1-hirudin fusion. Bacteria were grown to an optical density of 1.8 at 600 nm and induced with IPTG. The cells were concentrated by centrifugation, boiled, and electrophoresed (Thomas and Kornberg, 1975). The gel was stained with Coomassie brilliant blue (lower part). The upper part of the figure shows a densitometer scan of the gel track at 601 nm. The main peak represented 33.9% of the whole area.

the adjacent sequences, which may confer a favorable three-dimensional structure to the messenger RNA, thus enhancing its translation. The data presented in Figures 5 and 6 show a striking example of different protein expression controlled by the same promoter.

Hirudin could be strongly expressed when fused to the preceding coliphage λ C1 protein, which stabilized the hybrid polypeptide in the cell. Despite its extraordinary expression, the fusion protein resided in the bacteria in a soluble form and did not precipitate into crystalloid formations, which often accompany high-level expression (Ikehara et al., 1984). This greatly facilitated the purification process of the protein from the cell lysate.

Hirudin could be liberated from the composite polypeptide by cyanogen bromide cleavage. The α-thrombin-inhibiting activity of the product was considerable in both test systems, corresponding to about 4.8

mg/l of bacterial culture. Furthermore, purified biosynthetic hirudin had a specific activity similar to hirudin from the leech (Fortkamp et al., 1986).

Authentic hirudin from *Hirudo medicinalis* carries a sulfate residue at tyrosine 63. It has been demonstrated that the molecule retains part of its biological activity upon chemical desulfation (Chang, 1983). This fully agrees with our results that biosynthetic hirudin, which lacks the sulfate residue, also strongly inhibits α-thrombin.

ACKNOWLEDGMENTS

We thank U. Retzlaff, W. Weinig, and J. Weinmann for expert technical assistance. We are grateful to Professor W. W. Franke, Professor E. Fink, and Dr. R. Maschler for helpful discussions. We thank Dr. F. Stewart for a critical reading of the manuscript. We are indebted to E. Amann for his gift of plasmid pMF1. We thank the Plantorgan Werk KG, Bad Zwischenahn for the gifts of authentic hirudin and antihirudin antibody and for financial support.

REFERENCES

Adams, S. P., K. S. Kavka, E. J. Wykes, S. B. Holder, and G. R. Galluppi. (1983). Hindered dialkylamino nucleoside phosphite reagents in the synthesis of two DNA 51-mers. J. Am. Chem. Soc. 105:661.

Amann, E. J. Brosius, and M. Ptashne. (1983). Vectors bearing a hybrid trp-lac promoter useful for regulated expression of cloned genes in *Escherichia coli*. Gene 25:167.

Amann, E., M. Bröker, and F. Wurm. (1984). Expression of *Herpes simplex* virus type I glycoprotein C antigens in *Escherichia coli*. Gene 32:203.

Bolivar, F., R. L. Rodriguez, P. J. Greene, M. C. Betlach, H. L. Heynecker, H. W. Boyer, J. H. Crosa, and S. Falcow. (1977). Construction and characterisation of new cloning vehicles. II. A multipurpose cloning system. Gene 2:95.

Brent, R. and M. Ptashne. (1981). Mechanisms of action of the *lexA* gene product. Proc. Natl. Acad. Sci. USA 78:4204.

Buell, G., M. F. Schulz, G. Selzer, A. Chollet, N. R. Movva, D. Semon, S. Escanez, and E. Kawashima. (1985). Optimizing the expression in *E. coli* of a synthetic gene encoding somatomedin-C (IGF-I). Nucleic Acids Res. 13:1923.

Chang, J. Y. (1983). The functional domain of hirudin, a thrombin-specific inhibitor. FEBS Lett. 164:307.

Cohen, S. N., A. C. Y. Chang, and L. Hsu (1972). Non-chromosomal antibiotic resistance in bacteria: genetic transformation of *Escherichia coli* by R. factor DNA. Proc. Natl. Acad. Sci. USA 69:2110

DeBoer, H. A., L. J. Comstock, and M. Vasser (1983). The *tac* promoter: a functional hybrid derived from the *trp* and *lac* promoters. Proc. Natl. Acad. Sci. USA 80:21.

Dente, L., G. Cesareni, and R. Cortese (1983). pEMBL: a new family of single stranded plasmids. Nucleic Acids Res. 11:1645.

Dodt, J., H. P. Müller, U. Seemüller, and J. Y. Chang (1984). The complete amino acid sequence of hirudin, a thrombin specific inhibitor. FEBS Lett. 165:180

Fortkamp, E., M. Rieger, G. Heisterberg-Moutses, S. Schweitzer, and R. Sommer (1986). Cloning and expression in *Escherichia coli* of a synthetic DNA for hirudin, the blood coagulation inhibitor in the leech. DNA 6:511.

Goff, S. A. and A. L. Goldberg (1985). Production of abnormal proteins in *E. coli* stimulates transcription of *lon* and other heat shock genes. Cell 41:587.

Grosjeans, H. and W. Fiers (1982). Preferential codon usage in procaryotic genes: the optimal codon-anticodon interaction energy and the selective codon usage in efficiently expressed genes. Gene 18:199.

Hershko, A. and A. Ciechanover (1982). Mechanisms of intracellular protein breakdown. Ann. Rev. Biochem. 51:335.

Ikehara, M., E. Ohtsuga, T. Tokunaga, Y. Taniyama, S. Iwai, K. Kitano, S. Miyamoto, T. Ohgi, Y. Sakuragawa, K. Fujiyama, T. Ikari, M. Kobayashi, T. Miyake, S. Shibahara, A. Ono, T. Ueda, T. Tanaka, H. Baba, T. Miki, A. Sakurai, T. Oishi, O. Chisaka, and K. Matsubara (1984). Synthesis of a gene for human growth hormone and its expression in *Escherichia coli*. Proc. Natl. Acad. Sci. USA 81:5956.

Kempe, T., S. B. H. Kent, F. Chow, S. M. Peterson, W. I. Sundquist, J. J. L'Italien, D. Harbrecht, D. Plunkett, and W. J. DeLorbe (1985). Multiple-copy genes: production and modification of monomeric peptides from large multimeric fusion proteins. Gene 39:239.

Markwardt, F. (1970). Hirudin as an inhibitor of thrombin. In *Methods in Enzymology, Volume 19*, Perlmann, G. E. and L. Lorand (Eds.), Academic Press, New York, pp. 924-932.

McBride, L. J. and M. H. Caruthers (1983). An investigation of several deoxynucleoside phosphoramidites useful for synthesizing deoxyoligonucleotides. Tetrahedron Lett. 24:245.

Thomas, J. O. and R. D. Kornberg (1975). An octamer of histones in chromatin and free in solution. Proc. Natl. Acad. Sci. USA 72:2626.

Voller, A., D. Bidwell, and A. Bartlett (1976). Microplate enzyme immunoassays for the immunodiagnosis of virus infections. In *Manual of Clinical Immunology,* Rose, N. R. and H. Friedman (Eds.), American Society of Microbiology, Washington, D.C., pp. 506-512.

Yanisch-Perron, C., J. Vieira, and J. Messing (1985). Improved M13 phage cloning vectors and host strains: nucleotide sequences of the M13mp8 and pUC19 vectors. Gene 33:103.

23

Proteinase Inhibitors of the Leech *Hirudo medicinalis*

Ernst Fink
Plantorgan Arzneimittel, Bad Zwischenahn, Federal Republic of Germany
Ursula Seemüller, Johannes Dodt,* and Hans Fritz
University of Munich, Munich, Federal Republic of Germany

INTRODUCTION

Severe trauma, inflammation, or infection causes the activation of humoral cascade systems and various inflammatory cells, especially phagocytes (1). During such an inflammatory response, large amounts of thrombin and plasmin may be liberated, leading to disseminated intravascular coagulation (DIC). In addition, production of anaphylatoxins (e.g., C3a and C5a) due to complement activation leads to stimulation or even degranulation of polymorphonuclear (PMN) granulocytes (neutrophils), whereby neutral lysosomal proteinases, such as PMN elastase and cathepsin G, are released extracellularly. These lysosomal proteinases can inactivate various humoral factors, such as clotting factors V, VIII, and XIII) and degrade structural proteins (elastin, collagen) by unspecific digestion. The same holds true for bacterial proteinases, such as subtilisin.

Normally, potent proteinase inhibitors are present to prevent pathological reactions by excessively liberated proteinases (2). For example, antithrombin III (AT III) regulates the activity of thrombin, α_2-plasmin inhibitor (α_2PI) plasmin, and α_1-proteinase inhibitor

---

Present affiliation:
*Institut fur Organische Chemie und Biochemie, Tednische Hochschule Darmstadt, Darmstadt, Germany

(α_1PI) and α_2-macroglobulin (α_2M) inhibit the neutral lysosomal proteinases (besides the pancreatic proteinases) if released extracellularly. If, however, these endogenous inhibitors are extensively consumed during the inflammatory response (e.g., besides by complex formation with their target enzymes, by oxidation by oxidants produced during phagocytosis [α_1PI], and by unspecific degradation by lysosomal elastase [AT III, α_2PI]), the liberated, uncontrolled proteinases may contribute significantly to the development of DIC, lung dysfunction (adult respiratory distress syndrome or shock lung), or even multiple organ failure (1).

Interestingly, the medical leech *Hirudo medicinalis* produces several miniproteins that can strongly inhibit these proteinases (Fig. 1). Hirudin specifically reacts with thrombin. Bdellins are strong inhibitors of plasmin besides trypsin and sperm acrosin, and eglins are very potent antagonists of lysosomal PMN elastase and PMN cathepsin G, pancreatic chymotrypsin, and bacterial subtilisin (3). These inhibitors could be used, therefore, therapeutically to assist the endogenous inhibitor potential during an acquired proteinase-proteinase inhibitor imbalance.

Proteinase Inhibitors of the Medical Leech

Inhibitor	M_r	Inhibition of: Thrombin	Plasmin	Trypsin	Acrosin	PMN Elastase, PMN Cathepsin G, Chymotrypsin, Subtilisin
Hirudin	7 000	**+**	-	-	-	-
Bdellin A	6 000	-	**+**	**+**	**+**	-
Bdellin B	4 800	-	**+**	+	**+**	-
Eglin	8 000	-	(+)	-	-	**+**

+ very strong, + medium strong, (+) weak, - negative

Figure 1 Inhibitory specificity of proteinase inhibitor miniproteins from the leech *Hirudo medicinalis*. Enzymes of human (thrombin, plasmin, neutrophil PMN elastase, and cathepsin G), bovine (thrombin, trypsin, chymotrypsin), and porcine (plasmin, boar sperm acrosin) origin were used. "Very strong inhibition" means that the K_i values of the enzyme-inhibitor complexes are close to or below 1×10^{-10} mol/l.

THE PROTEINASE INHIBITORS OF THE MEDICAL LEECH

Hirudin and Isohirudins

The complete primary structure of hirudin is shown in Figure 2 (4). The molecule consists of a single polypeptide chain of 65 amino acid residues linked by three disulfide bridges, which form three loops comprised of 9, 13, and 18 amino acid residues, respectively. The six cysteines are located within only 34 residues in the N-terminal region of the molecule, thus probably causing a very rigid structure of this part of the polypeptide chain. Also remarkable is the accumulation of acidic residues at the C-terminal part of the molecule and especially of sulfated tyrosine in position 63. Removal of the sulfate group as well as cleavage of amino acids from the carboxy terminus (up to positions 63 and 60 or 57) by appropriate enzymes revealed that neither sulfated tyrosine nor the acidic residues close to the carboxy terminus are important or indispensable for the specific interaction with thrombin (5). Instead, the major contact or recognition area for thrombin seems to be the peptide portion between positions 40 and 48, the sequence of which corresponds to that in positions 149 to 157 in prothrombin (6); the latter sequence contains the cleavage site for thrombin.

During hirudin purification from leech extracts, another thrombin inhibitor could be separated, having an isoleucine instead of a valine residue at the amino terminus (5,7). Sequence analysis revealed that this newly discovered inhibitor form, designated hirudin PA, is a true isoform of hirudin: 80% of the amino acid residues are in identical positions (Fig. 3). Especially remarkable is the conserved sequence at the postulated recognition area for thrombin (positions 40 to 48) (8) and an extension in hirudin PA close to the carboxy terminus. Both inhibitors (and some isoforms) as well as their desulfated forms inhibit thrombin equally well, the dissociation equilibrium constants K_i of their complexes with thrombin being close to 3×10^{-11} mol/l (hirudin, hirudin PA) or 4×10^{-10} mol/l (desulfated hirudin) (5). Recently, another isoform of hirudin has been described, being more closely related to hirudin PA than to hirudin (cf. Fig. 3) (9). Obviously, hirudin occurs in several isoforms in the medical leech. This could be due to gene duplication and/or mutation during evolution, so that a family of genes has developed that encodes hirudin isoinhibitors. However, single isoinhibitors could be also associated with distinct leech populations (7).

The hirudins are highly effective thrombin inhibitors in vitro and in vivo. They react rapidly with their target enzyme in a 1:1 molar ratio. The hirudin-thrombin complex formed is very stable under normal conditions (i.e., in the physiological pH range). It can be dissociated by acidification or heating, whereby thrombin is denatured and hirudin is released in its active form (10). Due to their relatively low molecular masses of 7045 daltons (hirudin) and

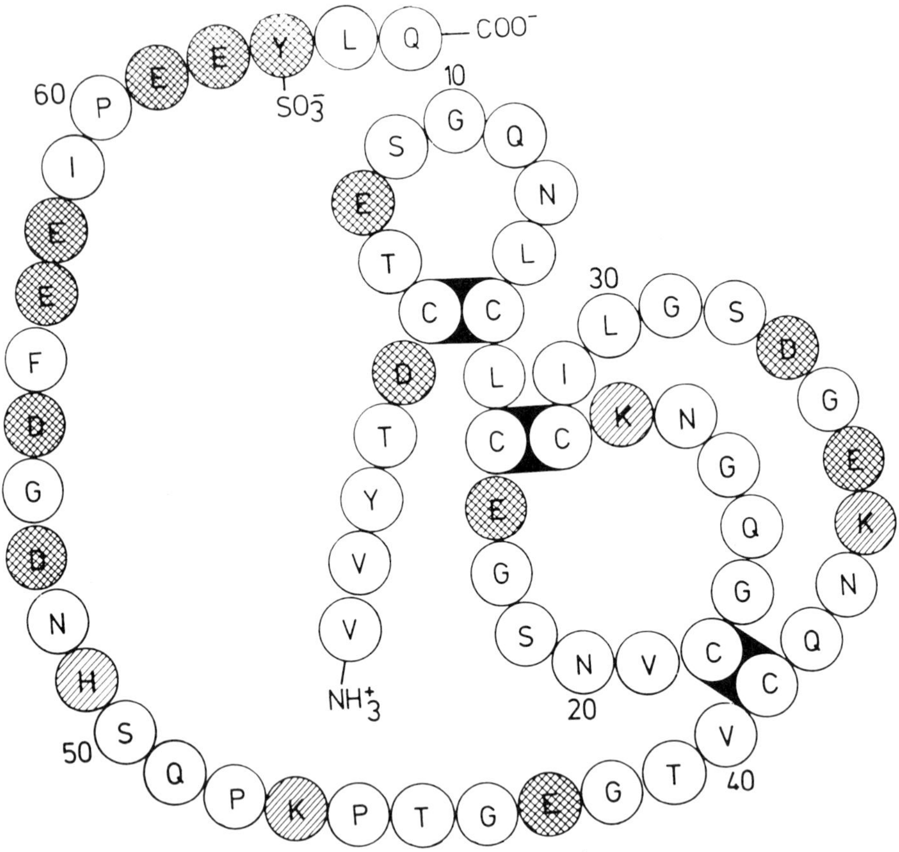

Figure 2 Primary structure of hirudin. Basic residues (H, K) and acidic residues (D, E, Y-SO_3) are indicated by distinct hatchings. (From Ref. 4.)

7087 daltons (hirudin PA), the hirudins diffuse rapidly between different compartments of an organism (10,11). This behavior as well as their high effectiveness in thrombin inhibition imply that suitable sensitive assays are available to follow the bioavailability of the hirudins after administration in vivo. For this purpose, we developed a very sensitive radioimmune bioassay (Fig. 4) (5). The sensitivity and assay range of this test, which recognizes exclusively inhibitorily active hirudins, is as low as 1 ng of hirudin per milliliter sample (i.e., approximately 0.002 ATE/ml) (cf. Fig. 4).

```
                     10                  20
Hirudin:      V V Y T D C T E S G Q N L C L C E G S N
Hirudin PA:   I T Y T D C T E S G Q N L C L C E G S N

                                30                  40
Hirudin:      V C G Q G N K C I L G S D G E K N Q C V
Hirudin PA:   V C G K G N K C I L G S Q G K D N Q C V

                                50                  60
Hirudin:      T G E G T P K P Q S H N D G D F E E I P
Hirudin PA:   T G E G T P K P Q S H N Q G D F E P I P

                    *                 *
Hirudin:      E E - Y L Q             Y = tyrosine-O-sulfate
                    *
Hirudin PA:   E D A Y D E
```

Figure 3 Amino acid sequences of hirudin and hirudin PA. Hirudin has a molecular mass of 7045 daltons, hirudin PA of 7087 daltons. Another isoform of hirudin sequenced very recently (9) differs from hirudin PA by the following amino acid exchanges: Asn 33 vs. Gln, Gly 36 vs. Asp, Asn 47 vs. Lys, Asn 53 vs. Gln, Glu 58 vs. Pro, and Glu 62 vs. Asp, which is followed directly (no Ala in position 63) by the three C-terminal residues of hirudin (Y*-L-Q). Furthermore, this inhibitor contains an elongation peptide at the amino terminus. (From Ref. 7.)

With this test, the plasma levels of hirudin in human subjects were followed after a single bolus injection of 1000 ATE/kg, and the effect on common clotting parameters was measured as a function of time (Fig. 5). Significant changes in these parameters could be observed in this individual up to 2 hours after administration of 3.5 mg hirudin. In addition, a clear dose-response relationship was observed between the amount of injected hirudin and the prolongation of partial thromboplastin time (PTT) and thrombin time (TT), the latter showing the most sensitive response (Fig. 6 and Table 1). Remarkably, most of the administered hirudin was eliminated within a relatively short time in the active form in urine (Table 1). The intravenous administration of the highly purified hirudin from leeches, a mixture of hirudin and hirudin PA, was tolerated by all individuals without any side effects.

Hirudin is rapidly eliminated from the circulation with a half-life ($t_{1/2}$) of approximately 0.6 to 1.5 hours, depending on the individual and amount of hirudin administered (see also the chapter by Markwardt elsewhere in this volume). This allows optimum control of hirudin

Radioimmuno-bioassay for Hirudin (H) with Thrombin (T)

$$\left.\begin{array}{l} H \underset{}{\overset{T}{\rightleftharpoons}} [H\text{-}T] \underset{}{\overset{IgG_T}{\rightleftharpoons}} [H\text{-}T]\text{-}IgG_T \\ {}^*H \qquad [{}^*H\text{-}T] \qquad\qquad [{}^*H\text{-}T]\text{-}IgG_T \end{array}\right\rangle \text{precipitate}$$

Purest Hirudin approx. 20 000 ATE/mg

Detection limits: 0.002 ATE ~ 1 ng

~ 0.04 ATE or ~ 2 ng/ml urine

~ 0.02 ATE or ~ 1 ng/ml plasma

Effective in vitro above approx.:	ATE/ml	ng/ml
• PTT (ellagic acid)	1	50
• platelet aggregation	0.5	25

Figure 4a Principle and detection limit of the radioimmune bioassay (RIBA) developed for the specific determination of inhibitorily active hirudin in biological samples. In the assay system, thrombin competes with hirudin (from the sample, H) and ^{125}I-hirudin (as tracer, *H) for complex formation. The mixture of complexes is precipitated by addition of thrombin-directed antibodies, and the amount of radioactivity present in the precipitate is counted. (For test evaluation see Figure 4b.) The high sensitivity of this RIBA is obvious by comparison with the lowest amounts of hirudin effective in inhibition of partial thromboplastin time (PTT, activated by ellagic acid) and platelet aggregation (see Table 1). The specific activity of hirudin (ATE/mg) may vary with the thrombin preparation applied to the inhibition assay. (From Ref. 5.)

plasma levels and, hence, the coagulation status simply by infusion. In contrast to heparin, hirudin does not need a cofactor to express its anticoagulant activity and might be, therefore, especially useful in cases with acquired antithrombin deficiency. Besides infusion, hirudin may also be administered subcutaneously, leading to measurable plasma levels in humans (Bichler et al., unpublished results). An interesting observation regarding local thrombotic events is the percutaneous absorption of hirudin after topical administration (Table 2).

So far, the major drawback to widespread clinical use of hirudin has been the limited amount available from natural sources. The successful production of hirudins by recombinant DNA technology by several groups will allow detailed studies in the near future on the antithrombotic effectiveness of this inhibitor (13).

PLATE 1

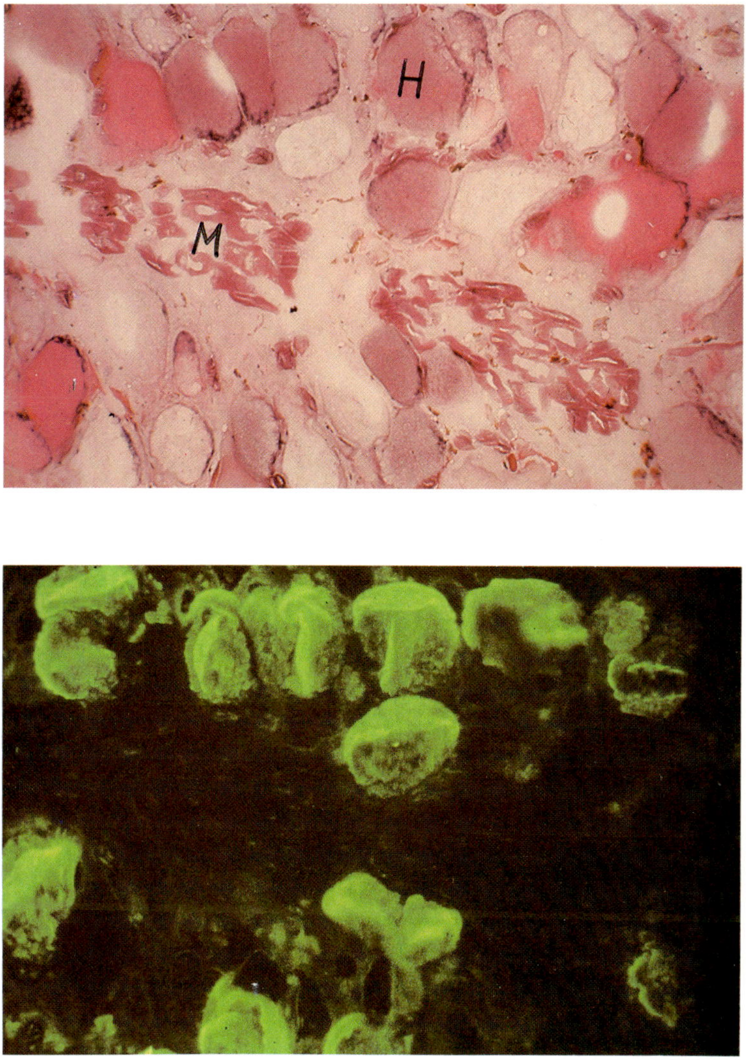

Figure 9 Allocation of immunofluorescent cells (lower part) to salivary gland cells of leech's throat (upper part) in parallel tissue sections. Immunofluorescence was observed only in areas binding hirudin-specific antibodies. The salivary gland cells were stained red with hemalum-eosin. H = salivary gland cells; M = muscle fibers. Magnification 131x. (From Ref. 27.)

PLATE 2

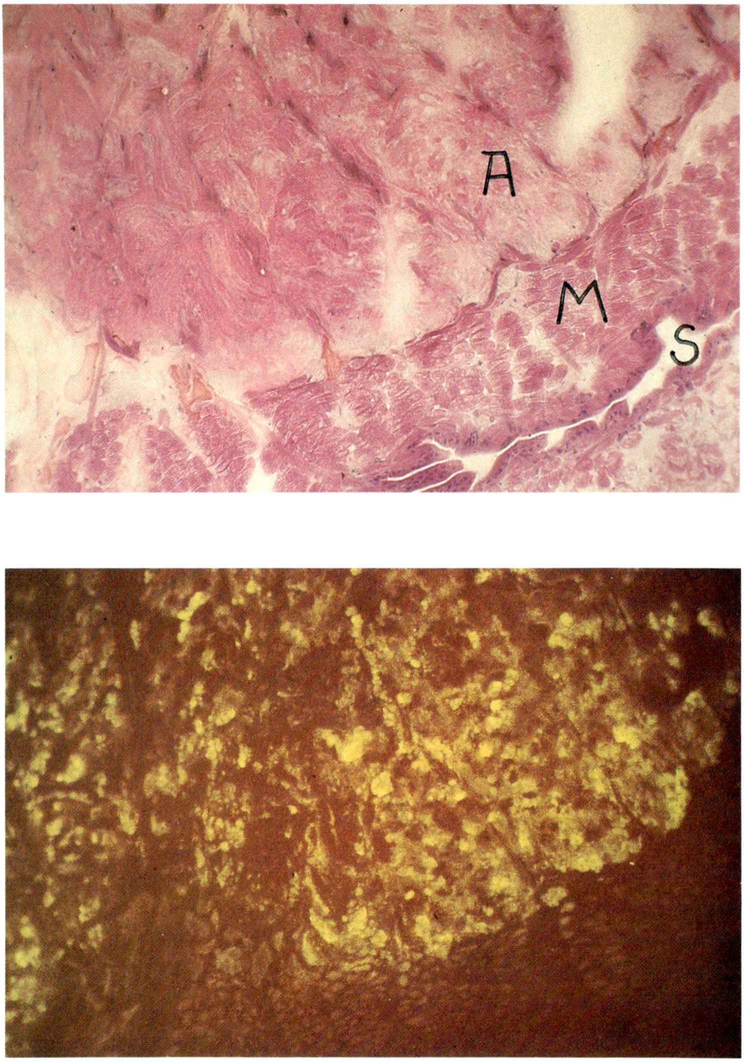

Figure 10 Allocation of immunofluorescence (lower part) to collecting ducts of salivary gland cells (upper part) in parallel tissue sections. Immunofluorescence was observed only in areas binding hirudin-specific antibodies. Histological identification was performed by staining with hemalum-eosin. A = collecting ducts; M = muscle fibers; S = cleft of oral cavity. Magnification 131x. (From Ref. 27.)

PLATE 3

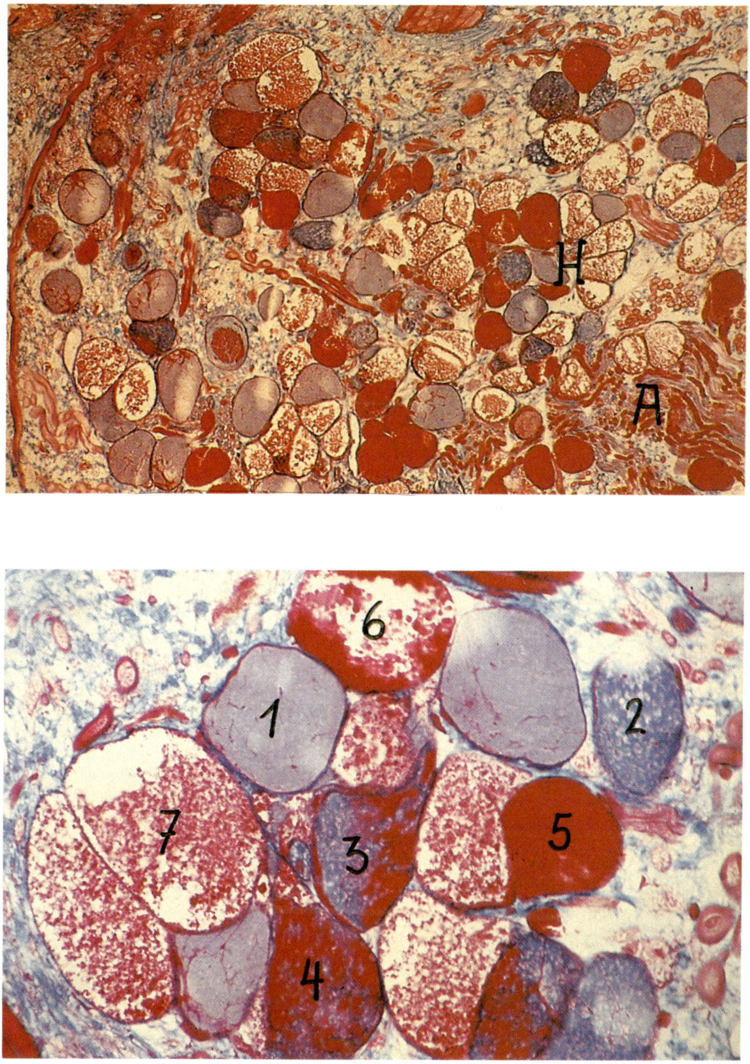

Figure 11 Salivary gland cells in the head part of the stained medical leech. The tissue sections were stained with Azan. The upper part shows a survey with salivary gland cells (H) and collecting ducts (A). Magnification 56x. The lower part represents salivary gland cells at different stages of development in higher magnification (338x). Histological staining allows discrimination of the following cell stages (colors): 1 = homogeneously bright-blue; 2 = granulated blue; 3 and 4 = mixed blue-red granulated; 5 = homogeneously deep-red; 6 = red-colored substances partly lost by secretion; 7 = fine-granulated red material left in "empty" cells. (From Ref. 27.)

PLATE 4

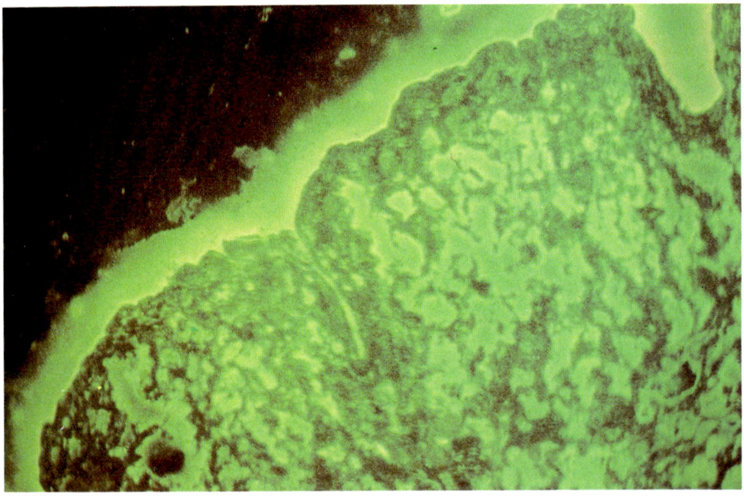

Figure 12 Localization of the bdellins and eglins in the skin area of the medical leech. With both bdellin- and eglin-specific antibodies intense fluorescence was observed in the saliva on the skin surface and beneath the skin border in mucus-producing cells (fully fluorescent) and in small blood vessels (dark lumen). Magnification 131x. (From Ref. 27.)

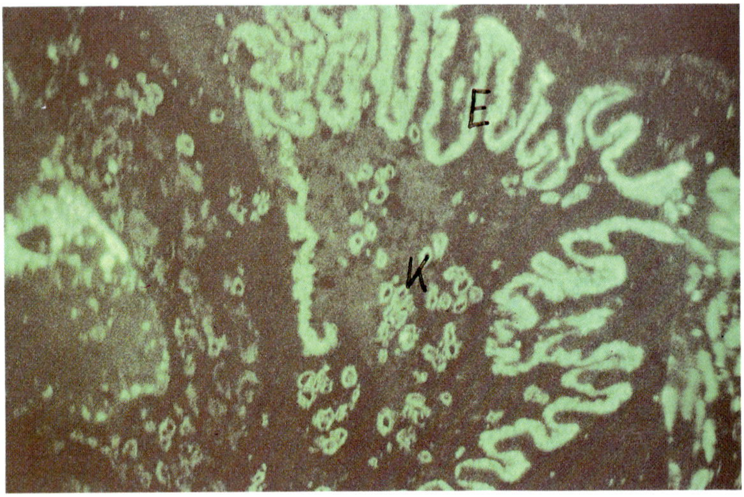

Figure 13 Localization of the bdellins and eglins in intestinal epithelium (E, strongly folded) and in the walls of small blood vessels (K). Intense fluorescence was observed with both bdellin- and eglin-specific antibodies. Magnification 131x. (From Ref. 27.)

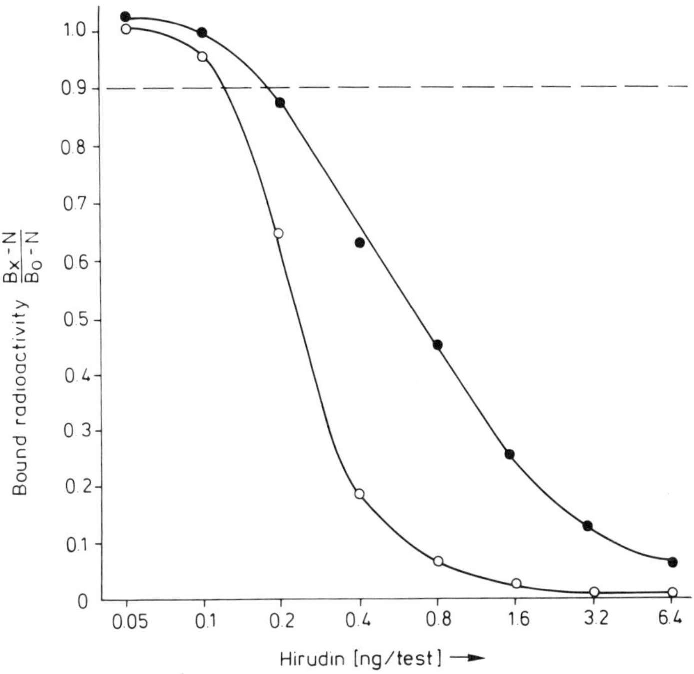

Figure 4b Typical standard curves for hirudin determination with the radioimmune bioassay (RIBA) (cf. Fig. 4a). Two methods were employed, differing in sensitivity: ●-● = tracer (*H) and hirudin sample (H) were mixed before addition of thrombin (T); ○-○ = hirudin sample was mixed with thrombin before addition of tracer; -- = this line marks the lower detection limit of the assay.

Bdellins

To date, two groups of trypsin-plasmin inhibitors bdellin A and bdellin B have been isolated from extracts of whole medical leeches (3,14). They are clearly distinguished by their chromatographic and electrophoretic behavior, as well as their amino acid compositions and molecular masses. Two typical representatives of each group are listed in Table 3. Their inhibitory specificity is restricted towards trypsin-like proteinases. Besides trypsin and sperm acrosin, porcine and human plasmin are strongly inhibited as equimolar enzyme-inhibitor complexes.

A partial amino acid sequence of a high molecular mass bdellin B-3 has been elucidated recently and comprises the inhibitorily active domain of Kazal-type structure (Fig. 7). Obviously, the bdellin B-3 molecule sequenced formerly (15) was a cleavage product of the high

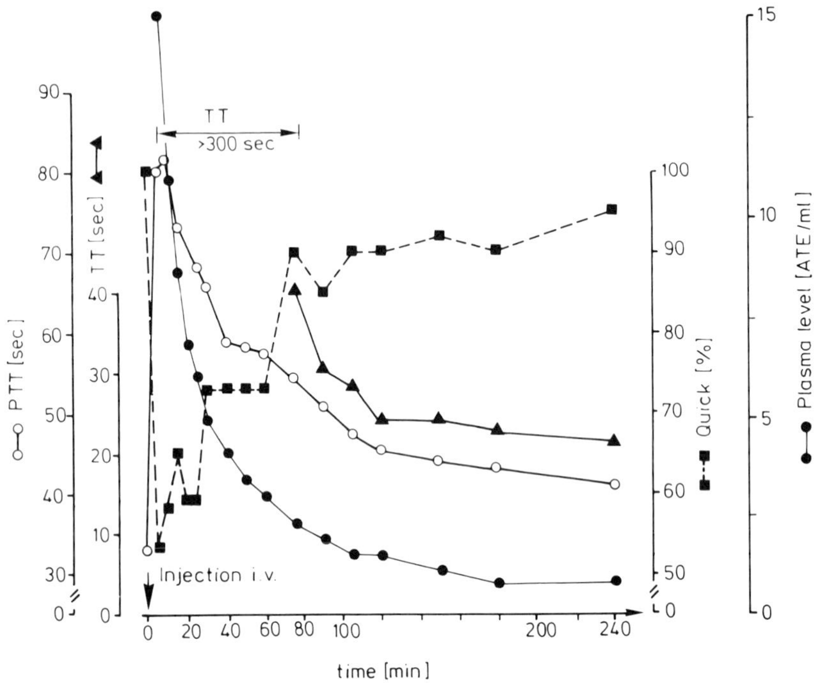

Figure 5 Plasma level of hirudin and its effect on PTT (partial thromboplastin time), TT (thrombin time), and Quick (prothrombin time) as a function of time after intravenous administration of the inhibitor in humans. Hirudin was injected as bolus of 1000 ATE/kg (i.e., approximately 3.5 mg highly purified inhibitor in a 70 kg individual). Hirudin, 1 ATE, corresponds to the inhibition of 1 NIH unit thrombin.

molecular mass form resulting from the use of affinity chromatography via insolubilized trypsin. The biological function of the bdellins is unknown. They occur in all areas of the body of the leech.

Eglins

Two isoforms of a miniprotein (eglin b and eglin c) with strong inhibitory activity against α-chymotrypsin, subtilisin, the neutrophil elastase, cathepsin G, as well as mast cell chymase have been isolated from extracts of whole leeches (16). They are composed of 70 amino acid residues with either histidine (eglin b) or tyrosine (eglin c) in position 35 (Fig. 8) (17,18). A pecularity among serine proteinase inhibitors is the lack of any cysteine residue and, hence, any stabilizing disulfide bridge in the inhibitor molecule (3,17).

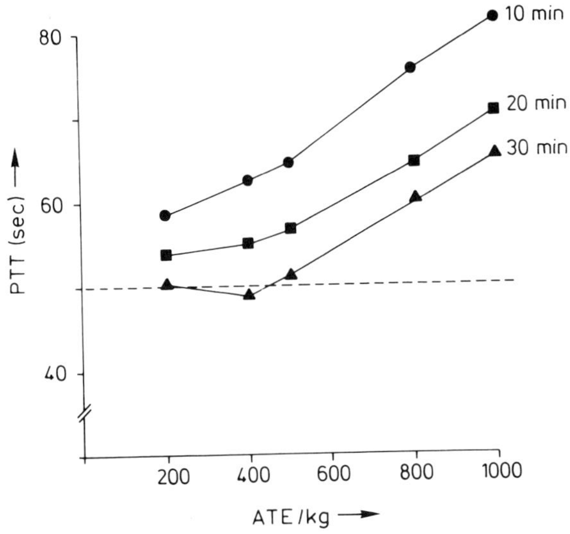

Figure 6 Effect of increasing amounts of intravenously injected hirudin (abscissa) on PTT (partial thromboplastin time) at certain time intervals (10, 20, 30 minutes) after inhibitor administration.

The residues at the reactive site peptide bond P_1-P_1' were identified as Leu and Asp in positions 45 and 46, respectively (3). Crystallographic data of the subtilisin-eglin complex obtained recently confirmed this observation and showed the extended contact area between eglin and subtilisin at a 1.2 Å resolution in great detail (19). Also interesting is the structural homology existing between eglin from leeches and inhibitors with similar specificity isolated from potatoes (44%) (20) and barley (33 and 36% (21).

Rapid and strong inhibition of lysosomal proteinases and especially of neutrophil elastase by eglin ($K_i < 10^{-10}$ mol/l) (22) suggested its use as antiinflammatory agent in experimental animal models. So far, a clear therapeutic effect could be demonstrated regarding the prevention of elastase-induced emphysema formation (23) and the development of shock lung during septicemia in pigs (24). These encouraging results justify the hope that eglin might become a therapeutically valuable antiinflammatory drug (25). The problem of the availability of sufficient inhibitor material (kg amounts) has been solved in the meantime by recombinant DNA technology. Rink and co-workers (26) succeeded in the expression of a synthetic eglin gene in *E. coli* in high yield (above 400 mg/l bacterial broth). The purified recombinant eglin (r-eglin) proved to be identical to authentic leech eglin, except that the α-amino group of the N-terminal threonine residue was

Table 1 Hirudin in Humans: Dose-Response Relationship and Renal Elimination

Bolus injection IV (ATE/kg)	1000	800	600	500	400	200
PT						
Max. value (%) up to (min)	59 (20)	50 (5)	70 (5)	64 (5)	69 (5)	69 (5)
Normal after (min)	60	30	<5	15	10	10
PTT						
Max. value (sec) up to (min)	82 (10)	80 (5)	60 (5)	65 (5)	67 (5)	64 (5)
Normal after (min)	90	75	30	30	30	30
TT						
Max. value (sec) up to (min)	40 (75)	54 (50)	59 (30)	56 (25)	49 (25)	40 (15)
Normal after (min)	120	105	60	60	50	30
Eliminated in urine (%)						
Up to 4 hr	46	71	96	83	83	70
Total	50	89	109	89	>83	88

Different amounts of hirudin, expressed in antithrombin units (ATE; 1 ATE corresponds to the inhibition of 1 NIH unit thrombin) were intravenously injected into different individuals. The clotting parameters prothrombin time, PTT (partial thromboplastin time), and TT (thrombin time) were measured in the time intervals shown in Figure 5. Maximum (max) values observed up to a certain time after injections were given, as well as the time intervals when normal values had been reached again. The amounts of hirudin eliminated in urine within 4 hours after injection and in total (up to 24 hours) are also listed.

Table 2 Percutaneous Absorption of Hirudin and Eglin Applied Topically*

Species/Model	Hirudin	Eglin	Urine	Test system
Pig skin	+†	+‡		E.I. and radiol.
Guinea pig	+§	+§	+	Immunological
Mouse	+	+	+	Radiol. and RIBA
Pig	+		+≠	Radiol. and RIBA
Human	+		+≠	RIBA

E.I. = enzyme inhibition (hirudin/thrombin, eglin/chymotrypsin); Immunological = by special skin test via sensibilization of mast cells by IgE antibodies; Radiol. = radioactivity after administration of labeled inhibitor; RIBA = radioimmune bioassay (cf. Fig. 3).

*For detailed penetration studies of pig skin, a pharmacological in vitro model (Sartorius chamber) was applied. In in vivo studies, ointments and gels containing hirudin plus eglin (Exhirud from Plantorgan) or hirudin alone were administered to all species listed. Details are given in several M.D. theses (dissertations), which may be obtained from the authors on request.

†7-12% and ‡0.5 to 12% penetrated within 48 hours (penetration is temperature and concentration dependent).

§Concentration maximum in serum after 70 minutes (immunol. test).

≠Only hirudin was estimated.

acetylated; this modification does not influence the inhibitory properties of eglin.

LOCALIZATION AND FUNCTION

The hirudins, bdellins, and eglins are not related immunologically. Specific antibodies against one of these three inhibitors do not cross-react with the other inhibitors. Antibodies were raised in rabbits using a special immunization protocol for each inhibitor type (27). Stimulation of hirudin antibody production was especially difficult and only nonprecipitating antibodies were obtained. In the meantime, Stöffler and associates (personal communication, April 1986) obtained precipitating hirudin antibodies in sheep.

Applying well-characterized, monospecific antibodies and the double-antibody technique (28), we were able to show the localization of each of the three inhibitors in the body of the leech (27). In agreement with the observation that hirudin is excreted with the saliva into the

Table 3 Amino Acid Compositions (Mol/Mol) and Molecular Masses of Bdellin B-3 and Bdellin A-4

	Bdellin	
Amino acid residue	B-3	A-4
Asx	5	7
Thr	4	3
Ser	2	3
Glx	6	5
Pro	0	3
Gly	4	4
Ala	4	4
Cys/2	6[a]	10[b]
Val	4	5
Met	0	1
Ile	0	1
Leu	2	1
Tyr	1	1
Phe	0	2
His	5	3
Lys	1	5
Arg	2	0
Trp	0	0
Total	45	58
M_r	4830	6224

[a]Presumably three disulfide bridges.
[b]Presumably five disulfide bridges.
Source: From Ref. 14.

wound after biting, this inhibitor was found exclusively in the salivary gland cells and their collecting ducts [Figs. 9-11 (Color Plates 1-3)]. Obviously, hirudin serves the leech to prevent blood clotting during and after sucking and, thus, to keep the animal mobile after feeding for a long period of time (several months).

The bdellins and eglins showed a localization very similar to each other. They are distributed over the whole body of the leech. The specific immunofluorescence was especially intense in the endothelial cell layers of blood lacunae and the epithelial mucous membranes of skin and intestine [Figs. 12 and 13 (Color Plate 4)]. It is tempting to speculate that these inhibitors exert a protective role against proteinases releases by invading organisms (skin) and/or endogenous digestive

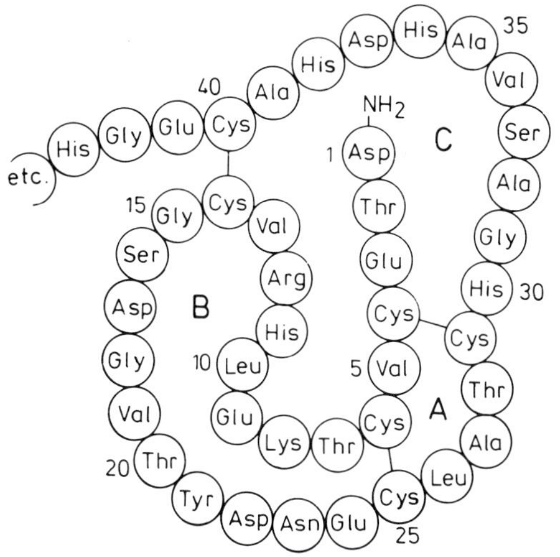

Figure 7 Model of the partial primary structure of bdellin B containing the inhibitorily active domain of bdellin B-3 (15). Bdellin B is extended in its high molecular mass form at the carboxy terminus by a polypeptide of the composition (calculated for the portion after Cys-40): 15 Asx, 28 Glx, 7 Gly, 1 Val, 28-29 His, 4 Lys. (From reference 3 and the Ph.D. thesis (dissertation) of H. Rehm [1985]: Isolierung und Charakterisierung eines langkettigen Proteinasen-Inhibitors vom Bdellintyp aus dem Blutegel Hirudo Medicinalis, University of Munich.)

```
                                    10                                      20
Thr-Glu-Phe-Gly-Ser-Glu-Leu-Lys-Ser-Phe-Pro-Glu-Val-Val-Gly-Lys-Thr-Val-Asp-Gln

                                    30                                      40
Ala-Arg-Glu-Tyr-Phe-Thr-Leu-His-Tyr-Pro-Gln-Tyr-Asp-Val-Tyr-Phe-Leu-Pro-Glu-Gly

                                    50                                      60
Ser-Pro-Val-Thr-Leu-Asp-Leu-Arg-Tyr-Asn-Arg-Val-Arg-Val-Phe-Tyr-Asn-Pro-Gly-Thr
                ‾‾‾‾‾‾‾
                                    70
Asn-Val-Val-Asn-His-Val-Pro-His-Val-Gly
```

Figure 8 Primary structure of eglin c. The inhibitor molecule is composed of 70 amino acid residues and contains no disulfide bridges. It has a molecular mass of 8099 daltons. Leu-Asp in positions 45 and 46 were found to comprise the reactive site peptide bond residues P_1-P_1'.

proteinases uses by the leech to degrade the sucked blood (intestine) and/or proteinases present in blood plasma and blood cells of the host (intestine). An identical distribution of the inhibitors was also found by biochemical means applied to dissected tissue material of leeches (27). Of further special interest is the mode of regulation of the various genes encoding the proteinase inhibitors of the medical leech.

ACKNOWLEDGMENT

The authors wish to express their sincere thanks to Uwe Christensen, Head of Plantorgan KG and to dedicate this chapter to him on the occasion of his 50th birthday. He has stimulated and enabled this research by generously supporting us with huge quantities of leech extracts as well as purified eglins and hirudins, and grants for PhD and MD theses and postdoctoral fellowships for many years. Due to his encouragement and support, the production of eglin and hirudin by recombinant DNA technology could be envisaged in good time so that these inhibitors are already available in kilogram (eglin) or gram (hirudin) amounts today. This development will also help to protect the medical leech in its natural surroundings in the future.

REFERENCES

1. Fritz, H., M. Jochum, K.-H. Duswald, H. Dittmer, and H. Kortmann. In *Selected Topics in Clinical Enzymology, Volume 2,* Goldberg, D. H. and K. Werner (Eds.). Walter de Gruyter & Co., Berlin, New York, pp. 305-328 (1984).
2. Travis, J. and G. S. Salvesen. Ann. Rev. Biochem. 52:655 (1983).
3. Seemüller, U., J. Dodt, E. Fink, and H. Fritz. In *Proteinase Inhibitors*, A. J. Barrett (Ed.). Elsevier Science Publishing, B.V., Amsterdam. In press.
4. Dodt, J., U. Seemüller, R. Maschler, and H. Fritz. Biol. Chem. Hoppe-Seyler 366:379 (1985).
5. Dodt, J. *Hirudine, die Thrombininhibitoren des Blutegels "Hirudo Medicinalis"*, PhD Thesis (Dissertation), University of Munich (1984).
6. Magnusson, S., T. E. Petersen, L. Sottrup-Jensen, and M. Clayes. In *Proteases and Biological Control*, E. Reich, D. B. Rifkin, and E. Shaw (Eds.). Cold Spring Harbor Laboratory, Cold Spring Harbor, pp. 123-149 (1975).
7. Dodt, J., W. Machleidt, U. Seemüller, R. Maschler, and H. Fritz. Biol. Chem. Hoppe-Seyler 367. In press.
8. Petersen, T. E., H. R. Roberts, L. Sottrup-Jensen, and S. Magnusson. In *Protides of the Biological Fluids, Volume 23,*

Proc. of the 23rd Colloquium, Peeters, H. (Ed.). Pergamon Press, Oxford, New York, pp. 145-149 (1976).
9. Harvey, R. P., E. Degryse, L. Stefani, F. Schamber, J. P. Cazenave, M. Courtney, P. Tolstosh, and J. P. Lecocq. Proc. Soc. Natl. Acad. Sci. USA 83:1084 (1986).
10. Baskova, I. P., D. U. Cherkesova, and V. V. Mosolov. Thromb. Res. 30:459 (1983).
11. Markwardt, F., G. Nowak, J. Stürzebecher, U. Griessbach, P. Walsmann, and G. Vogel. Thromb. Haemost. 52:60 (1984).
12. Bichler, J., U. Seemüller, R. Gemmerli, E. Fink, and H. Fritz. Arzn. Forsch./Drug. Res. Submitted.
13. Markwardt, F., J. Hauptmann, G. Nowak, Ch. Kleßen, and P. Walsmann. Thromb. Haemost. 47:226 (1982).
14. Fritz, H., M. Gebhardt, R. Meister, and E. Fink. In *Proceedings of the International Research Conference on Proteinase Inhibitors*, Fritz, H. and H. Tschesche (Eds.). Walter de Gruyter, Berlin, New York, pp. 271-280 (1971).
15. Krejci, K. and H. Fritz. FEBS Lett. 64:152 (1976).
16. Seemüller, U., M. Meier, K. Ohlsson, H.-P. Müller, and H. Fritz. Hoppe-Seyler's Z. Physiol. Chem. 358:1105 (1977).
17. Seemüller, U., M. Eulitz, H. Fritz, and A. Strobl. Hoppe-Seyler's Z. Physiol. Chem. 361:1841 (1980).
18. Chang, J.-Y., R. Knecht, R. Maschler, and U. Seemüller. Biol. Chem. Hoppe-Seyler 366:281 (1985).
19. Bode, W., E. Papamokos, D. Musil, U. Seemüller, and H. Fritz. EMBO J. In press.
20. Richardson, M. and L. Cossins. FEBS Lett. 5:11 (1974); 56:161 (1974).
21. Svendsen, I., S. Boisen, and J. Hejgaard. Carlsberg Res. Commun. 47:45 (1982).
22. Baici, A. and U. Seemüller. Biochem. J. 218:829 (1984).
23. Snider, G. L., P. J. Stone, E. C. Lucey, R. Breuer, J. D. Calore, T. Seshadri, A. Catanese, R. Maschler, and H.-P. Schnebli. Am. Rev. Respir. Dis. 132:1155 (1985).
24. Jochum, M., H. F. Welter, M. Siebeck, and H. Fritz. In *Proteinases in Inflammation and Tumor Invasion*, Tschesche, H. (Ed.). Walter de Gruyter & Co., Berlin. In press.
25. Schnebli, H. P., U. Seemüller, H. Fritz, R. Maschler, M. Liersch, J. L. Bodmer, G. D. Virca, E. C. Lucey, P. G. Stone, and G. L. Snider. In *Intracellular Protein Catabolism*, Khairallah, E. A., J. S. Bond, and J. W. C. Bird (Eds.). Alan R. Liss, Inc., New York, pp. 287-920 (1985).
26. Rink, H., M. Liersch, P. Sieber, and F. Meyer. Nucl. Acids Res. 12:6369 (1984).

27. Dietl, R. *Proteinasen-Inhibitoren und Proteinasen des Medizinischen Blutegels: Immunologie und Gewebelokalisation.* MD Thesis (Dissertation), University of Munich, 1985.
28. Larsson, L.-J. *Peptide Immunocytochemistry. Progress in Histochemistry and Cytochemistry, Volume 13.* Gustav Fischer Verlag, Stuttgart (1981).

24

Interaction of *Vipera ammodytes* Venom Proteinase Inhibitors with Trypsin and Kallikrein

Franc Gubenšek
"J. Stefan" Institute, E. Kardelj University, Ljubljana, Yugoslavia
Darko Kocjan
Lek Works, Ljubljana, Yugoslavia
A. Ritonja
"J. Stefan" Institute, E. Kardelj University, Ljubljana, Yugoslavia

INTRODUCTION

Among the large group of protein inhibitors of serine proteinases studied to date (1,2), relatively few were isolated from snake venoms. Those sequenced were inhibitors from the venoms of *Vipera russelli* (3), *Haemachatus haemachatus* and *Naja nivea* (4), *Dendroaspis polylepis polylepis* (5), *Bungarus fasciatus* (6), and *Vipera ammodytes* (7,8). The inhibitors are, however, present also in some other venoms of the Elapidae and Viperidae families of snakes (9). We have isolated three inhibitors of serine proteinases from the venom of the European long-nosed viper *Vipera ammodytes ammodytes*. Two (T.i.I and T.I.II) inhibit primarily trypsin and trypsin-like enzymes, whereas the third strongly inhibits only alpha-chymotrypsin (10,11). *Vipera ammodytes* and some other venom inhibitors inhibit kallikrein and plasmin in addition to trypsin, alpha-chymotrypsin, and elastase, and may thus interfere with hemostasis. The inhibition of these enzymes, however, does not seem to play an important role in the envenomation, where other more pronounced effects of venom components prevail. These inhibitors are interesting in studying the structure and function of enzymes and inhibitors. All venom inhibitors of serine proteinases are highly homologous to the Kunitz-bovine pancreatic trypsin inhibitor (BPTI) (12). They differ in specificities towards serine proteinases, depending not only on the reactive site residue P1 (2), but also on other residues that are in contact or nearly in contact with the enzyme (8,13).

The inhibition of kallikrein has been shown to be dependent on the presence of basic residues in at least one of the positions 17, 19, or 39, besides the reactive site residue in position 15 (14). The inhibitors from *Naja naja nivea* (NNV-II), *Hemacahatus hemachatus* (HHV-II), and *Vipera ammodytes* (T.i.I) venoms inhibit porcine pancreatic kallikrein with inhibition constants of 1.1, 8.3, and 30 nM, respectively (11,15). The inhibition of kallikrein by other venom inhibitors was studied in less detail. The same is also true for the inhibition of plasmin.

Circular dichroism (CD) spectra of proteins in the near-ultraviolet (UV) region between 250 and 330 nm are a sensitive reflection of the tertiary structure of a protein. Due to the high number of aromatic residues, particularly tryptophans, and disulfides that contribute to the CD in the near-UV region, the exact interpretation of the spectra is exceedingly difficult (16). The situation is somewhat simpler in the difference spectrum originating from the interaction between two proteins, where the dichroic absorption should result only from residues that are perturbed by this interaction. A preliminary study of the interaction of trypsin with BPTI has been reported (17). A careful CD study of the interaction of chymotrypsin with BPTI was already made in 1968 by Meloun and co-workers (18), and later of trypsin and BPTI by Rosenkranz (19). The crystal structures of complexes of BPTI with trypsin (20) and kallikrein (21), determined at 1.9- and 2.5-Å resolution, permitted a theoretical approach and helped to form a better understanding of the difference CD spectra originating from the formation of the complexes of BPTI and *Vipera ammodytes* venom inhibitor T.i.I with trypsin and kallikrein. The rotatory strength (RS) corresponding to the surface under the peak of CD absorption could be calculated from the crystallographic data, with the assumptions that the protein has the same structure in the crystal and in solution and that electric dipole-electric dipole coupling is the major source of the near-UV CD bands.

EXPERIMENTAL STUDIES

Vipera ammodytes venom trypsin inhibitor (T.i.I) was isolated as described previously (10). Bovine pancreatic trypsin inhibitor (BPTI, Trasylol from Bayer) and porcine pancreatic kallikrein were a gift from Professor H. Fritz from Munich. Beta-trypsin was prepared from trypsin (Sigma Chemical Company) (22). Its activity, determined by active site titration (23), was over 90%. Circular dichroism (CD) difference spectra were obtained by subtracting the spectrum of the complex from the sum of the spectra of a single enzyme and inhibitor. Spectra were recorded on a Jobin-Yvon dichrograph (Mark IV) attached to an Apple II E microcomputer. Spectra were normally recorded in a 1-cm silica cell at sensitivity 2×10^{-6} absorbance units/mm, at bandwidth 0.2 nm, and at 25°C. Eight to 10 spectra were recorded and averaged. Solutions of beta-trypsin were prepared in 0.05 M Tris-HCl buffer, pH 8.0, containing 0.02 M $CaCl_2$ and 0.1 M NaCl. Solutions of kallikrein were prepared in 0.1 M Tris-HCl buffer, pH 8.0. The concentration of proteins was always approximately 20 μM. Approximate rotatory strengths of near-UV transitions of tyrosine (^1L_b) and of tryptophan (^1L_a, ^1L_b) were calculated using Tinoco's expression for electric dipole coupling (24). We took into account tryptophanyl residues, tyrosyl residues, and peptide bonds as perturbing groups. The optical parameters used for the phenolic ring of tyrosine, the phenyl ring of phenylalanine, and the peptide bond were taken from Strickland (25). The indole parameters of the tryptophanyl residue were taken from the paper of Goux and co-workers (26).

RESULTS AND DISCUSSION

The difference CD spectrum of the complex of trypsin with *Vipera ammodytes* venom inhibitor T.i.I was compared with the spectrum of the analogous complex with BPTI (Fig. 1). It is evident that the two spectra differ particularly in the dichroic absorption of tryptophan (i.e., above 285 nm). The spectrum of T.i.I exhibits a considerably lower intensity in this region. Using the crystallographic data, we calculated the rotatory strengths (RS) for the electric dipole-electric dipole couplings in each protein and in the complex. The difference RS was then calculated in the same manner as the difference CD spectra. The calculated values for RS of tryptophans and tyrosines are presented in Table 1. The total sum of RS for tryptophan (3.75×10^{-40} esu) is in accord with the difference CD spectrum that shows a positive tryptophan band above 285 nm. A detailed analysis of the interactions between particular aromatic residues in the complex (Table 2) shows that a major contribution to tryptophan RS originates from the coupling of Trp 215 with Tyr 10I.

The second major contribution is from the interaction of Trp 141 with Tyr 35I. Additional proof that the most important contribution to the tryptophan RS does indeed result from the interaction of Trp 215 with Tyr 10I is the spectrum of the trypsin-T.i.I complex, where a considerable loss of Trp dichroic absorbance is evident. The primary structure of T.i.I (8) shows that Tyr 10I of BPTI is substituted by an Asp-residue in T.i.I, and consequently the major perturbant of Trp 215 is missing.

The tyrosine part of the experimental difference spectrum between 270 and 285 nm is not in such good agreement with the calculated RS. The detailed examination of contributions to RS from particular residues shows small contributions to both the positive and negative direc-

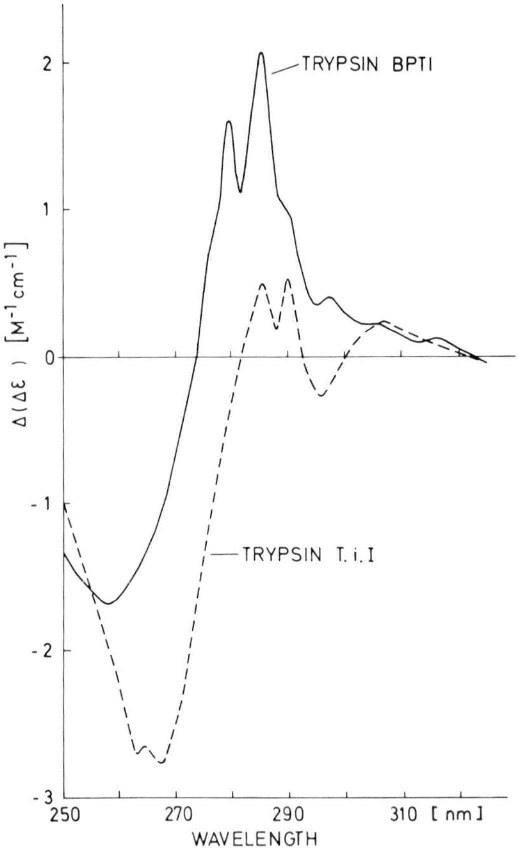

Figure 1 The difference CD spectra of complexes of trypsin with BPTI and with *Vipera ammodytes* venom trypsin inhibitor T.i.I.

tions; however, without a particularly strong single interaction. Both Trp and Tyr dichroic absorption seems to be superimposed on a broad negative cysteine band, with a minimum around 260 nm, which remained clearly expressed. In the trypsin-T.i.I inhibitor complex, the tyrosine intensity is even weaker.

The difference CD spectrum of the kallikrein-BPTI complex is similar to that of trypsin in the region of tryptophan absorption, although the intensity is weaker (Fig. 2). The tyrosine absorption is strongly positive, so that the negative, presumably disulfide band, visible in trypsin, practically disappears. Calculations of RS in the kallikrein-BPTI complex, based on the crystal structure (21), are presented in Table 3. Kallikrein crystallizes in two crystal forms,

Table 1 Summed Rotatory Strengths (in 10^{-40} esu) of Tryptophans and Tyrosines due to Electric Dipole-Electric Dipole Couplings in Trypsin, BPTI, and in Complex

	Tryptophan perturbation in		Tyrosine perturbation in		
Perturbant	Trypsin	Complex	Trypsin	BPTI	Complex
Amide (E)	46.52	41.54	1.24		2.72
Tyr (E)	17.90	13.12	-0.56		0.44
Trp (E)	-5.72	-7.43	4.43		2.22
Amide (I)		-0.58		-1.68	-0.92
Tyr (I)		8.30		-0.13	-0.19
Total	58.70	54.95	5.11	-1.81	4.27

Calculated difference RS = E + I - EI.
Tryptophan RS = 58.70 + 0 - 54.95 = 3.75.
Tyrosine RS = 5.11 - 1.81 - 4.27 = 0.03.

the main chains of which differ particularly in positions of external loops (27). The two crystal forms of kallikrein apparently differ even more in the positions of aromatic residues and, thus, different but strongly negative values of tryptophan RS are obtained in contrast to the measurement. A detailed analysis of the contributions of tryptophan RS resulting from coupling with inhibitor tyrosyl residues (Table 4) shows that two tryptophans, which are also present in trypsin (i.e., Trp 215 and Trp 141), contribute almost to the same extent in the kallikrein-BPTI complex. This means that the orientations of these residues in the crystal complexes of the two enzymes are practically identical. The contributions from the three tryptophans not present in trypsin only intensifies the total kallikrein tryptophan RS, yet the measured spectrum in the region of tryptophan absorption above 285 nm shows lower intensity than that of trypsin.

The tyrosine RS is, according to the calculated values, positive, and differs for the two crystal forms of kallikrein. In the experimental spectrum, the strongly positive dichroic absorption of tyrosines practically obscures the broad negative disulfide band. The examination of the contributions from particular pairs shows, similar to trypsin, numerous minor contributions, with the only exception being Tyr 99 and Tyr 63, which contribute over 0.25×10^{-40} esu.

The comparison of the spectrum of kallikrein-T.i.I complex with trypsin-T.i.I complex (Figs. 1 and 2) again shows a similar pattern. In both cases, the broad, negative, presumably disulfide band dominates. Tryptophans are more pronounced in trypsin and tyrosines in the kallikrein spectrum.

Table 2 Rotatory Strengths (in 10^{-40} esu) due to Electric Dipole-Electric Dipole Couplings of Tryptophans and Tyrosines of Trypsin in Its BPTI Complex

BPTI / Trypsin	Perturbing groups				
	Tyr 10I	Tyr 21I	Tyr 23I	Tyr 35I	
Trp 51	0.03(32.3)	0.48(32.0)	-0.10(42.3)	0.76(25.8)	
Trp 141	0.22(25.4)	-0.06(23.7)	-0.27(34.2)	2.24(20.8)	
Trp 215	4.06(12.4)	1.10(26.5)	0.82(30.1)	-0.36(14.0)	
Trp 237	0.63(27.6)	-0.49(32.7)	-0.01(40.7)	-0.77(22.9)	
Tot. Trp	4.94 +	1.03 +	0.44 +	1.87 =	8.29
Tyr 20	0.03(30.4)	0.03(34.9)	0.03(43.0)	-0.04(29.2)	
Tyr 29	-0.03(30.6)	0.04(33.4)	0.01(43.4)	0.02(26.7)	
Tyr 39	0.06(18.9)	0.22(10.5)	0.04(22.1)	-0.47(12.1)	
Tyr 59	0.05(21.1)	-0.11(19.7)	-0.05(28.7)	-0.22(13.6)	
Tyr 94	0.26(15.6)	-0.03(23.2)	0.07(29.3)	-0.10(11.6)	
Tyr 151	-0.08(16.9)	-0.06(16.4)	-0.03(24.4)	-0.15(14.3)	
Tyr 172	0.13(14.1)	0.07(29.6)	0.03(32.6)	0.06(17.9)	
Tyr 184	-0.05(26.0)	-0.02(36.4)	-0.04(41.9)	0.05(27.7)	
Tyr 228	-0.01(19.0)	-0.04(29.2)	-0.02(35.8)	0.03(18.8)	
Tyr 234	0.15(24.4)	0.00(33.0)	-0.03(39.9)	0.01(21.6)	
Tot. Tyr	0.51 +	0.10 +	0.01 +	-0.81 =	-0.19

Values in parentheses are distances in angstroms between the groups.

The major disagreement between the calculated RS and the observed dichroic absorption of tryptophan in kallikrein can only be explained by the greater mobility of aromatic residues of kallikrein in the kallikrein-BPTI complex, already noted by Bode and co-workers (27). Greater flexibility of kallikrein is also evident from larger differences between RS contributions within the kallikrein molecule in both free forms compared with the complex form. This difference is rather small in trypsin (Tables 1 and 3). We can assume that the greater flexibility of kallikrein results in a considerably different conformation of aromatic residues of the complex in solution, where the dichroic absorption was actually measured, compared with the conformation in the crystal from which the calculated RS was derived. Despite the fact that difference spectra involve fewer residues than did spectra of single proteins, their theoretical interpretation very often still remains difficult, particularly in the case where conformational changes occur in solution.

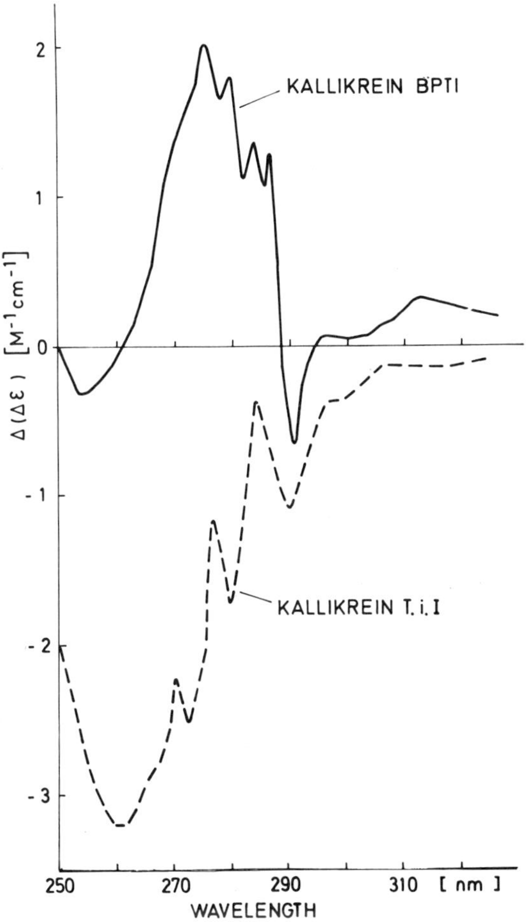

Figure 2 The difference CD spectra of complexes of porcine pancreatic kallikrein with BPTI and with *Vipera ammodytes* venom trypsin inhibitor T.i.I.

Table 3 Summed Rotatory Strengths (in 10^{-40} esu) of Tryptophans and Tyrosines due to Electric Dipole-Electric Dipole Couplings in Kallikrein, BPTI, and in Complex

	Tryptophan perturbation in			Tyrosine perturbation in			
Perturbant	Kall. I	Kall. II	Complex	Kall. I	Kall. II	BPTI	Complex
Amide (E)	0.80	11.59	42.75	4.95	2.90		3.69
Tyr (E)	-4.60	-1.48	22.31	-0.69	-0.62		0.47
Trp (E)	2.55	6.65	-4.32	0.27	5.60		-9.03
Amide (I)			2.29			-1.68	-3.15
Tyr (I)			11.92			-0.13	-0.88
Total	-1.25	16.76	74.95	4.53	7.88	-1.81	-8.90

Calculated difference RS = E + I - EI.
Tryptophan in kallikrein I: RS = -1.25 + 0 - 74.95 = -76.20.
Trytophan in kallikrein II: RS = 16.76 + 0 - 74.95 = -58.19.
Tyrosine in kallikrein I: RS = 4.53 - 1.81 + 8.90 = 11.62.
Tyrosine in kallikrein II: RS = 7.88 - 1.81 + 8.90 = 14.97.

Table 4 Rotatory Strengths (in 10^{-40} esu) of Tryptophans and Tyrosines due to Electric Dipole-Electric Dipole Couplings in BPTI Complex of Kallikrein

Kall. \ BPTI	Perturbing groups				
	Tyr 10I	Tyr 21I	Tyr 23I	Tyr 35I	
Trp 29	-0.25(31)	0.23(34)	0.09(44)	-0.39(22)	
Trp 51[a]	-0.18(32)	0.58(33)	0.01(43)	0.52(26)	
Trp 66	-0.17(30)	0.88(23)	0.05(35)	0.28(24)	
Trp 141[a]	0.26(26)	0.36(24)	-0.30(35)	2.01(21)	
Trp 209	1.09(28)	0.53(33)	0.52(42)	0.22(14)	
Trp 215[a]	4.87(12)	0.85(26)	1.06(30)	-0.50(22)	
Trp 237[a]	0.72(27)	-0.53(32)	-0.10(40)	-0.50(22)	
Total Trp	6.34 +	2.90 +	1.33 +	1.64 =	12.21
Tyr 34	-0.01(25)	-0.12(17)	-0.06(30)	0.12(18)	
Tyr 36	0.23(26)	0.06(15)	-0.01(27)	0.00(18)	
Tyr 63	-0.04(23)	0.02(20)	0.06(31)	0.27(15)	
Tyr 99	-0.36(10)	-0.07(24)	-0.01(27)	-0.16(12)	
Tyr 184[a]	0.09(26)	0.01(36)	0.00(40)	0.00(29)	
Tyr 228[a]	0.07(19)	-0.01(29)	0.00(36)	0.06(19)	
Tyr 234[a]	0.15(23)	-0.01(33)	0.01(40)	0.02(21)	
Total Tyr					-0.43

[a]Indicates residues present also in trypsin.
Values in parentheses are distances in angstroms between interacting groups.

ACKNOWLEDGMENTS

The authors are thankful to Professor R. H. Pain from the University of Newcastle for permitting the use of his dichrograph, to Professor R. Virden for his help with the instrument, to Professor H. Fritz from the University of Munich for the gift of proteins, and to Professor R. Huber and Dr. W. Bode from the Max Planck Institute for Biochemistry, Martinsried, for providing the crystallographic data. This work was supported by a Grant-in-Aid from The British Council, by the US-Yugoslav Joint Board on Technological Cooperation, and by the Research Community of Slovenia.

REFERENCES

1. Laskowski, M., Jr., M. Tashiro, M. W. Empie, S. J. Park, I. Kato, W. Ardelt, and M. Wieczorek. In *Proteinase Inhibitors: Medical and Biological Aspects*, Katunuma, N., H. Umezawa, and H. Holzer (Eds.), Japan Sci. Soc. Press, Tokyo; Springer-Verlag, Berlin, pp. 55-68, 1983.
2. Laskowski, M., Jr., and I. H. Kato, Ann. Rev. Biochem. 49:593 (1980).
3. Takahashi, H., S. Iwanaga, T. Kitagawa, Y. Hokama, and T. Suzuki. J. Biochem. (Tokyo) 76:721 (1974).
4. Hokama, Y., S. Iwanaga, T. Tatsuki, and T. Suzuki, J. Biochem. (Tokyo) 79:559 (1976).
5. Joubert, F. J. and D. J. Strydom, Eur. J. Biochem. 87:191 (1978).
6. Liu, C. S., T. C. Wu, and T. B. Lo. Int. J. Peptide Protein Res. 21:209 (1983).
7. Ritonja, A., B. Meloun, and F. Gubenšek. Biochim. Biophys. Acta 746:138 (1983).
8. Ritonja, A., B. Meloun, and F. Gubenšek. Biochim. Biophys. Acta 748:429 (1983).
9. Takahashi, H., S. Iwanaga, and T. Suzuki. Toxicon 12:193 (1974).
10. Ritonja, A., V. Turk, and F. Gubenšek. Eur. J. Biochem. 133:427 (1983).
11. Ritonja, A., V. Turk, and F. Gubenšek. In *Proteinases and Their Inhibitors*, Turk, V. and L. Vitale (Eds.), Mladinska Knjiga-Pergamon, Ljubljana, Oxford, pp. 339-343, 1981.
12. Kassel, B., M. Radicevic, M. J. Ansfield, and M. Laskowski, Jr. Biochem. Biophys. Res. Commun. 18:255 (1965).
13. Wachter, E., K. Deppner, H. Hochstrasser, K. Lempart, and R. Geiger. FEBS Lett. 119:58 (1980).
14. Fritz, H., E. Fink, and E. Truscheit. Fed. Proc. 38:2573 (1979).
15. Ritonja, A., R. Geiger, V. Turk, and F. Gubenšek. In *Recent Progress on Kinins*, Fritz, H., G. Dietze, F. Fiedler, and G. L. Haberland (Eds.), Birkhaeuser-Verlag, Basel, pp. 190-194, 1982.
16. Strickland, E. H. CRC Critical Reviews in Biochemistry 2:113 (1974).
17. Žerovnik, E., A. Ritonja, and F. Gubenšek, Period. Biol. 2:223 (1982).
18. Meloun, B., I. Fric, and F. Šorm. Coll. Czechoslovac Chem. Commun. 34:3127 (1969).
19. Rosenkranz, H. In *Bayer Symposium V "Proteinase Inhibitors,"* Fritz, H. (Ed.), Springer-Verlag, Berlin, pp. 458-462, 1974.

20. Huber, R., D. Kukla, W. Bode, P. Schwager, K. Bartels, J. Deisenhofer, and W. Steigemann. J. Mol. Biol. 89:73 (1974).
21. Chen, Z. and W. Bode. J. Mol. Biol. 164:283 (1983).
22. Strop, P. and D. Cechova. J. Chromat. 207:55 (1981).
23. Chase, T., Jr. and E. Shaw. Biochem. Biophys. Res. Commun. 29:508 (1967).
24. Tinoco, I. Adv. Chem. Phys. 4:113 (1962).
25. Strickland, E. H. Biochemistry 11:3465 (1972).
26. Goux, W. J., T. R. Kadesh, and T. M. Hooker. Biopolymers 15:977 (1976).
27. Bode, W., Z. Chen, K. Bartels, C. Kutzbach, G. Schmidt-Kastner, and H. Bartunik. J. Mol. Biol. 164:237 (1983).

25

The Action of Snake Venom Metalloproteinases on Plasma Proteinase Inhibitors

Lawrence F. Kress
Roswell Park Memorial Institute, Buffalo, New York

INTRODUCTION

Venoms from poisonous snakes of the Crotalidae and Viperidae families cause localized hemorrhage, tissue necrosis, edema, systemic coagulation defects, and death in most mammalian victims (1-3). The presence of proteolytic activity in these venoms is well documented (4-6), and some of the above symptoms result from the direct or indirect actions of venom proteinases. A number of the venom proteinases, of both the metallo and serine type, have been purified and shown to act on specific substrates such as factor X (7), prothrombin (8), fibrinogen (9-11), factor XIII (12), and other plasma proteins involved in coagulation and kinin release (5,6). Crotalid and viperid venoms also contain factors that cause localized hemorrhage by direct action on blood vessel walls (1). Hemorrhagic factors have been isolated from a number of these venoms and shown to be metalloproteinases (13-15).

The primary natural defense mechanism against the in vivo action of these venom proteinases and any endogenous proteinases activated by them is the victim's plasma proteinase inhibitor (PPI) system. The PPI system reacts with exogenous proteinases and also with endogenous proteinases involved in coagulation, fibrinolysis, complement formation, kinin liberation, and autodigestion (16,17). The PPI system typically functions by rapid or progressive formation of stoichiometric complexes between the inhibitor and the proteinase involved, and the complexes are then quickly cleared from the circulation (18,19). The inhibitors constitute approximately 10% of the total plasma proteins, and, when functioning normally, this system serves as an efficient regulator of proteolytic activity (16). However, several studies indicate that certain proteinases are not inactivated by the plasma proteinase inhibitors. Instead, these proteinases enzymatically inactivate the inhibitors without complex formation (20-22).

Investigations at Roswell Park Memorial Institute, Buffalo, New York, have been concerned with a systematic study of the effects of snake venom proteinases on human plasma proteinase inhibitors, following the initial report that human serum gradually lost all detectable inhibitory activity against trypsin and chymotrypsin during incubation with crotalid, viperid, and colubrid venoms, and that elapid (cobra) and hydrophid (sea snake) venoms were without effect (23).

This chapter summarizes the effects of venoms and purified venom metalloproteinases on the major human plasma proteinase inhibitors. Some recent results showing inhibition of venom metalloproteinases and hemorrhagic activity by opossum plasma proteinase inhibitors are also presented.

ENZYMATIC INACTIVATION OF HUMAN PLASMA PROTEINASE INHIBITORS

α_1-Proteinase Inhibitor (α_1-PI)

The crotalid, viperid, and colubrid venoms tested inactivated α_1-PI, whereas the elapid and hydrophid venoms lacked this activity (Table 1). The incubation of α_1-PI with a given amount of venom resulted in the gradual loss of inhibitory activity toward chymotrypsin, suggesting that the inactivation process involved enzymatic digestion of α_1-PI, rather than stoichiometric complex formation. The inhibitor inactivation could be readily stopped by the addition of EDTA to the reaction, indicating that venom metalloproteinases were involved. Phenylmethylsulfonyl fluoride was without effect on the reaction, showing that venom serine proteinases did not participate. The levels of inhibitor inactivation did not correlate with the level of caseinolytic activity within a given family of snakes. This suggested that venoms contained proteinases that preferentially attack native proteinase inhibitors, while showing only a limited activity toward typical denatured protein substrates such as casein.

Two metalloproteinases, designated proteinases I and II, were isolated from *Crotalus adamanteus* (Eastern diamondback rattlesnake) venom (24). Both proteinases I and II catalytically inactivated α_1-PI, and electrophoretic analysis revealed that intact α_1-PI (54 kDa) was cleaved by limited proteolysis to yield inactive α_1-PI (50 kDa) and a peptide (4 kDa), as shown diagrammatically in Figure 1. The gradual disappearance of intact α_1-PI coincided with a loss of inhibitory activity against trypsin and chymotrypsin. Sequence analysis (25) indicated that the inactivating cleavage had occurred at the Ala_{350}-Met

Table 1 Activity of Snake Venom Metalloproteinases on Human Plasma Proteinase Inhibitors

Venom source	Specific activity of venoms on individual inhibitors[a]						
	α_1-PI	AT III	AT III / Heparin	$C\bar{1}$-Inh	α_1-AC	α_2-AP	IαI
Crotalidae							
Crotalus adamanteus	50	0	42	5	12	0.2	0
C. atrox	44	0	35	438	3600	1.8	0
C. basiliscus	60	18	77	1280	126	7.6	0
Bothrops atrox	25	0	43	2	59	0.7	0
Trimeresurus flavoviridis	21	1	14	3	273	0.6	0
Agkistrodon piscivorus leukostoma	65	15	40	90	182	1.8	0
Lachesis muta	16	5	21	12	1675	6.6	0
Viperidae							
Bitis arietans	201	16	64	291	4520	5.6	0
Echis carinatus	7	4	17	31	760	0.4	0
Causus rhombeatus	13	30	323	4	35	0.3	0
Vipera ammodytes	NT	9	22	19	111	0.4	0
Colubridae							
Dispholidus typhus	22	12	52	87	6030	1.1	0
Elapidae							
Bungarus fasciatus	0	0	0	0	3	0	0
Naja naja kaouthia	0	0	0	0	2	0	0
Hydrophid							
Laticauda semifasciata	0	NT	NT	NT	0	NT	NT

[a]Specific activities are expressed as μg inhibitor inactivated/min/A_{280} unit of venom, calculated from the residual inhibitory activity after 30 minutes of incubation.
α_1-PI = α_1-proteinase inhibitor; AT III = antithrombin III; $C\bar{1}$-Inh = $C\bar{1}$-esterase inhibitor; α_1-AC = α_1-antichymotrypsin; α_2-AP = α_2-antiplasmin; IαI = inter-α-trypsin inhibitor; NT = not tested.

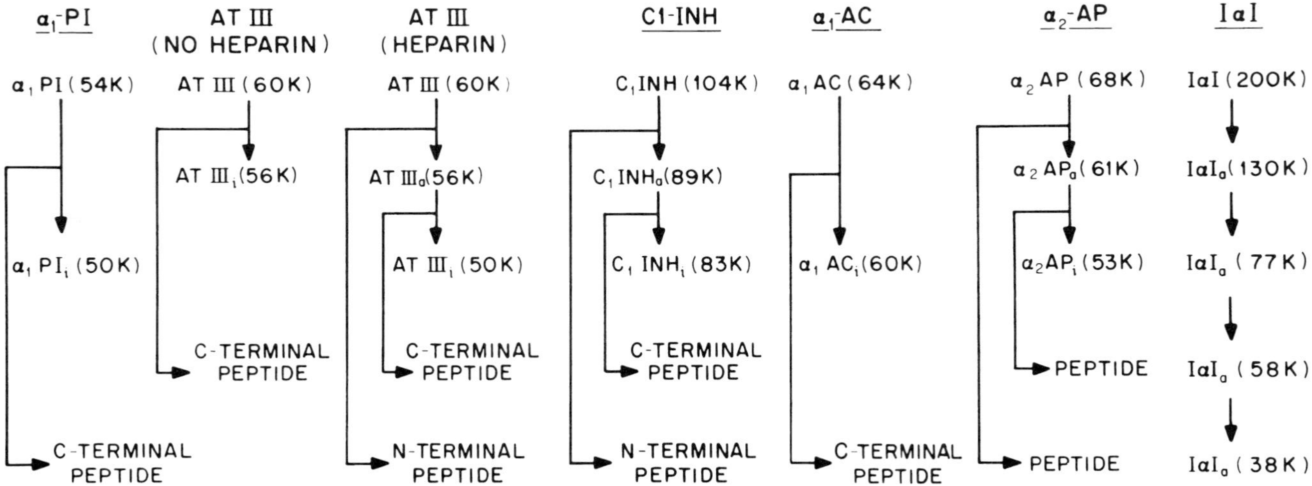

Figure 1 Schematic representation of human plasma proteinase inhibitor digestion by metalloproteinases, indicating the molecular weights and inhibitory activity (a = active, i = inactive) of the major products of the inhibitor digests and, where known, the origin of the peptide(s) released. The metalloproteinases used were: *C. adamanteus* proteinase II (α_1-PI, AT III, AT III/heparin); *C. atrox* α-proteinase followed by *P. aeruginosa* elastase (C$\bar{1}$-Inh); *C. atrox* α-proteinase (α_1-AC, IαI); and *C. basiliscus* crude venom (α_2-AP).

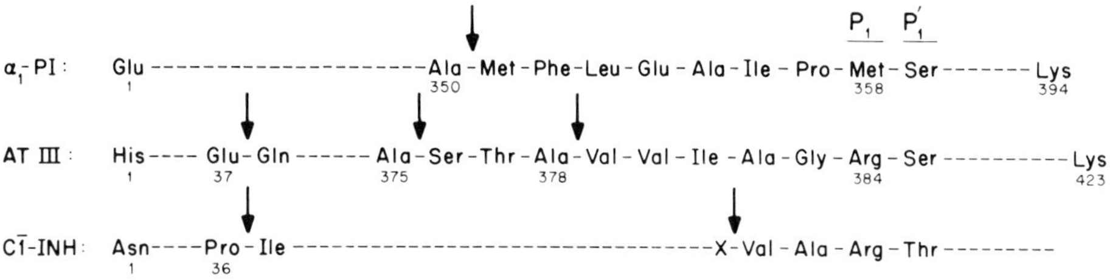

Figure 2 Location of inhibitor bonds cleaved during limited proteolysis by metalloproteinases. P_1-P_1' designates the inhibitor's reactive site. The arrows indicate the bond(s) cleaved by the following proteinases: in α_1-PI and AT III (heparin present), *C. adamanteus* proteinase II; in AT III (no heparin), only the Ala_{375}-Ser bond was cleaved; in C$\bar{1}$-Inh, *C. atrox* α-Proteinase (Pro_{36}-Ile) and *P. aeruginosa* elastase (X-Val).

bond, which is eight residues removed from the Met_{358}-Ser reactive site bond of α_1-PI, which is cleaved by trypsin or papain (Fig. 2) (20).

Antithrombin III (AT III)

The results for all the venoms tested on AT III are given in Table 1. In general, the venoms that react with antithrombin III digest the inhibitor comparatively slowly in the absence of heparin, and varying amounts of initial AT III activity still remained after 2 hours of incubation. In contrast, a rapid digestion leading to total inactivation of the inhibitor occurred within 2 hours when AT III was reacted with heparin prior to the addition of venom. The effect of heparin was most pronounced with several crotalid venoms, which produced essentially no loss of AT III activity in the absence of heparin (Table 1), but completely digested the inhibitor when heparin was present (26).

C. adamanteus proteinase II also inactivated AT III, and heparin caused several important differences in the overall reaction (27). Reaction of AT III (60 kDa) with proteinase II in the absence of heparin resulted in a slow loss of inhibitory activity against thrombin and the appearance of a band (56 kDa), corresponding to inactive inhibitor, and a peptide (Fig. 1). No further digestion products were detected. Carboxy-terminal analyses (Fig. 2) indicated that AT III inactivated in the absence of heparin was cleaved at the Ala_{375}-Ser bond. Dansylation of the inactivated AT III revealed that no cleavage had occurred in the amino-terminal region in the absence of heparin.

Incubation of AT III with heparin followed by catalytic amounts of proteinase II resulted initially in the conversion of intact inhibitor (60 kDa) into an intermediate species (56 kDa) that retained inhibitory activity against thrombin. Further proteolysis resulted in the gradual appearance of a third band (50 kDa) and the corresponding loss of inhibitory activity against thrombin. Electrophoretic analysis of the reaction products under nonreducing conditions revealed only one band (60 kDa), indicating that the cleavages had occurred within disulfide loops in the terminal regions of the inhibitor molecule. Amino- and carboxy-terminal sequence analyses of the inactivated AT III and amino-terminal analysis of the peptides released showed that cleavages had occurred at Glu_{37}-Gln, Ala_{375}-Ser, and Ala_{378}-Val (Fig. 2). Characterization of the peptide initially released during conversion of the intact inhibitor to the intermediate species indicated that the Glu-Gln cleavage occurred first and did not affect inhibitory activity against thrombin. The reactive site Arg_{384}-Ser bond, which is cleaved by thrombin during complex formation with AT III (28), was not cleaved by proteinase II either in the presence or the absence of heparin.

The results support the view that exposure to heparin causes a conformational change in AT III, which allows proteinase II to inflict a noninactivating cleavage in the amino-terminal region of the molecule, at a bond that was not accessible to the enzyme in the absence of heparin. The observation that venoms possess proteinases that selectively catalyze the inactivation of antithrombin III may have significance in terms of the in vivo functioning of other venom proteinases. For example, thrombocytin, a platelet-activating enzyme from *B. atrox* venom, is inactivated by complex formation with antithrombin III (29). In addition, several venom enzymes generate activated coagulation factors (5), which can in turn be inhibited by antithrombin III (30). The presence in venom of metalloproteinases that enzymatically inactivate antithrombin III would, therefore, complement the in vivo activity of other venom enzymes, thereby amplifying the disruption of the hemostatic balance noted with crotalid and viperid snakebites.

C$\bar{1}$-Esterase Inhibitor (C$\bar{1}$-Inh)

Incubation of C$\bar{1}$-Inh with crotalid, viperid, and colubrid snake venoms resulted in enzymatic inactivation of the inhibitor (Table 1). In contrast, C$\bar{1}$-Inh did not lose activity during incubation with elapid venoms; however, the intact inhibitor was gradually converted to an active species of 89 kDa during the incubation. With several elapid venoms, the active 89-kDa form was the reaction product, and no further proteolysis was noted after 2 hours of incubation (31).

A similar result was obtained during incubation of C$\bar{1}$-Inh with *Crotalus atrox* (Western diamondback rattlesnake) α-proteinase (E.C. 3.4.24.1), from which serine proteinase and hemorrhagic contaminants had been removed (32). Incubation of intact C$\bar{1}$-Inh (104 kDa) with α-proteinase resulted in formation of modified C$\bar{1}$-Inh (89 kDa), and no further cleavages occurred during prolonged incubation of C$\bar{1}$-Inh with α-proteinase. The modified inhibitor was fully active and could be obtained in homogeneous form by gel filtration. The amino-terminal sequence of modified C$\bar{1}$-Inh differed from the amino-terminal sequence of intact C$\bar{1}$-Inh. However, the first four residues of the new sequence were identical with residues 37 to 40 of C$\bar{1}$-Inh, indicating that the proteinase had cleaved the native inhibitor at the Pro_{36}-Ile bond in the amino-terminal region of the molecule (Fig. 2) (33).

Although several venom proteinases were able to inflict an inactivating cleavage in modified C$\bar{1}$-Inh (unpublished data), the best results were obtained with *Pseudomonas aeruginosa* elastase, and this metalloproteinase was chosen for studies aimed at delineating the C$\bar{1}$-Inh reactive site bond. Incubation of modified C$\bar{1}$-Inh (89 kDa) with *P. aeruginosa* elastase resulted in formation of inactive C$\bar{1}$-Inh (83 kDa) (Fig. 1). The amino-terminal sequence of the 83-kDa-inactivated C$\bar{1}$-Inh was identical to that of the 89-kDa-modified C$\bar{1}$-Inh, indicating that no further cleavage had occurred in the amino-terminal region of the inhibitor molecule. The sequence of the released peptide showed that the inactivating cleavage occurred at an X-Val bond (Fig. 2). The peptide overlapped the reactive site peptide released during complex formation with C$\bar{1}$-esterase, and established Arg-Thr as the reactive site P_1 and P'_1 residues of human C$\bar{1}$-Inh (33).

α_1-Antichymotrypsin (α_1-AC)

All venoms tested, except the hydrophid venoms, enzymatically inactivated α_1-AC (Table 1). Inhibitor inactivation by the elapid venoms required relatively large amounts of crude venom, and this accounted for the low specific activities (31). Several other elapid venoms also inactivated α_1-AC, indicating that these venoms contain low amounts of proteinases specific for native substrates such as α_1-AC (34). The α_1-AC was also readily inactivated by bacterial metalloproteinases and serine proteinases (35) and by human leucocyte elastase (unpublished data).

Incubation of *C. atrox* α-proteinase with α_1-AC (Fig. 1) resulted in limited proteolysis of native inhibitor (64 kDa) into inactive α_1-AC (60 kDa) and a peptide (unpublished). No structural analyses have been performed, but it seems likely that the inactivating cleavage occurred in the C-terminal region of α_1-AC near or at the reactive site Leu_{378}-Ser bond of the inhibitor (36).

α_2-Antiplasmin (α_2-AP)

Inhibitory activity against both plasmin and trypsin was lost at similar rates when α_2-AP was incubated with crude venoms (Table 1). The activity of the venom proteinase(s) on α_2-AP was generally much lower than that observed with the other plasma protein inhibitors that were inactivated. The inactivation reaction, however, is quite specific as judged by the lack of random proteolysis of α_2-AP, even after several hours of incubation with crude venoms. No α_2-AP activity was lost during incubation with elapid venoms, and no proteolysis of the inhibitor occurred (31). In this respect, α_2-AP resembles AT III rather than $C\bar{1}$-Inh, which is susceptible to elapid proteinases in the N-terminal region.

Incubation of α_2-AP with catalytic amounts of *C. basiliscus* venom resulted in cleavage of the intact inhibitor (68 kDa) to form a 61-kDa intermediate (Fig. 1). A second cleavage produced a 53-kDa product. Loss of activity correlated with appearance of the 53-kDa form, and it was concluded that the 61-kDa intermediate was active (31). Formation of an active intermediate (61 kDa) followed by α_2-AP inactivation was also observed with *P. aeruginosa* protease and elastase. However, with *C. adamanteus* proteinase II, only the active intermediate was generated and no α_2-AP inactivation occurred (unpublished observations). Again, no structural analyses have been performed on the α_2-AP digest products. However, by analogy to the results with AT III and $C\bar{1}$-Inh, it seems likely that the noninactivating cleavage occurred in the amino-terminal region, followed by an inactivating cleavage near or at the inhibitor-reactive site in the C-terminal region of the molecule.

PROTEOLYSIS OF INTER-α-TRYPSIN INHIBITOR (IαI)

Incubation of IαI with crotalid, viperid, colubrid, or elapid venoms resulted in random cleavage of the intact inhibitor (200 kDa) and formation of inhibitor fragments of 130, 77, 58, and 38 kDa, along with several minor products. The overall patterns of digestion varied among the venoms studied. However, a 77-kDa inhibitor cleavage-product was formed by all venoms tested, and this fragment was resistant to proteolysis even after a 24-hour incubation with the venoms (37).

A similar result (unpublished observations) was obtained when IαI was reacted with *C. atrox* α-proteinase (Fig. 1). This contrasted with the limited proteolysis of other plasma proteinase inhibitors previously noted with crude venoms and purified venom metalloproteinases, but is consistent with the previously observed susceptibility of IαI to proteolytic attack (38,39). No inactivation was noted since the inhibitor cleavage products all retained the ability to inhibit trypsin (Table 1). However, the IαI digest products did not inhibit venom serine-, or metalloproteinases.

A decrease in plasma IαI has been noted clinically following *C. atrox* snakebite, and cleavage of the inhibitor by endogenous proteinases was suggested (40). The present results indicate that direct hydrolysis of IαI by venom proteinases can also occur, and this effect would be amplified by subsequent attack on IαI by released endogenous proteinases. Since the precise physiological role of IαI has not yet been established, it is also possible that the venom digestion of IαI results in loss of inhibitory activity against other endogenous proteinases released at the site of the snakebite.

EFFECT OF α_2-MACROGLOBULIN (α_2-M) ON VENOM PROTEINASES

The inhibitor inactivations just mentioned could occur in vivo only if venom proteinases functioned in the presence of the inhibitor α_2-M, which has been shown to form complexes with members of all major categories of proteolytic enzymes (41). Incubation of several crotalid venoms with increasing amounts of α_2-M resulted in the loss of venom proteolytic activity on azocoll substrate. However, *C. adamanteus* venom retained 53% of its initial proteolytic activity in the presence of an excess of α_2-M. High molecular weight proteinases in this venom were shown to retain essentially all of their proteolytic activity in the presence of excess α_2-M, and are probably responsible for the residual proteolytic activity noted with *C. adamanteus*. *Bitis arietans* venom lost all proteolytic activity when incubated with α_2-M; the other viperid venoms studied retained significant amounts of proteolytic activity during incubation with the inhibitor. In the case of *Echis carinatus* venom, 48% of the original proteolytic activity remained after a short incubation (3 minutes) with α_2-M; this level was decreased to 20% when the incubation time was extended to 4 hours. A colubrid venom (*Dispholidus typus*) and an elapid venom (*Ophiophagus hannah*) also retained proteolytic activity when exposed to high amounts of α_2-M (42).

Thus, although some venoms exhibited the expected loss of proteolytic activity in the presence of α_2-M, others possess proteinases that either do not react with α_2-M or that form an enzyme-inhibitor complex only after long periods of incubation. In these instances, venom proteinases can retain activity in the presence of α_2-M. Also, the ability of α_2-M to form complexes with proteinases is retarded in the presence of substrate for the proteinase involved (43,44). Since venom proteinases can digest plasma proteinase inhibitors and other plasma components, the proteinases would be exposed in vivo to large excesses of potential substrates, and interaction with α_2-M would be decreased.

The hemorrhagic toxin of *C. adamanteus* venom, proteinase H, retained full activity in the presence of a measurable excess of human α_2-M, and no loss of α_2-M inhibitory activity was noted during the 20-hours incubation period (15). The data indicated that no proteinase-inhibitor complex was formed, and no enzymatic inactivation of α_2-M had occurred. These results are consistent with the observed occurrence of localized hemorrhage following rattlesnake envenomation (1). However, certain mammalian species (such as the opossum) do not exhibit localized hemorrhage following rattlesnake bite (45,46), and a possible mechanism to partially account for this is presented in the following section.

OPOSSUM SERUM INHIBITORS OF SNAKE VENOM METALLOPROTEINASES AND HEMORRHAGIC TOXINS

Fractionation of opossum serum and assay for ability to inhibit venom proteolytic activity have thus far resulted in the purification of an inhibitor designated "oprin" (opossum proteinase inhibitor) (47,48) and a hemorrhagic toxin inhibitor (HTI) (unpublished data). Oprin completely inhibited the metalloproteinases in *C. atrox* and *B. arietans* venoms, and caused a 60% decrease in *C. adamanteus* metalloproteinase activity. Experiments (not shown) indicated that oprin reacted stoichiometrically with *C. atrox* α-proteinase to form an inactive inhibitor-enzyme complex. Oprin is a major constituent of the opossum PPI system, which suggests a significant role in controlling the rapid influx of proteolytic activity accompanying envenomation, and in regulating endogenous metalloproteinase activity.

The effects of the opossum serum inhibitors on purified hemorrhagic toxins from *C. atrox* and *C. adamanteus* are shown in Figure 3. Oprin inhibits *C. atrox* HT-b (13) at a molar ratio of 1:1, but has no effect on *C. atrox* HT-a (13) or *C. adamanteus* proteinase H (15), even when a large excess of oprin is present. Opossum serum HTI, however, does inhibit HT-a and proteinase H, thereby preventing the hemorrhagic response. Neutralization of *C. atrox* venom hemorrhagic activity by opossum serum has previously been noted (49), but this is the first indication that the mechanism is due to formation of an inactive enzyme-inhibitor complex. In addition, the use of purified hemorrhagic toxins has indicated the presence of a previously overlooked inhibitor, HTI, the specificity of which is directed toward the most active hemorrhagic toxins from *C. atrox* and *C. adamanteus* venoms.

OVERVIEW

The following generalizations can be made with respect to venom proteinase inactivation of α_1-PI, AT III, α_1-AC, α_2-AP, and $C\bar{1}$-Inh. These inhibitors were all inactivated with catalytic amounts of venom proteinases. No stable proteinase-proteinase inhibitor complexes were detected, and no random proteolysis of the inhibitors occurred. Bond cleavage appeared to be quite specific, with only one or two high molecular weight inhibitor-derivatives being formed. The C-terminal peptide cleavage products in the case of α_1-PI, AT III, and $C\bar{1}$-Inh

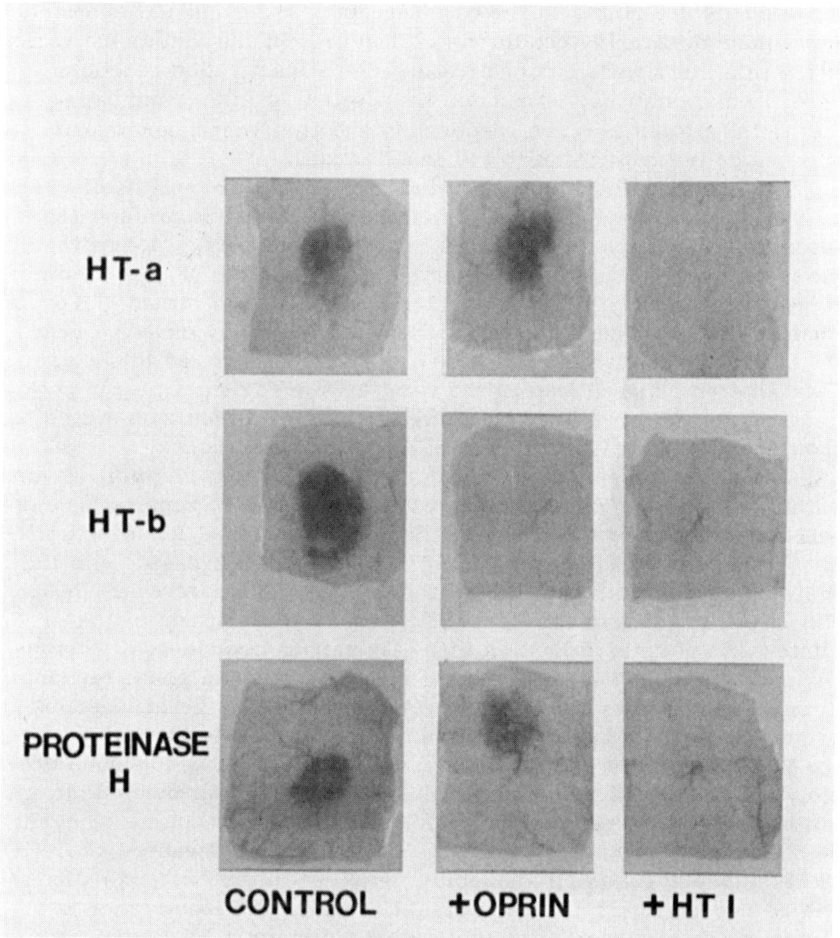

Figure 3 Inhibition of snake venom hemorrhagic toxins by inhibitors isolated from opossum serum. Oprin (opossum proteinase inhibitor) and HTI (opossum serum hemorrhagic toxin inhibitor) were incubated with the following venom hemorrhagic toxins: *C. atrox* HT-a and HT-b (13) and *C. adamanteus* proteinase H (15). Aliquots (100 μl) were injected subcutaneously into mice, and the hemorrhagic responses photographed after 5 hours. The controls shown are the responses for the hemorrhagic toxins alone; oprin and HTI controls (not shown) elicited no hemorrhagic response.

remained tightly bound to the core molecule, and could be separated only under denaturing conditions. Cleavages in the N-terminal regions of the inhibitors were noninactivating. The inactivation reactions were, in all instances, terminated by addition of EDTA, indicating that metalloproteinases are responsible and that venom serine proteinases do not contribute to PPI inactivation.

These studies have demonstrated that limited proteolysis of PPI by venom metalloproteinases is an effective method for probing the reactive site regions of the native inhibitor molecules. Interestingly, the inactivating cleavages all occurred near, but not at, the inhibitor-reactive site bond. Recently, bacterial enzymes and human leukocyte elastase have also been shown to inactivate AT III by cleavage near the inhibitor-reactive site. These observations have led to the proposal that an exposed loop in the reactive site region of plasma proteinase inhibitors has been conserved during evolution as a switch to allow inhibitor inactivation by specific proteolysis (50).

The physiological significance of PPI inactivation following envenomation has not yet been fully established. However, reports have appeared, indicating that human PPI levels were altered following snakebite. Circulating α_2-M, α_2-AP, AT III, and IαI decreased, and the acute phase reactant α_1-PI increased (40,51). The decreases in α_2-M and α_2-AP can be ascribed predominantly to consumption of the inhibitors via complex formation with endogenous proteinases, whereas the increased α_1-PI level is part of the victim's acute phase response to snakebite. IαI is known to be randomly digested by endogenous proteinases (39). The measurements all reflect systemic changes in the PPI system, and indicate a significant amount of endogenous proteolytic activity following snakebite. Whether these proteolytic events are initiated in part by venom metalloproteinases remains to be evaluated. No reports have appeared as yet concerning localized alterations of PPI function following snakebite. Further in vivo determinations will be needed to establish the extent to which PPI inactivation is a factor in the development of the localized symptoms associated with crotalid, viperid, and colubrid envenomation.

ACKNOWLEDGMENTS

This research was supported by Grant no. HL22996 from the National Institutes of Health, Bethesda, Maryland. The technical assistance of Joseph Catanese and Mary Hufnagel is gratefully acknowledged.

REFERENCES

1. Ohsaka, A. In *Handbook of Experimental Pharmacology, Volume 52*, Lee, C.-Y. (Ed.), Springer-Verlag, New York, p. 480 (1979).

2. Russell, F. E. In *Handbook of Experimental Pharmacology, Volume 52*, Lee, C.-Y. (Ed.), Springer-Verlag, New York, p. 978 (1979).
3. Seegers, W. H. and C. Ouyang. In *Handbook of Experimental Pharmacology, Volume 52*, Lee, C.-Y. (Ed.), Springer-Verlag, New York, p. 684 (1979).
4. Jimenez-Porras, J. M. Clin. Toxicol. 3:389 (1970).
5. Tu, A. T. *Venoms: Chemistry and Molecular Biology*, Wiley-Interscience, New York (1977).
6. Iwanaga, S. and T. Suzuki. In *Handbook of Experimental Pharmacology, Volume 52*, Lee, C.-Y. (Ed.), Springer-Verlag, New York, p. 61 (1979).
7. DiScipio, R. G., M. A. Hermodson, and E. W. Davie. Biochemistry 16:5253 (1977).
8. Walker, F. J., W. G. Owen, and C. T. Esmon. Biochemistry 19:1020 (1980).
9. Markland, F. S., Jr. Methods Enzymol. 45:223 (1976).
10. Stocker, K. and G. H. Barlow. Methods Enzymol. 45:214 (1976).
11. Nolan, C., L. S. Hall, and G. H. Barlow. Methods Enzymol. 45:205 (1976).
12. Kirby, E. P., S. Niewarowski, K. Stocker, C. Ketner, E. Shaw, and T. M. Brudzynski. Biochemistry 18:3564 (1979).
13. Bjarnason, J. B. and A. T. Tu. Biochemistry 17:3395 (1978).
14. Bjarnason, J. B. and J. W. Fox. Biochemistry 22:3770 (1983).
15. Kurecki, T. and L. F. Kress. Toxicon 23:657 (1985).
16. Heimburger, N. In *Proteases and Biological Control, Volume 2*, Reich, E., D. B. Rifkin, and E. Shaw (Eds.), Cold Spring Harbor Laboratory, New York, p. 367 (1975).
17. Collen, D. Thromb. Haemost. 43:77 (1980).
18. Ohlsson, K. and C.-B. Laurell. Clin. Sci. Mol. Med. 51:87 (1976).
19. Imber, M. J. and S. V. Pizzo. J. Biol. Chem. 256:8134 (1981).
20. Johnson, D. A. and J. Travis. Biochem. J. 163:639 (1977).
21. Brower, M. S. and P. C. Harpel. J. Biol. Chem. 257:9849 (1982).
22. Jochum, M., S. Lander, N. Heimburger, and H. Fritz. Hoppe-Seyler's Zeit. Physiol. Chem. 362:103 (1981).
23. Kress, L. F. and E. A. Paroski. Biochem. Biophys. Res. Commun. 83:649 (1978).
24. Kurecki, T., M. Laskowski, Sr., and L. F. Kress. J. Biol. Chem. 253:8340 (1978).
25. Kress, L. F., T. Kurecki, S. K. Chan, and M. Laskowski, Sr. J. Biol. Chem. 254:5317 (1979).
26. Kress, L. F. and J. J. Catanese. Biochim. Biophys. Acta 615:178 (1980).
27. Kress, L. F. and J. J. Catanese. Biochemistry 20:7432 (1981).

28. Jornvall, H., W. W. Fish, and I. Bjork. FEBS Lett. 106:358 (1979).
29. Niewiarowski, S., E. P. Kirby, T. M. Brudzynski, and K. Stocker. Biochemistry 18:3570 (1979).
30. Stead, N., A. P. Kaplan, and R. D. Rosenberg. J. Biol. Chem. 251:6481 (1976).
31. Kress, L. F., J. J. Catanese, and T. Hirayama. Biochim. Biophys. Acta 745:113 (1983).
32. Kruzel, M. and L. F. Kress. Anal. Biochem. 151:471 (1985).
33. Salvesen, G., J. J. Catanese, L. F. Kress, and J. Travis. J. Biol. Chem. 260:2432 (1985).
34. Kress, L. F. and M. E. Hufnagel. Comp. Biochem. Physiol. 77B:431 (1984).
35. Kress, L. F. Acta Biochim. Polon. 30:159 (1983).
36. Morii, M. and J. Travis. J. Biol. Chem. 258:12749 (1983).
37. Catanese, J. J. and L. F. Kress. Comp. Biochem. Physiol. 80B: 507 (1985).
38. Steinbuch, M. Methods Enzymol. 45:760 (1976).
39. Dietl, T., W. Dobrinski, and K. Hochstrasser. Hoppe-Seyler's Z. Physiol. Chem. 360:1313 (1979).
40. Bagley, E. A., K. Lonberg-Holm, B. V. Pandya, and A. Z. Budzynski. Electrophoresis 4:238 (1983).
41. Starkey, P. M. and A. J. Barrett. In *Proteinases in Mammalian Cells and Tissues*, Barrett, A. J. (Ed.), North-Holland, New York (1977).
42. Kress, L. F. and J. J. Catanese. Toxicon 19:501 (1981).
43. Werb, Z., M. C. Burleigh, A. J. Barrett, and P. M. Starkey. Biochem. J. 139:359 (1974).
44. Schapira, M., C. F. Scott, A. James, L. D. Silver, F. Kueppers, H. L. James, and R. W. Colman. Biochemistry 21:567 (1982).
45. Kilmon, J. A., Jr. Toxicon 14:337 (1976).
46. Werner, R. M. and J. A. Vick. Toxicon 15:29 (1977).
47. Kress, L. F. and J. J. Catanese. Fed. Proc. 44:1431 (1985).
48. Kress, L. F., J. J. Catanese, and L. G. Sheflin. J. Cell. Biochem. Suppl. 10A:264 (1986).
49. Menchaca, J. M. and J. C. Perez. Toxicon 19:623 (1981).
50. Carrell, R. W. and M. C. Owen. Nature 317:730 (1985).
51. Budzynski, A. Z., B. V. Pandya, R. N. Rubin, B. S. Brizuela, T. Soszka, and G. J. Stewart. Blood 63:1 (1984).

26

Purification and Double Chain Structure of Anticoagulant Factor from Five-Pace Snake *(Agkistrodon acutus)* Venom

Yang Lu, Wan-zhi Huang, Chun Wang, and Zi-xian Lu*
University of Science and Technology, Hefei, Anhui, Republic of China

INTRODUCTION

Ouyang and Teng (1972) demonstrated that the venom of hundred-pace snake (*Agkistrodon acutus*) had a coagulant action at high concentration and an anticoagulant action at a low concentration in vitro. By means of DEAE-Sephadex A-50 column chromatography, Cheng and Ouyang (1967) obtained 12 protein peaks from the crude venom and found that anticoagulant activity was located in peaks 1, 6, and 7, with the highest in peak 6. Ouyang and co-workers purified an anticoagulant principle (ACP) from the venom of hundred-pace snake (Ouyang and Teng, 1972) and from *Trimeresurus gramineus* (Ouyang and Yang, 1975). It was revealed that ACP, protein in nature, possessed distinct anticoagulant activity both in vitro and in vivo (Ouyang and Teng, 1978). Sugihara and associates (1980) purified two anticoagulant factors (A1 and A2) from the venom of *Agkistrodon acutus* from Taiwan and found that both contained two peptide chains. Ouyang and co-workers suggested that there might be reversible binding of ACP to some activating factors for prothrombin, and this might interfere with the coagulation process. After studying the inhibition of the conversion of prothrombin to thrombin by the prothrombinase complex, Teng and Seegers (1981) concluded that the anticoagulant principle of hundred-pace snake, AAV-Xal, binds with the light chain of factor Xa on the sites that are necessary for

Present affiliation: Shanghai Institute of Biochemistry, Chinese Academy of Sciences, Shanghai, Republic of China phospholipid function. This is one category of anticoagulant factors from snake venom.

Boffa and Boffa (1973) studied the anticoagulant factors of venoms of Viperidae and found that these factors inhibited the activities of platelet phospholipid. Ouyang (1978), Evans (1980), Lyubimtseva (1983), Condrea (1982), Li (1984), Shukla (1981), Ruan (1984) and their co-workers isolated basic phospholipase A (bPLA) from *Trimeresurus mucrosquamatus*, *Naja nigricollis*, *Vipera ursini renardi*, Russell's viper, *Agkistrodon halys blomhoffii*, and *Agkistrodon halys Pallas* and found that these bPLAs also showed anticoagulant effects that were parallel to the phospholipase activity. Some of these results indicated that the anticoagulant activity of the PLA was in part due to binding of PLA to phospholipid and in part a lytic action on phospholipid by the enzyme. This is another category of anticoagulant factors from snake venom.

In the present work, an ACF was purified from five-pace snake (*Agkistrodon acutus*) venom, and its structure was studied. The results demonstrated that ACF is an anticoagulant factor similar to the A1 and A2 purified by Sugihara and his co-workers, but not to bPLA.

MATERIALS AND METHODS

Materials

The lyophilyzed venom of five-pace snake (*Agkistrodon acutus*) was purchased from Qimen Snake-Bite Hospital of Anhui. Sephadex was purchased from Pharmacia. Standard proteins for molecular weight calibration for electrophoresis and gel filtration chromatography were bovine serum albumin (BSA), ovalbumin, chymotrypsinogen, and cytochrome C. Blue Dextran-2000 and acetyl-tyrosine ethyl ester (ATEE) were made by Dongfeng Biochemicals Factory of the Shanghai Institute of Biochemistry.

DEAE-Sephadex Column Chromatography

Two g of the crude venom was applied to a column (3.2 × 100 cm) of DEAE-Sephadex, as described previously (Xu et al., 1981). The anticoagulant activity by plasma prothrombin time was tested in every five tubes, each of which contained 6 ml of eluate. Elution was monitored on a LKB Uvicord at 280 nm and 4°C.

Sephadex G-75 Column for Further Purification of the ACF

Fraction 7 was further purified using gel filtration successively on 1.8 × 100 cm and 1.8 × 50 cm columns of Sephadex G-75. For each column, 0.15 M NaCl was used for elution. The flow rate was 8 ml/hr, and every 3 ml was collected. The effluent was monitored at 280 nm and 4°C with a LKB Uvicord.

Disc Electrophoresis

Polyacrylamide gel electrophoresis (PAGE) with a 7.5% gel was used in Tris-Gly buffer at pH 8.9 for 4 hours at room temperature, 5 mÅ/ tube. Coomassie brilliant blue R-250 (0.1%) in 10% acetic acid was used for staining the proteins and 10% acetic acid for destaining.

Molecular Weight Determination by SDS-PAGE and Gel Filtration

SDS-PAGE was performed in phosphate buffer (pH 7.1) containing 0.1% SDS. Bovine serum albumin (BSA), ovalbumin, chymotrypsinogen, and cytochrome C were used as molecular weight standards. A gel filtration method for molecular weight measurement was performed according to the procedure of Andrews (1964). A Sephadex G-100 column, 1.6 × 50 cm, was equilibrated and eluted with 0.15 M NaCl. The void volume and total inner volume were measured using Blue Dextran-2000 and N-acetyl-L-tyrosine ethyl ester, respectively. The calibration curve was made by BSA, ovalbumin, trypsin, and cytochrome C as standards.

The disulfide bond(s) of ACF were oxidized by the method of Hirs (1956). A mixture of 9 ml formic acid with a ml of H_2O_2 was kept at room temperature for 1.5 hours, and 0.2 ml of the mixture was dropped onto 1.5 mg ACF and incubated at room temperature for 1 hour. One ml of water was then added to stop the reaction, and the mixture was dried in a vacuum.

Chemical Analysis

An LKB 4430 amino acid analyzer was used for the amino acid analysis after 24 hours of hydrolysis of the sample. The N-terminal amino acid and sequence of ACF were determined using the method of Chang (1980). Carbohydrate was measured using the method of Dubois and colleagues (1956).

Blood-Clotting Activity Measurement

Whole blood-clotting time was measured according to the procedure of Lee and White (1913). Calcium-clotting time was measured as described by Biggs and MacFarlane (1962). Plasma prothrombin time (PPT) was measured according to Quick (1938). The rabbit plasma was prepared by mixing rabbit blood with 1/9 of 0.1 M sodium oxalate or 3.8% sodium citrate and centrifuged at 2000 rpm for 12 minutes. Rabbit brain thromboplastin (factor III), which we prepared ourselves (Biggs and MacFarlane, 1962), was dissolved in 0.15 M NaCl to 0.05%.

Enzyme Activities Measurements

Caseinolytic activity was measured according to Rick (1963). Arginine esterase activity was measured by N-α-tosyl-L-arginine methyl ester as described by Hummel (1959). L-aminoacid oxidase activity was tested following Wellner's method (1971). Activities of 5'-nucleotidase and nucleotide pyrophosphatase were measured according to Heppel and Hilmol (1955) and Lindberg and Ernster (1956), respectively. Marinetty's method (1965) was used for determining phospholipidase A activity.

Toxicity Measurements

Hemorrhagic activity was measured as described by Kondo and associates (1960), and toxicity was determined according to the method of Litchfield and Wilcoxon (1949).

RESULTS

Purification and Purity of ACF

Chromatography on DEAE-Sephadex yielded 12 fractions (Fig. 1), as reported previously (Xu et al., 1981). The anticoagulant activity was mainly distributed in fractions 6 and 7 (more in fraction 7). Fraction 7 was further purified by chromatography and rechromatography on Sephadex G-75, finally yielding a single peak. The active fraction was lyophilized. As judged by PAGE, ACF was homogeneous (Fig. 2). The anticoagulant activity of pure ACF was 32 times higher than that of the crude venom. The isoelectric point of ACF was 5.3 as measured by isoelectric focusing.

Chemical Structure and Molecular Weight of ACF

The amino acid composition of ACF is given in Table 1. It contains a high content (28%) of acidic amino acid. The carbohydrate content is 7.9% on a weight basis. The N-terminal of ACF is Asp, and only a single N-terminal was observed. The N-terminal sequence measured by DABITC-PTH method was Asp-Asp-Val-Val ····.

On SDS-PAGE with 2-mercaptoethanol, a single band appeared. Its molecular weight was 11,600 daltons, as judged by the calibration curve (Fig. 3). The ACF on gel filtration was homogeneous with a molecular weight of 23,400 daltons (Fig. 4).

Electrophoresis of ACF after treatment in 8 M urea and in 1% SDS gave single bands of the same mobility (Fig. 5a). After oxidation by performic acid, ACF revealed a single band of smaller molecular weight than native ACF in PAGE at pH 8.9 (Fig. 5b). In gradient, PAGE-oxidized ACF revealed a single band with a molecular weight of about

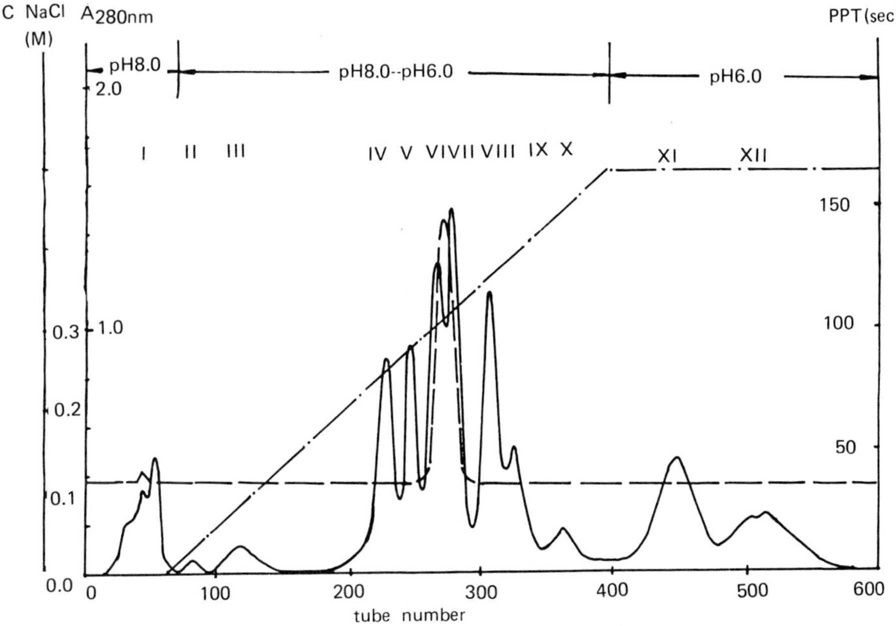

Figure 1 Elution profile of crude venom of five-pace snake on DEAE-Sephadex A-50, 6 ml per tube, and 24 ml/hr. Gradients changed as indicated in the top of the figure. —— =eluted curve; ---- = PPT activity; — · — · — · = NaCl gradient.

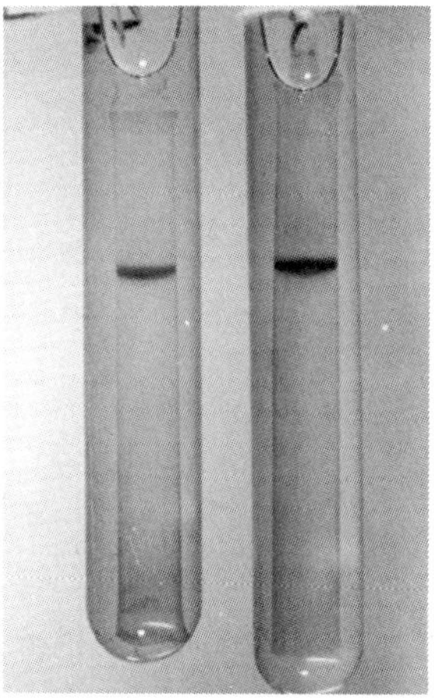

Figure 2 Disc electrophoresis (7.5%, pH 8.9) of ACF. Stained with Coomassie brilliant blue R-250 in 10% acetic acid. Both tubes contained the same sample, but the amount on the right was twice that on the left.

Table 1 Comparison of Amino Acid Content of Some Anticoagulant Factors

	Asp	Thr	Ser	Glu	Gly	Ala	Half Cys	Val	Met	Ile	Leu	Tyr	Phe	Lys	His	Arg	Pro	Trp
A. acutus (Anhui, China)	22	10	22	36	13	15	6	10	3	12	8	8	14	17	5	3	1	—
A. acutus (Taiwan, China)	17	8	16	23	10	10	9	8	3	6	9	7	10	14	5	5	—	—
T. gramineus (Taiwan, China)	18	7	6	12	11	7	15	8	3	7	8	7	4	17	3	6	7	4

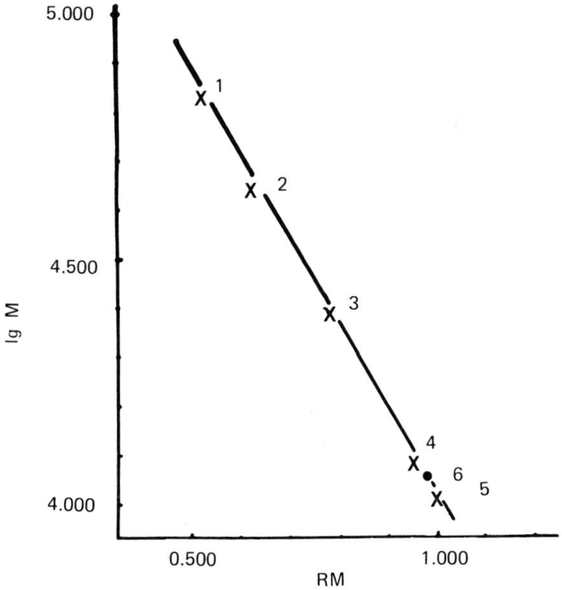

Figure 3 Estimation of molecular weight of ACF by SDS-PAGE under reducing conditions; 0.1% SDS and 10% gel at pH 7.1. RM = relative electrophoresis mobility. Standard proteins: 1 = BSA; 2 = ovalbumin; 3 = chymotrypsinogen; 4 = cytochrome C; 5 = Pinellia trypsin inhibitor; 6 = ACF.

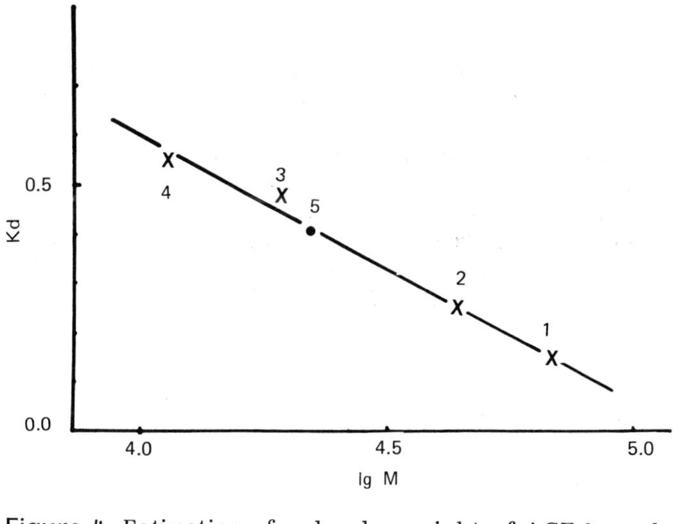

Figure 4 Estimation of molecular weight of ACF by gel filtration on Sephadex G-100. Kd was the distribution constant of gel filtration. 1 = BSA; 2 = ovalbumin; 3 = trypsin; 4 = cytochrome C; 5 = ACF.

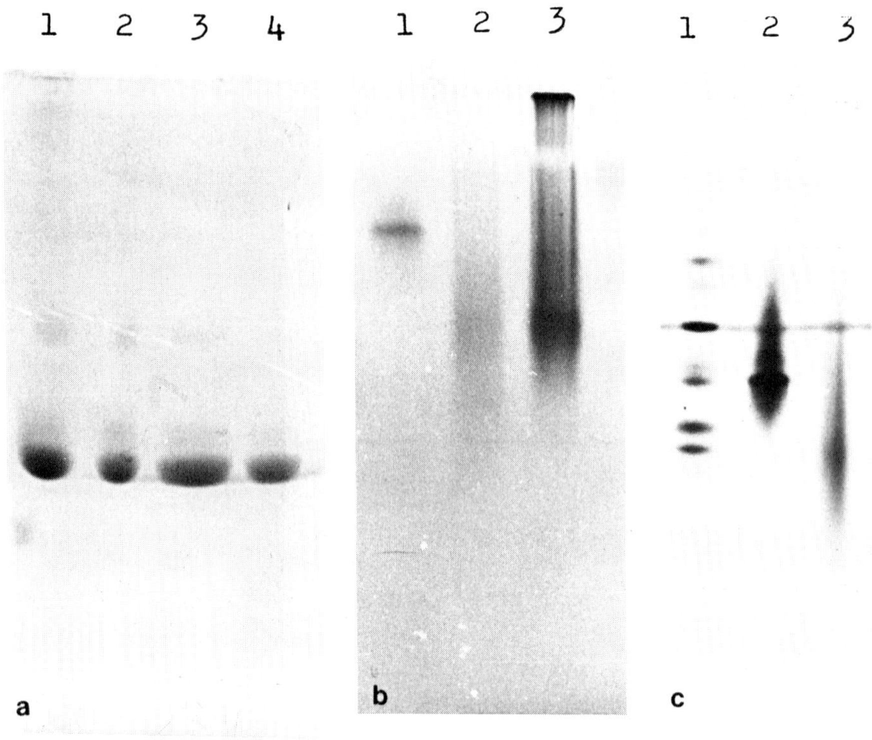

Figure 5 PAGE of split peptide chain(s) of ACF. a = Gradient (12% to 18% acrylanide) PAGE, column 1 and 2, ACF dissolved in 1% SDS, 3, 4, in 8 M urea; b = PAGE (14% gel) at pH 8.9, column 1 and 3, native ACF and performic acid oxidized ACF, respectively, and column 2, a diluted sample of oxidized ACF; c = Gradient (12/30) PAGE, column 1, standard proteins including phosphorylase b (MW 94,000), albumin (MW 67,000), ovalbumin (MW 43,000), carbonic anhydrase (MW 30,000), trypsin inhibitor (MW 20,100), and α-lactalbumin (MW 11,400), 2 and 3, native and performic acid treated ACF, respectively.

12,000 daltons, while native ACF gave a single band with a molecular weight of about 24,000 daltons (Fig. 5c).

Biological Activities of ACF

These results demonstrated that ACF did not show any caseinolytic, esterolytic, L-aminoacid oxidase, 5'-nucleotidase, nucleotide pyrophosphatase, phospholipase A, hemorrhagic, or lethal activity. In contrast, the crude venom displayed caseinolytic (20.6 μg Tyr$\cdot$min$^{-1}\cdot$mg^{-1}) and esterolytic (3500 μM$\cdot$min$^{-1}\cdot$mg^{-1}) activities and peak 7 had mild hemorrhagic and lethal activities (LD_{50} > 1 mg/mg body weight).

DISCUSSION

Until now, all the anticoagulant factors from snake venoms studied could be classified into two categories. One does not possess enzyme activities, and the second possesses activity of basic phospholipase A. It was suggested that the anticoagulant factors of the first category prolong the blood-clotting time through a reversible binding with some plasma factors (Ouyang and Yang, 1975; Ouyang and Teng, 1978; Sugihara et al., 1980; Teng and Seegers, 1981), while bPLA lengthens the blood-clotting time by hydrolysis of phospholipids or by inhibition of the phospholipids' function (Ouyang and Teng, 1978; Evans et al., 1980; Lyubimtseva, 1983; Condrea et al., 1982; Li et al., 1984; Shukla and Hanahan, 1981; Ruan et al., 1984).

In this chapter, we report the isolation of pure ACF from the venom of five-pace snake (*Agkistrodon acutus*) from southern Anhui Province, China. It does not indicate any kind of proteolytic, esterolytic, or phospholipase A activities or certain other enzymatic activities. These results suggest that ACF belongs to the first category of anticoagulant factors. ACF is composed of amino acids with a high content of acidic residues. These properties are similar to that of the ACP purified by Ouyang from *Agkistrodon acutus* (Taiwan) and *Trimeresurus gramineus*. There are, however, some points of difference between ACF from Anhui and ACP and A1 and A2 from Taiwan in amino acid composition (Table 1) and in carbohydrate content, which in the case of A1 and A2, amounted to only 0.31%. Xu and co-workers (1981) demonstrated that there were some differences in the hemorrhagic components between the snake venoms from the two *Agkistrodon acutus* snakes studied.

Sugihara and associates (1980) reported that both of the two anticoagulant factors, A1 and A2, the molecular weights of which are 25,800 and 24,000, respectively, had two peptide chains. The molecular weights of these peptide chains from A1 are 14,000 and 12,800 and those from A2 are 13,300 and 10,500, denoting that the peptide chain structure of our ACF is different from A1 and A2 in that we find only one kind of peptide chain in the ACF molecule. Under the conditions that keeps the molecule intact, such as in 1% SDS or 8 M urea solutions without reducing reagent, the molecular weight of ACF is 23,400 daltons, while under the conditions that disulfide bonds are destroyed by reducing or oxidizing reagents, the molecular weight is 11,600 daltons. The former is just twice of the latter one, and it is concluded that ACF is composed of two identical peptide chains linked together by disulfide bond(s). The results of N-terminal sequence determination, indicating only Asp-Asp-Val-Val-, support this conclusion.

CONCLUSIONS

An anticoagulant factor (ACF) was purified from the venom of five-pace snake (*Agkistrodon acutus*) from southern Anhui, China, on DEAE-Sephadex A-50 and Sephadex G-75. Sixty-five mg of pure ACF was obtained from 2 gm of crude lyophilyzed venom. It showed distinct anticoagulant action in tests of whole blood-clotting time, calcium-clotting time, and plasma prothrombin time. ACF was homogeneous as judged by polyacrylamide gel electrophoresis (PAGE) at pH 8.9. The anticoagulant activity of ACF was 32 times higher than that of the crude venom. The isoelectric point was 5.3, and the molecular weight was 11,600 daltons by SDS-PAGE under reducing conditions and 23,400 daltons by a gel filtration method. This suggested a two-peptide chain structure for the ACF molecule. It was observed that these two chains were identical and linked together through disulfide bond(s). The ACF was composed of 205 amino acid residues, of which 28% were acidic, with the N-terminal sequence, Asp-Asp-Val-Val-, and 7.9% carbohydrate. The ACF showed neither caseinolytic, esterolytic, L-amino acid oxidase, 5'-nucleotidase, nucleotide pyrophosphatase, or phospholipase A activities, nor hemorrhagic or lethal activities.

ACKNOWLEDGMENT

We are grateful to Professor X. Xu, Department of Biology, University of Science and Technology of China, Professor C. G. Ruan, Medical College of Suzhou, and Professor C. W. Chi, Shanghai Institute of Biochemistry, for their invaluable advice about blood coagulation and enzymatic activity measurements. We also thank Professor L. S. Gong, Department of Biology, University of Science and Technology of China, for the gift of 5'-UMP, and Mr. R. H. Yu, Miss Y. Xu, and Miss X. T. Xi for their help in the daily experiments and chemical analyses.

REFERENCES

Andrews, P. (1964). Estimation of the molecular weight of protein by Sephadex gel-filtration, Biochem. J. 91:222.

Biggs, R. and R. G. MacFarlane (1962). *Human Blood Coagulation and Disorders, Edition 3,* Blackwell Scientific, Oxford.

Boffa, G. A. and M. C. Boffa (1973). Analytical chromatographic and immunochemical characterization of ACP factor from Viper aspis venom (aspis inhibitor). C. R. Soc. Biol. 107:654.

Chang, J. Y. (1980). Amino-terminal analysis of polypeptide using dimethylaminoazobenzene isothiocyanate. Anal. Biochem. 102:384.

Cheng, H. C. and C. Ouyang. (1967). Isolation of coagulant and anticoagulant principles from the venom of *Agkistrodon acutus*. Toxicon 4:235.

Condrea, E., C. C. Yang, and P. Rosenberg (1982). Additional evidence for a lack of correlation between anticoagulant activity and phospholipid hydrolysis of snake venom. Thromb. Haemost. 47:298.

Dubois, M., K. A. Gills, P. A. Reberrs, and E. Smith (1956). Colorimetric method for determination of sugar and related substance. Anal. Chem. 28:350.

Evans, H. J., R. Fraabson, and G. D. Oureshi (1980). Isolation of anticoagulant proteins from cobra (*Naja nigricollis*) venom. Identity with phospholipases A_2. J. Biol. Chem. 255:3793.

Heppel, L. A. and R. L. Hilmol (1955). 5'-nucleotides. Methods Enzymol. 2:546.

Hirs, C. H. W. (1956). The oxidation of ribonuclease with performic acid. J. Biol. Chem. 219:611.

Hummel, C. W. (1959). A modified spectrophotometric determination of chymotrypsin, trypsin and thrombin. Can. J. Biochem. Physiol. 37:393.

Kondo, H., S. Kondo, H. Ikezawa, R R. Murata, and A. Ohsaka. (1960). Quantitative method for determination of hemorrhagic activity of Habu snake venom. Jpn. J. Med. Sci. Biol. 13:43.

Lee, R. L. and P. D. White (1913). Am. J. Med. Sci. 145:495.

Li, Y. S., G. F. Liu, Q. C. Wang, G. C. Tu, and Y. L. Ran (1984). Purification and physicochemical properties of phospholipase A_2 from Vipera russelli siamensis from Fujian, China. 5th Cong. Chinese Biochem. Soc. Symp. 1:316 (in Chinese).

Lindberg, O. and L. Ernster (1956). Determination of organic phosphorus compounds by phosphate analogs. Methods Biochem. Anal. 3:1.

Litchfield, J. R. and F. Wilcoxon (1949). A simplified method of evaluating close-effect experiments. J. Pharm. Exp. Ther. 96:99.

Lyubimtseva, G. E. (1983). Study of Viper ursini renardi ch. venom V. Phospholipase A_2 with anticoagulant properties. Khim. Prir. Soedin. 3:399.

Marinetty, G. (1965). The action of phospholipase A on lipoproteins. Biochim. Biophys. Acta 98:554.

Ouyang, C. and C. M. Teng (1972). Purification and properties of the anticoagulant principle of *Agkistrodon acutus* venom. Biochim. Biophys. Acta 278:155.

Ouyang, C. and F. Y. Yang (1975). Purification and properties of the anticoagulant principle of *Trimeresurus gramineus* venom. Biochim. Biophys. Acta 386:479.

Ouyang, C. and C. M. Teng (1978). In vivo effects of the purified thrombin-like and anticoagulant principle of *Agkistrodon acutus* (hundred-pace snake) venom. Toxicon 16:583.

Ouyang, C., C. M. Teng, Y. C. Chen, and S. C. Lin (1978). Purification and characterization of the anticoagulant principle of *Trimeresurus mucrosquamatus* venom. Biochim. Biophys. Acta 541: 394.

Quick, A. J. (1938). The nature of the bleeding in jaundice. J. Am. Med. Assoc. 110:1658.

Rick, W. (1963). In *Methods of Enzymatic Analysis*, Bergmeyer, H.-U. (Ed.), Academic Press, New York, p. 800.

Ruan, C. G., D. W. Cheng, H. Chen, X. F. Wu, and Y. C. Chen. (1984). Anticoagulant effect of phospholipase A from *Agkistrodon halys*, 5th Cong. Chinese Biochem. Soc. Symp. 1:320.

Shukla, S. D. and D. F. Hanahan (1981). Differences in the pattern of attack of acidic, neutral and basic phospholipases A_2 of *Agkistrodon halys blomhoffii* on human erythrocyte membranes. Arch. Biochem. Biophys. 209:668.

Sugihara, H., T. Nikai, T. Naakamura, H. Daido, and T. Tanake (1980). Enzymochemical studies on snake venoms III. Purification of hemorrhagic principle Ac_4-proteinase in the venom of *Agkistrodon acutus*. Yakugaku Zasshi 100:811.

Teng, C. M. and W. H. Seegers (1981). *Agkistrodon acutus* snake venom inhibitors prothrombinase complex formation. Thromb. Res. 23:255.

Wellner, D. (1971). L-amino acid oxidase (snake venom). Methods Enzymol. 17B:597.

Xu, X., C. Wang, J. Liu, and Z. X. Lu (1981). Purification and characterization of hemorrhagic components from *Agkistrodon acutus* (hundred-pace snake) venom. Toxicon 19:663.

IV

PLATELET ACTIVATION AND AGGREGATION

27

Platelet-Aggregating Noncoagulant Snake Venom Fractions

An Overview

Kenneth M. Brinkhous and Scott V. Smith
University of North Carolina School of Medicine, Chapel Hill, North Carolina

INTRODUCTION

In recent years, many venom factors exhibiting various effects on platelets and the hemostatic mechanism have been isolated. These include "novel" platelet activators without coagulant activity. The first description of the coagulant effects of venoms has been attributed to a 1737 publication by Geoffrey and Hunauld (1). They observed that the blood of cats and dogs bitten by vipers did not coagulate. Reviews of the early history of venoms and the study of their hemostatic properties have been published by Leake (2) and Boquet (3). Suspicion that venoms had a platelet effect apparently came from reports of thrombocytopenia in some envenomated patients. It was not until the 1960s that interest focused on the action of certain crude venoms and venom components on platelets. During the period from 1965 to 1973, several platelet-active whole venoms were identified. After this, many noncoagulant platelet-aggregating venom fractions were identified and purified with the use of biochemical techniques for high-resolution separation of proteins.

Table 1 lists the venoms recognized during the transitional period from 1965 to 1973 as having potent platelet effects. In 1965, Davey and Lüscher (4) described a powerful platelet-aggregating action of crude *Trimeresurus purpureomaculata* venom when tested on a suspension of washed human blood platelets. The venom also caused a platelet nucleotide release reaction. Marked contraction of the platelet aggregates was observed. The platelet-aggregating substance was

Table 1 Recognition of Snake Venoms with Potent Platelet Aggregation/Release Effects

Venom source	Date	Reference
Trimeresurus purpureomaculata	1965	Davey and Lüscher (4)
Crotalus terrificus	1966	Markwardt, Barthel, Glusa, and Hoffmann (5)
Trimeresurus okinavensis	1967	Davey and Lüscher (6)
Bitis nasicornis	1970	MacKay, Ferguson, and McNicol (8)
Bitis arietans	1973	Phillips, Weiss, and Christy (12)

nondialyzable, heat stable at 56°C for 30 minutes, inhibited by hirudin and heparin, and its action was Ca^{2+}-dependent. The crude venom had limited thrombin-like activity. Other venoms, which initiated a platelet release reaction, were observed with little or no platelet-aggregating activity. In 1966, Markwardt and co-workers (5) tested 29 venoms on isolated rabbit platelets for platelet release of serotonin, histamine, and ADP. Several *Crotalus* and *Naja* venoms were most effective, with venoms of two *Bothrops* species being moderately active in causing serotonin release. The venom of a South American rattlesnake, *C. terrificus*, was especially potent. This was confirmed during 1967 and 1968 by Davey and Lüscher (6,7), who identified an additional venom, *T. okinavensis*, which caused a strong platelet nucleotide release reaction. The nucleotide-releasing activity of the two *Trimeresurus* and the *C. terrificus* venoms was separated functionally from their esterolytic and coagulant activities by treatment with diisopropylfluorophosphate (DFP). A noncoagulant platelet-active venom fraction was obtained by electrophoresis in 1% agarose gels. During the period from 1970 to 1973, it was reported that venoms of two members of the *Bitis* genus, *B. nasicornis* and *B. arietans*, also had significant platelet-activating effects. *B. nasicornis* crude venom increased platelet adhesivity in whole blood and in platelet-rich plasma (PRP) (8). Partial purification of this platelet-active factor was obtained. Four fractions obtained by passage of crude venom through a carbomethylcellulose column caused platelet aggregation and enhanced platelet ADP reactivity. The platelet effects of *B. arietans* venom have been studied both in vivo and in vitro (9-12). Patients demonstrated thrombocytopenia (9,10) and a coagulopathy. In vivo effects were also observed in animal models injected with the crude venom.

Rats (9) and baboons (11) were used. Thrombocytopenia and coagulopathy developed. Heparin treatment was not beneficial in reversing thrombocytopenia in either animal model. In vitro studies with human PRP demonstrated a dose-dependent, irreversible platelet aggregation that occurred in the presence of the anticoagulants, citrate, EDTA, and heparin (11,12).

This chapter will cover the major noncoagulant platelet-active venom factors that have been isolated in the period beginning in 1969. These factors represent a heterogeneous group of substances obtained from several genera of snakes, which appear to act in diverse ways in bringing about platelet aggregation and activation. These various factors will be considered in chronological order based on the time when they were first reported.

TRIMERESURUS OKINAVENSIS VENOM PLATELET-AGGREGATING FACTOR

In 1969, Davey and Esnouf (13) reported the isolation of a potent platelet-active fraction from the venom of *T. okinavensis*, the Okinawan pit viper, himehabu (6). This was the first highly purified fraction of the noncoagulant platelet-active type to be reported. Purification of 55 to 100-fold was achieved by ion exchange chromatography using a TEAE-cellulose column with stepwise elution with a phosphate buffer, followed by gel filtration on Bio-Gel P-200 or 8% agarose (Table 2). The purified fraction was a protein-polysaccharide complex of high molecular weight. After Pronase digestion, it had molecular weight of approximately 400,000 daltons. It was shown to consist of 10% peptides, 85% neutral sugars, and 5% hexosamines. The platelet-active factor was distinct from the proteolytic, esterolytic, and coagulant materials also present in the venom. The fraction possessed no demonstrable enzymatic activity. It caused irreversible platelet aggregation of human PRP and platelet release of ADP. The action was Ca^{2+}-dependent for aggregation, but not for nucleotide release. The platelet-aggregating activity was impaired by various inhibitors. These included EDTA, adenosine, and the thiol reagents N-ethylmaleimide and p-chloromercuribenzenesulfonic acid. The minimal concentration of the venom factor causing platelet aggregation was 10^{-6} mg/ml.

CONVULXIN

In 1970 in her thesis, Prado-Franceschi (14) described an extremely potent platelet activator of high purity prepared from *Crotalus durissus terrificus* venom. The substance had been named convulxin in 1967 (15), based on the observation that intravenous injection of the substance in mice caused transient tachypnea and apnea and convul-

Table 2 Purification of Platelet-Aggregating Activity from the Venom of *T. okinavensis*

Method of purification	Fold-purification relative to crude venom
A. 1. TEAE-cellulose chromatography[a]	28
2. Bio-Gel P-200 filtration	55
B. 1. Pronase digestion and Bio-Gel P-200 filtration	45
2. TEAE-cellulose chromatography[a]	85
3. Agarose (SAG-8) filtration	100

[a]Stepwise elution with Tris phosphate buffers. Eight major protein fractions were obtained; 80% of the platelet-aggregating activity eluted with 0.32 M phosphate in the seventh fraction.

sions. Earlier, Markwardt and co-workers had identified a platelet-releasing activity in this venom (5), which may have been due to convulxin. Convulxin has subsequently been isolated from three subspecies of the South American rattlesnake, *C. durissus*: *cascavella*, *terrificus*, and *collilineatus*.

Convulxin, when gel-filtered on Sephadex G-75, with an exclusion limit for globular proteins of about 80,000 daltons, was present in the void volume. This indicated that convulxin had a high molecular weight (16). Several purification methods have been reported (16-21). The strategy of purification was embodied in three steps. First was a preliminary step of solubilization of convulxin partially freed of other venom proteins. Second was a chromatographic step using either Sephadex G-75 or Sephadex G-100, with an exclusion limit of about 150,000 daltons. The active material appeared in the void volume freed of smaller molecular weight materials that eluted in the included volume of the columns. In the third step, the void volume containing the high molecular weight material was subjected to molecular sieving using either Sephadex G-200 with an exclusion limit of about 600,000 daltons, or Sepharose 6B or 4B, with exclusion limits of $4\text{-}20 \times 10^6$ daltons. Two apparently homogeneous preparations of convulxin were obtained, one from *C. d. terrificus* venom (18), the other from *C. d. cascavella* venom (19). The latter preparation has been studied in some detail (19-22). Convulxin is an acid-soluble glycoprotein with a carbohydrate content of about 5%. Two forms of the active native protein were isolated from venom, with apparent molecular weights of approximately 144,000 and 300,000 daltons at acidic pH. By urea dissociation, a still smaller form with convulxin activity of apparent molecular weight of 72,000 was isolated. The latter is believed to be the monomeric form, with the native forms present in venom being the dimer and tetramer, respectively. With negative-staining electron microscopy, convulxin exhibited a rosette pattern 150 Å in diameter. Based on the electron microscopic studies, a molecular model of convulxin was developed. In this model, the rosette consisted of a ring of eight protomers (estimated molecular weight of 36,000). A pair of rosettes formed a highly organized structure, resembling two doughnuts stuck together, which when viewed from the side gave an appearance of a doublet rod-like structure consisting of 16 protomers in all (20). On reduction of the monomer, molecular weight 72,000 daltons, two subunits present in equal proportion were identified by SDS polyacrylamide gel electrophoresis and by gel filtration (21). The α-subunit had an apparent molecular weight of 12,600 daltons, the β subunit, 13,500 daltons. The convulxin monomer is thus postulated to have a structure of $\alpha_3\beta_3$, joined by disulfide bonds.

Convulxin activity was assayed with platelet aggregation tests using guinea pig PRP (16), rabbit PRP (17), or washed rabbit platelets (20). Convulxin action is prompt, causing platelet aggregation in guinea pig PRP in 1 to 5 seconds. It had no coagulant action. Activity was destroyed when the purified convulxin preparations were heated to 90°C for 3 minutes (17). Lyophilized preparations were stable (17,20). *Crotalus* antivenin, an equine antisera, neutralized convulxin activity. However, a *Bothrops* antivenin and an antiserum to crotoxin (a neurotoxic principle from *C. terrificus* venom) did not (18).

The action of convulxin on platelets appeared to be a direct one, since it aggregated washed platelets. This aggregation was potentiated by addition of fibrinogen. Aggregation was both dose-dependent and Ca^{2+}-dependent. Chelating agents such as EDTA prevented the platelet aggregation/activation response. Prostacyclin inhibited convulxin-induced aggregation (16), as did indomethacin if a low dose of convulxin was used. Convulxin-stimulated platelet release of ATP, thromboxane A_2, and acetyl glyceryl ether phosphorylcholine (AGEPC) (16,23). Convulxin was not species-specific in its activity. It aggregated the platelets from man, rabbit, guinea pig, rat, and dog. Electron microscopy of convulxin-treated platelets revealed loose platelet aggregates with degranulation, especially of the dense bodies, but without disruption of the cytoplasmic platelet membrane (23). Although the basic mechanism of action of convulxin on platelets is unknown, it has been suggested that its action may be analogous to the platelet-aggregating effect of collagen in some respects, but unlike that of thrombin. On a weight basis, convulxin was estimated to be 150 times more potent than thrombin in its platelet effect, and 1000 times more potent than collagen.

The striking effect of intravenous injection of purified convulxin preparations was an immediate and profound thrombocytopenia, with a slow return of platelet counts to the normal range. Circulating platelet aggregates were observed. The thrombocytopenia was accompanied by a moderate leukopenia. No sequestration of either platelets or leukocytes could be found in the capillary beds of the lungs or kidneys, nor in the sinusoids of the liver or spleen of rabbits (17). Other systemic manifestations of convulxin injections were hypotension, respiratory distress with apnea, tachypnea and bronchoconstriction (16), and convulsions. The severity of these disturbances was dose dependent, which was the basis of in vivo bioassays (18). Differences in the effects of convulxin injections in various animals have been observed. For example, convulsions were observed in cats, guinea pigs, and mice, but not in dogs or rabbits (16-18). Also, there have been differences in in vivo effects with different high-purity preparations of approximately the same potency. One preparation prepared from *C. d. terrificus* given to mice caused respiratory distress in low doses, with convulsions in higher doses. Another high-purity preparation made from *C. d. cascavella* venom caused only respiratory distress, with no convulsions occurring regardless of dosage level. Whether this difference in in vivo effect is fundamental or dependent on the degree of purification or some other factor is unknown (18,20). The respiratory and central nervous system effects are most likely a result of platelet thromboembolism. Intravenous injection of convulxin-induced platelet aggregates formed in vitro in rabbits produced the same systemic manifestations as did injections of purified convulxin, except for the thrombocytopenia (17). Also, production of a thrombocytopenic state by injection of an antiplatelet antiserum made the animal refractory to the effects of convulxin injections.

COBRA VENOM FACTOR

During 1971 to 1972, Dodds and Pickering reported studies on the effects of an anticomplementary cobra venom factor (CVF) on blood platelets (24,25). The cobra factor used was that previously isolated from *Naja haja* venom (26). In vitro studies demonstrated macroscopic aggregation of washed guinea pig platelets in 7 to 8 minutes, with release of platelet nucleotide and platelet factors 3 and 4. The CVF failed to cause clotting of plasma or purified fibrinogen solutions. There appeared to be a species-specificity in platelet effects, with human platelets being least reactive, guinea pig platelets most reactive, and dog and rabbit platelets giving an intermediate response. Clot retraction in thrombin-induced clots in guinea pig, dog, rabbit, and human PRP was interfered with by CVF (24). Similar results with human PRP had been reported earlier (27). The in vivo effects of intraperitoneal injections in guinea pigs were complex. There was no thrombocytopenia (25). Hemolysis, evidence of a coagulopathy, and a delayed hyperfibrinogenemia occurred. It is suggested that CVF acts directly on the platelet surface membrane, an action perhaps analogous to that on red cells with the associated hemolysis.

THROMBOCYTIN

In 1977, Niewiarowski and associates (28) described the isolation of a new platelet-aggregating protein from the venom of *Bothrops atrox marajoensis*. The factor aggregated platelets directly and caused platelet release of platelet factor 4 and serotonin. Thrombocytin promoted the clot-retracting effect of platelets. Thus, nonretracting clots of PRP, produced with the thrombin-like enzyme of the venom, retracted rapidly if thrombocytin was present. Thrombocytin was purified to homogeneity from crude venom by precipitation with sodium salicylate, solubilization in ammonium formate solution, pH 3.0, and then chromatography on a heparin-agarose (Sepharose 4B) column (29,30).

Thrombocytin is a serine proteinase. It is a single-chain glycoprotein with a molecular weight of 36,000 daltons, containing an estimated 5.6% carbohydrate (29). It was purified approximately 160 times in comparison with crude venom. Thrombocytin acts via direct platelet activation, probably through a thrombin-like action, requiring neither prothrombin nor factor X. It was less active on washed platelets, however, than it was on PRP. Thrombocytin was only about 15% as active as thrombin in causing aggregation of PRP. It demonstrated amidolytic activity on synthetic peptide substrates. Other effects of this fraction are activation of factor XIII and increasing factor VIII procoagulant activity. It has a trace of coagulant activity and degrades the α-fibrinogen chain. Inhibition of these actions could be obtained using soybean trypsin inhibitor, heparin, DFP, prostaglandin E_1 (PGE_1), antithrombin III with heparin, amitriptylene, cytochalasin B, and several arginine chloromethylketones. Inhibition could not be achieved with indomethacin or hirudin. It has been suggested that the platelet release reaction induced by thrombocytin is a phenomenon secondary to the aggregation of platelets (31).

BOTROCETIN (VENOM COAGGLUTININ)

In 1978, Read and co-workers (32) reported the isolation of a platelet-aggregating factor from the venoms of several species of *Bothrops* snakes. Its platelet-aggregating effect was dependent on the presence of plasma von Willebrand factor (vWF). Botrocetin had no coagulant or evident enzymatic activity. Its activity was not species specific. It has become a valuable reagent for the bioassay of vWF. Botrocetin is discussed in detail elsewhere in this volume and is mentioned here only in its historical context.

CROTALUS VENOM PHOSPHOLIPASE A_2 PLATELET-AGGREGATING FACTOR

In 1979, Heinrich and Beckmann (33), reported on the platelet-aggregating effect of phospholipase A_2 isolated form the venom of *C. terrificus terrificus*. The crude venom caused severe neurotoxic effects. This phospholipase A_2 in a dose range of 10 to 50 μg/ml induced a concentration-dependent aggregation of platelets with a release reaction. Higher doses of 100 μg/ml produced platelet lysis. Smaller doses in the range of 1 to 5 μg/ml had no apparent effect on platelets, although platelets so treated became refractory to the aggregating action of collagen. These actions were inhibited by EDTA, PGE_1, and apyrase. It is believed that its action on platelets is due to release of arachidonate acid from the platelet membrane phospholipids, followed by prostaglandin and thromboxane synthesis. High concentrations of the factor induce platelet lysis due to phospholipid breakdown in the platelet membrane.

PLATELET AGGREGOSERPENTIN

In 1979, Ouyang and Teng (34) described a purified platelet aggregation principle with phospholipase activity that was isolated initially from the venom of the Formosan viper, *T. mucrosquamatus*. They called this principle "platelet aggregoserpentin" (PAS). It was without coagulant activity. Subsequent work suggests that there may be several different aggregoserpentins. This subject is discussed elsewhere in this book.

THROMBOLECTIN

In 1980, Gartner and associates (35) described the first lectin isolated from snake venom. It was obtained from *B. atrox* venom by affinity chromatography, using asialofetuin-Sepharose 4B. The thrombolectin was discovered when it was found that high concentrations of thrombocytin (28) agglutinated formaldehyde-fixed bovine erythrocytes. This activity was separated from the serine proteinase activity of thrombocytin, and was demonstrated to be a lectin by specific inhibition of erythrocyte agglutination with saccharides. Lactose and a lactose analog blocked the agglutination reaction.

Table 3 Lectins from Snake Venoms

Lectin	Venom source	Reference
Thrombolectin	*Bothrops atrox* (fer-de-lance)	35
Bushmaster lectin	*Lachesis muta*	36
Jameson's-mamba lectin	*Dendroaspis jamesonii*	36
Southern copperhead lectin	*Agkistrodon contortrix contortrix*	37
Western diamondback rattlesnake lectin	*Crotalus atrox*	37
Western cottonmouth lectin	*Ancistrodon piscivorous leukostoma*	37

Thrombolectin also caused platelet aggregation. The authors suggested that lectins were significant platelet-aggregating components of venom (35). This aspect of thrombolectin actions needs further study. It is believed that thrombolectin needs two or more sugar-binding sites to cross-link or agglutinate platelets or other cells. If lactose inhibits thrombolectin-induced platelet aggregation as it does red cell agglutination, the unbound sugar in solution competes for binding sites on the lectin molecule, thus making them unavailable for interaction with sugar molecules bound to the platelet cell surface. Thrombolectin preparations as purified by chromatography appeared to be homogeneous. It was a dimer with an approximate molecular weight of 28,000. No carbohydrate could be detected. Its action as a hemagglutinin was Ca^{2+} dependent.

Since the identification of this venom lectin, a number of other snake venom lectins have been identified (Table 3) (36,37). Except for the mamba lectin, they resemble thrombolectin in several aspects. They have the same molecular weight (28,000 daltons) and the same Ca^{2+}-dependent hemagglutination, specifically inhibited by lactose. Data on the platelet-aggregating activity of these lectins have not yet been published (38).

CROTALOCYTIN

In 1980, Schaimer and co-workers (39) described a new platelet-aggregating factor isolated from the venom of the timber rattlesnake, *C. horridus horridus*. They named this factor crotalocytin. The authors had postulated that a separate platelet-active factor was present in the venom, in addition to the well recognized thrombin-like enzyme, after studying a patient bitten by this species of rattlesnake. Defibrination and a very severe thrombocytopenia were present. On testing the crude venom, it exhibited a platelet-aggregating activity that was much more powerful than was its fibrinogen-clotting action, with the ratio being about 1:300.

Crotalocytin was purified to near homogeneity. The crude venom was subjected to several purification steps, including Sephadex G-100 gel filtration, low ionic strength protein precipitation, and DEAE-A50 Sephadex chromatography. The activity of the purified crotalocytin was about 36 times greater than that of the crude venom. The isolated venom fraction was identified as a single-chain polypeptide with a molecular weight of 55,000 daltons. No carbohydrate could be found. The purified fraction possessed proteolytic and amidolytic activity. It was inhibited by DFP and appeared to be a serine proteinase (40).

The platelet-aggregating effect of crotalocytin was tested on PRP and gel-filtered platelets. Its aggregating effect was Ca^{2+}-dependent, being inhibited by EDTA. Simultaneously with aggregation, there was ATP secretion from the platelets and, in addition, depletion of dense granules and α-granules as judged by release of serotonin, low affinity PF-4, and fibrinogen. The aggregation was not potentiated by added fibrinogen. The fraction was active on chymotrypsin-treated platelets, but not on formalin-fixed platelets. The platelet-aggregating activity of crotalocytin was about 100 times more potent than thrombocytin. Crotalocytin was inactivated by commercial antivenin.

Aspirin-treated platelets were refractory to the aggregating action of crotalocytin, and aggregation of isolated normal platelets was inhibited by prostaglandin E_1 (PGE_1). It was suggested that prostaglandin intermediates and thromboxane were involved in the crotalocytin-induced aggregation/secretion reaction. Also, apyrase partially inhibited secretion, indicating that ADP may be additionally required for the platelet action of crotalocytin. The authors suggested that the action of crotalocytin on platelets is similar to that of low doses of thrombin and collagen.

SUMMARY AND CONCLUSIONS

Nine different platelet-active venom factors without coagulant activity have been described since 1969. They represent a heterogeneous group of compounds, both in composition and in function. Table 4 displays selected data regarding the source, biochemical characteristics, and action of these factors. More work is needed to provide a functional classification of these factors, especially in view of the many recent advances in platelet physiology and biochemistry. Mono-

Table 4 Summary of the Noncoagulant Platelet-Active Snake Venom Fractions

Fraction (and year)	Venom source(s)	Active substance	Degree of purity[a]	Enzymatic activity	MW (daltons)	Potency[b] (mg/ml)	Mode of action	Ca^{2+} dependent
T. okinavensis venom platelet-aggregating factor (1969)	*T. okinavensis*	Protein-polysaccharide complex	55-100X	None	400,000	$1 \cdot 10^{-6}$	Resembles collagen and thrombin in PRP	Yes
Convulxin (1970)	*C. d. cascavella* *C. d. terrificus* *C. d. collilineatus*	Acid-soluble glycoprotein	20X	None	300,000 (tetramer)	$7 \cdot 10^{-6}$ (1000X collagen)	Direct (resembles collagen)	Yes
Cobra venom factor (1971)	*N. haja*	Anticomplementary protein						Yes
Thrombocytin (1977)	*B. atrox marajoensis*	Single-chain glycoprotein	160X	Serine protease	36,000	$3.6 \cdot 10^{-3}$	Direct (thrombin-like?)	Yes
Botrocetin (1978)	*B. jararaca*	Protein	c. 200X	None	26,500		Via vWF	No
Crotalus venom phospholipase A_2 platelet-aggregating factor (1979)	*C. terrificus terrificus*	Protein (enzyme)		Phospholipase			Phospholipase activity	Yes
Platelet aggrego-serpentin (1979)	*T. mucrosquamatus* *T. gramineus*	Glycoprotein	c. 400X	Phospholipase	62,000 43,000	$1 \cdot 10^{-5}$	Direct (via AGEPC)	Yes
Thrombolectin (1980)	*B. atrox*	Lectin; no carbohydrate	410X	None	28,000 (dimer)	$1.5 \cdot 10^{-3}$	Cross-linkage between sugars	Yes
Crotalocytin (1980)	*C. horridus horridus*	Single-chain polypeptide	36X	Serine protease	55,000		Direct (resembles low doses of collagen and thrombin)	Yes

[a]In relation to the crude venom.
[b]Minimal amount causing platelet aggregation.

clonal antibodies, for example, could be used to identify the platelet membrane receptors as well as the plasma or other platelet components that participate in the aggregation/release response brought about by these venom factors. These compounds provide a group of natural platelet-active substances that could be valuable as platelet probes. Of the group, only botrocetin has found a place in clinical application, being used for the bioassay of von Willebrand factor. With so many different platelet-aggregating factors with different mechanisms of action and from snakes from many continents, all interfering with mammalian platelet function, these factors must be important for the ecology of the snake.

REFERENCES

1. Geoffrey and Hunauld. Memoire dans lequel on examine si l'Huile d'Olive est un spécifique contra la morsure des Viperes. In *Histoire de L'Academie Royale des Sciences*, Royal Academy of Sciences, Paris, p. 183 (1737).
2. Leake, C. D. Development of knowledge of venoms. In *Venomous Animals and Their Venoms*, Bücher, W., E. E. Buckley, and V. Deulofeu (Eds.), Academic Press, New York pp. 1-12 (1968).
3. Boquet, P. History of snake venom research. In *Snake Venoms, Handbook of Experimental Pharmacology, Volume 52*, Lee, C-Y. (Ed.), Springer-Verlag, New York, pp. 1-14 (1979).
4. Davey, M. G. and E. F. Lüscher. Actions of some coagulant snake venoms on blood platelets. Nature 207:730 (1965).
5. Markwardt, F., W. Barthel, E. Glusa, and A. Hoffmann. Über die freisetzung biogener Amine aus Blutplättchen durch tierische Gifte. Naunyn-Schmiedebergs Arch. Exp. Pathol. Pharmak. 252:297 (1966).
6. Davey, M. G. and E. F. Lüscher. Actions of thrombin and other coagulant and proteolytic enzymes on blood platelets. Nature 216:857 (1967).
7. Davey, M. G. and E. F. Lüscher. Release reactions of human platelets induced by thrombin and other agents. Biochim. Biophys. Acta 165:490 (1968).
8. MacKay, N., J. C. Ferguson, and G. P. McNicol. Effects of the venom of the rhinoceros horned viper (*Bitis nasicornis*) on blood coagulation, platelet aggregation, and fibrinolysis. J. Clin. Pathol. 23:789 (1970).
9. Phillips, L. L., H. J. Weiss, L. Pessar, and N. P. Christy. Effects of puff adder venom on the coagulation mechanism. I. In vivo. Toxicon 11:423 (1973).
10. Warrell, D. A., L. D. Ormerod, and McD. N. Davidson. Bites by puff-adder (*Bitis arietans*) in Nigeria, and value of antivenom. Br. Med. J. 4:697 (1975).

. Brink, S. and J. G. Steytler. Effects of puff-adder venom on coagulation, fibrinolysis, and platelet aggregation in the baboon. S.A. Med. Tydskr. 48:1205 (1974).
. Phillips, L. L., H. J. Weiss, and N. P. Christy. Effects of puff adder venom on the coagulation mechanism. II. *In vitro*. Thromb. Diath. Haemorrh. 30:499 (1973).
. Davey, M. G. and M. P. Esnouf. The isolation of a component of the venom of *Trimeresurus okinavensis* that causes the aggregation of blood platelets. Biochem. J. 111:733 (1969).
. Prado-Franceschi, J. *Estudo Sobre Convulxina*, Doctoral Thesis, Institute of Biology, State University of Campinas, Brazil (1970).
. Vital Brazil, O., J. Prado-Franceschi, and E. Waisbich. Ciencia Cultura 19:658 (1967).
. Vargaftig, B. B., J. Prado-Franceschi, M. Chignard, J. Lefort, and G. Marlas. Activation of guinea pig platelets induced by convulxin, a substance extracted from the venom of *Crotalus durissus cascavella*. Eur. J. Pharmacol. 68:451 (1980).
. Prado-Franceschi, J., D. Q. Tavares, R. Hertel, and A. Lobo de Araújo. Effects of convulxin, a toxin from rattlesnake venom, on platelets and leukocytes of anesthetized rabbits. Toxicon 19:661 (1981).
. Prado-Franceschi, J. and O. Vital Brazil. Convulxin, a new toxin from the venom of the South American rattlesnake *Crotalus durissus terrificus*. Toxicon 19:875 (1981).
. Marlas, G. Purification and preliminary structure of a potent platelet aggregating glycoprotein isolated from the venom of *Crotalus durissus cascavella*. Toxicon 20:289 (1982).
. Marlas, G., D. Joseph, and C. Huet. I. Isolation and electron microscope studies of a potent platelet-aggregating glycoprotein from the venom of *Crotalus durissus cascavella*. Biochim. 65:405 (1983).
. Marlas, G., D. Joseph, and C. Huet. II. Subunit structure of a potent platelet-activating glycoprotein isolated from the venom of *Crotalus durissus cascavella*. Biochim. 65:619 (1983).
. Vargaftig, B. B. Platelet activation by non-coagulant snake venom components. Toxicon 20:279 (1982).
. Vargaftig, B. B., D. Joseph, F. Wal, G. Marlas, M. Chignard, and L. G. Chevance. Convulxin-induced activation of intact and of thrombin-degranulated rabbit platelets: specific crossed desensitization with collagen. Eur. J. Pharmocol. 92:57 (1983).
. Dodds, W. J. and R. J. Pickering. Purified cobra venom factor: effect on blood platelets. Proc. Soc. Exp. Biol. Med. 140:429 (1972).
. Dodds, W. J. and R. J. Pickering. The effect of cobra venom factor on hemostasis in guinea pigs. Blood 40:400 (1972).

26. Pickering, R. J., M. R. Wolfson, R. A. Good, and H. Gewurz. Passive hemolysis by serum and cobra venom factor: a new mechanism inducing membrane damage by complement. Proc. Natl. Acad. Sci. USA 62:521 (1969).
27. Taylor, F. B., Jr. and H. J. Muller-Eberhard. Qualitative description of factors involved in the retraction and lysis of dilute whole blood clots and in the aggregation and retraction of platelets. J. Clin. Invest. 49:2068 (1970).
28. Niewiarowski, S., E. P. Kirby, and K. Stocker. Thrombocytin—a novel platelet activating enzyme from *Bothrops atrox* venom. Thromb. Res. 10:863 (1977).
29. Kirby, E. P., S. Niewiarowski, K. Stocker, C. Kettner, E. Shaw, and T. M. Brudzynski. Thrombocytin, a serine protease from *Bothrops atrox* venom. 1. Purification and characterization of the enzyme. Biochemistry 18:3564 (1979).
30. Niewiarowski, S., E. P. Kirby, T. M. Brudzynski, and K. Stocker. Thrombocytin, a serine protease from *Bothrops atrox* venom. 2. Interactions with platelets and plasma-clotting factors. Biochemistry 18:3570 (1979).
31. Kubisz, P., A. Arabi, F. Seghier, and S. Cronberg. Investigations on the effect of thrombocytin on platelets. Thromb. Res. 33:225 (1984).
32. Read, M. S., R. W. Shermer, and K. M. Brinkhous. Venom coagglutinin: an activator of platelet aggregation dependent on von Willebrand factor. Proc. Natl. Acad. Sci. USA 75:4514 (1978).
33. Heinrich, D. and S. Beckmann. Phospholipase A_2-induced platelet aggregation, release and lysis. Thromb. Haemost. 42:246 (1979) (Abstract).
34. Ouyang, C. and C. M. Teng. The action mechanism of the purified platelet aggregation principle of *Trimeresurus mucrosquamatus* venom. Thromb. Haemost. (Stuttg.) 41:475 (1979).
35. Gartner, T. K., K. Stocker, and D. C. Williams. Thrombolectin: a lectin isolated from *Bothrops atrox* venom. FEBS Lett. 117:13 (1980).
36. Ogilvie, M. L. and T. K. Gartner. Identification of lectins in snake venoms. J. Herpetol. 18:285 (1984).
37. Gartner, T. K. and M. L. Ogilvie. Isolation and characterization of three Ca^{2+}-dependent β-galactosides—specific lectins from snake venoms. Biochem. J. 224:301 (1984).
38. Gartner, T. K. Personal communication (1985).
39. Schmaier, A. H., W. Claypool, and R. W. Colman. Crotalocytin: recognition and purification of a timber rattlesnake platelet aggregation protein. Blood 56:1013 (1980).
40. Schmaier, A. H. and R. W. Colman. Crotalocytin: characterization of the timber rattlesnake platelet activating protein. Blood 56:1020 (1980).

28

Botrocetin and von Willebrand Factor

Kenneth M. Brinkhous, Scott V. Smith, and Marjorie S. Read
University of North Carolina School of Medicine, Chapel Hill, North Carolina

INTRODUCTION

Botrocetin and von Willebrand Factor

A factor known as botrocetin or venom coagglutinin in the venoms of several snakes, mainly in the *Bothrops* genus, causes platelet aggregation in a few seconds when added to platelet-rich plasma. The aggregation reaction is dependent on the presence in plasma of the macromolecular adhesive protein known as von Willebrand factor (vWF) (1). The recognition and testing of the venom factor botrocetin has been related entirely to its interaction with this plasma protein and platelets. Hence, a brief summary of the present understanding of the vWF protein will be presented first.

The vWF protein is so named because of its reduction or absence in plasma in the inherited autosomal bleeding disorder known as von Willebrand disease (vWD). In normal plasma, vWF is complexed with factor VIII to form the factor VIII-vWF complex. Factor VIII is the protein that is deficient in hemophilia-A plasma, also called antihemophilic factor (AHF). The two proteins, vWF and factor VIII, can be measured by immunological or bioassay methods. Both polyclonal and monoclonal antibodies have been used for the immunoassays. The proteins so determined are known as vWF antigen (vWF:Ag) and factor VIII antigen (F.VIII:Ag), respectively, according to the nomenclature recently adopted by the International Committee on Thrombosis and Haemostasis (2). The biologic activity of the two types of protein can be determined in vitro by platelet agglutination and clotting tests, respectively. The biologic activity of the vWF protein is designated as vWF with an indication of the "activator" used. The factor VIII clotting activity is known as F.VIII:C, the C referring to "coagulant."

The vWF activity is determined in vWF-dependent platelet agglutination tests. In 1972, ristocetin, an antibiotic, was recognized as the first activator of vWF-dependent platelet agglutination in human plasma (3). The vWF activity elicited with ristocetin is designated as vWF/ristocetin. The old nomenclature for vWF/ristocetin was F.VIIIR:RiCoF, with the R in F.VIIIR, a synonym for vWF, denoting "related" and the RiCoF standing for ristocetin cofactor. Venom coagglutinin, now called botrocetin, was discovered in 1978 as a second "activator" of the vWF-dependent platelet agglutination (1). The vWF activity determined with botrocetin is designated as vWF/botrocetin. The vWF of certain animal plasmas, mainly ruminants and the pig, can also be assayed by direct agglutination of human platelets without an activator. When vWF is so determined, it is designated vWF/PaggF or vWF/PAF, the latter abbreviations, PAggF or PAF, referring to platelet-aggregating factor (4). The PAF used in this sense must be distinguished from platelet-activating factor, or acetyl glyceryl ether phosphorylcholine (AGEPC), also designated PAF.

Plasma vWF exists in plasma as a series of multimers of different sizes. The molecular weight of the vWF subunit is about 200,000, with an array of multimers being formed, the largest being about 12 $\times 10^6$ (5). Factor VIII in variable amounts is attached to the multimers of vWF in normal plasma by noncovalent bonds to form the factor VIII-vWF complex. In hemophilic plasma, which lacks factor VIII, the vWF exists as an array of vWF macromolecules, not as the macromolecular complex as in normal plasma. The vWF acts as a carrier for factor VIII. When isolated factor VIII free of vWF is administered to hemophilic subjects, the F.VIII is rapidly complexed in the circulation to its carrier, the endogeneous vWF (6).

The other functions of vWF are related to the platelet and platelet-related hemostatic mechanisms, adhesion and aggregation. Adhesion of platelets to damaged vascular surfaces is largely regulated by vWF. In normal subjects with vWF, platelets attach to the damaged intima devoid of endothelium and promptly spread to form a platelet monolayer cover. In vWD subjects without vWF, this does not occur (7). The vWF appears to interact with collagen and fibronectin on vascular surfaces. The vWF-dependent platelet aggregation is a major mechanism in the formation of the hemostatic platelet plug, the control of bleeding, as well as in the formation of platelet thrombi. The platelet membrane receptor for vWF-induced aggregation is glycoprotein Ib (8) as well as the fibrinogen-ADP receptor, glycoprotein IIb/IIIa (9). Current evidence indicates that in platelet aggregation induced with ristocetin (8) and botrocetin (10), the vWF interacts with the glycoprotein Ib receptor. Since fixed and fixed-lyophilized platelets can be used with ristocetin or botrocetin to measure vWF, the term platelet agglutination is often used for this reaction, indicating it is a platelet surface reaction without the requirement for metabolically active platelets.

Discovery of Botrocetin

Two circumstances contributed to the discovery of botrocetin. One was the availability of a male dog with mild vWD, characterized by mild bleeding, occurring especially after trimming of the toenails. The plasma factor VIII was irregularly low. We planned to establish a colony of these dogs by a systematic breeding program (11). The other circumstance was the finding that ristocetin, the only activator of vWF-dependent platelet agglutination known at that time, was inactive with dog plasma. Hence, canine plasma vWF could not be measured. This situation prevented us from identifying heterozygous animals to establish an effective and efficient breeding program. Since venoms were known to contain a large number of different proteins that affected the hemostatic system, we started a systematic search of all available venoms for any that might cause platelet agglutination of normal platelet-rich plasma, but limited or no aggregation of von Willebrand dog plasma. Venoms of 73 different snakes were examined in this respect. Five of these venoms, *Bothrops alternatus*, *B. jararaca*, *B. medusa*, *B. neuwiedii*, and *Bitis arietans*, showed the greatest activity, although marginal activity was demonstrated for other venoms (1). *B. jararaca* venom has been used mainly for subsequent studies.

The venom or *Bothrops* factor was originally called venom coagglutinin (VCA) (1), since it was found to be a noncoagulant platelet-agglutinating agent that was a cofactor for vWF activity. Because its action was similar in many respects to ristocetin, it was later named botrocetin, emphasizing both the genus of snake from which the venom was obtained and the resemblance of its action to that of ristocetin.

BIOASSAYS FOR BOTROCETIN AND VON WILLEBRAND FACTOR

Assay Methodology: Macroscopic and Aggregometric

The bioassays of both botrocetin and vWF/botrocetin are based on determination of the rate of platelet agglutination in standardized platelet aggregation tests. The main components of the tests are vWF, botrocetin, and platelets, plus buffer to maintain a pH of 7.4. Two general procedures are used: (a) macroscopic testing in which the time required for the appearance of visible platelet clumps is determined, the platelet aggregation time; and (b) aggregometry testing in which the slope of the aggregometry curve is used as an index of the rate of platelet aggregation. Two types of macroscopic testing are used: (a) a flocculation macrotiter test, and (b) a tap tube test. The values obtained with these two macroscopic tests are similar. Replicate values of ± 10% are obtained by experienced operators. Aggregometry testing provides a permanent tracing of the pattern of platelet aggregation. An example with canine plasmas is shown in Figure 1. The rate of aggregation is ascertained by identifying the steepest part of the curve, constructing a tangent to that part of the curve, and then determining the slope of the tangent line. In vWD plasma, botrocetin had no effect.

Platelet Preservation

In bioassays for botrocetin and vWF/botrocetin, the rate of platelet agglutination can vary with different fresh platelet preparations, even if all other components of the test mixtures are the same. In addition to the difficulty in quality control from day to day with fresh platelets, they must be prepared daily. Two discoveries have resulted in the availability of stable and reliable platelet preparations for the bioassay procedures. The first was the finding that (para)formaldehyde-fixed platelets retained the receptor activity for vWF-dependent platelet agglutination (12). In fact, the fixed platelets usually were more reactive than fresh platelets in the tests. The second discovery was that the fixed platelets could be lyophilized and evenly resuspended without spontaneous platelet aggregation (13). The lyophilized platelets retained their full ability to aggregate in the vWF-dependent platelet aggregation tests. A photograph of a batch of lyophilized platelets is shown in Figure 2. These platelets, when resuspended, are discrete and are aggregated in 5 to 7 seconds in a standard test mixture with vWF plus botrocetin or ristocetin. The lyophilized platelets stored at -20°C appear to be stable for years.

Botrocetin Bioassay

The bioassay of botrocetin is based on the observation that the rate of platelet aggregation under standard conditions with a constant amount of vWF is dependent on the concentration of botrocetin (14). The rate of macroscopic platelet aggregation as a function of botrocetin concentration is demonstrated in Table 1. A unit of botrocetin is defined as that amount in 1 ml which will cause platelet aggregation in 14 seconds in a standard four-part macroscopic test system, in which the concentration of vWF is 1 unit/ml. This unit was derived by determining the range of small changes in botrocetin concentration that caused maximal changes in platelet aggregation times (14). In order to assay the potency of a botrocetin preparation, it is diluted to give samples that cause platelet aggregation in the range of 12 to 16 seconds. From Table 1, which relates platelet aggregation time to

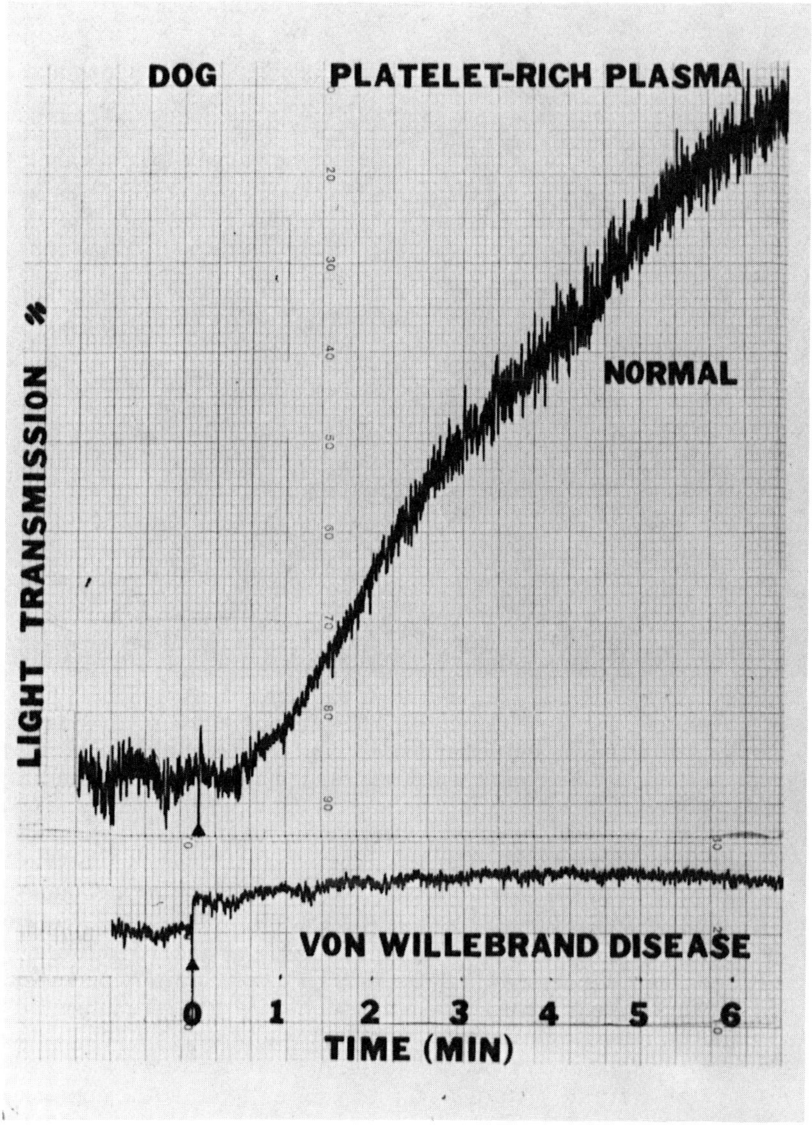

Figure 1 Aggregometry curves of platelet-rich normal and vWD dog plasmas after the addition of botrocetin (arrows). Final botrocetin concentraiton was 6 units/ml. In normal plasma, progressive aggregation, as judged by increased light transmission, follows a short lag phase. In vWD plasma, no change in light transmission occurs, with no platelet aggregation as confirmed by phase microscopy.

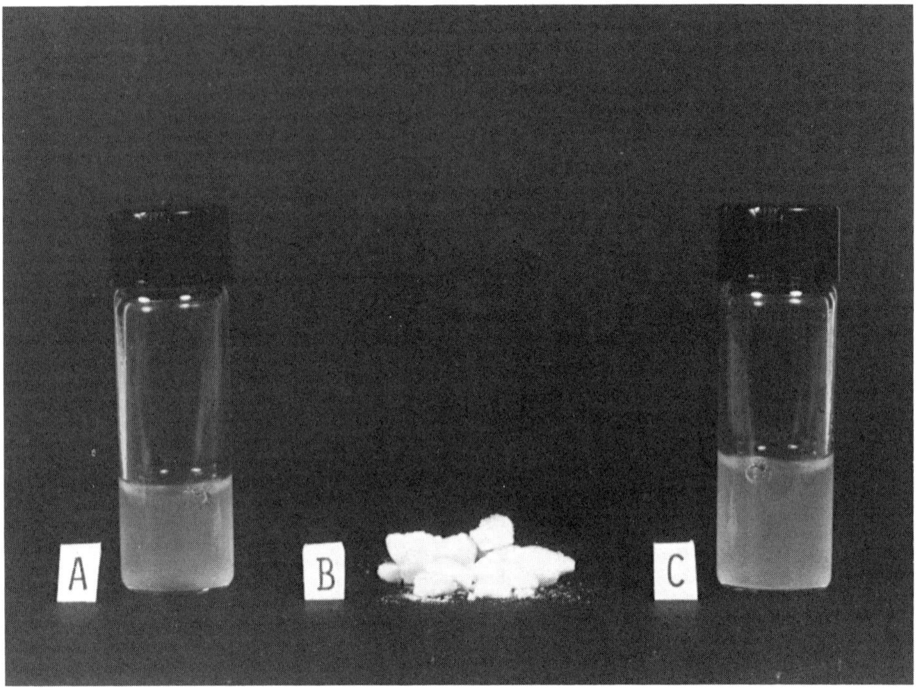

Figure 2 Steps in the preparation of lyophilized platelets. The platelets are fixed (12) (A), then lyophilized (13) (B), and stored at -20°C. Before use, the lyophilized platelets are resuspended in imidazole buffer, pH 7.35 (C).

concentration of botrocetin, the potency of the preparation is determined by multiplying the reciprocal of the total dilution of the preparation by the conversion factor in the table. For example, if an unknown botrocetin preparation diluted 1:128, with a further dilution of 1:4 in the test mixture, gives a platelet aggregation time of 13 seconds, the unitage is 128 × 4 × 1.16 or 594 botrocetin units/ml.

The ability to assay botrocetin on the basis of its functional activity was pivotal in further studies of this venom factor. An accurate assay procedure for vWF/botrocetin was dependent on use of botrocetin preparation of known potency, as were studies on its isolation, properties, and in vivo effects.

von Willebrand Factor Bioassay with Botrocetin

Botrocetin offers a test reagent for the bioassay of vWF with either the macroscopic or aggregometric methodology (15), and can be used with

Table 1 Conversion Factors for Bioassy of Botrocetin

Platelet aggregation time (sec)	Botrocetin (units/ml)
12.0	1.35
12.5	1.25
13.0	1.16
13.5	1.08
14.0	1.00
14.5	0.92
15.0	0.86
15.5	0.78
16.0	0.75

platelet-rich plasma, isolated fresh platelet preparations, suspensions of formaldehyde-fixed platelets, or lyophilized-fixed platelets after resuspension. A unit of vWF is that amount present in 1 ml of normal human reference plasma. The vWF values in human plasmas obtained with botrocetin and ristocetin were found to be comparable (15,16). If the distribution of the array of multimeric forms of vWF is abnormal, with a reduction or absence of the higher molecular weight forms, the apparent vWF/ristocetin concentration is lower than vWF/botrocetin values (17,18). Calculations to determine the amount of vWF/botrocetin and vWF/ristocetin are carried out in the same way. For both macroscopic and aggregometric testing, the rate of platelet aggregation of a test sample is compared with that of a normal reference plasma to which is assigned a value of 1 unit vWF/botrocetin and 1 unit vWF/ristocetin per ml. Each plasma is serially diluted. With macroscopic testing, the platelet aggregation times for each diluted sample of normal and test plasmas are plotted on semilog graph paper, as illustrated in Figure 3. The vWF values for the test plasmas are determined by locating points on the reference plasma curve that represent the same platelet aggregation time as in the test plasma, and read off the corresponding vWF concentration on the abscissa. With aggregometric testing, the ratio of the slopes of the line tangent to the steepest part of the aggregometric curve for the test plasma to that of reference plasma provides the vWF unit value per ml of the test plasma (19).

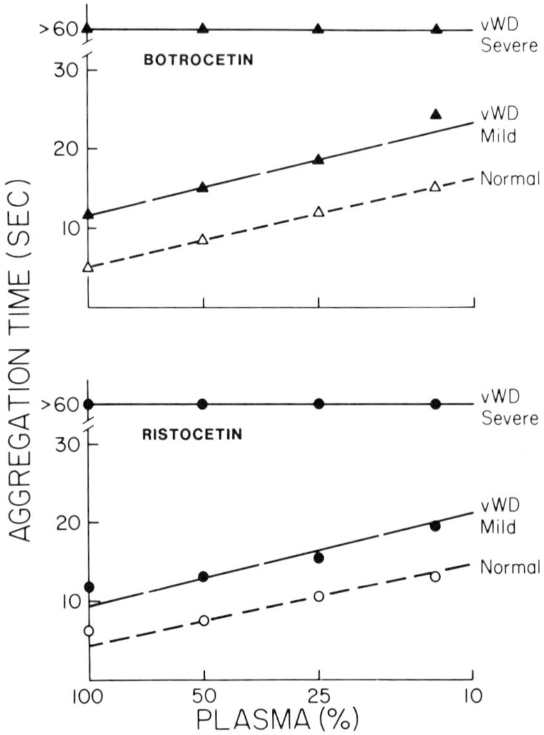

Figure 3 Bioassay of human vWF in patients with heterozygous (mild) and homozygous (severe) vWD using botrocetin (upper plot) and ristocetin (lower plot) as the "activator." The *final* concentration of botrocetin in the test mixture was 12.5 units/ml; of ristocetin it was 0.6 mg/ml. The vWF/botrocetin values are 0.28 units/ml for the mild vWD subject and less than 0.05 units/ml for the severe vWD subject. The corresponding vWF/ristocetin values were 0.28 units/ml and less than 0.05 units/ml.

ISOLATION AND CHARACTERIZATION OF BOTROCETIN

The crude venoms containing botrocetin activity also contain thrombin-like enzymes (1). The fibrinogen-clotting time is used as an index of the concentration of thrombin-like enzymes. Serially diluted crude venom samples are used. In our experience, the clotting time with diluted venom samples is usually much slower than the rate of platelet aggregation. In one study with *B. jararaca* venom, when the clotting time was 2 minutes, the platelet aggregation time was 7 to 8 seconds. Thus screening of venoms for botrocetin activity can usually be carried

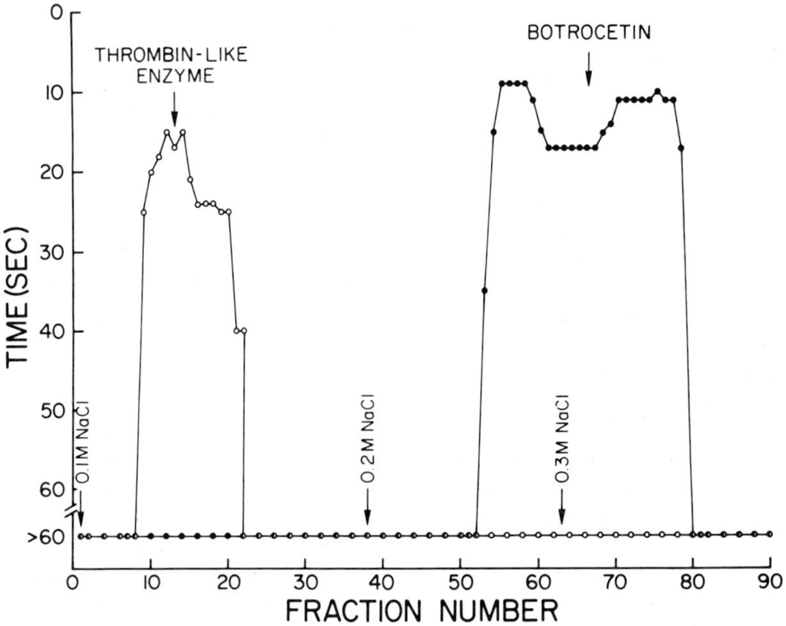

Figure 4 Isolation of botrocetin on a DEAE-cellulose column using *B. jararaca* venom. The column was eluted starting with 0.1 M buffered NaCl solution, followed by stepwise increases in the molarity of NaCl. The thrombin-like enzyme was determined by the clotting time of a fibrinogen solution. Botrocetin activity was assessed by the macroscopic platelet aggregation time.

out without undue interference with fibrin clotting. In bioassays of plasmas, vWF/botrocetin freed of thrombin-like enzyme is needed. Successful separation of these two activities in venom was accomplished by DEAE-cellulose chromatography (Fig. 4). The thrombin-like enzyme eluted with the starting buffer, whereas the botrocetin activity was recovered using stepwise elution with salt solutions of increasing molarity. The elution fractions containing botrocetin are assayed for content of the platelet-aggregating activity and are lyophilized, with storage at -20°C or 4°C. The preparations are stable for over 12 months.

Botrocetin as isolated by ion exchange chromatography has been partially characterized. The fractions containing botrocetin lost none of their activity on dialysis with a casing with a cutoff for molecules 12 kDa or greater in size. After polyacrylamide gel electrophoresis, the botrocetin fractions were found to be heterogeneous. One major band in the range of 25 to 29 kDa and two very faint minor bands stained with Coomassie brilliant blue. Recovery of the active fraction from an unstained polyacrylamide gel localized the aggregating activity to the major staining band only. Further studies on slab gels indicate the approximate molecular weight of botrocetin to be 26.5 kDa. Botrocetin appears to be a protein, as indicated by the Coomassie blue stain, and to contain no detectable carbohydrate, as indicated by the periodic acid-Schiff reagent. Botrocetin activity was stable at room temperature for several hours, but at 60°C, most of the activity (78%) was lost in 2 hours. This sensitivity to heat segregates botrocetin from heat-stable venom proteins such as phospholipase A_2, arginine ester hydrolase, thrombin-like enzyme, and factor V activator (20). No specific proteolytic activity of botrocetin has been identified. Substrates used were casein, TAMe (tosyl-L-arginine methyl ester), fibrinogen, and BAEE (benzoyl-L-arginine ethyl ester). Protease inhibitors, including trasylol, DFP (diisopropylfluorophosphatase), benzamidine hydrochloride, and PMSF (phenylmethylsulfonyl fluoride) did not interfere with botrocetin-induced aggregation. Also, botrocetin was fully active in the presence of the sulfhydryl inhibitors, PCMB (p-chloromercurobenzoic acid), MMC (methyl mercuric chloride), NEM (n-ethyl maleimide), and DTNB (5,5'-dithiobis-2-nitrobenzoic acid), implying that its aggregating activity probably does not involve SH groups. Likewise, botrocetin activity is unaffected by incubation with EDTA or by the presence of EDTA or citrate in the plasma during bioassays. This contrasts with such venom metalloproteins as phospholipases and venom endopeptidases, which require calcium for their activity. Heparin in doses sufficient to prevent blood from clotting does not interfere with the action of botrocetin. Tryptic digestion resulted in a nearly complete loss of platelet-aggregating activity.

MECHANISMS OF ACTION OF BOTROCETIN

Role of von Willebrand Factor

The essentiality of von Willebrand factor for the platelet-aggregating action of botrocetin was shown in two ways. One was the lack of platelet aggregation in platelet-rich plasma and in mixtures of platelet-free plasma and platelets if the plasma source was from a subject with severe vWD (Fig. 1). This was true with plasmas obtained from humans, pigs, and dogs with homozygous vWD. The other was the inhibitory effects of polyclonal anti-vWF antibodies in preventing or decelerating botrocetin-induced platelet aggregation with normal plasma (1). Table 2 illustrates the effect of antibody neutralization with normal plasma (1). Table 2 also shows the effect of antibody neutralization of vWF on vWF/botrocetin activity. Three different anti-vWF antisera were used, as follows: (a) a human antihuman vWF antibody occurring in a multitransfused

Table 2 Antibody Neutralization of vWF/Botrocetin

Antibody	vWF source (normal plasma)	Human platelet aggregation time (sec)			
		Botrocetin		Control[a]	
		Ab	No Ab	Ab	No Ab
Human anti-human vWF	Human (1:2)	>120	11.0	Neg.	Neg.
Goat anti-human vWF (1:20)	Human (1:4)	>120	15.0	Neg.	Neg.
Rabbit anti-porcine vWF (1:8)	Canine (1:16)	>120	20.0	Neg.	Neg.

[a]No botrocetin

homozygous vWD subject, (b) an antihuman vWF antibody raised in a goat, and (c) an antiporcine vWF antibody raised in a rabbit. Each antibody inhibited vWF/botrocetin platelet aggregation. These antibodies likewise inhibited vWF-dependent aggregation initiated with ristocetin (1).

The distribution of the multimeric forms of vWF in plasma influences the rate of vWF-induced platelet aggregation initiated by the activators, ristocetin and botrocetin (17,18). Ristocetin is known to react with only the high molecular weight multimers of vWF (5). Three plasma vWF preparations in which the high molecular weight multimers were greatly decreased or absent were tested with both activators. These vWF preparations were variant vWD type-IIA plasma, a high-purity commercial factor VIII concentrate, and supernatant plasma after preparation of cryoprecipitate (18). With comparable levels of vWF:Ag in all three preparations, vWF/ristocetin was not detectable, whereas the test for vWF/botrocetin was positive. From these and similar experiments, it was concluded that vWF/botrocetin correlated better with vWF:Ag values than did vWF/ristocetin. Further studies are needed to determine if the lowest molecular weight multimers of vWF are fully reactive with botrocetin. In addition, these studies suggest that the site of action of botrocetin on vWF may be different from the ristocetin-reactive site.

There are two additional studies that suggest that ristocetin and botrocetin may react with different domains on the vWF molecule. In one study, a patient with benign gammopathy and acquired von Willebrand syndrome possessed a monoclonal antibody that bound to vWF (21). This binding antibody, however, did not inhibit vWF/botrocetin in platelet aggregation and inhibited vWF/ristocetin only slightly. These data suggest that the antibody bound to an epitope that was coterminous, at least partially, with the ristocetin-reactive domain on the vWF molecule, whereas the epitope was separate from the botrocetin-reactive domain. Thus, the ristocetin- and botrocetin-reactive domains on the vWF molecule may be distinct (21). In another study, vancomycin inhibited vWF/ristocetin, but not vWF/botrocetin, suggesting vancomycin and ristocetin bound to the same domain on vWF, which was different from the botrocetin-reactive domain (22). It is unknown if botrocetin reacts directly with vWF to produce an activated form that will agglutinate platelets, or if it mediates this reaction in some other manner.

Role of Platelet Receptors

The membrane proteins that are concerned with vWF binding to platelets and presumably with vWF-dependent aggregation of platelets are glycoproteins Ib (GPIb) and IIb/IIIa (5). For ristocetin-induced aggregation of platelets, it is believed that GP Ib is needed. A comparison of the effects of variations in the membrane glycoprotein composition on vWF/ristocetin and vWF/botrocetin activity is shown in Table 3. In Glanzmann's thrombasthenia in which GP IIb/IIIa is missing or impaired and in which GP Ib is intact, both ristocetin and botrocetin caused platelet aggregation. In Bernard-Soulier syndrome (BSS), with lack of platelet GP Ib, ristocetin-induced platelet aggregation does not occur. However, botrocetin does aggregate BSS platelets, although less rapidly than normal platelets (22). In platelets exposed to a purified metalloprotein from *Serratia marcescens*, in which GP Ib is lost (23), vWF-dependent aggregation is absent with ristocetin, but positive with botrocetin (10). A monoclonal anti-GP Ib Ab caused inhibition of both vWF/ristocetin (8) and vWF/botrocetin (10). From these and other data, it appears that the GP Ib membrane receptor is needed for botrocetin action, although impairment or lack of GP Ib is more effective in preventing ristocetin aggregation than that due to botrocetin.

USES OF vWF BIOASSAYS WITH BOTROCETIN

Human Studies

The vWF/botrocetin bioassays are employed to measure the levels of vWF in plasma and plasma concentrates, as well as platelet vWF (15,16, 21). Table 4 shows a comparison of results using both botrocetin and ristocetin in assaying plasma vWF. Normal plasmas and plasmas collected in the study of families with classic vWD were used. Good agreement between the two types of vWF bioassays was observed (15).

Table 3 Comparison of vWF/Botrocetin and vWF/Ristocetin Activities Using Normal and GP Ib or GP IIb/IIIa Receptor-Deficient Platelets

Platelets	Platelet receptor: Ib	Platelet receptor: IIb/IIIa	Platelet aggregation: Botrocetin	Platelet aggregation: Ristocetin	References
Normal	Present	Present	++	++	15
Glanzmann's thrombasthenia	Present	Absent	++	++	22
Bernard-Soulier syndrome	Absent	Present	+	-	22
Serratia marcenscens (treated)	Absent	Present	+	-	10,23

++ = aggregation
+ = reduced aggregation
- = no aggregation

Table 4 Comparison of Bioassays of vWF/Botrocetin and vWF/Ristocetin in Normal Plasmas and Plasmas from Affected Members of vWD Families

Plasmas			
Subject	Sex	vWF/Botrocetin (%)	vWF/Ristocetin (%)
Normal subjects			
A.B.	M	100	100
J.S.	F	100	100
K.K.	F	73	78
R.B.	M	100	100
B.B.	M	124	122
N.C.	F	133	122
J.P.	F	83	89
R.W.	M	94	101
C.H.	M	78	63
C.S.	F	100	115
Mean ± SD		98.5 ± 13.2	99.0 ± 12.4
vWD (Heterozygotes)			
L.C.	F	36	53
M.N.	F	21	24
K.K.	M	50	56
vWD (Homozygotes)			
V.K.	F	<5	<5
B.B.	F	<5	<5
F.M.	M	<5	<5

The combined use of botrocetin and ristocetin vWF bioassays is valuable in studying variant vWD and platelet-type vWD. In variant vWD, type-IIA, with lack of high molecular weight vWF multimers, as indicated above, botrocetin-induced platelet aggregation is positive and correlates well with vWF:Ag levels (17,18). Botrocetin has been proposed for use in measuring the sensitivity of platelets to aggregate in platelet-type vWD. This disease is believed to be due to an intrinsic platelet abnormality, with plasma vWF near normal levels or moderately reduced. Early studies demonstrated that in this disease, binding of patients' vWF to platelets and platelet aggregation in patients' platelet-rich plasma (PRP) were induced with low concentrations of ristocetin, too low to cause either vWF binding to normal platelets or aggregation of normal platelet-rich plasma (24,25). Recently, Takahashi and coworkers (26) demonstrated that concentrations of botrocetin too low to cause aggregation in normal PRP caused pronounced aggregation of patient's PRP in platelet-type vWD. By using serially diluted botrocetin as activator, the difference in platelet reactivity in patients' platelets and normal platelets is well demonstrated, especially if plasma vWF is near normal levels.

The botrocetin-vWF-platelet-aggregating system may also be used to identify and quantitate circulating anticoagulants or antibodies to vWF (1,21). These antibodies may develop in multitransfused patients with severe vWD or may develop spontaneously in autoimmune von Willebrand syndrome (21). The test is performed by mixing samples of test plasma, serially diluted, with normal plasma and determining the highest dilution of plasma that prolongs the botrocetin-induced aggregation of platelets (21).

It has been proposed that botrocetin may be used to harvest platelets from blood or platelet-rich plasma with the platelet aggregates sedimenting rapidly. Deaggregation can be brought about with dextran sulfate. Botrocetin can also be used to aggregate megakaryocytes in harvested suspensions of cells of bone marrow aspirates to assist in their isolation from other cells.

Animal Studies

The first use of botrocetin in animal plasma was for the assay of vWF to determine the phenotypes in an inbred strain of vWD dogs, whether homozygous, heterozygous, or normal (1,11). Botrocetin was a necessity since dog plasma was found to be ristocetin-resistant. A systematic study was then made to determine the applicability of botrocetin for the assay of vWF in other animal plasmas (27). Table 5 summarizes the data on the reactivity of botrocetin with plasmas from different animals. The tests were carried out with platelets from four species of animals and with human platelets. In addition to the different species of platelets used in this study, it was also demonstrated that mixtures

Table 5 Reactivity of vWF of Various Animal Plasmas with Botrocetin and Heterologous and Homologous Platelets: Comparison with vWF Reactivity in Ristocetin and PAggF Tests

Plasma/species source	Platelet aggregation/platelet source				
	Human	Dog	Pig	Horse	Cow
Human	+(+)	+	+	+	+
Chimpanzee	+(+)	+	+	+	+
Rhesus	+(+)	+	+	+	+
Dog	+	+	+	+	+
Rat	+	+	+	+	+
Guinea pig	+	+	+	+	+
Pig	+(++)	+	+	+	+
Rabbit	+	+	+	+	+
Cow	+(++)	+	+	+	+
Sheep	+(++)	+	+	+	+
Goat	+(++)	+	+	+	+
Horse	+(+)	+	+	+	+
Cat	+	+	+	+	+

+ = Aggregation with botrocetin, negative in ristocetin and PAggF tests. +(+) = Aggregation with botrocetin and ristocetin, negative in PAggF tests. +(++) = Aggregation in botrocetin, ristocetin, and PAggF tests.

of homologous platelets and plasma from all of these species reacted well with botrocetin (27).

IN VIVO EFFECTS OF BOTROCETIN

Induced von Willebrand Syndrome

The effects of intravenous administration of botrocetin have been determined in dogs, pigs, and rats. Botrocetin preparations of known potencies were used in these studies, since the effects are dose-dependent. The results obtained from botrocetin administration in a normal dog (14) are illustrated in Figure 5. After injection of botrocetin, there

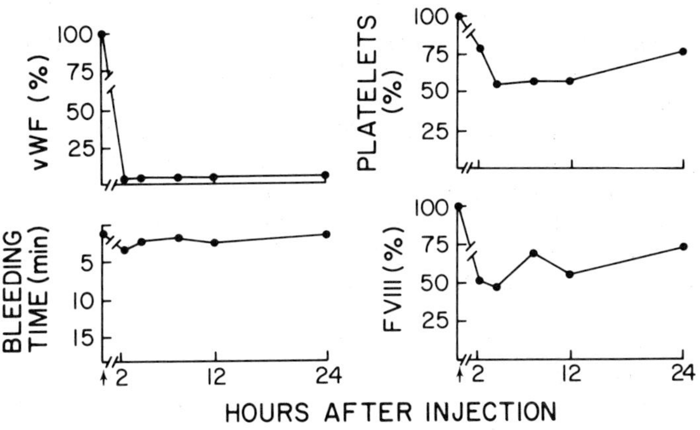

Figure 5 Induction of von Willebrand syndrome in a normal dog by administration of botrocetin (86 units/kg, arrow). Preinfusion values of vWF, platelets, and F.VIII were assigned as 100%, and postinfusion values represent changes from these baselines. The injections appeared innocuous, with no overt changes in the animal.

is a rapid loss of active vWF in the circulation, with undetectable levels of vWF/botrocetin. The vWF:Ag is also greatly depressed, but not as severely as is vWF/botrocetin. There is a concurrent but less severe depression in the level of F.VIII:C. The bleeding time remains normal. There is also a transient thrombocytopenia, with near normal platelet counts being restored in minutes or a few hours at most, depending on the dosage of botrocetin. Following a single injection of a sufficient dose of botrocetin (as in Fig. 5), there is nearly total loss of vWF from the plasma for 24 hours. Complete recovery occurs in a few days. By giving repeated doses of vWF at intervals, reduced levels of vWF, vWF:Ag, and F.VIII:C persist for several days. This has been demonstrated in rats, in which a chronic state of vWF deficiency lasting for several days was produced by repeated small injections over a period of time (28). The animals of each species had no overt signs or symptoms of any type with injection of botrocetin, suggesting that except for the plasma changes, botrocetin is innocuous.

The botrocetin appeared to cause activation and utilization of vWF, with its prompt removal from the circulation. Because most of the F.VIII:C is complexed to vWF in plasma, it could be passively removed from the circulation, along with vWF. The findings of decreased levels of plasma vWF, vWF:Ag, and F.VIII:C following administration of botrocetin to normal animals is somewhat analogous to the acquired vW syndrome (vWS) occurring in humans. For this reason, we called the condition produced by botrocetin injections in animals "induced vWS." Since botrocetin is reactive with the vWF of all animals tested, it would appear that this animal model of vWS could be produced in a wide variety of animals.

Acquired von Willebrand syndrome (vWS) has been reported many times in man (21). This syndrome is characterized not only by reduced plasma levels of the various factors of the F.VIII-vWF complex, but often by a long bleeding time and hemorrhagic diathesis as well. The syndrome may be of several types. In the autoimmune type of vWS, an anti-vWF antibody either neutralizes vWF or binds to vWF, with removal of the antigen-antibody complex from the circulation, depleting the plasma of this macromolecule and its attached F.VIII:C. This syndrome is very different from the botrocetin-induced syndrome, where there appears to be a massive consumption or utilization of plasma vWF. In some patients, vWS is a complication of a malignancy, most often hematopoietic, in which no anti-vWF antibodies have been demonstrated. The immune status of these patients is uncertain, and some patients may have a nonimmune type of vWS (29). These patients could have a vWF consumption-type of vWS analogous to the induced animal syndrome, possibly triggered by release of a botrocetin-like physiologic factor that results in vWF consumption.

Refractoriness of Animals with vWD

The effect of administration of botrocetin to animals with vWD has also been tested. Pigs and dogs with vWD from the Chapel Hill colony were used. The homozygous or severely affected animals of each species have no detectable vWF in their plasmas, with greatly reduced F.VIII:C and a long bleeding time. The heterozygous animals have mean plasma vWF levels of about 50% of normal, with normal or slightly reduced plasma F.VIII:C. The bleeding time is normal. Being of autosomal inheritance, the disorder occurs in both males and females. When botrocetin was administered to these animals, the response was very different from that observed in normal animals (30). Also, the homozygous animals responded differently from the heterozygous animals. In the heterozygous animals, the plasma vWF levels became undetectable with small doses of botrocetin, too small to produce more than a mild reduction in plasma vWF in normal animals. Plasma F.VIII:C was likewise greatly reduced and there was a transient mild thrombocytopenia. The recovery in the heterozygous animals was slow, suggesting that their cellular synthesis of vWF was quantitatively impaired or their stores of vWF reduced. It was postulated that small injections of botrocetin might help identify those carriers with near normal levels of vWF and F.VIII:C, since a disproportionate decrease in plasma vWF occurs compared with normal subjects. In the homozygous vWD animals, botrocetin injections were without any detectable effect. There was no thrombocytopenia.

SUMMARY AND CONCLUSIONS

Botrocetin, originally called venom coagglutinin, is a potent noncoagulant platelet-aggregating material present in the venoms of several snakes, particularly in certain members of the *Bothrops* genus. Its platelet-aggregating activity is dependent on the presence of the macromolecular adhesive plasma protein, von Willebrand factor. Botrocetin can be largely separated from the thrombin-like enzyme in crude venom by ion exhange chromatography. It has been further purified to near homogeneity. Its molecular weight is about 26,500. It is a heat-labile protein (60°C) without demonstrable enzyme activity. A method of botrocetin bioassay has been developed and a unit of botrocetin defined, based on the rate of platelet aggregation with a constant amount of vWF. Although the mechanism of its action in causing platelet aggregation is uncertain, preliminary studies suggest it causes activation of vWF, which in turn binds to the platelet receptor glycoprotein Ib. The venom factor mediates the binding of vWF to platelets. The domain on the vWF molecule that interacts with botrocetin appears to be different from the one that interacts with ristocetin.

Botrocetin appears to be a unique probe for the detection and measurement of vWF and in vivo manipulation of plasma vWF levels. It is reactive with a wide range of varying-sized vWF multimers, unlike other probes for vWF, ristocetin, and the PAggF reaction, which are interactive with high molecular weight vWF multimers only. Furthermore, unlike other probes of vWF, botrocetin is reactive with vWF from all species of animals tested. Botrocetin is a valuable reagent for the clinical laboratory assessment of the quantitative and qualitative status of plasma vWF. In addition to platelets in blood, botrocetin also causes aggregation of megakaryocytes in bone marrow cell suspensions, and may provide a method for the isolation of these cells. An induced von Willebrand syndrome, in which the factor VIII-vWF complex is greatly reduced or depleted, occurs when purified botrocetin is given intravenously in dogs, pigs, and rats. The severity of the syndrome is dose dependent.

ACKNOWLEDGMENT

This work was supported in part by a Grant no. HR-01648-39 from the National Institutes of Health, Bethesda, Maryland.

REFERENCES

1. Read, M.S., R. W. Shermer, and K. M. Brinkhous. Venom coagglutinin: an activator of platelet aggregation dependent on von Willebrand factor. Proc. Natl. Acad. Sci. USA 75:4514 (1978).

2. Marder, V. J., P. M. Mannucci, B. G. Firkin, L. W. Hoyer, and D. Meyer. Standard nomenclature for factor VIII and von Willebrand factor: a recommendation by the International Committee on Thrombosis and Haemostasis. Thromb. Haemost. In press.
3. Howard, M. A. and B. G. Firkin. Ristocetin—a new tool in the investigation of platelet aggregation. Thromb. Diath. Haemost. 26:362 (1971).
4. Griggs, T. R., H. A. Cooper, W. P. Webster, R. H. Wagner, and K. M. Brinkhous. Plasma aggregating factor (bovine) for human platelets: a marker for study of antihemophilic and von Willebrand factors. Proc. Natl. Acad. Sci. USA 70:2814 (1973).
5. Zimmerman, T. S., Z. M. Ruggeri, and C. A. Fulcher. Factor VIII/von Willebrand factor. Progr. Hematol. 13:279 (1983).
6. Brinkhous, K. M., H. Sandberg, J. B. Garris, C. Mattsson, M. Palm, T. Griggs, and M. S. Read. Purified human factor VIII:C: comparative hemostatic response after infusions into hemophilic and von Willebrand disease dogs. Proc. Natl. Acad. Sci. USA 82:8752 (1985).
7. Reddick, R. L., T. R. Griggs, M. A. Lamb, and K. M. Brinkhous. Platelet adhesion to damaged coronary arteries: comparison in normal and von Willebrand disease swine. Proc. Natl. Acad. Sci. USA 79:5076 (1982).
8. Coller, B. S., E. I. Peerschke, L. E. Scudder, and C. A. Sullivan. Studies with a murine monoclonal antibody that abolishes ristocetin-induced binding of von Willebrand factor to platelets: additional evidence in support of GPIb as a platelet receptor for von Willebrand factor. Blood 61:99 (1983).
9. Timmons, S., M. Kloczewiak, and J. Hawiger. ADP-dependent common receptor mechanism for binding of von Willebrand factor and fibrinogen to human platelets. Proc. Natl. Acad. Sci. USA 81:4935 (1984).
10. Read, M. S. Unpublished data.
11. Brinkhous, K. M., M. S. Read, R. L. Reddick, and T. R. Griggs. Pathophysiology of platelet aggregating von Willebrand factor: applications of the venom coagglutinin vWF assay. Ann. NY Acad Sci. 370:191 (1981).
12. Allain, J. P., H. A. Cooper, R. H. Wagner, and K. M. Brinkhous. Platelets fixed with paraformaldehyde: a new reagent for assay of von Willebrand factor and platelet aggregating factor. J. Lab. Clin. Med. 85:318 (1975).
13. Brinkhous, K. M. and M. S. Read. Preservation of platelet receptors for platelet aggregating factor/von Willebrand factor by air drying, freezing, or lyophilization: new stable platelet preparations for von Willebrand factor assays. Thromb. Res. 13:591 (1978).

14. Brinkhous, K. M., D. S. Barnes, J. Y. Potter, and M. S. Read. Von Willebrand syndrome induced by a *Bothrops* venom factor: bioassay for venom coagglutinin. Proc. Natl. Acad. Sci. USA 78:3230 (1981).
15. Brinkhous, K. M., and M. S. Read. Use of venom coagglutinin and lyophilized platelets in testing for platelet-aggregating von Willebrand factor. Blood 55:517 (1980).
16. Furlan, M., B. A. Perret, R. Felix, and E. A. Beck. Quantitative und qualitative Beurteilung des faktor VIII/von Willebrand—Proteins bei verschiedenen formen der von Willebrandschen krankheit. Beziehung zwischen molekulgrosse, botrocetin-cofaktor und latex-antigen. Schweiz. Med. Wochenschr. 114:1391 (1984).
17. Howard, M. S., H. H. Salem, K. B. Thomas, L. Hau, J. Perkin, M. Coghlan, and B. G. Firkin. Variant von Willebrand's disease type B—revisited. Blood 60:1420 (1982).
18. Brinkhous, K. M., M. S. Read, W. A. Fricke, and R. H. Wagner. Botrocetin (venom coagglutinin): reaction with a broad spectrum of multimeric forms of factor VIII macromolecular complex. Proc. Natl. Acad. Sci. USA 80:1463 (1983).
19. Brinkhous, K. M. and M. S. Read. Fixed platelets: their use in platelet aggregation tests. In *Platelets, Methods in Enzymology*, Hawiger, J. (Ed.), Academic Press, New York (in press).19.
20. Iwanga, S. and T. Suzuki. Enzymes in snake venom. In *Snake Venoms*, Lee, C-Y. (Ed.), Springer-Verlag, Berlin, p. 61 (1979).
21. Fricke, W. A., K. M. Brinkhous, J. B. Garris, and H. R. Roberts. Comparison of inhibitory and binding characteristics of an antibody causing acquired von Willebrand syndrome: an assay for von Willebrand factor binding by antibody. Blood 66:562 (1985).
22. Howard, M. A., J. Perkin, H. H. Salem, and B. G. Firkin. The agglutination of human platelets by botrocetin: evidence that botrocetin and ristocetin act at different sites on the factor VIII molecule and platelet membrane. Br. J. Haemat. 57:25 (1984).
23. Cooper, H. A., W. P. Bennett, G. C. White II, and R. H. Wagner. Hydrolysis of human platelet membrane glycoproteins with a *Serratia marcescens* metalloprotease: effect on response to thrombin and von Willebrand factor. Proc. Natl. Acad. Sci. USA 79:1433 (1982).
24. Weiss, H. J., D. Meyer, R. Rabinowitz, G. Pietu, J. P. Girma, W. J. Vicic, and J. Rogers. Pseudo-von Willebrand's disease: an intrinsic platelet defect with aggregation by unmodified human factor VIII/von Willebrand factor and enhanced adsorption of its high-molecular-weight multimers. N. Engl. J. Med. 306:326 (1982).
25. Miller, J. L. and A. Castella. Platelet-type von Willebrand's disease: characterization of a new bleeding disorder. Blood 60:790 (1982).

26. Takahashi, H., R. Nagayama, A. Hattori, and A. Shibata. Botrocetin- and polybrene-induced platelet aggregation in platelet-type von Willebrand disease. Am. J. Hematol. 18:179 (1985).
27. Read, M. S., J. Y. Potter, and K. M. Brinkhous. Venom coagglutinin for detection of von Willebrand factor activity in animal plasmas. J. Lab. Clin. Med. 101:74 (1983).
28. Sanders, W. E., Jr. and K. M. Brinkhous. Unpublished data.
29. Roussi, J. H., L. L. Houbouyan, R. Alterescu, B. Franc, and A. F. Goguel. Acquired von Willebrand's syndrome associated with hairy cell leukaemia. Br. J. Haemat. 46:503 (1980).
30. Brown, W. H., M. S. Read, D. S. Barnes, T. C. Flynn, W. E. Sanders, and K. M. Brinkhous. Response of von Willebrand disease pigs and dogs to venom coagglutinin. Fed. Proc. 41:705 1982).

29

Properties of a Potent Platelet Aggregation Inhibitor from *Echis carinatus* Snake Venom

Che-ming Teng and Yunn-hwa Ma
National Taiwan University College of Medicine, Taipei, Taiwan

INTRODUCTION

Hemorrhagic snake (Viperidae or Crotalidae) venoms are mixtures of proteins affecting blood vessels, plasma coagulation, and platelet functions. Procoagulant, anticoagulant, and fibrinolytic principals have been purified and utilized in laboratory tests, in clinical uses, and in studies of the mechanism of hemostasis (Morita et al., 1976; Seegers and Ouyang, 1979; Teng and Seegers, 1981; Ouyang et al., 1982).

In recent years, some venom principals affecting platelet aggregation were purified, and their chemical natures and action properties were studied (Table 1). Among the aggregating agents, there are proteinases with procoagulant activity, or nonenzymatic glycoproteins which induce aggregation independently of ADP and thromboxane pathways. Coagglutinin was claimed to induce aggregation or agglutination associated with von Willebrand factor. Cardiotoxin, a membrane-active peptide isolated from *Elapidae* snake venoms, could potentiate the aggregation by enhancing the activation of endogeneous phospholipase A_2 and, thus, increasing thromboxane formation.

On the other hand, there are venom enzymes inhibiting platelet aggregation (e.g., phospholipase A_2, 5'-nucleotidase, and α-fibrinogenase). However, some phospholipase A_2 inhibits platelet aggregation independently of its enzymatic activities. In this chapter, we report a nonenzymatic inhibitor purified from *Echis carinatus* snake venom, which inhibits the aggregation induced by various inducers at a common pathway by interfering with the interaction between fibrinogen and its specific receptors on the platelet membrane.

Table 1 Classification and Properties of Purified Venom Proteins Affecting Platelet Functions

Classification	Venom	Active principle and properties	References
I. Aggregating protein			
Enzymatic	*B. atrox*	Thrombocytin, coagulant, serine proteinase	Kirby et al., 1979
	C. h. horridus	Crotalocytin, serine proteinase	Schmaier and Colman, 1980
	E. carinatus	Ecarin, prothrombin activator	Teng et al., 1985
Nonenzymatic	*T. mucrosquamatus*	Glycoprotein, noncoagulant, ADP-independent pathway	Teng et al., 1981
	T. gramineus	Glycoprotein, noncoagulant, ADP-independent pathway	Ouyang and Huang, 1983c
	C. d. cascavella	Convulxin, noncoagulant, PAF formation	Vargaftig, 1982
	Bothrops	Coagglutinin, von Willebrand factor-dependent	Read et al., 1978
II. Potentiating protein	*N. naja atra*	Cardiotoxin, membrane-active peptide-enhancing thromboxane formation	Teng et al., 1984b
III. Inhibitory protein			
Enzymatic	*T. mucrosquamatus* *V. russelli* *N. naja atra*	Phospholipase A_2, PBPB-sensitive, also causes aggregation of albumin-free platelets	Ouyang and Huang, 1984 Teng et al., 1984 Tend et al., 1986
	A. rhodostoma *T. gramineus* *T. mucrosquamatus*	α-fibrinogenase, specifically α(A)-chain splitting	Ouyang et al., 1983a Huang et al., 1984 Ouyang et al., 1979
	T. gramineus	5'-Nucleotidase, ADP hydrolysis	Ouyang and Huang, 1983a
Nonenzymatic	*T. gramineus* *A. halys*	Phospholipase A_2, PBPB-insensitive	Ouyang and Huang, 1983b Ouyang et al., 1983b
	E. carinatus	Carinatin, glycoprotein, fibrinogen receptor antagonism	Teng, et al.. 1985

MATERIALS AND METHODS

Materials

Platelet aggregation inhibitor (carinatin) was purified from the venom of *E. carinatus* (Ouyang et al., 1985), and was homogeneous as judged by the disc electrophoresis. Collagen (bovine achilles tendon, Sigma Chemical Company) was homogenized in 25 mM acetic acid. Thrombin (bovine, Parke & Davis Company) was dissolved in 50% glycerol to give a stock solution of 100 NIH units/ml. Ionophore A-23187 and platelet-activating factor (PAF) were obtained form Calbiochem-Behring Company. Rabbit fibrinogen (90% clottable), sodium arachidonate, adenosine diphosphate (ADP), thimerosal, purified luciferase-luciferin, indomethacin, apyrase, α-chymotrypsin, and bovine serum albumin were purchased from Sigma Chemical Company.

Platelet Preparations

Platelet-rich plasma and washed platelets were prepared as previously described (Teng et al., 1984a, 1984b). Chymotrypsin-treated platelets were prepared by the method of Niewiarowski and co-workers (1981) by incubation of washed platelets with α-chymotrypsin (8 U/ml) at room temperature for 15 minutes, and then washed with Tyrode's solution. The medium for platelet suspension was Tyrode's solution of the following composition (mM): NaCl (136.8), KCl (2.8), $NaHCO_3$ (11.9), $MgCl_2$ (1.1), NaH_2PO_4 (0.33), $CaCl_2$ (1.0), and glucose (11.2). Apyrase (0.1 mg/ml) and bovine serum albumin (3.5 mg/ml) were added.

Measurement of Aggregation and ATP Release

Platelet aggregation was measured by the turbidimetric method (Cross, 1964) at 37°C and 900 rpm using a Payton Lumi-aggregometer (Model 1020). The percentage of aggregation was calculated as described previously (Teng et al., 1984a). For measuring ATP release, the bioluminescence method was used. Luciferase-luciferin was added to the platelet suspension 1 minute before the addition of the inducer. The gain of the fluorescent intensity was adjusted to about 80% of the maximal deflection. The area under the curve (cm^2) was calculated to measure the relative fluorescent intensity.

Thromboxane B_2 Assay

At 6 minutes after the addition of the inducer, EDTA (2 mM) and indomethacin (50 μM) were added to platelet suspension. After the centrifugation, thromboxane B_2 content of the supernatant was measured with radioimmunoassay kits (New England Nuclear Company), according to the procedure described by the manufacturer.

RESULTS AND DISCUSSION

We purified platelet aggregation inducer and inhibitor from *E. carinatus* venom (Ouyang et al., 1985). The former induced platelet aggregation through prothrombin activation, while the latter was devoid of any enzymatic activity possessed by the crude venom. The venom inhibitor (carinatin) was an acidic glycoprotein with a molecular weight of 6800 daltons. Upon heat treatment at 90°C in acidic medium, it was stable for 4 hours. Carinatin decreased the elasticity of the whole blood clot as measured by thromboelastograph, but did not prolong the coagulation time of plasma.

Carinatin inhibited the aggregation of washed platelets induced by collagen, thrombin, ionophore A-23187, arachidonic acid, ADP, and platelet-activating factor (PAF) (Fig. 1). Hence, this inhibitory effect was relatively nonspecific, and the IC_{50} (concentration causing 50% inhibition of aggregation) was estimated to be around 10 μg/ml for all aggregation inducers. This inhibition occurred only in authentic aggregation, because the agglutination of formaldehyde-fixed platelets induced by polylysine was not affected. The effect of carinatin on platelets was nonenzymatic because its incubation with platelets did not increase the inhibitory effect. Furthermore, it could be washed out from platelets, and further addition of carinatin caused the same inhibitory action.

The formation of thromboxane B_2 in platelets caused by various inducers was not significantly affected by carinatin (Table 2). This indicates that carinatin may not affect the formation of thromboxanes or prostaglandin endoperoxides. In fact, carinatin caused further inhibition of collagen-induced platelet aggregation in the presence of indomethacin (50 μM) which was sufficient to block the cyclo-oxygenase activity. Inhibition of aggregation was not due to a defect in the release of ADP from platelets upon the activation by the inducers because ATP release, except that induced by collagen, was not affected by carinatin (Table 2). Furthermore, carinatin still inhibited the aggregation of ADP-refractory platelets and thrombin-degranulated platelets (Teng et al., 1985). Hence, the pathway of ADP also could not have been involved. In addition to thromboxane A_2 and ADP, PAF was proposed to be the other possible pathway (Vargaftig et al., 1981) for inducing platelet aggregation. However, PAF-induced aggregation was inhibited by carinatin with a similar IC_{50} as other inducers. Thus, carinatin may affect aggregation at the common pathway after the activation of platelets, but not at a specific pathway or the aggregation process. The first possible mechanism we considered for this nonspecific effect was inhibition of calcium influx. However, the in-

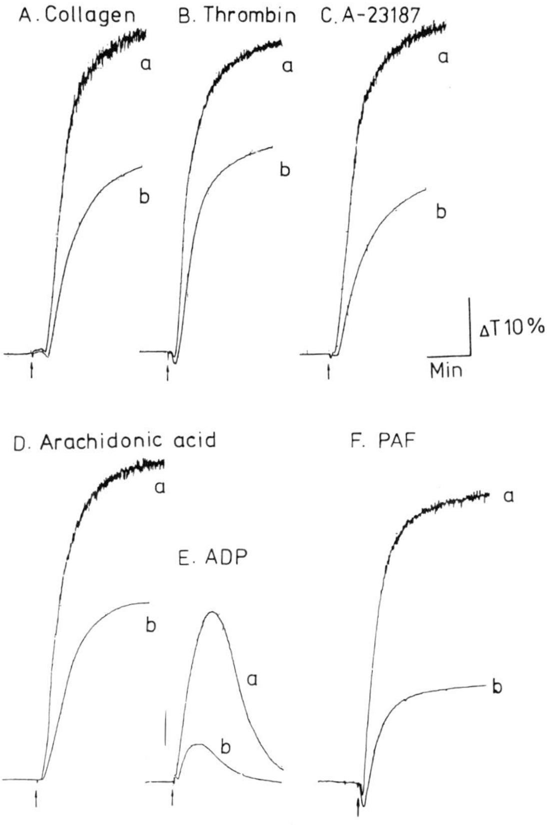

Figure 1 Inhibitory action of carinatin on the aggregation of washed rabbit platelets. Aggregation agent was added at the arrow: collagen (A, 10 μg/ml), thrombin (B, 0.1 U/ml), ionophore A-23187 (C, 2 μM), arachidonic acid (D, 20 μM), ADP (E, 20 μM), PAF (F, 10 ng/ml). (a) control; (b) carinatin (10 μg/ml) was added 1 minute prior to the aggregating agent.

crease of calcium concentration from 1 to 30 mM could not counteract the inhibitory action, even when ionophore A-23187 was added. Neither was its inhibitory effect due to the membrane-stabilizing effect because only the aggregation, not the lysis, of platelets caused by thimerosal was inhibited by carinatin.

The inhibitory effect of carinatin on the aggregation of platelet-rich plasma induced by collagen was four times less potent than that in washed platelets (Fig. 2). The IC_{50} for platelet-rich plasma was calculated to be 40 μg/ml, whereas that for washed platelets was 10

Table 2 Effect of Carinatin on Aggregation, Thromboxane B_2 Formation, and ATP Release Induced by Collagen, Thrombin, Ionophore A-23187, Arachidonic Acid, ADP, and Platelet-Activating Factor (PAF) in Washed Rabbit Platelets

	Aggregation (%)	Thromboxane B_2 (ng/3 × 10^8 platelets)	ATP release*
Control			
Collagen	87.2 ± 1.8 (8)	164.9 ± 33.8 (4)	35.0 ± 5.4 (6)
Thrombin	88.9 ± 1.1 (12)	132.8 ± 40.8 (8)	24.2 ± 3.4 (5)
A-23187	86.7 ± 1.7 (6)	186.7 ± 62.8 (7)	31.1 ± 5.6 (5)
Arachidonic acid	75.2 ± 5.2 (5)	73.9 ± 12.1 (5)	32.1 ± 6.2 (6)
ADP	53.5 ± 5.0 (4)	8.7 ± 0.6 (4)	—
PAF	87.6 ± 1.4 (5)	37.5 ± 5.1 (5)	—
With carinatin			
Collagen	50.6 ± 3.2 (10)‡	173.8 ± 22.9 (3)	19.2 ± 3.7 (6)†
Thrombin	53.2 ± 2.8 (12)‡	220.1 ± 60.4 (8)	19.2 ± 6.9 (5)
A-23187	60.7 ± 3.6 (7)‡	225.8 ± 80.4 (6)	38.1 ± 9.5 (5)
Arachidonic acid	31.9 ± 3.7 (5)‡	90.6 ± 42.1 (4)	35.3 ± 7.9 (6)
ADP	20.5 ± 5.8 (5)‡	7.6 ± 0.7 (4)	—
PAF	56.2 ± 3.9 (6)‡	51.9 ± 11.0 (4)	—

*Area under the curve (cm^2).
†$p < 0.05$, ‡$p < 0.001$ compared with the respective control.
Values are expressed as means ± SE, and experimental numbers are in parentheses. Final concentration: 10 μg/ml (collagen), 0.1 U/ml (thrombin), 2 μM (ionophore A-23187), 20 μM (arachidonic acid), 20 μM (ADP), 10 ng/ml (PAF), 10 μg/ml (carinatin).

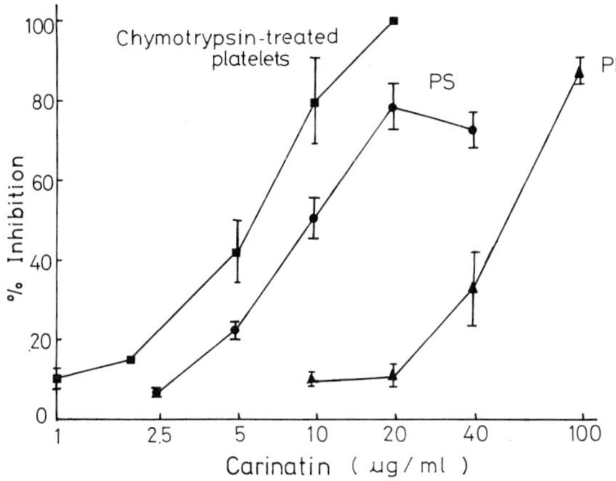

Figure 2 Dose-response curves of carinatin in the aggregation of platelet-rich plasma, platelet suspension, and chymotrypsin-treated platelets. Aggregating agent was collagen (10 µg/ml) in platelet-rich plasma (PRP) or platelet suspension (PS), or it was fibrinogen (0.3 mg/ml) in chymotrypsin-treated platelets. Data are expressed as the means ± SE (n = 3 to 8).

µg/ml. This gave us a hint that some components in plasma had an inhibitory action on the effect of carinatin. Fibrinogen had an antagonizing effect on carinatin in collagen-induced aggregation of washed platelets. This explained its weaker inhibition on the aggregation of platelet-rich plasma. The antagonism by fibrinogen was a competitive one against carinatin (Fig. 3). Because carinatin did not possess any fibrinogenolytic activity, its inhibitory effects on platelet aggregation was not similar to those produced by α-fibrinogenases (Ouyang et al., 1979, 1983a; Huang et al., 1984). Albumin (3.5 mg/ml) did not have any effect on carinatin. However, a high concentraiton of albumin (40 mg/ml) had an antagonizing effect on carinatin, but was still much weaker than that caused by fibrinogen (0.4 mg/ml). Thus, fibrinogen is the major protein antagonizing the effect of carinatin in plasma.

There are reports that the fibrinogen-platelet interaction is very important for the aggregation process (Tangen et al., 1971; Niewiarowski et al., 1972), especially the ADP-induced platelet aggregation. It has also been reported that fibrinogen receptors could be exposed after the treatment of platelets with chymotrypsin (Greenberg et al., 1979). The properties of exposed receptors were the same as those induced by ADP (Niewiarowski et al., 1981). A low concentration of fibrinogen could induce aggregation of these chymotrypsin-treated platelets. Carinatin caused more marked inhibition of the aggregation

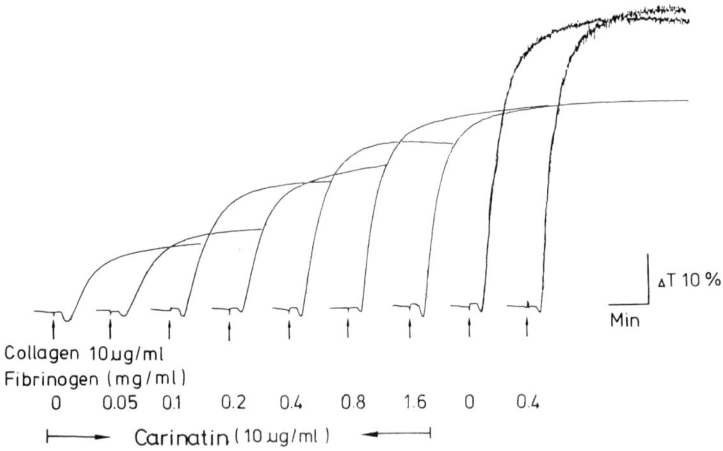

Figure 3 Antagonism by fibrinogen of the inhibitory action of carinatin in washed platelets. Fibrinogen (0.05 to 1.6 mg/ml) and/or carinatin (10 μg/ml) were added to platelet suspension 1 minute before the addition of collagen (10 μg/ml, arrow).

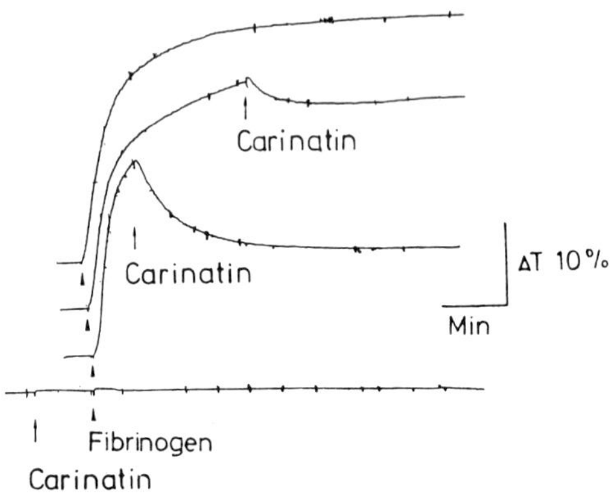

Figure 4 Effect of carinatin on the fibrinogen-induced aggregation of chymotrypsin-treated platelets. Carinatin (10 μg/ml) was added 1 minute before or 0.5, 2 minutes after the addition of fibrinogen (0.2 mg/ml).

of these modified platelets (Fig. 2). Pretreatment with carinatin (10 to 20 μg/ml) completely inhibited the aggregation. It also deaggregated the clumped platelets induced by fibrinogen. The earlier the addition of carinatin, the greater the extent of deaggregation (Fig. 4). In addition to ADP, it is still uncertain whether the fibrinogen-platelet interaction is involved in the aggregation of platelets caused by other inducers (Plow and Marguerie, 1980; Bennett et al., 1981; Harfenist et al., 1982; Morinelli et al., 1983). Because carinatin inhibited platelet aggregation induced by various inducers, we suggest that the fibrinogen-platelet interaction is a common final pathway of the aggregation process. Carinatin may be a valuable tool for this study.

ACKNOWLEDGMENTS

This work was supported by a Grant no. NSC74-0412-B002-55 from the National Science Council of the Republic of China.

REFERENCES

Bennett, J. S., G. Vilaire, and J. W. Burch (1981). A role for prostaglandins and thromboxanes in the exposure of platelet-fibrinogen receptors. J. Clin. Invest. 68:981.

Cross, M. J. (1964). Effect of fibrinogen on the aggregation of platelet by adenosine diphosphate. Thromb. Diath. Haemorrh. 12:524.

Greenberg, J. P., M. A. Packham, M. A. Guccione, E. J. Harfenist, J. L. Orr, R. L. Kinlough-Rathbone, D. W. Perry, and J. F. Mustard (1979). The effect of pretreatment of human or rabbit platelets with chymotrypsin on their responses to human fibrinogen and aggregating agents. Blood 54:753.

Harfenist, E. J., M. A. Guccione, M. A. Packham, R. L. Kinlough-Rathbone, and J. F. Mustard (1982). Arachidonate induced fibrinogen binding to thrombin degranulated platelets is independent of released ADP. Blood 59:956.

Huang, T. F., J. H. Chang, and C. Ouyang (1984). Characterization of hemorrhagic principles from *Trimeresurus gramineus* snake venom. Toxicon 22:45.

Kirby, E. P., S. N. Niewiarowski, K. Stocker, C. Kettner, E. Shaw, and T. M. Budzynski (1979). Thrombocytin: a serine protease from Bothrops atrox venom. 1. Purification and characterization of the enzyme. Biochemistry 18:3564.

Morinelli, T. A., S. Niewiarowski, E. Kornecki, W. R. Figures, Y. Wachtfogel, and R. W. Colman (1983). Platelet aggregation and exposure of fibrinogen receptors by prostaglandin endoperoxide analogues. Blood 61:41.

Morita, T., S. Iwanaga, and T. Suzuki (1976). The mechanism of bovine prothrombin by an activator isolated from *Echis carinatus* venom and characterization of the new active intermediates. J. Biochem. (Tokyo) 79:1089.

Niewiarowski, S., A. Z. Budzynski, T. A. Morinelli, T. M. Budzynski, and G. J. Stewart (1981). Exposure of fibrinogen receptor on human platelets by proteolytic enzymes. J. Biol. Chem. 256: 917.

Niewiarowski, S., E. Regoeczi, and J. F. Mustard (1972). Platelet interaction with fibrinogen and fibrin: comparison of the interaction of platelets with that of fibroblasts, leukocytes and erythrocytes. Ann. NY Acad. Sci. 201:72.

Ouyang, C. and T. F. Huang (1983a). Inhibition of platelet aggregation by 5'-nucleotidase purified from *Trimeresurus gramineus* snake venom. Toxicon 21:491.

Ouyang, C. and T. F. Huang (1983b). Potent platelet aggregation inhibitor from *Trimeresurus gramineus* snake venom. Biochim. Biophys. Acta 757:332.

Ouyang, C. and T. F. Huang (1983c). A potent aggregation inducer from *Trimeresurus gramineus* snake venom. Biochim. Biophys. Acta 761:126.

Ouyang, C., H. I. Yeh, and T. F. Huang (1983b). A potent platelet aggregation inhibitor purified from *Agkistrodon halys* (mamushi) snake venom. Toxicon 21:797.

Ouynag, C. and T. F. Huang (1984). Effect of the purified phospholipase A_2 from snake and bee venoms on rabbit platelet function. Toxicon 22:705.

Ouyang, C., L. T. Hwang, and T. F. Huang (1983a). α-Fibrinogenase from *Agkistrodon rhodostoma* (Malayan pit viper) snake venom. Toxicon 21:25.

Ouyang, C., Y. H. Ma, H. C. Jih, and C. M. Teng (1985). Characterization of the platelet aggregation inducer and inhibitor from *Echis carinatus* snake venom. Biochim. Biophys. Acta 841:1.

Ouyang, C., C. M. Teng, and Y. C. Chen (1979). Properties of fibrinogen degradation products produced by α- and β-fibrinogenases of *Trimeresurus mucrosquamatus* snake venom. Toxicon 17:121.

Ouyang, C., C. M. Teng, and T. F. Huang (1982). Characterization of the purified principles of Formosan snake venoms which affect blood coagulation and platelet aggregation. J. Formosan Med. Assoc. 81:781.

Plow, E. F. and G. A. Marguerie (1980). Participation of ADP in the binding of fibrinogen to thrombin-stimulated platelets. Blood 256:553.

Read, M. S., R. W. Shermer, and K. M. Brinkhous (1978). Venom coagglutinin: an activator of platelet aggregation dependent on von Willebrand factor. Proc. Natl. Acad. Sci. USA 75:4514.

Schmaier, A. H. and R. W. Colman (1980). Crotalocytin: characterization of Timber rattlesnake platelet activating protein. Blood 56:1020.

Seegers, W. H. and C. Ouyang (1979). Snake venoms and blood coagulation. In *Handbook of Experimental Pharmacology, Volume 52, Snake Venoms*, Lee, C. Y. (Ed.), Springer-Verlga, Berlin, Heidelberg, New York, p. 684-750.

Tangen, O., H. J. Berman, and P. Murphy (1971). Gel-filtration A new technique for separation of blood platelets from plasma. Thromb. Diath. Haemorrh. 25:268.

Teng, C. M., Y. H. Chen, and C. Ouyang (1984a). Biphasic effect on platelet aggregation by phospholipase A purified from *Vipera russellii* snake venom. Biochim. Biophys. Acta 772:393.

Teng, C. M., W. Jy, and C. Ouyang (1984b). Cardiotoxin from *Naja naja atra* snake venom: a potentiator of platelet aggregation. Toxicon 22:463.

Teng, C. M., Y. P. Kuo, L. G. Lee, and C. Ouyang (1986). Effect of cobra venom phospholipase A_2 on platelet aggregation in comparison with those produced by arachidonic acid and Lysophosphatidylcholine. Thromb. Res. 44:875.

Teng, C. M., K. K. Liao, J. P. Wang, H. S. Lin, and C. Ouyang (1981). Ultrastructural changes and release reaction of platelets induced by an aggregation inducer purified from *Trimeresurus mucrosquamatus* (Formosan Habu) snake venom. Toxicon 19:121.

Teng, C. M., Y. H. Ma, and C. Ouyang (1985). Action mechanism of the platelet aggregation inducer and inhibitor from *Echis carinatus* snake venom. Biochim. Biophys. Acta 841:8.

Teng, C. M. and W. H. Seegers (1981). Activation of Factor X and thrombin zymogens in 25% sodium citrate solution. Thromb. Res. 22:213.

Vargaftig, B. B. (1982). Platelet activation by non-coagulant snake venom components. Toxicon 20:279.

Vargaftig, B. B., M. Chignard, and J. Benveniste (1981). Present concepts on the mechanisms of platelet aggregation. Biochem. Pharmacol. 30:263.

30

Activation of Human Platelets Induced by the Aggregoserpentin of *Trimeresurus mucrosquamatus* Venom (TMVA)

Jennifer Hong Chen,* Haiying Wan, and Changgeng Ruan
Thrombosis and Haemostasis Research Unit, Suzhou Medical College, Suzhou, Republic of China

INTRODUCTION

Recently, several platelet-activating principles from Crotalidae venoms have been purified and characterized (1,2). Some are diisopropylfluorophosphate (DFP)-sensitive serine proteinases, such as thrombocytin (3) and crotalocytin (4). Others are DFP-resistant and nonenzymatic proteins, such as the platelet aggregoserpentin from *Trimeresurus mucrosquamatus* (PAS-TM) (5), the platelet aggregoserpentin from *Trimeresurus gramineus* (PAS-TG) (2), and convulxin, a purified platelet-activating factor from *Crotalus durissus terrificus* (1).

We studied the platelet aggregation induced by the *Trimeresurus mucrosquamatus* venom (from the Hunan Province) and found that the actions of this venom were thromboxane A_2-independent (6). This is different from the actions of PAS-TM, which was dependent on prostaglandin synthesis and whose actions were inhibited by aspirin (7). Using DEAE-Sephadex A-50 column chromatography and gel filtration, the *Trimeresurus mucrosquamatus* venom (from the Hunan Province) aggregoserpentin (TMVA) was purified. It is a glycoprotein with a molecular weight of about 68,000 daltons by SDS polyacrylamide gel electrophoresis, but the carbohydrate content (7%) is much less than PAS-TM (Wu, X. et al., unpublished data). In this chapter, we attempt to elucidate the mechanism of TMVA-induced platelet aggregation.

---

* *Present affiliation*: Stanford University, Stanford, California

MATERIALS AND METHODS

The TMVA was provided by Dr. Y. Xiong (Kunming Institute of Zoology, Academia Sinica, Kunming, China). Arachidonic acid, ADP, creatine phosphate (CP), and creatine phosphokinase (CPK) were purchased from Sigma Chemical Company, St. Louis, Missouri. The PAF-acether and A-23187 were obtained from Calbiochem, La Jolla, California, and ristocetin was from Lundbeck. Thrombin was purchased from Tianjin Biochemical Manufactory, China. Tetramethyl pyrazine (TP) and mepacrine were provided by Pharmacology Department, Suzhou Medical College, Suzhou, China. Monoclonal antibodies to human platelet, AN-51 (8) and SZ-2 (9), were prepared at our laboratory.

Platelet aggregation was performed by the turbidimetric method using an aggregometer (PAM Model 2, Jiangsu, China), as described previously (6). Platelet-rich plasma (PRP) was prepared by conventional centrifugation of human citrated blood. In the case of thrombin-treated experiments, human platelets were washed free from their plasma and resuspended in a modified Tyrode's solution (2.6 mM KCl, 1.05 mM $MgCl_2$, 137 mM NaCl, 1.3 mM $CaCl_2$, 10 mM Tris, 5.05 mM glucose, pH 7.4) containing 0.25% bovine serum albumin. Platelet counts were adjusted to 300,000 per μl.

Degranulated platelets were prepared according to the method essentially described by Teng and co-workers (10). A suspension of washed human platelets was prewarmed to 37°C and mixed with thrombin (0.3 U/ml) at 37°C in the presence of EDTA (0.2 mM). After incubation for 6 minutes, this mixture was centrifuged at 10,000 × *g* for 10 minutes, and the platelet pellets were washed twice and resuspended in Tyrode's solution.

Electron microscopic studies were performed according to White (11). The platelets were added to an equal volume of 0.125% glutaraldehyde and centrifuged for 3 minutes at 15,000 *g* in an Eppendorf centrifuge. The samples were subsequently fixed and embedded by standard procedures. Thin sections were stained with uranyl acetate and lead citrate before study in an Hitachi H-500 electron microscope.

Serotonin (5HT) release was measured after prelabeling citrated PRP with [^{3}H]-5HT (5-hydroxy [G-^{3}H] tryptamine creatine sulfate, 10 Ci/mM (from Shanghai Institute of Nuclear Science, Academia Sinica) for 30 minutes at 37°C. Labeled PRP was stirred in an aggregometer tube with different inducers for 3 minutes. The reaction was stopped by adding ice-cold EDTA (0.1 M) and centrifuging for 30 seconds at 15,000 *g* in an Eppendorf centrifuge. The results were expressed as percent release of total ^{3}H-5HT in the platelets. When using unlabeled PRP, serotonin was fluorometrically measured by the formation of a fluorophore after condensation with orthophthaldialdehyde (12).

Thromboxane (TX) B_2 was measured by radioimmunoassay according to the procedure of Fitzpatrick (13), using an iodinated tracer of

TXB_2 (14). The antiserum used for the assay was kindly donated by Dr. Maclouf (Unité 150 INSERM, Hôpital Lariboisière, Paris, France).

RESULTS AND DISCUSSION

Activation of Human Platelets Induced by TMVA

The minimal concentration of the TMVA necessary to induce human platelet aggregation was 0.2 μg/ml, about 1/20 of that of the crude venom (6). The aggregation rate of human platelets was proportional to the concentration of TMVA, up to 20 μg/ml (Fig. 1). The TMVA also induced a dose-dependent release of serotonin from human platelets (Table 1).

Serotonin was measured fluorometrically after condensation with O-phthaldialdehyde.

Independence of Release Reaction

In the presence of EDTA (0.2 mM), thrombin (0.3 U/ml) caused a release reaction, but did not induce aggregation of human platelets. Under these conditions, about 98% of the serotonin was released from thrombin-treated platelets, and most of the dense bodies and α-granules were depleted as examined by electron microscopy (Fig. 2). Thrombin-degranulated platelets could be aggregated by TMVA at higher concentrations (30 μg/ml) in the presence of fibrinogen (0.3%), but failed to respond to ADP, thrombin itself, or TMVA in the absence of fibrinogen (Fig. 3). Like PAS-TM (10) and PAS-TG (2), TMVA-induced platelet aggregation appears to be independent of the release reaction.

Independence of TXA_2 Formation

The TMVA not only induced platelet aggregation, but also the formation of TXB_2, the stable metabolite of TXA_2. The effect of TMVA on the formation of TXB_2 was the same as that of PAF, but less potent than that of arachidonic acid (Fig. 4a). When the formation of TXB_2 was inhibited by aspirin (10 mM), the platelet aggregation induced by arachidonic acid (100 μg/ml) was abolished, but aspirin did not inhibit the platelet aggregation induced by TMVA and PAF (Fig. 4b). This activity of TMVA is different from that of PAS-TM, whose activity was completely inhibited by aspirin (5 mM) and related to prostaglandin synthesis (7).

Pretreatment of platelets with the combination of aspirin (10 mM) and an ADP-removing system, CP/CPK (5 mM/10 U/ml), also did not prevent the TMVA (10 μg/ml)- and PAF (3×10^{-7} M)-induced platelet aggregation (Fig. 5). Therefore the mechanism for platelet aggrega-

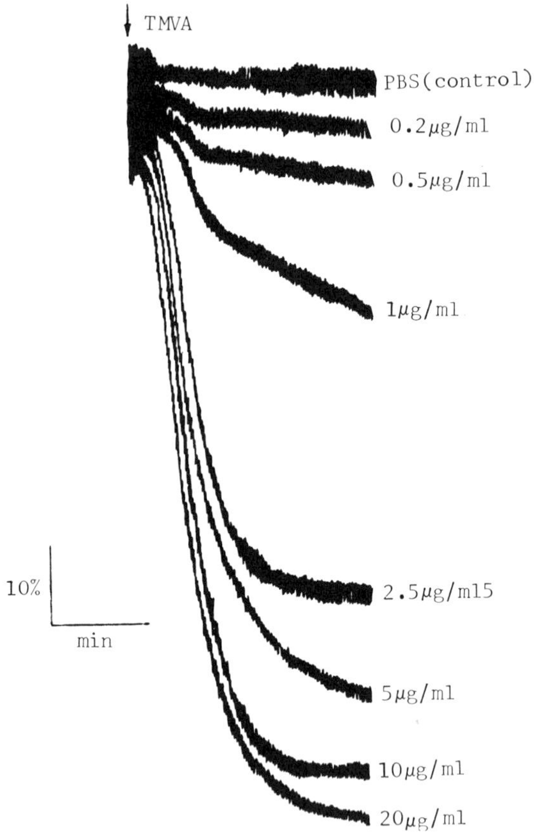

Figure 1 TMVA-induced platelet aggregation. Platelet aggregation was expressed by light transmission.

Table 1 Effect of TMVA on Platelet Aggregation and 5-HT Release Reaction

TMVA (μg/ml)	% Aggregation	% Release
1	24 ± 7	–
5	44 ± 12	23 ± 16
10	63 ± 7	61 ± 13
20	74 ± 8	72 ± 10

Values represent means ± SD (n = 6).

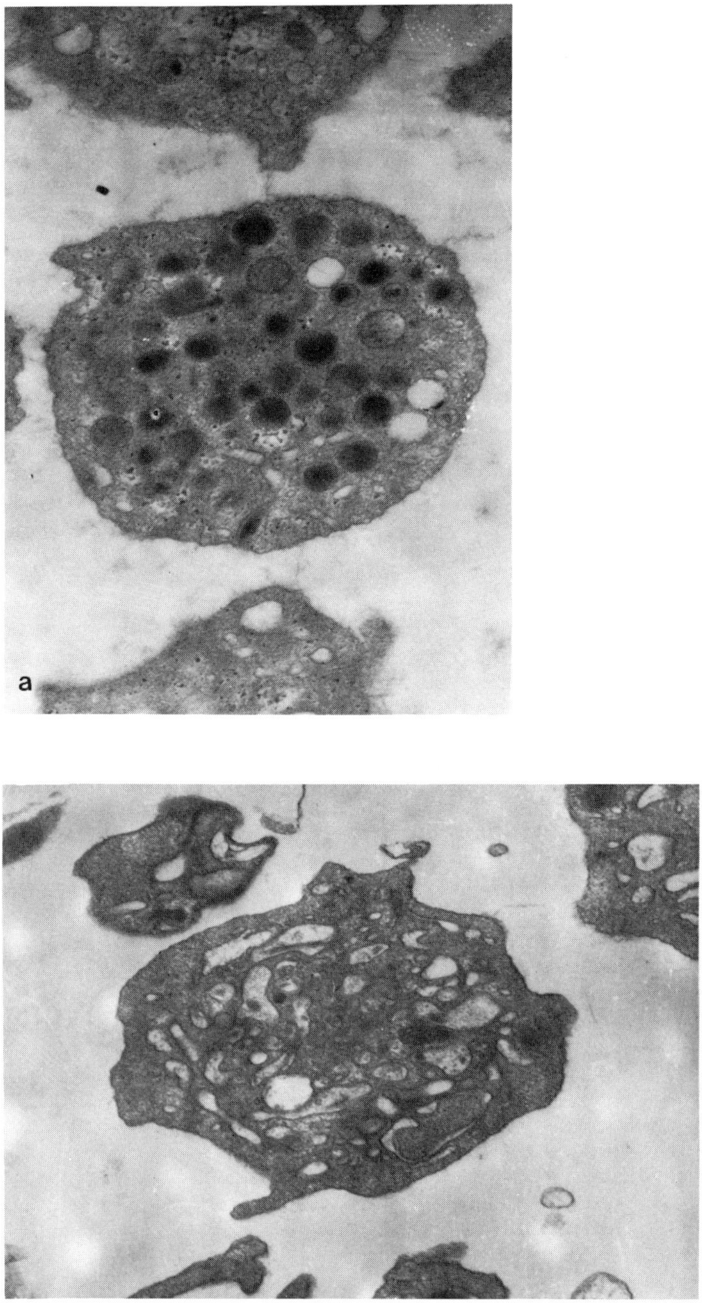

Figure 2 (a) Thin section of human platelets at 37°C for 6 minutes. Magnification: 15,000 X. (b) Degranulated platelets from a sample exposed to 0.3 u/ml of thrombin at 37°C for 6 minutes in the presence of EDTA (0.2 mM). Magnification 12,000 X.

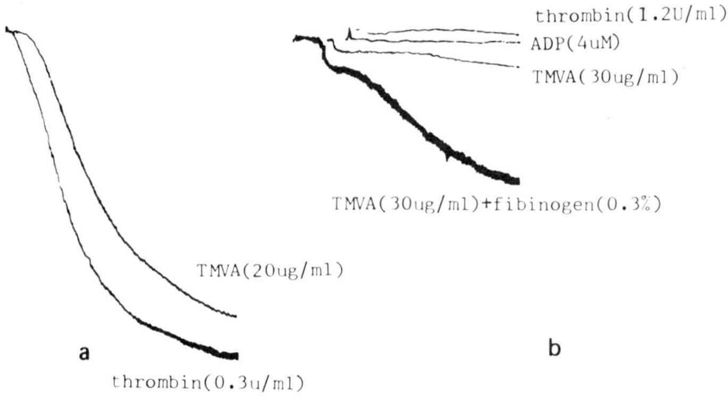

Figure 3 Effect of TMVA on aggregation of the thrombin-treated (degranulated) platelets. (a) Control platelets. (b) Degranulated platelets.

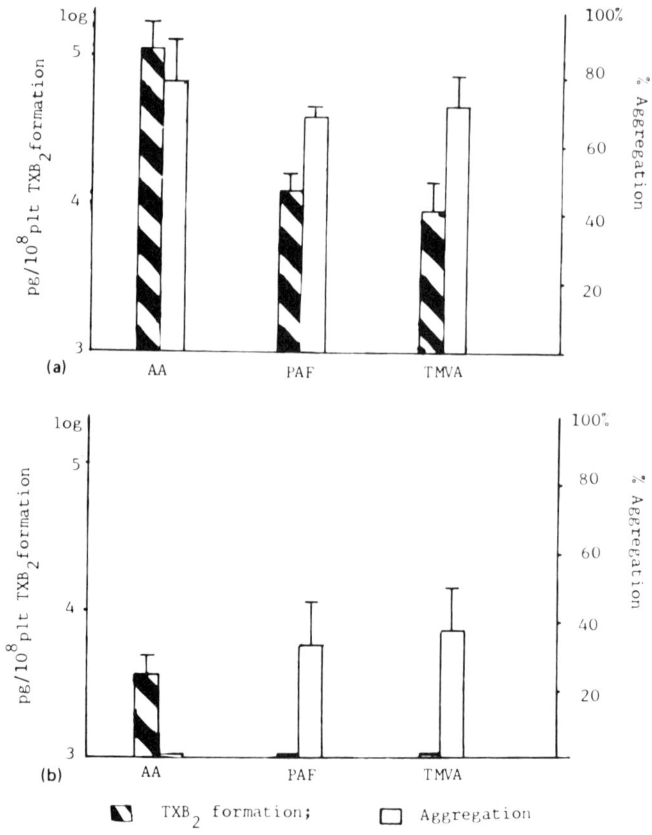

Figure 4 (a) Effects of inducers on TXB_2 formation and platelet aggregation. AA = 100 μg/ml; PAF = 3×10^{-7} M; TMVA = 10 μg/ml. (b) Effect of aspirin (10 mM) on TXB_2 formation and platelet aggregation. AA = 100 μg/ml; PAF = 3×10^{-7} M; TMVA = 10 μg/ml.

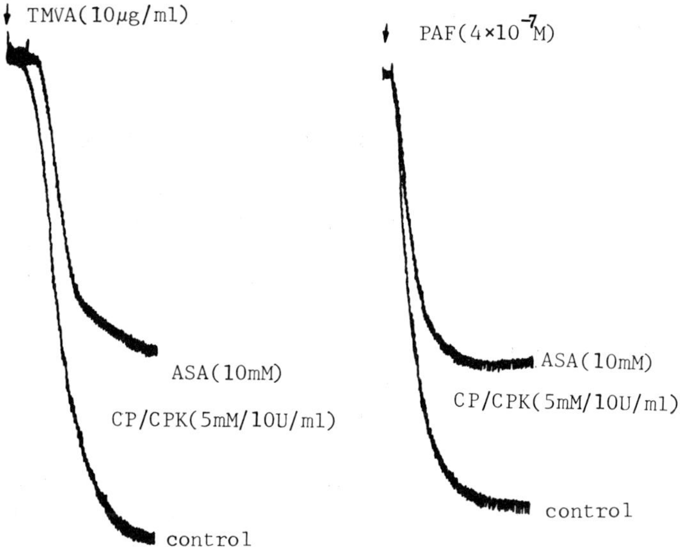

Figure 5 Effect of aspirin, creatine phosphate, and creatine phosphokinase on platelet aggregation.

tion with TMVA does not entail ADP, release reaction, or thromboxane. This was consistent with PAS-TG (2) and convulxin (1).

Common Pathways for TMVA and PAF

Mepacrine (25-100 μM) inhibited TMVA (10 μg/ml)-induced platelet aggregation and release reaction in a dose dependent manner (Table 2). Tetramethyl pyrazine (TP), a compound isolated from Chinese herbs (15), also inhibited TMVA (10 μg/ml)-induced platelet aggregation and release reaction, and this inhibition was dose-dependent (1.5 to 6 mM) Table 3). In the case of PAF-induced aggregation and release reaction, mepacrine and TP had a similar effect, but A-23187 was not susceptible to these inhibitors (Tables 2 and 3). Mepacrine and TP also inhibited the formation of TXB_2 induced by TMVA (data not shown). Mepacrine is a phospholipase A inhibitor, and the action of TP on human platelets appears to be mediated via inhibition of phospholipase C (unpublished data). Our present experiments do not indicate the TMVA-PAF relationships, but suggest that there are probably some common pathways for TMVA and PAF (e.g., a thromboxane A_2-independent mechanism via activation of phospholipase A_2 and/or phospholipase C for producing calcium flux during platelet stimulation) (16).

Table 2 Effect of Mepacrine on Platelet Aggregation and $^3H[5HT]$ Release Reaction

Inducer	Mepacrine (μM)	% Aggregation	% Release
TMVA	0	64 ± 8 (n = 3)	58 ± 11 (n = 3)
(10 μg/ml)	25	31 ± 13 (n = 3)	15 ± 2 (n = 3)
	50	18 ± 5 (n = 3)	14 ± 2 (n = 3)
	100	10 ± 6 (n = 3)	11 ± 6 (n = 3)
PAF	0	57 ± 2 (n = 3)	46,33
(1.82×10^{-7} M)	25	13,28	12,10
	50	10,16	13,13
	100	5,8	11,0
A_{23187}	0	67 ± 13 (n = 3)	67 ± 17 (n = 3)
(15 μM)	25	53 ± 14 (n = 3)	62 ± 14 (n = 3)
	50	69 ± 18 (n = 3)	67 ± 10 (n = 3)
	100	56,48	72,77

Values represent means ± SD.

Table 3 Effect of Tetramethyl Pyrazine (TP) on Platelet Aggregation and $[^3H]5HT$ Release Reaction

Inducer	TP (mM)	% Aggregation	% Release
TMVA	0	64 ± 8 (n = 3)	58 ± 11 (n = 3)
(10 μg/ml)	1.5	49 ± 16 (n = 3)	40 ± 9 (n = 3)
	3	38 ± 6 (n = 3)	29 ± 10 (n = 3)
	6	7 ± 3 (n = 3)	10 ± 3 (n = 3)
PAF	0	62 ± 14 (n = 3)	45 ± 10 (n = 3)
(1.82×10^{-7} M)	1.5	35 ± 20 (n = 3)	18 ± 11 (n = 3)
	3	26 ± 10 (n = 3)	3,0
	6	0	0
A_{23187}	0	61,62	69 ± 17 (n = 3)
(15 μM)	1.5	63 ± 14 (n = 3)	62 ± 14 (n = 3)
	3	68,57	64,90
	6	49	92

Values represent means ± SD.

Common Sites of Action for TMVA, PAF, and Collagen

We used two monoclonal antibodies to human platelet membrane glycoprotein I_b (GPI_b), An-51 and SZ-2, to study the effect of these antibodies on TMVA-induced platelet aggregation. The AN-51 specifically inhibited ristocetin-induced platelet aggregation (8). The SZ-2 not only inhibited platelet aggregation induced by ristocetin, but also platelet aggregation induced by collagen, and the epitope for SZ-2 is related to the platelet collagen receptor (17).

The platelet aggregation induced by TMVA (10 μg/ml) and PAF (3×10^{-7} M) was not inhibited by AN-51. However, SZ-2 dose-dependently (5 to 10 μg/ml) inhibited TMVA- and PAF-induced platelet aggregation (Fig. 6).

Recently, Vargaftig and co-workers (18) demonstrated the specific cross-desensitization of convulxin with collagen, which appears to involve more basic processes, such as common sites of action. Our results suggest that TMVA, PAF, and collagen may share common sites of action at the platelet membrane level, and one of these sites is probably related to GPI_b and recognized by monoclonal antibody SZ-2.

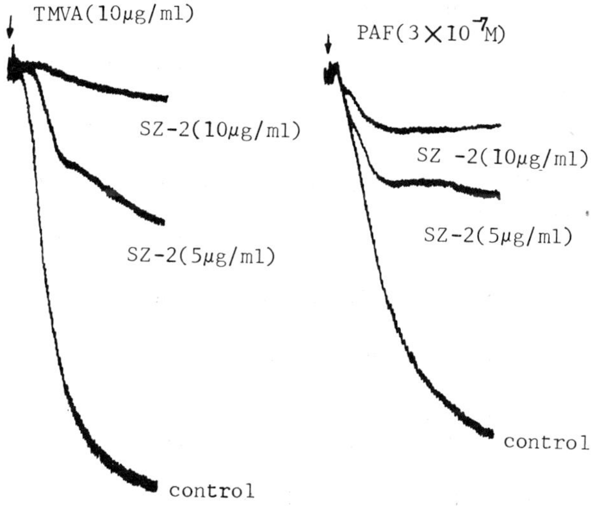

Figure 6 Effect of monoclonal antibody SZ-2 on platelet aggregation induced by TMVA (10 μg/ml) and PAF (3×10^{-7} M).

SUMMARY

In these experiments, TMVA induced the concentration-dependent aggregation of human platelets, which was accompanied by the formation of TXA_2 and the release of serotonin from the challenged platelets. However, the platelet aggregation induced by TMVA was independent of the release reaction and the formation of TXA_2. The actions of TMVA on human platelets were different from those of ADP, arachidonic acid, thrombin and A-23187, but similar to that of PAF and dependent on the activation of phospholipase A_2 and/or C. Collagen, TMVA, and PAF may share common sites of action on the platelet membrane.

ACKNOWLEDGMENT

This work was supported in part by the Science Fund of the Chinese Academy of Sciences, Beijing (Project no. 82-172).

REFERENCES

1. Vargaftig, B. B. Toxicon 20:279 (1982).
2. Ouyang, C. and T-F. Huang. Biochim. Biophys. Acta 761:126 (1983).
3. Kirby, E. P., S. Niewiarowski, K. Stocker, C. Kettner, E. Shaw, and T. M. Brudzynski. Biochemistry 18:3564 (1979).
4. Schmaier, A. H., W. Claypool, and R. W. Colman. Blood 56:1013 (1980).
5. Ouyang, C., J. P. Wang, and C. M. Teng. Biochim. Biophys. Acta 630:246 (1980).
6. Ruan, C., X. Xi, H. Wan, P. Li, Z. Gu, W. Wan, Y. Xiong, C. Yang, and X. Chen. Xoo. Res. 4:245 (1983).
7. Ouyang, C. and C. M. Teng. Thromb. Haemost. 41:475 (1979).
8. Ruan, C., G. Tobelem, A. J. McMichael, L. Drouet, Y. Legrand, L. Degos, N. Kieffer, H. Lee, and J. P. Caen. Br. J. Haemat. 49:511 (1981).
9. Ruan, C., X. Xi, P. Li, Q. Wu, H. Wan, X. Du, and J. Gu. Chinese J. Hematol. 6:130 (1985).
10. Teng, C. T., K. K. Liao, J. P. Wang, H. S. Lin, and C. Ouyang. Toxicon 19:121 (1981).
11. White, J. G. Blood 31:604 (1968).
12. Komesu, N. N. and J. H. Thompson. Eur. J. Pharmacol. 16:248 (1971).
13. Fitzpatrick, F. A. Methods Enzymol. 86:286 (1982).
14. Maclouf, J., M. Pradel, P. Pradelles, and F. Dray. Biochim. Biophys. Acta 431:139 (1976).
15. Beijing Institute of Pharmaceutical Industry. Natl. Med. J. China 7:420 (1977).

16. Nichols, W. L., J. M. Gerrard, and P. Didisheim. In *Recent Advances in Blood Coagulation*, Poller, L. (Ed.), Churchill-Livingstone, Edinburgh, pp. 1-39 (1981).
17. Ruan, C., X. Du, X. Xi, P. A. Castaldi, and M. C. Berndt. Thromb. Haemost. 54:124 (1985).
18. Vargaftig, B. B., D. Joseph, F. Wal, G. Marlas, M. Chignard, and L. G. Chevance. Eur. J. Pharmacol. 92:57 (1983).

V
HEMORRHAGIC FACTORS

31
Hemorrhagic Proteases from Snake Venoms

Review

Anthony T. Tu
Colorado State University, Fort Collins, Colorado

BRIEF DESCRIPTION OF HEMORRHAGE

Snakes of Crotalidae and Viperidae families induce hemorrhage and myonecrosis on envenomation (1-4). Hemorrhage is the rupture of the capillary endothelium, causing blood components to penetrate the surrounding tissues besides exuding through openings (Fig. 1). Recently, there has been considerable progress in studies on snake venom-induced hemorrhagic action. The present chapter reviews the current status of experimental venom hemorrhagic research.

The onset of hemorrhage is very fast, and the hemorrhagic lesion can be observed 2 minutes after the injection of *Crotalus atrox* venom. Intradermal, subcutaneous, and intramuscular injections of venom tend to produce a localized hemorrhagic lesion (Figs. 2 and 3). Intravenous injection of venom tends to produce systemic hemorrhagic lesions in many organs and tissues (Fig. 4). Of course, the difference is not that clear, as a high-dose injection of venom by the intramuscular route can also induce systemic lesions.

DIFFERENT HEMORRHAGIC TOXINS

There are many hemorrhagic toxins that have been isolated from snake venoms. Usually, a venom contains more than one hemorrhagic toxin (HT) (5,6). For instance, *C. atrox* venom is known to contain seven hemorrhagic toxins with different isoelectric points, molecular weights, and amino acid compositions. *C. horridus horridus* venom was found to con-

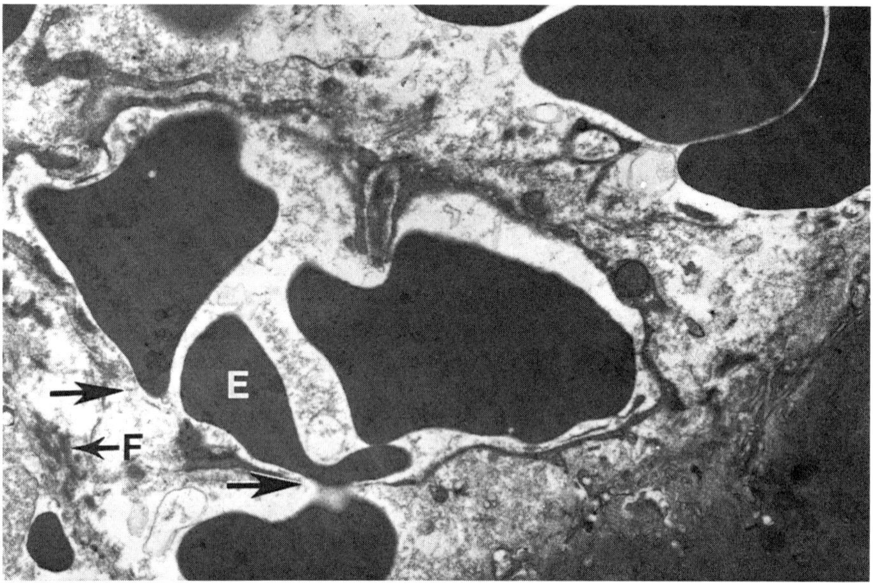

Figure 1 Hemorrhage induced by HT*a* of *Crotalus atrox*. Electron micrograph shows the breakage of endothelial cells. Note thinness of endothelium and presence of large gaps (arrows) through which erythrocytes (E) are escaping. Note also the presence of fibrin (F). (Reprinted from Ref. 35, with permission of the author and the *American Journal of Pathology*.)

tain at least three distinct hemorrhagic toxins (7). The molecular weights of these toxins range from 20,000 to 83,000. With different physical and chemical properties, it is quite convincing that each toxin is different from the other. However, it is better to confirm the difference of each toxin by other independent methods. For this purpose, immunodiffusion was employed to differentiate the various hemorrhagic toxins, especially for those toxins isolated from the same venom. As can be seen from Figure 5A, both HT*a* and *f* responded to *C. atrox* antibody, but the respective precipitation lines did not fuse together, indicating that the two toxins were immunologically distinct and, thus, different toxins. The difference between HT*a*, *c*, and *f* was clearly demonstrated in immunodiffusion using antibody against HT*c* (Fig. 5B). HT*c* responded very well to its own antibody and formed a precipitation line, while HT*a* and *f* did not show any reaction. From this evidence, it is clear that each hemorrhagic toxin is a unique protein distinctly different from each other.

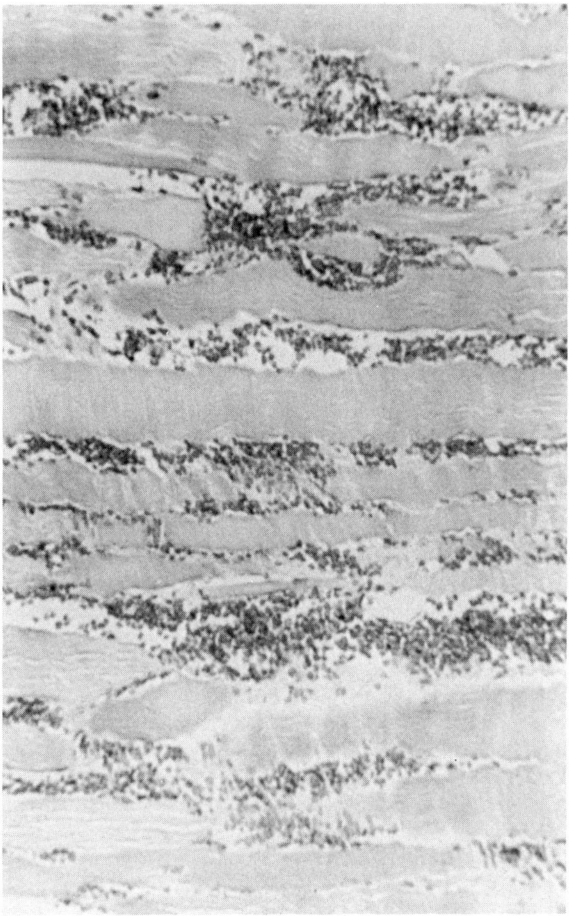

Figure 2 Hemorrhage in muscle induced by intramuscular injection of HT*f* of *C. atrox* venom. The histograph was obtained from Dr. T. Nikai.

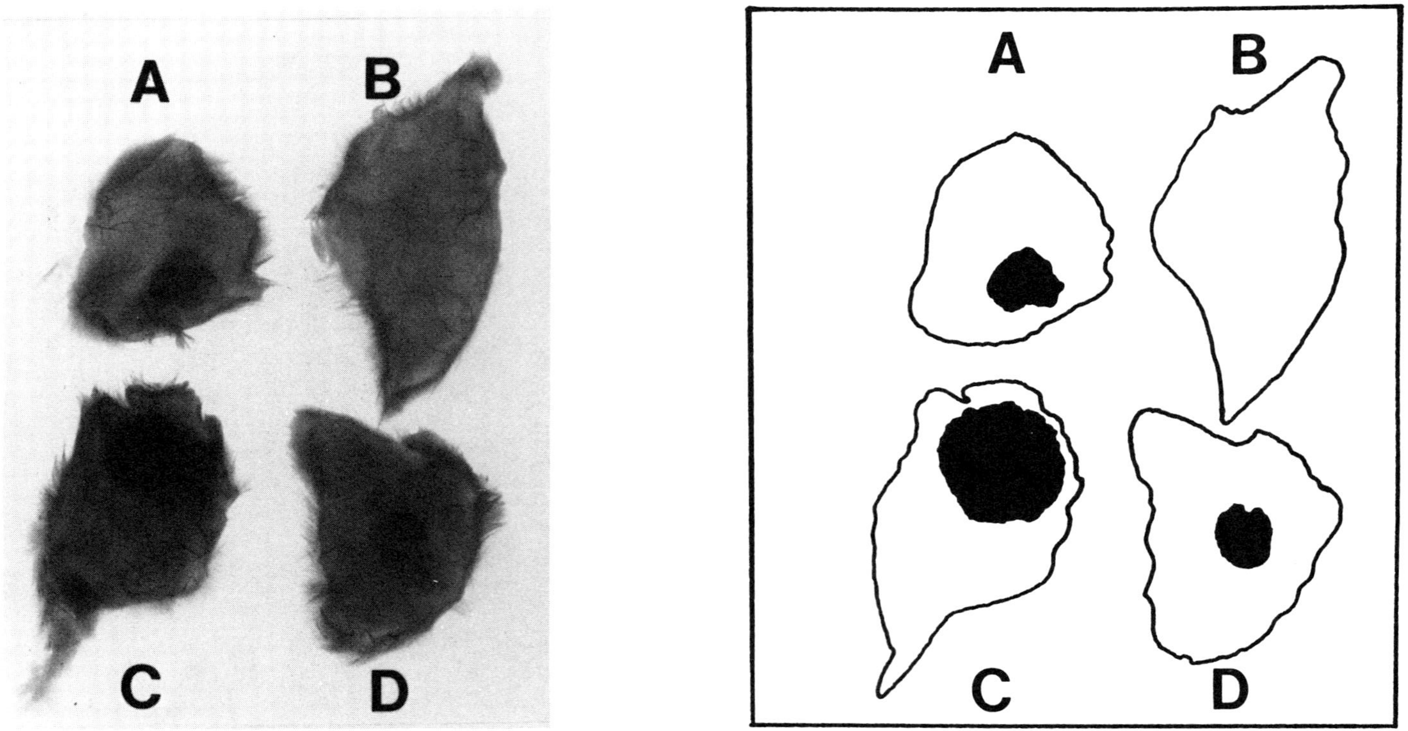

Figure 3

PRESENCE OF ZINC

As early as 1919, Delezenne (8) observed the presence of zinc in snake venoms. In addition to zinc, magnesium and calcium were reported in Crotalidae venoms (9). Although magnesium and calcium can be removed by dialysis, zinc is more firmly bound to snake venoms (10). Venom hemorrhagic and proteolytic activities were known to be eliminated with a chelating agent, suggesting that some metal ions were essential for these activities (11,12).

Actual analysis of zinc was done on several purified proteolytic enzymes and hemorrhagic toxins from snake venoms. They show that many of them contained 1 mole of zinc per mole of protein (Table 1). Some venom proteolytic enzymes contained 1 mole of zinc, but their hemorrhagic activity was never tested. Some hemorrhagic toxins were found to be inactivated, but their zinc content was never analyzed. In the future, Table 1 may become more complete as more investigators analyze their samples for proteolytic and hemorrhagic activities, and zinc content.

PROTEOLYTIC AND NONPROTEOLYTIC

Since snake venoms with high proteolytic activity were also strongly hemorrhagic, it was suspected that venom proteinases were responsible for hemorrhagic action. Moreover, EDTA eliminated both hemorrhagic and proteolytic activities, strengthening the idea of hemorrhagic toxins being proteolytic enzymes. However, this was not proved definitely

Figure 3 Hemorrhage induced by AC_1 protease (A) from *Trimeresurus flavoviridis* venom. No hemorrhage was induced by injecting hyaluronidase alone (B). Hemorrhage was enahnced by mixing AC_1 proteinase and hyaluronidase (C). No enhancement by mixing AC_1 proteinase and BSA (D). (Reprinted form Ref. 59, with permission of the author and Pergamon Press.)

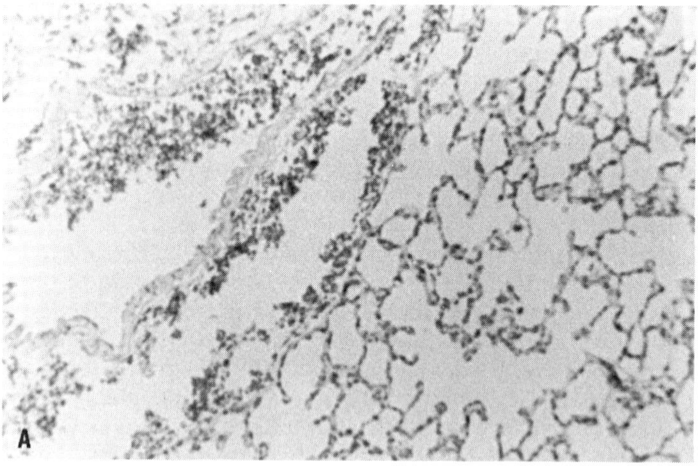

Figure 4 Systemic hemorrhage induced in mice by injecting HT*f* of *C. atrox* venom intravenously. (A) Lung; (B) heart; (C) liver. (Reprinted from Ref. 6, with permission of the author and Academic Press.)

until several hemorrhagic toxins were isolated and their proteolytic activities were actually demonstrated (5). These hemorrhagic proteinases are summarized in Table 2.

As of now, 33 hemorrhagic toxins have been isolated from various snake venoms and all have been shown to possess proteolytic activity. However, earlier works on three snake venoms showed hemorrhagic toxins to be nonproteolytic (Table 3). They were HR-I from *A. halys blomhoffii* venom and HR2a and HR2b from *T. flavoviridis* venom (13, 14). Three hemorrhagic toxins were isolated from *Vipera palestinae* venom and two had proteolytic activity, whereas the third was negative against casein (15). The substrate used in both cases was casein, following the Kunitz method for assay. Kunitz's method is based on the measurement of acid-soluble peptide fragments after trichloroacetic acid (TCA) precipitation. If the proteolysis fragments were acid-insoluble, they might not be detected. Moreover, casein is a protein not very sensitive to hydrolysis by some proteinases. Although HR2a and HR2b were shown to be nonproteolytic, they hydrolyzed basement

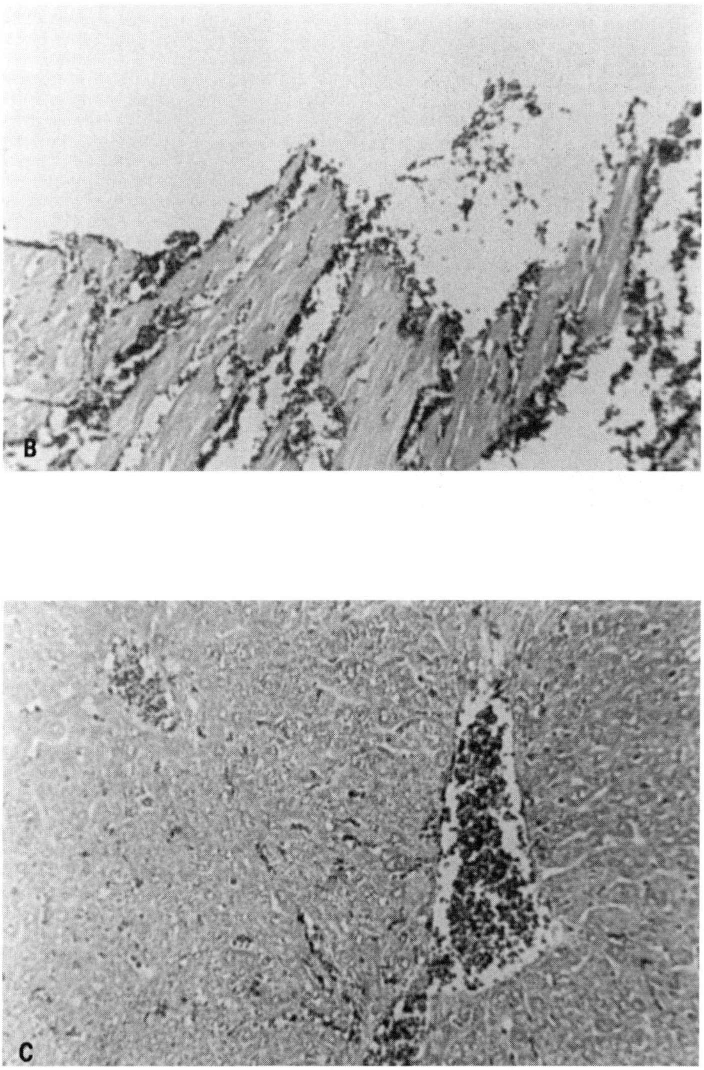

Figure 4 (*continued*)

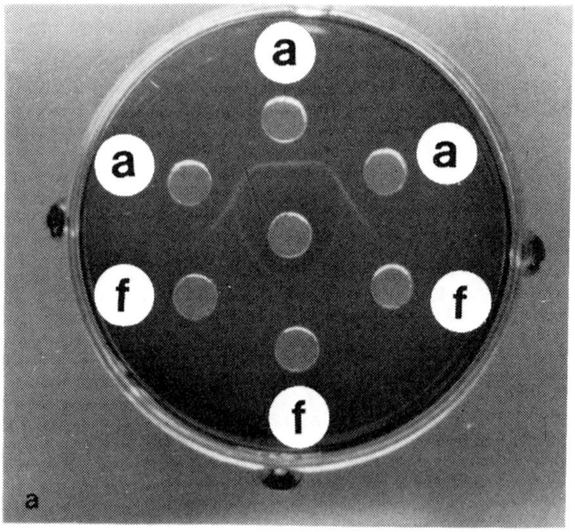

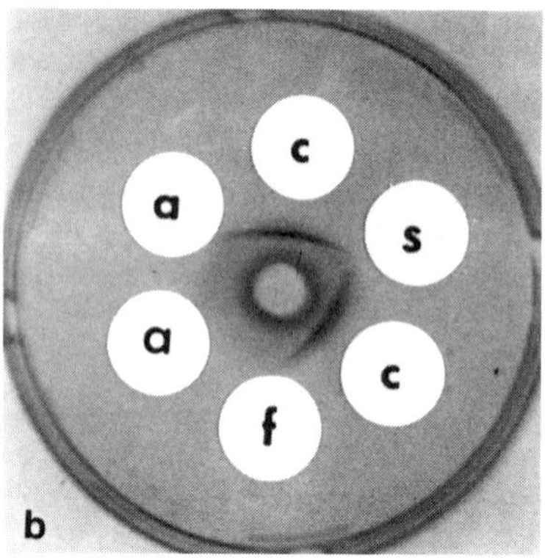

Figure 5 Different hemorrhagic toxins from the same venom are immunologically distinct proteins. (a) Precipitin lines HT*a* and HT*f* indicate that they do not fuse together. Antibody in the center well is *C. atrox* antivenin. (b) HT*a* and HT*f* do not react with anti-HT*c* (in the center well). S = saline control. (Reprinted from Ref. 6, with permission of the author and Academic Press.)

Table 1 Zinc Content and Properties of Purified Hemorrhagic Toxins and Proteolytic Enzymes Isolated from Snake Venoms

Venoms (origin)	Name	Mol Zn/mol protein	Molecular weight (daltons)	Biological activity tested	Reference
Agkistrodon acutus (Formosa)	AC_1 proteinase	1.15	24,000	Hemorrhagic, proteolytic	24
A. contortrix mokasen (USA)	Fibrinogenase	1.0	22,900	Proteolytic	20
A. piscivorus leucostoma (USA)	Leucostoma peptidase A	0.98	22,500	Proteolytic	21
Crotalus atrox (USA)	Hemorrhagic toxin				
	a	0.99	68,000	Hemorrhagic, proteolytic	5
	b	0.82	24,000	Hemorrhagic, proteolytic, myotoxic	5
	c	0.86	25,000	Hemorrhagic, proteolytic	5
	d	0.86	24,000	Hemorrhagic, proteolytic	5
	e	1.03	25,700	Hemorrhagic, proteolytic	5
	f	1.15	64,000	Hemorrhagic, proteolytic	6
C. horridus horridus (USA)	HP-IV	1.0	56,000	Hemorrhagic, proteolytic	14
Trimeresurus flavoviridis (Japan)	Proteinase	1.0	22,000	Hemorrhagic, proteolytic	22

Table 2 List of Purified Hemorrhagic Toxins with Proteolytic Activities

Venom	Name	References
Agkistrodon acutus	AC_1-proteinase	23, 24
	AC_2-proteinase	25
	AC_3-proteinase	26
	AC_4-proteinase	27
	AC_5-proteinase	28
	Aa HI	29
	A_a-toxin I	30, 31
	A_a-toxin II	30, 31
	A_a-toxin III	30, 31
A. halys blomhoffii	HR-I	32
	HR-II	33
Bitis arietans	Hemorrhagic principle	34
Bothrops neuwiedi	NHF_a	35
Crotalus atrox	HT*a*	5, 35, 36
	HT*b*	5, 19, 35
	HT*c*	5
	HT*d*	5
	HT*e*	5, 35
	HT*f*	6
	HT*g*	37
C. horridus horridus	Hemorrhagic proteinase	38, 39
Trimeresurus flavoviridis	HR1	40
	HR2	40
	$HR2_a$	41
	$HR2_b$	41
T. gramineus	HR1	42
	HR2	42
T. mucrosquamatus	Mucrotoxin *a*	43
	a	44
	b	44
Vipera palestinae	HR1	15
	HR2	15

Table 3 Hemorrhagic Toxins Claimed to be Nonproteolytic

Venom	Name	Substrate	Reference
Agkistrodon halys blomhoffii	HR-I[a]	Casein (-)	45
Trimeresurus flavoviridis	HR2a[b] HR2b[b]	Casein (-)	14
Vipera palaestinae	HR3	Casein (-)	15

[a]Proteolytic with other substrates (16).
[b]Proteolytic with glomerular basement membrane (41).

membrane. The basement membrane is a supporting tissue surrounding the capillaries. It consists of collagen and other proteins and glycoproteins. Hydrolysis of basement membrane by HR2a and HR2b was a good indicator that these hemorrhagic toxins did possess proteolytic activity.

It seems that the key to solving this conflicting fact is to isolate the hemorrhagic toxin claimed to be nonproteolytic and reexamine its proteolytic activity. This was exactly what was done (16). HR-I from *A. halys blomhoffii* was isolated, and a number of substrates were used for reexamination of proteolytic activity. When casein was used, the protein was not hydrolyzed, as the original investigators found. However, when all other substrates were used, all were hydrolyzed by HRI including dimethylcasein.

Dimethylcasein assay with trinitrobenzenesulfonic acid (TNBS) is particularly very useful. With dimethylsulfate, all amino groups were dimethylated. When the substrate was hydrolyzed, a new N-terminal should be generated, and this amino group was detected with the reagent TNBS (Fig. 6). Oxidized insulin B-chain, glucagon, and other substrates of known amino acid sequences were hydrolyzed. The fragments were isolated by high performance liquid chromatography (HPLC), and from the amino acid compositions of fragments, it was possible to determine the sites of cleavage.

From this evidence, it was clear that HR-I hitherto considered to be a nonproteinase was, indeed, a proteolytic enzyme. It is thus important to employ different substrates and different proteolytic assay methods to determine whether a hemorrhagic toxin is proteolytic or not. Similarly, HR2a and HR2b obtained from *T. flavoviridis* venom were also proven to be proteolytic (17).

Specificity of other hemorrhagic toxins was also shown by many investigators using different substrates. Some examples are shown in Tables 4 and 5.

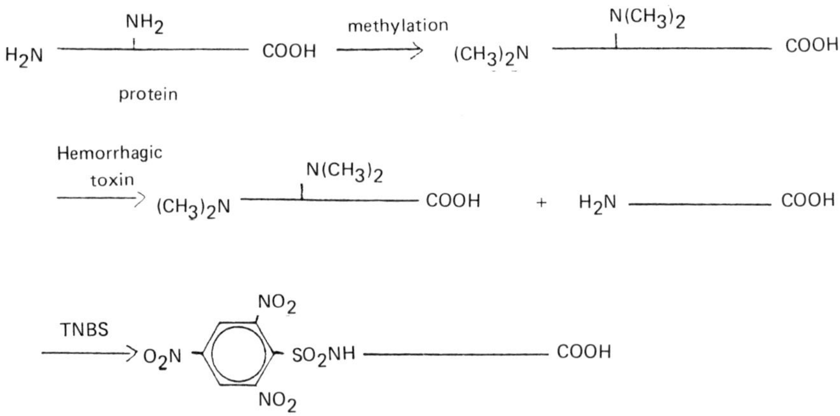

Figure 6 The equation shows dimethylcasein as a substrate for proteolysis and its detection with TNBS.

ANTICOAGULANT PROPERTY OF HEMORRHAGIC TOXINS

Snake venoms, especially those of the Crotalidae and Viperidae families, are known to contain both coagulants and anticoagulants that can be separated by fractionation. Many coagulants were isolated, and the mechanism of action of some coagulants was elucidated. For instance, arvin, crotalase, reptilase, and many other thrombin-like enzymes and fibrinogenases were extensively studied.

However, the mechanism of many anticoagulants has not been so well studied. In the past, hemorrhagic toxins were thought to be different venom components from anticoagulants. As part of the proteolytic specificity of hemorrhagic toxins, we found that fibrinogen was hydrolyzed. However, unlike those from thrombin-like enzymes, none of the hemorrhagic toxins formed clots from fibrinogen (6,18,19). This suggests that many hemorrhagic toxins are fibrinogenases. From a reduced fibrinogen slab electrophoretic pattern, some hemorrhagic toxins hydrolyzed the Aα-chain first, with the Bβ-chain being hydrolyzed later (Fig. 7). On the other hand, some hemorrhagic toxins hydrolyzed the γ-chain only (18). The difference from thrombin-like enzymes is that hemorrhagic toxins do not produce a clot, whereas thrombin-like enzymes produce a clot. The reason for this difference is that thrombin-like enzymes hydrolyzed Aα- and/or Bβ-chains and produced the respective fibrinopeptide A and/or fibrinopeptide B. The sites of cleavage for hemorrhagic toxins are apparently different from those of thrombin-line enzymes.

In Table 6, all fibrinogenases hydrolyzing Aα-, Bβ, or γ-chains of fibrinogen are summarized. Some formed clot (thrombin-like enzymes)

Table 4 Peptide Bonds of Glucogon Cleaved by Various Proteinases

1 His-Ser-Gln -Gly-Thr(5)-Phe-Thr-Ser -Asp-Tyr(10) -Ser-Lys -Tyr -Leu -Asp(15) -Ser -Arg -Arg- Ala-Gln(20)-Asp -Phe -Val-Gln -Trp(25) -Leu -Met -Asn-Thr(29)

	His-Ser	Ser-Gln	Gln-Gly	Gly-Thr	Thr-Phe	Phe-Thr	Thr-Ser	Ser-Asp	Asp-Tyr	Tyr-Ser	Ser-Lys	Lys-Tyr	Tyr-Leu	Leu-Asp	Asp-Ser	Ser-Arg	Arg-Arg	Arg-Ala	Ala-Gln	Gln-Asp	Asp-Phe	Phe-Val	Val-Gln	Gln-Trp	Trp-Leu	Leu-Met	Met-Asn	Asn-Thr
Hemorrhagic toxin *b* (60)					▲					▲					▲						▲				▲			
Thermomycolin (61)			▲		▲	▲		▲				▲	▲	▲	▲	▲	▲			▲		▲		▲	▲	▲	▲	
Aspergillopeptidase (62)			▲			▲		▲						▲		▲											▲	

Table 5 Peptide Bonds of Insulin B-Chain Cleaved by Various Proteinases

Sequence (residues 1–30): Phe-Val-Asn-Gln-His-Leu-$CySO_3H$-Gly-Ser-His-Leu-Val-Glu-Ala-Leu-Tyr-Leu-Val-$CySO_3H$-Gly-Glu-Ar-Gly-Phe-Phe-Tyr-Thr-Pro-Lys-Ala

▲ = bond cleaved

	Asn3-Gln4	Gln4-His5	Leu6-$CySO_3H$7	$CySO_3H$7-Gly8	Gly8-Ser9	Ser9-His10	His10-Leu11	Leu11-Val12	Val12-Glu13	Glu13-Ala14	Ala14-Leu15	Leu15-Tyr16	Tyr16-Leu17	Leu17-Val18	Val18-$CySO_3H$19	$CySO_3H$19-Gly20	Gly23-Phe24	Phe24-Phe25	Phe25-Tyr26	Tyr26-Thr27	Lys29-Ala30
Hemorrhagic toxin *b* (60)	▲						▲						▲				▲				
Hemorrhagic toxin *e* (63)	▲					▲					▲										
Thermomycolin (60)		▲								▲		▲						▲			
Alkaline protease (*Aspergillus flavus*)	▲	▲		▲					▲	▲		▲	▲			▲		▲	▲		▲
Elastase (porcine) (65)						▲			▲		▲				▲		▲				
Subtilisin Carlsberg (66)		▲				▲		▲				▲	▲	▲						▲	
Papain (67)	▲		▲	▲	▲					▲	▲			▲		▲			▲	▲	

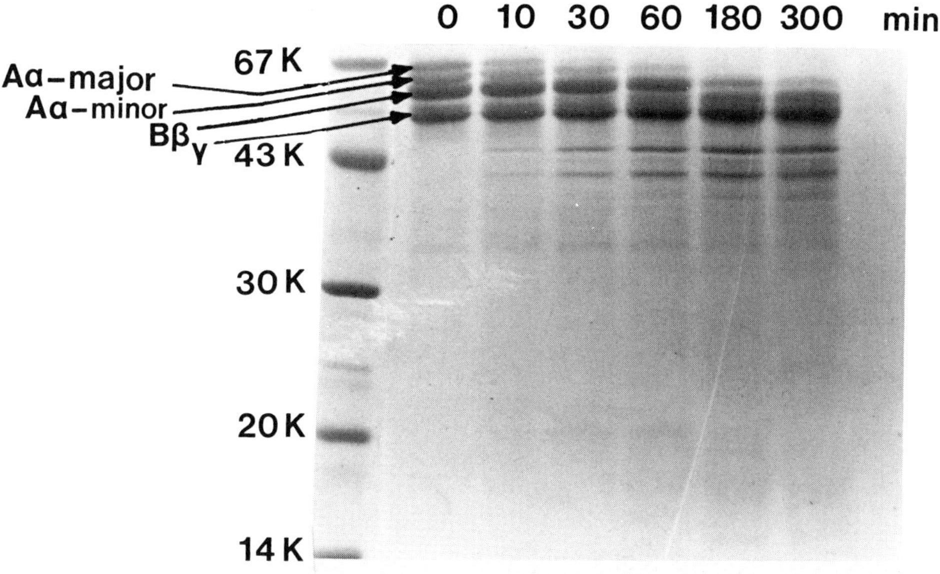

Figure 7 Hydrolysis of fibrinogen by HR-I from *A. halys blomhoffii* venom. The hemorrhagic toxin did not hydrolyze casein; therefore, it was reported to be nonproteolytic. However, it was shown recently that HR-I hydrolyzed many proteins, including fibrinogen. (From Ref. 16.)

and some did not (anticoagulant proteinases). Therefore, the relationship of snake venom proteinases and hemorrhagic toxins can be summarized as follows:

1. Snake venom of Crotalidae contains a variety of proteolytic enzymes.
2. Some proteinases produce hemorrhage on envenomation and some do not.
3. Practically all snake hemorrhagic toxins have proteolytic activity and destroy fibrinogen, rendering it unable to form a clot. They are anticoagulant.
4. Some proteinases form a clot by degrading fibrinogen. They are thrombin-like enzymes.
5. Some proteinases are nonhemorrhagic, but hydrolyze fibrinogen.

Snake venom proteases may be divided into several groups based on their ability to form a clot and cause hemorrhage:

Table 6 Fibrinogen Degradation by Snake Venom Enzymes

Venom	Name	Molecular weight (daltons)	Effect on Aα	Bβ	γ	Hemorrhagic activity	Reference
A. Clot formed							
—	Thrombin	33,700	+	+	-	-	46
Agkistrodon contortrix contortrix	Coagulant	100,000	-	+	-	Not determined	47, 48
A. halys pallas	Thrombin-like	43,000	+	+	-	Not determined	49
A. rhodostoma	Arvin (ancrod)	33,000	+	-	-	-	50
Bothrops jararaca	Reptilase I	29,000	+	-	-	Not determined	51
	Reptilase II	31,400	+	-	-	Not determined	51
	Batroxobin		+	-	-	Not determined	47
Crotalus adamanteus	Crotalase	32,700	+	-	-	Not determined	52
C. horridus horridus	Thrombin-like	43,000	+	-	-	Not determined	53
Echis carinatus	Ecarin		+	+	+	Not determined	54
B. No clot formed							
Agkistrodon contortrix mokasen	Fibrinogenase		+	-	-	Not determined	20
A. halys blomhoffii	HR-I		+	+	-	+	16
Cerastes cerastes	F_4		+	+	-	-	55
Crotalus atrox	HT*b*	24,000	+	+	-	+	19
	HT*f*	64,000	-	-	+	+	6
	Fibrinogenase	31,000	+	+	-	-	24
	Fibrinogenase	22,900	+	+	-	Not determined	56
C. horridus horridus	HP-IV	56,000	+	+	-	+	38
Naja nigricollis	F1	58,000	+	-	-	Not determined	57
T. mucrosquamatus	Factor a		-	+	-	+	44
	Factor b		+	-	-	+	44
	α-Fibrinogenase	22,000	+	-	-	Not determined	58
	β-Fibrinogenase	25,500	-	+	-	Not determined	58

1. Thrombin-like enzymes that produce a clot without hemorrhage.
2. Anticoagulant proteinases.
   a. No clot produced, but cause hemorrhage.
   b. No clot produced and no hemorrhage.

Since the anticoagulant proteinases in this context are fibrinogenases, then these two enzyme groups can be rearranged in terms of their fibrinogenase, clotting, and hemorrhagic activities as follows:

1. Fibrinogenase with clot formation, but without hemorrhage.
2. Fibrinogenase without clot formation and with hemorrhage.
3. Fibrinogenase without clot formation and without hemorrhage.

ACKNOWLEDGMENT

This work was supported by Grant no. GM15591 from the National Institutes of Health, Bethesda, Maryland.

REFERENCES

1. Tu, A. T. *Venoms: Chemistry and Molecular Biology*, John Wiley, New York (1977).
2. Ownby, C. L. In *Rattlesnake Venoms: Their actions and Treatment*, Tu, A.T. (Ed.). Marcel Dekker, New York (1982).
3. Stringer, J. M., R. A. Kainer, and A. T. Tu. Am. J. Pathol. 67:127 (1972).
4. Ownby, C. L., R. A. Kainer, and A. T. Tu. Am. J. Pathol. 76:401 (1974).
5. Bjarnason, J. B. and A. T. Tu. Biochemistry 17:3395 (1978).
6. Nikai, T., N. Mori, M. Kishida, H. Sugihara, and A. T. Tu. Arch. Biochem. Biophys. 231:309 (1984).
7. Sullivan, J. A., E. Farr, and C. R. Geren. Toxicon 17:269 (1979).
8. Delezenne, C. Ann. Inst. Pasteur 33:68 (1919).
9. Suzuki, T. The Snake 2:75 (1970).
10. Frederich, C. and A. T. Tu. Biochem. Pharmacol. 20:1549 (1971).
11. Flowers, H. H. Toxicon 1:131 (1963).
12. Goncher, C. R. and H. H. Flowers. Toxicon 2:139 (1964).
13. Oshima, G., T. Omori-Satoh, S. Iwanaga, and T. Suzuki. J. Biochem. 72:1483 (1972).
14. Takahashi, T. and A. Ohsaka. Biochim. Biophys. Acta 207:65 (1970).
15. Ovadia, M. Toxicon 16:479 (1978).
16. Nikai, T., E. Oguri, M. Kishida, H. Sugihara, N. Mori, and A. T. Tu. Int. J. Biochem. In press (1986).
17. Nikai, T. Personal communication (1985).

18. Tu, A. T. In *Zinc and its Role in Biology and Nutrition*, Sigel, H. (Ed.). Marcel Dekker, New York (1983).
19. Komori, Y., S. Hagihara, and A. T. Tu. Biochim. Biophys. Acta 829:127 (1985).
20. Moran, J. B. and C. Geren. Biochim. Biophys. Acta 659:161 (1981).
21. Spiekerman, A. M., K. K. Fredericks, F. W. Wagner, and J. M. Prescott. Biochim. Biophys. Acta 293:464 (1973).
22. Hagihara, S. Acta Med. Univ. Kagoshima 16:127 (1974).
23. Nikai, T., H. Sugihara, and T. Tanaka. Yakugaku Zashi 97:507 (1977).
24. Nikai, T., H. Ishizaki, A. T. Tu, and H. Sugihara. Comp. Biochem. Physiol. 72C:103 (1982).
25. Sugihara, H., T. Nikai, T. Kawaguchi, and T. Tanaka. Yakugaku Zashi 98:1523 (1978).
26. Sugihara, H., T. Nikai, H. Umeda, and T. Tanaka. Yakugaku Zashi 99:1161 (1979).
27. Sugihara, H., T. Nikai, G. Kinoshita, and T. Tanaka. Yakugaku Zashi 100:855 (1980).
28. Mori, N., T. Nikai, and H. Sugihara. Toxicon 22:451 (1984).
29. Zhang, J., Z. Chen. Y. He, and X. Xu. Toxicon 22:931 (1984).
30. Xu, X. and Z. Y. Yu. Chung-Kuo Ko Hsueh Chi Shu Ta Hsueh Pao 10:105 (1980).
31. Xu, X., C. Wang, J. Liu, and Z. Lu. Toxicon 19:633 (1981).
32. Suzuki, T. Mem. Inst. Butantan 33:519 (1966).
33. Mebs, D. and F. Panholzer. Toxicon 20:509 (1982).
34. Mandelbaum, F. R., M. T. Assakura, and A. P. Riechl. Toxicon 22:193 (1984).
35. Ownby, C. L., J. Bjarnason, and A. T. Tu. Am. J. Pathol. 93:201 (1978).
36. Tu, A. T., T. Nikai, and J. O. Baker. Biochemistry 20:7004 (1981).
37. Nikai, T. Personal communication.
38. Civello, D. J., M. L. Duong, and C. R. Geren. Biochemistry 22:749 (1983).
39. Civello, D. J., H. R. Allen, T. T. Lee, and C. R. Geren. Toxicon 22:235 (1984).
40. Ohsaka, A., H. Ikezawa, H. Kondo, S. Kondo, and N. Uchida. Br. J. Exp. Pathol. 41:478 (1960).
41. Ohsaka, A., M. Just, and E. Habermann. Biochim. Biophys. Acta 323:415 (1973).
42. Huang, T. F., J. H. Chang, and C. Ouyang. Toxicon 22:45 (1984).
43. Sugihara, H., M. Moriura, and T. Nikai. Toxicon 21:247 (1983).
44. Nikai, T., N. Mori, M. Kishida, Y. Kato, C. Takenaka, T. Murakami, S. Shigezano, and H. Sugihara. Biochim. Biophys.

Acta 838:122 (1985).
45. Oshima, G., T. Omori-Satoh, S. Iwanaga, and T. Suzuki. J. Biochem. 72:1483 (1972).
46. Ewart, M. R., M. W. C. Hatton, J. M. Basford, and K. S. Dodgson. Biochem. J. 118:603 (1970).
47. Stocker, K., H. Fischer, and J. Meier. Toxicon 20:265 (1982).
48. Dyr, J. E., B. Blombäck, and F. Kornalik. Thromb. Res. 30: 185 (1983).
49. Guan, L. F., C. W. Chi, and M. Yuan. Thromb. Res. 35:301 (1984).
50. Blombäck, B. Ark. Kemi 12:321 (1958).
51. Holleman, W. H., and L. J. Weiss. J. Biol. Chem. 251:1663 (1976).
52. Markland, F. S. and H. Pirkle. In *Chemistry and Biology of Thrombin*. Lundbland, R. L., J. W. Fenton, II, and K. G. Mann (Eds.). Ann Arbor Science, Ann Arbor (1977).
53. Shu, Y. Y., J. B. Moran, and C. R. Geren. Biochim. Biophys. Acta 748:236 (1983).
54. Fortova, H., J. E. Dyr, Z. Vodrazka, and F. Kornalik. J. Chromatogr. 259:473 (1983).
55. Daoud, E., F. El-Asmar, and A. T. Tu. Unpublished data (1986).
56. Sapru, Z. Z., A. T. Tu, and G. S. Bailey. Biochim. Biophys. Acta 474:225 (1983).
57. Evans, H. J. Biochim. Biophys. Acta 802:49 (1984).
58. Ouyang, C., C. M. Teng, and Y. C. Chen. Biochim. Biophys. Acta 481:622 (1977).
59. Tu, A. T. and R. H. Hendon. Comp. Biochem. Physiol. 76B:377 (1983).
60. Hagihara, S., Y. Komori, and A. T. Tu. Comp. Biochem. Physiol. 82C:21 (1985).
61. Stevenson, K. J. and G. M. Gaucher. Biochem. J. 151:527 (1975).
62. Nordwing, A. and W. F. Jahn. Eur. J. Biochem. 3:519 (1968).
63. Bjarnason, J. B. and J. W. Fox. Biochemistry 22:3770 (1983).
64. Turkova, J. and O. Mikes. Biochim. Biophys. Acta 198:386 (1970).
65. Narayanan, A. S. and R. A. Anwar. Biochem. J. 114:11 (1969).
66. Johansen, J. T., M. Ottensen, I. Svendsen, and G. Wybrandt. Comp. Rend. Trav. Lab. Carlsberg 36:365 (1986).
67. Johansen, J. T. and M. Attesen. Comp. Rend. Trav. Lab. Carlsberg 36:265 (1968).

32

Isolation and Characterization of the Weakly Acidic Hemorrhagins from Timber Rattlesnake Venom

Yu-yan Shu
Guangxi Medical College, Nanning, Republic of China
H. Raymond Allen* and Collis R. Geren
University of Arkansas, Fayetteville, Arkansas

INTRODUCTION

Hemorrhagins are common components of venoms of snakes belonging to the Crotalidae family. Rattlesnake bites in general cause severe local hemorrhage and tissue damage to the extent that permanent damage is often a problem (1). Only a small number of rattlesnake venom hemorrhagins have been well characterized (2-6). These have all been described as metal-dependent proteinases.

Since the original report by Geren and co-workers (7) that timber rattlesnake, *Crotalus horridus horridus*, venom contained at least three distinct hemorrhagins, two of these have been purified and further characterized. The most abundant of these was characterized by Geren and associates (8,9) as being a highly specific metal-dependent proteinase. The most potent of these (hemorrhage size/μg protein) was described by Geren and Young (10) in a preliminary report as also being a metal-dependent proteinase, but it differed dramatically from the first hemorrhagin described by Geren and associates (8,9). In this chapter, the most recently studied hemorrhagins from timber rattlesnake venom are described. They differ from the first two in being smaller and apparently lacking any essential proteolytic activity.

MATERIALS AND METHODS

Lyophilized timber rattlesnake venom was from the Miami Serpentarium; ion exchange resins and 3 mm chromatography paper were from Whatman;

*Present affiliation: Louisiana State University Medical Center, New Orleans, Louisiana electrophoresis materials, molecular sieve resins, and isoelectric focusing ampholytes were from Bio-Rad; acetylated cytochrome c standards for isoelectric focusing were from United States Biochemical; neuraminidase, type V, was from Sigma Chemical Company; hide powder azure was from Calbiochem Behring; Spectra/Por dialysis tubing, 12,000 to 14,000 molecular weight cutoff was from Fisher; and Centriflo membrane cones were from Amicon. Unless otherwise stated, all other chemicals were from Sigma Chemical Company.

Timber rattlesnake venom was fractionated by DEAE-cellulose chromatography, as previously described by Geren and co-workers (7). When necessary, protein solutions were concentrated with CF-25 Centriflo membrane cones. Protein concentrations were estimated by absorbance at 280 nm, assuming an extinction coefficient of 1.0 for a 1.0 mg/ml solution.

All bioassays were conducted with either Tex:(ICR) albino or C3H mice. Subcutaneous (sc) injection in the fatty area between the shoulder blades of a maximum volume of 0.2 ml was used for hemorrhage studies. The MHD (minimum hemorrhagic dose) was defined as that amount of material that would give a 1-cm hemorrhage 5 hours after venom (or purified component) challenge.

Clotting assays were accomplished at 37°C. Each assay consisted of 300 μl of 5 mg/ml fibrinogen (95% clottable) in 20 mM Tris-buffered saline, pH 7.4, and 12 μg of suspected coagulant. The fibrinogen was preincubated for 5 minutes at 37°C prior to addition of the coagulant. Each assay was tilted every 15 seconds, and clot formation was detected visually.

TAME (tosyl-L-arginine methyl ester)- and BAEE (benzoyl-L-arginine ethyl ester)-hydrolyzing activities were determined spectrophotometrically by the methods of Hummel (11) and Schwert and Takenaka (12), respectively. The hydrolysis of casein was determined as described by Laskowski (13). The hydrolysis of hide power azure, elastin congo red, and keratin azure were determined essentially by the method described by Rinderknecht and associates (14) for the hydrolysis of hide powder azure. These assays have been described in more detail elsewhere (9). Neuraminidase was assayed using the sialic acid determination method of Warren (15).

The oxidized A- and B-chains of bovine insulin as well as bee venom melittin were used as potential substrates for BII_t. One mg of each in 1.0 ml of buffer was incubated with 45 μg of the hemorrhagic fraction (added in 15 μg aliquots at 2-hour intervals) for 24 hours at 37°C. Incubation of the A-chain of insulin was in 20 mM NH_4HCO_3, pH 8.5. The B-chain of insulin and melittin were incubated with BII_t in 20 mM $NaHCO_3$, pH 9.5. Dansylation, hydrolysis, and thin-layer chromatography were as previously described (8).

Sodium dodecyl sulfate (SDS) polyacrylamide gel electrophoresis was accomplished by the method of Weber and Osborn (16), whereas the discontinuous method was that of Ornstein (17) and Davis (18). Visualization was either by Coomassie blue or silver stain (19). Schiff's carbohydrate stain (20) was also used. Cross-linked hemoglobin standards were routinely included. Isoelectric points were determined as previously described (8).

Thin-layer gel (TLG) chromatography was accomplished with a Pharmacia TLG apparatus. Swollen G-200 Sephadex superfine in 50 mM sodium acetate, pH 6.0, was spread on a 20- × 40-cm plate to a thickness of 0.6 mm.

Protein bands were dissected from electrophoresis gels and the protein extracted and destained, as described by Gibson and Gracy (21). Amino acid compositions of proteins were determined using a microcomputer-controlled micropore amino acid analyzer (22). Detection was by the ninhydrin reaction and was monitored at 548 nm. Hydrolysis of samples was with constant boiling HCl at 110°C for 24 hours (23).

RESULTS

Lyophilized timber rattlesnake venom was fractionated on DEAE-cellulose as described by Geren and co-workers (7). Although that study (7) reported the third hemorrhagic activity of whole venom to be centered in the fractions labeled III_a, the adjacent II_t fraction was found to be the fraction of choice from which to isolate the remaining hemorrhagin(s) from timber rattlesnake venom. The minimum hemorrhagic dose (1-cm lesion) for II_t in C3H mice was 13 μg.

The hemorrhagic activity from II_t was further purified by chromatography on DEAE-Sephadex A50, as described by Figure 1. As obtained from this chromatographic analysis, this material (named BII_t) had a minimum hemorrhagic dose of 8.5 μg. Low concentrations of BII_t showed a sharp doublet on discontinuous polyacrylamide gel electrophoresis (Fig. 2). Larger amounts appeared to migrate as a single band. Densitometric scans of the lower concentration gels showed the two proteins to be present in approximately equal amounts. Both bands were positive with Schiff's stain. Scanning of such gels revealed the bands to be of equal intensity. Carbohydrate content was estimated at 10% as compared with the staining intensities of four different concentrations of fetuin. Incubation of 100 μg of BII_t with 10 μg of neuraminidase in 40 mM of sodium acetate, pH 5.0, for 1 hour at 37°C had no effect on the hemorrhagins' migration in discontinuous gel electrophoresis. No condensation of the two bands was observed, nor did the R_f values of the stained bands change as compared with the electrophoretic patterns of BII_t incubated without neuraminidase. Ten micrograms of neuraminidase incubated with 120 μg of mucin under the same conditions catalyzed the release of 2.5 μg of N-acetylneu-

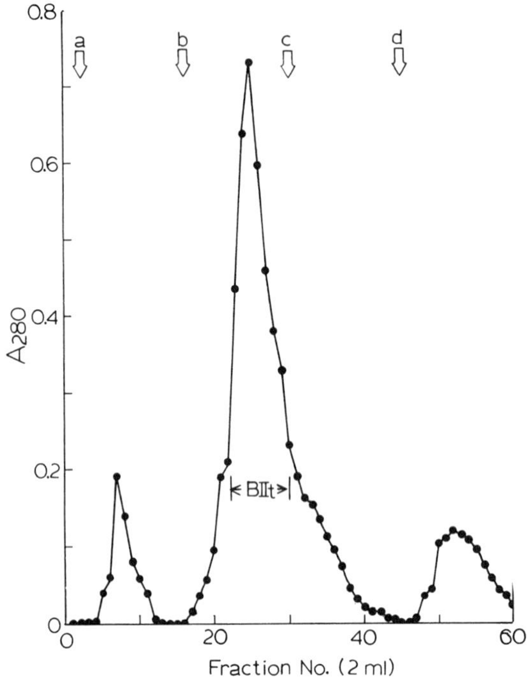

Figure 1 Purification of hemorrhagins. A 1- × 17-cm glass column of DEAE-Sephadex A50 was equilibrated with 5 mM of sodium acetate, pH 6.0. Six mg of the II_t fraction from timber rattlesnake venom was applied in 3 ml of 5-mM sodium acetate. As indicated by the arrows, the column was successively eluted with 30 ml volumes of (a) 5 mM sodium acetate, pH 6.0; (b) 50 mM sodium acetate, pH 6.0; (c) 50 mM sodium acetate, pH 7.0; and (d) 50 mM sodium acetate, pH 6.0, containing 50 mM sodium chloride. Chromatography was at 4°C.

raminic acid after 30 minutes and 4.5 μg after 1 hour of incubation. Two closely migrating bands were also apparent on the isoelectric focusing of BII_t on polyacrylamide gels (Fig. 2). The isoelectric points of the two bands were 5.67 ± 0.03 and 5.93 ± 0.06 as compared with the migration of acetylated cytochrome c standards. The amino acid compositions of the two protein bands sliced from the discontinuous electrophoresis gels are given in Table 1. These results indicate that the two proteins are very similar. Slight differences in Glx and Arg account for the difference in electrophoretic migration of the two proteins. Additional aliquots of BII_t were separated by discontinuous electrophoresis, and the resulting gels were scanned at 280 nm rather

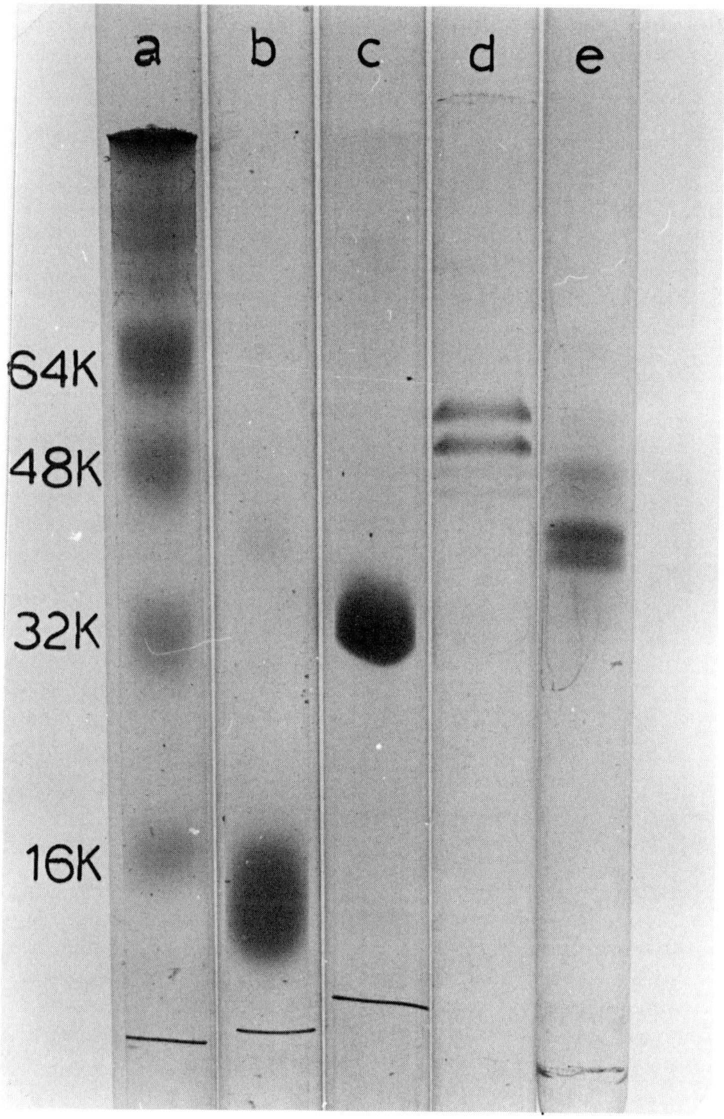

Figure 2 Electrophoresis and isoelectric focusing gels of BII_t. The gel on the left (a) contained cross-linked hemoglobin standard, with K representing 1000 daltons. The remaining gels all contained 20 μg of BII_t with (b) showing SDS electrophoresis of reduced BII_t; (c) SDS electrophoresis of nonreduced BII_t; (d) isoelectric focusing of BII_t; and (e) discontinuous electrophoresis of BII_t.

Table 1 Amino Acid Compositions of Hemorrhagins Separated by Discontinuous Gel Electrophoresis

Amino acid	Faster band[a]	Slower band
Asx	19.6 ± 0.2	19.6 ± 0.2
Thr	10.8 ± 0.1	11.0 ± 0.0
Ser	19.3 ± 0.0	21.2 ± 0.1
Glx	26.3 ± 0.2	28.1 ± 0.4
Pro	10.8 ± 0.5	10.0 ± 0.7
Gly	30.8 ± 0.0	28.8 ± 0.7
Ala	12.3 ± 0.0	12.4 ± 0.0
Cys	3.0 ± 0.0	2.1 ± 0.5
Val	10.4 ± 0.0	11.3 ± 0.4
Met	2.7 ± 0.0	2.8 ± 0.1
Ile	7.4 ± 0.0	7.0 ± 0.2
Leu	12.6 ± 0.0	12.5 ± 0.1
Tyr	5.0 ± 0.0	5.8 ± 0.0
Phe	9.4 ± 0.2	10.2 ± 0.2
His	5.1 ± 0.0	5.3 ± 0.2
Lys	13.1 ± 0.3	13.0 ± 0.1
Arg	8.8 ± 0.0	5.9 ± 0.2

[a]As shown in Figure 2, BII_t separates into two bands on electrophoresis on a discontinuous system. The faster and slower band designations indicate their relative migration. Bands were dissected from the gels and protein extracted, destained, and analyzed as described earlier. The composition of the bands was calculated based on 22,000. All values are the average of two determinations.

than stained to determine protein location. Duplicate gels were run with 20-, 30-, and 40-μg amounts of BII_t. Using the scans as guides, the two bands were dissected from the gels, crushed with a glass on glass homogenizer, and each extracted into 250 μl of 20 mM Tris-buffered saline, pH 7.4, by stirring for 30 minutes at 4°C. Both peaks produced similar hemorrhages, indicating that both were hemorrhagic. Recovery of activity from the gels was poor. Hemorrhages were 5 mm in diameter, whereas the hemorrhage produced by 20 μg of BII_t (the maximum amount expected to be extracted from each band) produces a lesion in excess of 2 cm. The two hemorrhagins present in BII_t migrated as a single band on SDS electrophoresis gels (nonreducing conditions) with an apparent molecular weight of 22,100 ± 600 as compared with cross-linked hemoglobin standard. Reduction resulted in a single faster-migrating species of a 12,900- ± 300-daltons (Fig. 2). The stained protein bands were sectioned from these gels and their amino acid compositions determined to ascertain whether or not reduction was resulting in the loss of lower molecular weight polypeptide(s) that would not be detectable on gels under the described conditions. The compositions of the reduced and nonreduced proteins were the same, and the amounts of amino acid residues recovered from each preparation were equal, indicating no loss of low molecular weight fragments. Thin-layer gel chromatography of BII_t on Sephadex G-200 also failed to resolve the two hemorrhagins. They migrated as a single spot, with an apparent molecular weight of 25,000 ± 500 as compared with the migration of bovine serum albumin, ovalbumin, B-lactoglobulin, and myoglobin.

The two proteins proved very difficult to separate by any technique other than the discontinuous electrophoresis described above. A variety of other ion exchange and molecular sieve materials were examined, with no improvement in separation. As the two proteins were very difficult to separate in quantity and their hemorrhagic and amino acid compositions were very similar, their properties were examined without separation (i.e., as BII_t).

Figure 3 compares the effect of incubating BII_t at 4°C with dialyzing it against its own solvent buffer for the same length of time and against its buffer containing 20 mM EDTA. The EDTA dialysis destroyed all hemorrhagic activity. Long-term dialysis against buffer also eventually destroyed this activity. Ultraviolet spectra of BII_t before and after dialysis showed no change, so that loss of protein through the dialysis tubing was not the cause. Dialysis of inactive material overnight against the solvent buffer with added 50 mM of $ZnCl_2$ and 50 mM of $CaCl_2$ resulted in no recovery of hemorrhagic activity. Aliquots of BII_t containing 17 and 34 μg of protein were incubated with 20 mM of EDTA in 20 mM of Tris-buffered saline for 15 minutes, then injected subcutaneously into mice (250 μl total volume). Two mice were used for each amount. This short-term incubation with

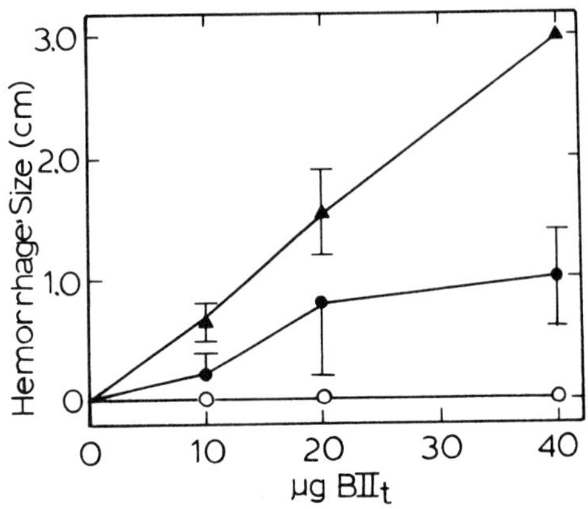

Figure 3 Effect of dialysis alone and dialysis against EDTA on the hemorrhagic activity of BII_t. Half-ml aliquots of BII_t (0.34 mg/ml in 50 mM sodium acetate, pH 6.0) were incubated overnight at 4°C (▲), dialyzed overnight at 4°C against 500 ml of 50 mM sodium acetate, pH 6.0 (●), or dialyzed overnight at 4°C against 500 ml of 50 mM sodium acetate, pH 6.0 containing 20 mM of EDTA (○).

EDTA reduced hemorrhage size 68% as compared with that induced by BII_t incubated in the absence of EDTA. Neither incubation of BII_t with 50 mM of N-ethylmaleimide for 1 hour at room temperature nor with 20 mM of iodoacetamide for 1 hour at room temperature caused any loss of hemorrhagic activity. Incubation with 40 mM of phenylmethylsulfonyl fluoride for 2 minutes with agitation also caused no loss of hemorrhagic activity. In all cases, the preparations were dialyzed for 2 hours against 0.1 M Tris, pH 7.4, prior to injection as the modification reagents are themselves irritants. The time chosen was such as to allow little loss of hemorrhagic activity due to dialysis itself. The data obtained with phenylmethylsulfonyl fluoride are included in Table 2 to show typical results and experimental design. Incubation of BII_t with 20 mM of dithiothreitol for 1 hour at room temperature followed by 1 hour incubation with 50 mM iodoacetamide resulted in total loss of hemorrhagic activity. Four animals were injected with 10 μg of the treated BII_t, four with 20 μg, and four with 40 μg. No hemorrhages were observed in any of the treated animals.

The original II_t fraction was examined for its ability to catalyze the hydrolysis of hide powder azure, keratin azure, and elastin congo red by the procedures described. The maximum amount of II_t assayed was 40 μg per 1.4 ml of assay. No hydrolysis of any of these insoluble substrates was observed after 5 hours of incubation at 37°C.

Table 2 Effect of Phenylmethylsulfonyl Fluoride on the Hemorrhagic Activity of BII_t[a]

Treatment	Amount of BII_t injected (μg)	Hemorrhagic size (cm)
None	10	0.8 × 0.5
	20	2.0 × 3.0
Incubation with acetonitrile[b]	10	0.2 × 0.5
	10	0.5 × 0.5
	20	2.0 × 3.0
Incubation with acetonitrile and PMSF[c]	10	0.2 × 0.2
	10	0.5 × 0.5
	20	2.0 × 2.0

[a]Hemorrhages were determined as described in the section on Materials and Methods.
[b]Four parts of 1.0 mg/ml BII_t in 50 mM of sodium acetate, pH 6.0, were incubated with one part of acetonitrile for 2 minutes with mixing.
[c]Four parts of 1.0 mg/ml BII_t in 50 mM of sodium acetate, pH 6.0, were incubated with one part of 200 mM of phenylmethylsulfonyl fluoride in acetonitrile for 2 minutes with mixing. Precipitated PMSF was removed by centrifugation. Samples were also dialyzed prior to injection.

BII_t catalyzed the hydrolysis of TAME (tosyl-L-arginine methyl ester) with an activity of 11.4 ± 0.8 units/mg and BAEE (benzoyl-L-arginine ethyl ester) with an activity of 17.0 ± 1.4 units/mg. Incubation of BII_t with phenylmethylsulfonyl fluoride under the same conditions that caused no loss of hemorrhagic activity (see Table 2) caused an 80% decrease in esterolytic activity. Reduction of BII_t with dithiothreitol followed by alkylation with iodoacetamide under the same conditions that totally inactivated the hemorrhagic activity had no effect on either BAEE or TAME hydrolysis.

BII_t does catalyze the hydrolysis of casein in a time- and concentration-dependent manner. Reduction and alkylation of BII_t (same conditions as described earlier) resulted in a preparation that retained 93% of its hydrolytic activity, but none of its hemorrhagic activity. Also, BII_t that had been subjected to discontinuous gel electrophoresis followed by extraction of protein from the gels retained only esterolytic and hemorrhagic activities.

BII_t has no apparent anticoagulant activity, but is coagulant with purified fibrinogen. Twelve μg of BII_t caused a clot in 4 minutes using the procedure described here. Treatment of BII_t with penylmethylsulfonyl fluoride (as described in Table 2) resulted in material that would not cause fibrinogen to clot, even with 2 hours of incubation. As shown in Table 2, this same material retained its ability to cause hemorrhage. This treated BII_t also caused the appearance of no new N-termini when incubated with the A- and B-chains of insulin and with melittin.

DISCUSSION

Two weakly acidic hemorrhagins with similar hemorrhagic potencies and subunit and amino acid compositions have been isolated from timber rattlesnake venom (8-10). The current hemorrhagins have, to date, displayed no proteolytic action that is essential for hemorrhagic activity. The caseinolytic action observed has definitely been shown to be unnecessary for hemorrhage induction. Likewise, the coagulant activity has been shown not to be involved. The coagulant activity is probably due to slight contamination of BII_t with the thrombin-like coagulant, which we earlier reported as being present in fraction III (24). The caseinolytic activity was also most probably a minor contaminant, as it was separated from the hemorrhagic activity by electrophoresis and was unaffected by incubation with dithiothreitol. It should come as no surprise that very small amounts of these contaminant proteins should copurify with any individual venom protein because the whole venom contains a variety of such hydrolytic entities. Although the hemorrhagic activity can be destroyed with no effect on the esterolytic activity, the converse cannot be stated absolutely. Only 80% of the TAME- and BAEE-hydrolyzing activities were lost with incubation with phenylmethylsulfonyl fluoride. Nevertheless, these results, combined with the earlier report by Geren and co-workers (7), showing the BAEE- and TAME-hydrolyzing activities separating on either side of the starting material used for this isolation of hemorrhagins strongly suggest the esterases to be minor contaminants.

It is possible that these two hemorrhagins are the result of some proteolytic action on a single protein in the whole venom. If so, this occurs prior to our receipt of the venom, since incubation of the whole venom at 37°C for 24 hours results in no change in electrophoretic patterns (25).

It is quite possible, contrary to the earlier hypothesis by Naki and co-workers (6), that not all hemorrhagins from Crotalid venoms are very specific metal-dependent proteinases. BII_t does appear to be metal-dependent, but does not appear to be a proteinase.

ACKNOWLEDGMENTS

This work was supported in part by Grants GM 24173, RCDA ES 00052, and BRSG RR07101 from the National Institutes of Health, Bethesda, Maryland. Y.-Y. Shu was supported in part by the Fulbright College of Arts and Sciences of the University of Arkansas, Fayetteville, Arkansas.

REFERENCES

1. Tu, A. T. In *Venoms: Chemistry and Molecular Biology*. Wiley, New York (1977).
2. Fabiano, R. J. and A. T. Tu. Biochemistry 20:21 (1981).
3. Bjarnason, J. B. and A. T. Tu. Biochemistry 17:3395 (1978).
4. Ohsaka, A., M. Just, and E. Habermann. Biochim. Biophys. Acta 323:415 (1973).
5. Xu, X., C. Wang, J. Liu, and Z. Lu. Toxicon 19:633 (1981).
6. Nakai, T., H. Ishizacki, A. T. Tu, and H. Sugihara. Comp. Biochem. Physiol. 72C:103 (1982).
7. Sullivan, J. A., E. Farr, and C. R. Geren. Toxicon 17:269 (1979).
8. Civello, D. J., L. H. Duong, and C. R. Geren. Biochemistry 22:749 (1983).
9. Civello, D. J., J. B. Moran, and C. R. Geren. Biochemistry 22:755 (1983).
10. Geren, C. R. and E. F. Young. Fed. Proc. 42:1810 (1983).
11. Hummel, B. C. W. Can. J. Biochem. Physiol. 37:1393 (1959).
12. Schwert, G. W. and Y. Takenaka. Biochim. Biophys. Acta 16:570 (1955).
13. Laskowski, M. Methods Enzymol. 2:26 (1955).
14. Rinderknecht, H., M. C. Geokas, P. Silverman, and B. J. Haverback. Clin. Chim. Acta 21:197 (1968).
15. Warren, L. J. Biol. Chem. 234:1971 (1959).
16. Weber, K. and M. Osborn. J. Biol. Chem. 244:4406 (1969).
17. Ornstein, L. Ann. N.Y. Acad. Sci. 121:321 (1964).
18. Davis, B. J. Ann. N.Y. Acad. Sci. 121:404 (1964).
19. Merril, C. R., D. Goldman, S. A. Sedman, and M. H. Ebert. Science 211:1437 (1980).
20. Zacharias, R. M., T. E. Zell, J. H. Morrison, and J. J. Woodlock. Anal. Biochem. 30:148 (1969).
21. Gibson, D. R. and R. W. Gracy. Anal. Biochem. 96:352 (1979).
22. Durham, B. and C. R. Geren. Anal. Biochem. 116:311 (1981).
23. Spackman, D. H., W. H. Stein, and S. Moore. Anal. Chem. 30: 1190 (1958).
24. Shu, Y.-Y., J. B. Moran, and C. R. Geren. Biochim. Biophys. Acta 748:236 (1983).
25. Moran, J. B. and C. R. Geren. Toxicon 17:237 (1979).

33

Characterization of Substrate Site Mapping of Two Hemorrhagic Zinc Proteases from *Crotalus atrox* Venom

Jon B. Bjarnason
University of Iceland, Reykjavik, Iceland
Jay W. Fox
University of Virginia Medical School, Charlottesville, Virginia

INTRODUCTION

One of the most striking consequences of crotalid and viperid envenomation is hemorrhage. In less severe situations, hemorrhage is usually observed to be localized in cutaneous and subcutaneous tissues at the envenomation site. Yet, in extreme cases, hemorrhage can be evasive throughout a substantial area of the involved extremity and can sometimes be observed in organs such as the heart, kidneys, brain, and lungs.

The initial event of snake-induced hemorrhage appears to be the disruption of the pericellular collagenous basement membrane and other connective tissue collagens by these proteolytic toxins with high specificities toward peptide bonds of the connective tissue proteins. It is, therefore, relevant to explore the proteolytic specificities that these hemorrhagic toxins possess and in what manner they act on potential connective tissue substrates.

Five hemorrhagic toxins have been isolated from the venom of *Crotalus atrox* and have been characterized as metalloproteinases in which their native forms contain zinc as the metal (1,2). The hemorrhagic and proteolytic activities of these toxins are associated with their zinc content (1,2). Since the discovery of the catalytically essential zinc in the hemorrhagic toxins of *C. atrox* (1), other hemorrhagic toxins have also been found to contain zinc that is essential for activity (3-5). The metal is likely involved either in substrate interaction or in catalysis, or both. To further our understanding of the catalytic action of the hemorrhagic proteinases and to delineate the mechanism of hemorrhage, it is important to examine the substrate specificities, the active site environments, and mechanisms of inhibition of these enzymes.

The purpose of this investigation was to determine the peptide bond specificities of hemorrhagic toxins c and d (Ht-c and Ht-d) on various peptide substrates and to probe the active site environments of the toxins using specifically designed synthetic peptides as well as known inhibitors of other metalloproteinases.

MATERIALS AND METHODS

Lyophilized crude venom was purchased from the Miami Serpentarium. Hemorrhagic toxins c and d were isolated from the crude venom by a previously published method (1) with Tris (hydroxymethyl aminomethane)-glycinate disc gel and sodium dodecyl sulfate (SDS) disc gel electrophoresis as the criteria of purity. The oxidized B-chain of bovine insulin was purchased from the Serva Chemical Company and purified to homogeneity by reverse phase high performance liquid chromatography (HPLC). Boc-(*tert*-butyloxycarbonyl)amino acids and solid-phase peptide synthesis resins were from Peninsula Company. Abz-Ala-Gly-Leu-Ala-Nba was from Enzyme Systems Products, Livermore, California. All other reagents and solvents were of analytical grade or higher quality.

Molecular Weights, Isoelectric Points, and Amino Acid Compositions

The molecular weights of the toxins were determined by SDS polyacrylamide gel electrophoresis. The toxins and standards were treated with sample buffer containing SDS and 2-mercaptoethanol for 5 minutes at 100°C and then run on polyacrylamide gels of varying acrylamide concentrations (5, 8, 10, and 14%) (6). Protein concentrations were determined by the method of Bradford (7).

Isoelectric focusing was performed in Iso Gel agarose (FMC Corporation) at 15°C. Ampholytes of pH range 3.5 to 9.5 supplemented with pH range 9 to 11 were used to form the gradient. Iso Gel pI markers (FMC Corporation) were used to observe the pH gradient.

The amino acid compositions of the reduced and carboxyamidomethylated toxins were determined on a Beckman 121 amino acid analyzer (8). Samples (50 μg) of the toxins were dissolved in 500 μl of 5.7 N HCl containing 0.3% phenol and 0.2% β-mercaptoethanol. The tubes were evacuated, sealed, and then hydrolyzed at 110°C for 24, 48, and 72 hours. All hydrolyses were performed in triplicate for each hydrolysis time.

Spectroscopy

A JASCO spectropolarimeter (Model J-41C) was used to obtain the circular dichroic (CD) spectra at 25°C. The proteins were dissolved in 10 mM Tris, pH 7.9, containing 2 mM of $CaCl_2$ and 100 mM of NaCl. Multiple scans were made and average spectra were constructed by a microprocessor.

Immunology

Nine-week-old Balb/c male mice were injected subcutaneously with either 20 μg of Ht-c or 20 μg of Ht-d in Freund's complete adjuvant. Four weeks later, the mice were boosted subcutaneously with 20-μg each of antigen in Freund's incomplete adjuvant. One week later, the mice were bled and the resultant sera assayed for cross-reactivity on Ouchterlony plates.

Peptide Synthesis

All peptides were manually synthesized via solid-phase methodology (9). Following treatment of the peptide resin by hydrogen fluoride (HF), the crude peptide product was desalted by gel filtration on Sephdex G-10 and purified by HPLC on a Waters μBondapak reverse-phase C-18 column (30 × 0.8 cm; 10 μm), using a 0.1% TFA:$CNCH_3$ gradient system. The purified product was assayed by thin-layer chromatography (TLC) in two solvent systems: analytic HPLC on a Waters C-18 reverse-phase μBondapak column (30 × 0.4 cm; 10 μm) and amino acid analysis.

Specificity Studies

For the digestion of the oxidized insulin B-chain, 0.76 mg of the peptide was dissolved in 0.675 ml of 10 mM Tris-HCl, pH 8.1, and then incubated with 25 μl of either Ht-c or Ht-d (1 mg/ml) at 37°C. At timed intervals, 75-μl aliquots were withdrawn from the digestion mixture, mixed with 10 μl of glacial acetic acid to stop the reaction, and then frozen until analysis. Seventy-five μl of this mixture was then injected onto a Waters C-18 μBondapak column (30 × 0.8 cm) equilibrated with 0.1% TFA (buffer A). The column was developed with a gradient of 0 to 60% buffer B, containing 0.1% TFA in $CNCH_3$ at 1%/min with a flow rate of 1.0 ml/min. Peak fractions were collected, evaporated to dryness, and hydrolyzed in 5.7 N HCl at 110°C for 24 hours, followed by amino acid analysis. From the amino acid composition of the peak fractions and the known insulin chain sequence, the sites of cleavage were deduced.

To determine the kinetic constants for the cleavage by the toxins of the insulin peptide analogs, 2.0-ml reaction mixtures of each peptide were incubated at 37°C with the toxin in a 0.05-M sodium phosphate buffer at pH 7.5, such that the proteinase concentration was 0.52 μM and the peptide concentration range was 10 to 1000 μM. At timed intervals, 200-μl aliquots from the digestion mixture were taken, added to 300 μl of H_2O, and then reacted with 500 μl of a fluorescamine solution (30 mg/100 ml HPLC grade acetone). The fluorescence of the mixture was recorded at $\lambda_{excitation}$ = 390 nm and $\lambda_{emission}$ = 475 nm on a Perkin Elmer MPF-44A fluorescence spectrofluorometer. In order to convert change of fluorescence into absolute molar velocities, standard curves of fluorescence versus concentration of substrates and products before and after exhaustive digestion by the toxins were constructed for each peptide substrate (10). The rate of hydrolysis was linear past the amount of peptide hydrolysis necessary for obtaining accurate measurements of initial velocities. Aliquots of these digestions were subjected to HPLC separation and isolation and amino acid analyses as just described in order to ascertain the sites of cleavage.

The kinetic parameters Km, kcat, and kcat/kM were calculated from the velocity versus substrate concentration curve constructed from the digestion data for each synthetic peptide assayed using a nonlinear curve-fitting computer program (11). To further map the substrate binding sites of the two enzymes, the kinetic constants were determined for the cleavage of a series of synthetic peptides. The cleavage reactions were performed in a like manner as described for the insulin analog peptides.

Inhibition Studies

The inhibitory potential of various peptides and hydroxamic acid amino acids was examined with Ht-c and Ht-d. The fluorogenic substrate used for these assays was Abz(2-aminobenzoyl)-Ala-Gly-Leu-Ala-Nba (4-nitrobenzylamide). Fluorescence intensity versus concentration of substrates and hydrolysis products curves were constructed according to the procedure of Castillo and co-workers (12). The reactions were monitored at $\lambda_{excitation}$ = 340 nm and $\lambda_{emission}$ = 415 nm on a FOCI manual spectrofluorometer to obtain values of initial velocities. Inhibitor concentrations used were 10^{-2} to 10^{2} times the K_I values. All reactions were performed in triplicate. The K_I values were calculated from Dixon plots of the reactions in the presence of inhibitors (13).

RESULTS

Chemical Properties

Hemorrhagic toxins c and d were isolated from the venom of *C. atrox* and were demonstrated to be homogeneous by SDS and Tris-glycinate polyacrylamide gel electrophoresis (PAGE). From isoelectric focusing, the isoelectric points of Ht-c and Ht-d was determined to be 6.2 and 6.1, respectively. Thus Ht-d is slightly more acidic than Ht-c and might, therefore, be expected to have a higher negative charge than Ht-c at any given pH. This is manifested in the higher mobility of Ht-d on Tris-glycinate polyacrylamide gel electrophoresis, as well as its tighter binding to DEAE-anionic exchange resin during the isolation of the toxins (1). The molecular weights of both toxins were determined to be 24,000 and identical as far as could be detected using SDS gel electrophoresis and the Ferguson plot method for the molecular weight determination.

Amino Acid Composition

The amino acid analyses performed on the reduced and carboxyamidomethylated toxins show that the compositions are nearly identical. A difference of one residue is observed for six amino acids; all others are equal for both toxins. Most noteworthy of these differences is an additional His residue in the more basic hemorrhagic toxin c and an additional Ser residue in hemorrhagic toxin d. A difference in one residue was also observed for Ala, Val, Ile, and Leu. The reduced and alkylated toxins were found to contain two carboxymethylcysteine residues, whereas the nonreduced and alkylated toxins contained two half-cystines and no carboxylmethylcysteine residues. It was, therefore, concluded that each toxin contained one disulfide bond. According to the amino acid composition, Ht-c contains 218 residues, whereas Ht-d contains 216 residues, 56 of which are acidic residues or amidated forms thereof.

Immunochemical Comparisons

Antiserum raised in mice against Ht-c was assayed on double diffusion Ouchterlony plates for cross-reactivity with Ht-c, Ht-d, and Ht-e, another hemorrhagic proteinase from *C. atrox* venom (1,2). As seen in Figure 1, no cross-reactivity was observed between Ht-c antiserum and Ht-e. Ht-c antiserum did cross-react with both Ht-c and Ht-d. No spurs were observed in the precipitin lines between Ht-c and Ht-d and, therefore, the two proteinases appear to share identical antigenic structures. This experiment also demonstrated that Ht-e does not share any antigenic sites with Ht-c.

Circular Dichroism Spectroscopy

The circular dichroic (CD) spectra of the two proteins in the peptide regions are seen in Figure 2, and the aromatic region is seen in Figure 3. Overall, the CD spectra for Ht-c and Ht-d appear almost identical

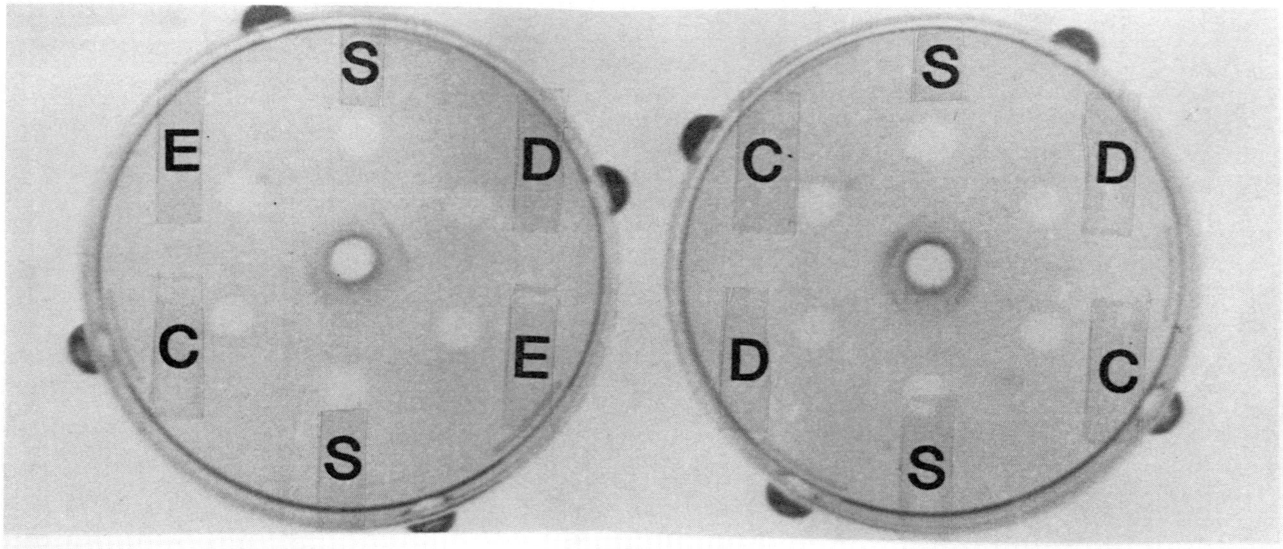

Figure 1 Double-diffusion Ouchterlony plates of Ht-c and Ht-d. Antiserum against Ht-c was placed in the middle well, hemorrhagic toxins C, D, and E were in the outer wells (left plate), and hemorrhagic toxin C and D were in the outer wells of right plate. S = saline control.

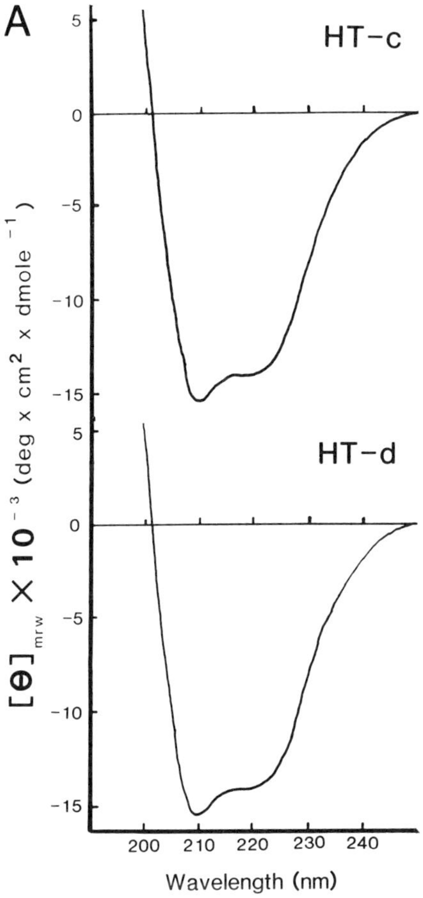

Figure 2 Comparison of the circular dichroic spectra of Ht-c and Ht-d in the peptide region.

with respect to minima positions and ellipticities. In the peptide region of both spectra, two minima are observed: 209 and 222 nm, with mean residue weight ellipticities of approximately -15,450 and -14,200, respectively. Both spectra indicate that the two proteins' secondary structures are approximately 40% α-helical (14,15). The CD spectra in the aromatic region of Ht-c and Ht-d are likewise nearly identical, with minima observed at 262, 267, 276, 284, and 292 nm with molar ellipticities of -8,500, -11,000, -11,500, -12,500, and -9,000, respectively. Also, one can observe a broad minimum centered around 310 nm. The minima at 262 and 264 nm correlate with phenylalanine CD

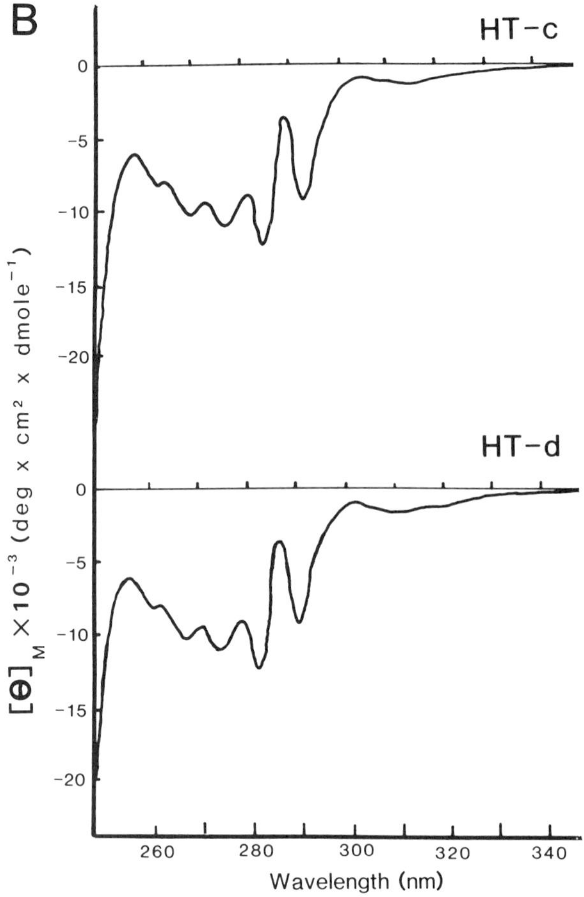

Figure 3 Comparison of the circular dichroic spectra of Ht-c and Ht-d in the aromatic region.

(16). The minima at 276 and 284 nm are probably due to CD from tyrosyl residues, and the 292-nm minimum is most likely due to contributions from tyrosyl and/or tryptophanyl residues (17). The broad, shallow, negative band centered around 310 nm is likely the result of CD due to the disulfide in each protein.

The major differences between the spectra of the two toxins in the aromatic region can be observed in the region between 275 and 292 nm. The spectrum of Ht-d shows a shoulder at 278 nm on the 276-nm peak, which is absent on the Ht-c spectrum. On the other hand, the spectrum of Ht-c has shoulders at 282 and 288 nm on the 284-nm peak and a shoulder at 290 nm on the 292-nm peak, all of which are absent on the Ht-d spectrum. This is indicative of subtle differences in the conformations of the two toxins.

Sites of Cleavage

The sites of cleavage of oxidized insulin B-chain by HT-c and Ht-d were deduced from the amino acid composition of digestion fragments observed at various time intervals. Both enzymes cleaved the B-chain at identical positions and order. Furthermore, both enzymes digested the chain at approximately identical rates. Analysis of the peptide fragments and their order of appearance allowed for deduction of the sites and order of insulin B-chain cleavage by both proteinases (Fig. 4). Both enzymes cleave the B-chain at five positions. Four of the hydrolyzed bonds contained a leucine residue at the P_1' position.* The first bond of insulin B-chain to be cleaved contained an alanine residue at the P_1 position, indicating a preference for a smaller residue at that position, in addition to leucine at the P_1' position. The only bond cleaved that did not have a P_1' leucine was the Gly_{23}-Phe_{24} bond.

Insulin Peptide Analogs

The results of the cleavage studies on the truncated insulin B-chain primary cleavage site peptides (termed insulin analogs) are shown in Table 1. The site of peptide bond cleavage was confirmed by amino acid analysis of the digestion products. Although the rates of cleavage were slow, both enzymes demonstrated a preference for extended substrates, as evidenced by the kcat/Km values. Most interestingly, the shortest peptide, NH_2-Ala-Leu-Tyr-Leu, was not cleaved by either enzyme. Unlike thermolysin, neither enzyme cleaved the synthetic substrate furylacryloyl-Gly-Leu-NH_2 (FAGLA).

Acetylated Peptides

The substrate binding subsites of the two enzymes were probed via cleavage studies using various synthetic peptides that contained different residues at several key positions within their structures. All of these synthetic peptides were acetylated at their amino terminals. For both enzymes, the rates of cleavage of these peptide substrates were appreciably greater than those of the unacetylated insulin analogs (Table 2). The peptide Ac(acetyl)-Val-Ala-Leu-Leu-Ala-COOH (pep-

---

*The nomenclature used for individual amino acid residues (P_1, P_1', P_2', etc.) of an inhibitor or substrate and for subsites (S_1, S_1', S_2', etc.) of the enzyme is that of Schechter and Berger (1967) see Ref. 40.

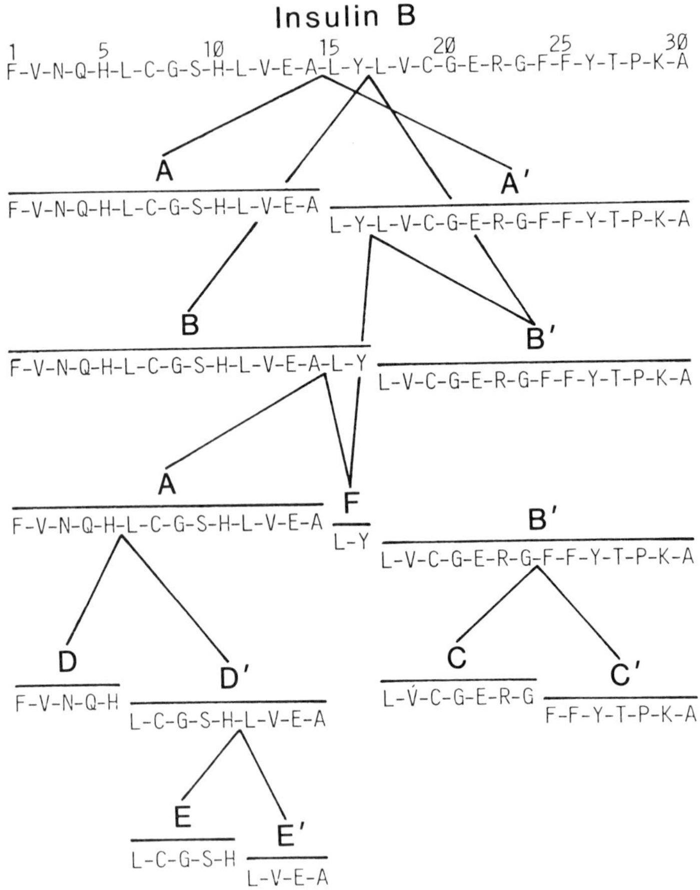

Figure 4 Digestion peptides of oxidized insulin B-chain by Ht-c or Ht-d. Progression of peptides produced was determined by amino acid analysis of peptides isolated by HPLC.

tide no. 1) gave the largest kcat/Km value of all the peptides assayed for both Ht-c and Ht-d (except the fluorogenic substrate). The enzymes were sensitive to substitutions at the substrate sites from position P_3 to position P'_2. The substitution of a glycyl residue for an alanyl residue at site P'_3 did not seem to have much affect on the kcat/Km value (peptide no. 2 vs. 3).

The enzymes greatly favored the aliphatic leucyl residue at position P'_2 in a substrate compared with a glycyl residue (peptide no. 1 vs. 3). Ht-c did not demonstrate any significant kinetic differences

Table 1 Kinetic Parameters of Cleavage of Insulin B-Chain Peptide Analogs by Ht-c and Ht-d

Substrate site	Km × 10^4 (M)		kcat × 10^4 (min^{-1})		kcat/Km (min^{-1}, M-1)	
P_4 P_3 P_2 P_1 P'_1 P'_2 P'_3	Ht-c	Ht-d	Ht-c	Ht-d	Ht-c	Ht-d
NH_2 – L–V–E–A–L–Y–L	1.1	0.32	118	43	107	134
NH_2 – V–E–A–L–Y–L	1.2	0.23	71	30	60	130
NH_2 – E–A–L–Y–L	2.9	1.4	52	25	18	18
NH_2 – A–L–Y–L[a]			NDC[b]	NDC[b]		

[a]Inhibition measured (see Table 3).
[b]NDC = no detectable cleavage.

Table 2 Kinetic Parameters of Cleavage of Synthetic Peptides by Ht-c and Ht-d

Peptide no.	Substrate site $P_3\ P_2\ P_1\ P'_1\ P'_2\ P'_3$	Km × 10^4 (M)		kcat (min^{-1})		kcat/Km × 10^{-2} ($min^{-1}\ M^{-1}$)	
		Ht-c	Ht-d	Ht-c	Ht-d	Ht-c	Ht-d
1	Ac—V—A—L—L—A	6.2	6.2	8.5	8.3	137	132
2	Ac—V—A—L—G—G	4.5	3.1	0.46	0.5	10.2	16.1
3	Ac—V—A—L—G—G	3.1	3.1	0.47	0.6	15.2	19.4
4	Ac—A—A—L—G—A	7.1	7.1	0.30	0.3	4.23	4.23
5	Ac—V—A—A—L—G—A	6.2	6.9	5.9	5.9	95.2	85.5
6	Ac—A—V—L—G—A[a]			NDC[b]	NDC		
7	Ac—V—G—L—G—A	2.1		0.46		21.9	
8	Ac—V—A—F—G—A	5.0		1.1		22.0	
9	Ac—E—A—L—G—A	7.6		1.0		13.8	
10	Ac—V—E—L—G—A[a]			NDC	NDC		
11	Ac—V—K—L—G—A[a]			NDC	NDC		
12	Ac—K—A—L—G—A[a]			NDC	NDC		
13	Ac—V—A—I—G—A[a]			NDC	NDC		
14	AB_z—A—G—L—A—NBA	3.1		145		4680	

[a]Inhibition assayed for, but not detected.
[b]NDC = no detectable cleavage.

when a phenylalanyl residue was substituted for the leucyl residue (peptide no. 3 vs. 8). However, it is very interesting to note that when an isoleucyl residue was placed at the P_1' site (peptide no. 13), no cleavage was detected nor did that peptide inhibit the enzymes, suggesting that peptide no. 13 does not bind to the enzymes' substrate-binding sites.

The enzymes both showed preference for the small aliphatic alanyl residue or a glycyl residue at the P_1 site. Substitution of a valyl residue at this position (peptide no. 6) resulted in a noncleaved peptide, which also did not act as an inhibitor. Additionally, the placement of a positively charged lysyl residue or a negatively charged glutamyl residue at the P_1 site gave rise to unhydrolyzed peptides (peptide nos. 10 and 11).

The bulky aliphatic valyl residue at position P_2 was preferred over the smaller alanyl residue (peptide no. 3 vs. 4). Moving the glutamyl residue from the P_1 site to the P_2 site (peptide no. 10 to 9) gives rise to a cleavable peptide, apparently indicating that a negatively charged residue is tolerable at the P_2 site. However, moving the positively charged lysyl residue from the P_1 site to the P_2 (peptide no. 11 to 12) does not give a cleavable product nor any detectable binding as assayed with inhibition measurements.

Extending the peptide substrate to include a valyl residue at the P_3 site (peptide no. 5) raises the kcat/Km compared with the value for peptide no. 4, suggesting that the enzymes prefer an extended substrate with a bulky residue at the P_3 site. This is corroborated by the kinetic data from the fluorgenic peptide no. 14 containing the 2-aminobenzoyl group at the amino terminal of the peptide.

Inhibition Studies

The results of the inhibition studies on Ht-c by hydroxamic acid amino acids, two chloromethyl esters, one synthetic peptide, and phosphoramidon are seen in Table 3. The hydroxamic acid amino acids containing an aliphatic side chain all exhibit competitive inhibition of the enzyme in the 10^{-4} M range. The L-Leu hydroxamic acid appears to be a slightly better inhibitor than the other hydroxamic acids. It is plausible that the primary mechanism of inhibition is the binding of the hydroxamate function to the enzyme's zinc atom, with some degree of specificity being conferred on the inhibitors by the amino acid side chain, as has previously been suggested for these types of inhibitors with thermolysin (18).

The hydroxamic acid inhibitors thus proved to be reasonably good inhibitors of the hemorrhagic proteinases, albeit with a rather low degree of specificity contributed by the single amino acid residue bound to the hydroxamate moiety. Nishino and Powers (18) referred to the Z-Gly-Gly-NHOH hydroxamic acid, having a K_I value of 94×10^{-5} M,

Table 3 K_I Values for Inhibitors of Ht-c

Inhibitor	$K_I \times 10^5$ (M)[a]
L-Leu-NHOH	15 ± 10
L-Phe-NHOH	43 ± 21
L-Val-NHOH	48 ± 15
L-Ile-NHOH	92 ± 44
MeO-Suc-Gly-Leu-Phe-CH_2Cl	43 ± 12
L-Leu-CH_2Cl	75 ± 24
NH_2-Ala-Leu-Tyr-Leu	25 ± 4
Phosphoramidon	33 ± 18

[a]Substrate used in these experiments was Abz-Ala-Gly-Leu-Ala-Nba.

as a nonspecific hydroxamic acid inhibitor of thermolysin. The L-Ile-NHOH hydroxamic acid inhibitor with Ht-c similarly had a K_I value of 92×10^{-5} M, also indicative of little or no specificity contributed by the L-Ile amino acid moiety of the inhibitor. This is in agreement with the observation that the L-Ile containing acetylated peptide no. 13 (Ac-Val-Ala-Ile-Gly-Ala-COOH) was neither a substrate nor an inhibitor of the toxins (see Table 2). On the other hand, the L-Leu-NHOH hydroxamic acid had a K_I value of 15×10^{-5} M with Ht-c as compared with 19×10^{-5} M found with thermolysin (18). This is likely due to a contribution to binding by the L-Leu-residue in the hydroxamate and, therefore, gives rise to a certain degree of specificity.

The chloromethyl esters did not irreversibly inhibit the enzymes and, in turn, were not particularly good inhibitors. However, the extended peptide chain chloromethyl inhibitor (MeO-Suc-Gly-Leu-Phe-CH_2Cl) did appear to be a relatively better inhibitor than L-Leu-CH_2Cl, suggesting the enhancement of inhibition with extended enzyme-binding sites.

The peptide NH_2-Ala-Leu-Tyr-Leu-COOH, which was not cleaved by the enzymes (Table 1), acted as a competitive inhibitor. Finally, phosphoramidon, (N-L-rhamnopyranosyl-oxyhydroxyphosphinyl)-L-leucyl-L-tryptophan, which is a potent inhibitor of thermolysin (K_I = 2.8×10^{-8} M) (19), also inhibits Ht-c, although not nearly as well as it does thermolysin.

DISCUSSION

The most outstanding local effects of rattlesnake envenomation are edema, hemorrhage, and myonecrosis. In cases of severe poisoning, these effects can lead to massive tissue destruction and subsequent dysfunction or loss of the affected extremity.

Western diamondback rattlesnake (*Crotalus atrox*) venom contains at least five proteolytic, hemorrhagic toxins, two of which have relatively low hemorrhagic activities as demonstrated by their high minimum hemorrhagic doses (1). These particular proteinases (Ht-c and Ht-d) show striking similarities in their physical and biochemical characteristics. Their molecular weights appear identical by SDS-PAGE (24,000 daltons), the isoelectric points of Ht-c and Ht-d are nearly identical (6.2 and 6.1, respectively), and their amino acid compositions show only minor differences. As evidenced by the CD spectra in both the peptide and aromatic regions, the solution conformations of the two toxins also appear to be similar. In addition, the antigenic structures of the two toxins seem to be identical, although different from the other *C. atrox* hemorrhagic proteinases. Thus, it is apparent that the two hemorrhagic proteinases Ht-c and Ht-d are very similar enzymes.

Although the two hemorrhagic toxins Ht-c and Ht-d are quite similar, they are definitely biochemically distinct from the other three hemorrhagic toxins (Ht-a, b, and e from *C. atrox* venom). This is clearly demonstrated by differences in amino acid compositions, isoelectric points, and CD spectra, as well as with the immunochemical comparisons (1,2,20).

An increasing number of reports are appearing in the literature on proteinases from *Crotalus atrox* venom involving various biological effects. In addition to the five hemorrhagic toxins, there are reports of one collagenase (21), two kallikrein-like enzymes (22), and four anticoagulant proteinases (23) that have all been purified and extensively characterized. Four anticomplimentary factors have also been purified, although not extensively characterized (24), and two fibrinolytic enzymes (25) and three caseinolytic proteases (26) have been isolated but not purified or characterized extensively, although one of the caseinolytic proteinases was purified further and some of its properties determined (27-29). This adds up to a total of 21 proteinases observed in the venom of *C. atrox* thus far. Most likely some of these proteinases are identical reducing their number somewhat, although such comparisons cannot be made with much confidence for those proteinases that have not been extensively characterized.

However, as previously suggested by Pandya and Budzynski (23), hemorrhagic toxin b and anticoagulant proteinase I are most likely one and the same enzyme. All available data support this supposition. It was also suggested that α-proteinase (27) and the fibrinolytic proteinase of Mr 21,500 (25) are identical to the aforementioned proteinases. We are in agreement with these authors on the suggestion that α-proteinase is most likely identical to Ht-b and proteinase I, judging from the available data as well as additional data from our laboratories on the cleavage specificities of Ht-b (soon to be published). Similarities with the fibrinolytic proteinase of Mr 21,500 are, of course, much more doubtful since only very scant information is available on that proteinase. It also appears quite probable that the anticoagulant proteinase II (23) and the kallikrein-like enzymes E-I and E-II (22) are identical enzymes. This is manifested in their amino acid compositions, molecular weights, arginine esterase activities, and along with other observations. On the other hand, the hemorrhagic toxins presently being studied appear to be dissimilar form all the other proteinases observed so far from *C. atrox* venom, as well as all other known venom proteinases (20) and nonvenomous zinc proteinases reported to date.

In summary, these two proteinases are very similar zinc enzymes, with only minor differences observed in their biochemical properties. However, Ht-c and Ht-d are not identical to any other known hemorrhagic or nonhemorrhagic venom or nonvenom-derived proteinase isolated thus far. This is exemplified by the comparison of their chemical and physical characteristics with other proteinases.

As an approach toward understanding the substrate-specificities of Ht-c and Ht-d, the sites and order of cleavage of peptide bonds in oxidized insulin B-chain were examined. The overall pattern of cleavage on the insulin B-chain is similar to several hemorrhagic and nonhemorrhagic snake venom proteinases, as well as two nonsnake venom proteinases (thermolysin and elastase), both of which can produce hemorrhage in vivo (Table 4). It is of interest to note that the majority of the snake venom proteinases have a rather limited specificity for this substrate, cleaving only three to five peptide bonds in the chain. In addition, one proteinase, hemorrhagic proteinase IV, from *C. h. horridus*, cleaves at only one position. The Ala_{14}-Leu_{15} cleavage site is common to all the proteinases except Protease A from *Bothrops jararaca* venom. Most of these proteases also cleave the His_{10}-Leu_{11} and His_5-Leu_6 bonds. Many cleave at the Tyr_{16}-Leu_{17} and Gly_{23}-Phe_{24} bonds as well. It is apparent that the cleavage sites of some of these proteinases are determined by the size and structure of the substrate and not confined to only the two amino acids comprising the peptide bond that is cleaved.

In spite of the similarities of cleavage specificities of these 18 proteases, no two or more have exactly identical cleavage patterns except the isoenzymes Ht-c and Ht-d, the basic hemorrhagic toxin Ht-b from *C. atrox* venom and the partially purified α-proteinase fraction, also from *C. atrox* venom. Ht-c and Ht-d are seen to be nearly identical in their insulin B-chain cleavage pattern to the pattern of the weakly hemorrhagic proteinase bothropasin from *B. jararaca* venom. The

Table 4 Sites of Cleavage of Proteolytic Hemorrhagic Toxins and Other Venom and Nonvenom Proteinases on the Oxidized B-Chain of Insulin

Phe[1]–Val–Asn–Gln–His[5]–Leu–Cys(SO_3^-)–Gly–Ser–His[10]–Leu–Val–Glu–Ala–Leu[15]–Tyr–Leu–Val–Cys(SO_3^-)–Gly[20]–Glu–Arg–Gly–Phe–Phe[25]–Tyr–Thr–Pro–Lys–Ala[30]

	Phe1–Val2	Asn3–Gln4	His5–Leu6	Ser9–His10	His10–Leu11	Val12–Glu13	Ala14–Leu15	Leu15–Tyr16	Tyr16–Leu17	Leu17–Val18	Val18–Cys19	Gly20–Glu21	Arg22–Gly23	Gly23–Phe24	Phe24–Phe25	Phe25–Tyr26	Tyr26–Thr27
Hemorrhagic toxin e (*C. atrox*)[a]		▲		▲			▲										
Hemorrhagic toxins c and d (*C. atrox*)[b]			▲		▲		▲		▲					▲			
Hemorrhagic toxin b (*C. atrox*)[c]			▲		▲		▲		▲					▲			
Hemorrhagic toxin a (*C. atrox*)[c]		▲	▲		▲		▲		▲								
Hemorrhagic factor 2 (*B. jararaca*)[a]					▲		▲		▲						▲		
Bothropasin (*B. jararaca*)[d]			▲		▲		▲		▲						▲		
Hemorrhagic protease IV (*C. h. horridus*)[e]							▲										
HR-I, proteinase b (*A. halys blomhoffi*)[a]			▲	▲	▲		▲							▲			
Proteinase a (*A. halys blomhoffi*)[a]			▲	▲			▲										
Proteinase c (*A. halys blomhoffi*)[a]					▲		▲							▲			
α-Proteinase, partially pure (*C. atrox*)[a]			▲		▲		▲		▲								
H_2 Protease (*T. flavoviridis*)[a]		▲			▲		▲										
Proteinase II (*C. adamanteus*)[a]	▲		▲		▲		▲	▲	▲								
Peptidase A (*A. piscivorus leucostoma*)[a]	▲		▲		▲		▲					▲		▲	▲		
Thermolysin[a]			▲		▲		▲		▲					▲	▲		
Elastase[a]				▲		▲	▲			▲	▲			▲			
Protease A (*B. jararaca*)[a]													▲			▲	▲

[a]See reference 20. [b]Present investigation. [c]Fox and Bjarnason, unpublished data. [d]See reference 5. [e]See reference 41.

difference in their patterns is that Ht-c and Ht-d cleave at the Gly_{23}-Phe_{24} bonds, whereas bothropasin cleaves at the Phe_{24}-Phe_{25} bonds. Thermolysin cleaves at all the same peptide bond positions in the insulin B-chain as Ht-c and Ht-d, with the addition of the Phe_{24}-Phe_{25} bond.

Thermolysin, like the metalloendopeptidase angiotensin-converting enzyme and carboxypeptidases A and B, contains a zinc ion in the active site that interacts with the carbonyl oxygen of the scissle peptide bond (30,31). Many synthetic inhibitors of this family of enzymes have been designed to contain a ligand, which coordinates with the zinc in the active site of the enzyme (32-34). Other inhibitors utilize a leaving group, such as a halide, which upon nucleophilic attack by an active site residue leaves the inhibitor, which then forms a covalent bond to the enzyme. These ligands and active groups are generally also coupled to a peptide or amino acid conferring a specificity to the inhibitor which is then capable of interacting in a competitive manner with the substrate binding site(s) of the enzyme (18,32,35,36). Due to similarities in the mechanism of action of these zinc metalloproteinases, an inhibitor of one of these enzymes may inhibit other enzymes of the same class when modified to conform to the enzymes' individual specificity requireemtns (34,37,38). Accordingly, substrates and inhibitors for Ht-c and Ht-d were designed and synthesized based on the primary sites of cleavage of the insulin B-chain by these toxins, such that the character of the substrate-binding sites of these enzymes could more clearly be defined.

From the kinetics of the hydrolysis of the insulin B-chain peptide analogs (Table 1), it is evident that this particular group of peptides was cleaved so slowly that they must be considered extremely poor substrates. The smallest peptide of this group, the tetrapeptide NH_2-Ala-Leu-Tyr-Leu-COOH, showed no measurable cleavage. In fact, it proved to be an inhibitor of the toxins ($K_I = 2.5 \times 10^{-4}$ M) when Abz-Ala-Gly-Leu-Ala-Nba was used as substrate. The slow cleavage of the insulin B-chain analogs could perhaps be attributed to the free amino and/or carboxyl termini of these peptides. This is plausible in light of the fact that the tetrapeptide is not cleaved, whereas the kcat and kcat/Km for the other peptides increase with the increase in peptide chain length. Also, in the case of thermolysin, peptides containing either a free amino terminus at the P_1 site or a free carboxyl terminus at the P_1' serve as potent inhibitors (30). Furthermore, from our results, we note that N-acetylated synthetic peptides, which are structurally similar to the insulin B-chain analog peptides, show significantly accelerated cleavages (Table 2).

The fluorogenic substrate Abz-Ala-Gly-Leu-Ala-Nba is actually a very good substrate for Ht-c and Ht-d, having the highest kcat/Km values for synthetic substrates tested thus far with these toxins. This particular substrate was cleaved at the Gly-Leu peptide bond, suggestive of similarities with the vertebrate collagenases that cleave interstitial collagen at Gly-Ile or Gly-Leu bonds (39). This is further supported by our preliminary experiments, which have shown that hemorrhagic toxins from *C. atrox* discriminately digest native interstitial collagen as well as native, type IV collagen and type V collagen.

A generalized view of the enzymes' substrate binding sites (S_3 through S'_3) can be approached from the cleavage data of the synthetic peptides. The enzymes seem to prefer hydrophobic residues in their S'_1 and S'_2 sites. However, the interesting exception is seen with peptide no. 13, which has an isoleucyl residue at its P'_1 site. Apparently, the enzymes cannot accept an isoleucyl residue at the S'_1 sites since peptide no. 13 does not act as a competitive inhibitor toward the enzymes. The S'_3 sites of the enzymes do not demonstrate any significant preference for either a glycyl or alanyl residue. The S_1 sites in the enzymes are rather selective in that they do not cleave peptides containing charged residues (glutamyl, lysyl) or bulky aliphatic side chains (valyl) at that position as contrasted with thermolysin (30). The S_2 sites of the enzymes seem to prefer a bulky aliphatic side chain like valyl over a smaller side chain (Ala) and will not cleave a peptide with positively charged lysyl residue at that position. Finally, it appears that the enzymes have an extended substrate-binding site in that the placement of large, bulky groups at the P_3 position of a substrate enhances catalysis compared with similar but truncated peptides.

As previously stated, it has been shown that an inhibitor of one of the zinc metalloproteinases may serve as an inhibitor of other enzymes of the same class when modified to conform to their individual specificity requirements. Therefore, we undertook a study on the effectiveness of hydroxamic acid type of inhibitors against the peptidase activity of Ht-c. Hydroxamic acid amino acids and dipeptide hydroxamic acids have been shown to be good inhibitors against thermolysin (18). The L-Leu-NHOH inhibitor with thermolysin had a K_I of 190 μM as compared with our data with Ht-c, having a K_I with this inhibitor of 150 μM. The dipeptide and tripeptide hydroxamic acid inhibitors were seen to be even more effective against thermolysin, with values for K_I in the low μM range (18). This suggests that future inhibitors of the hydroxamic acid class developed against these venom proteinases be lengthened to include more than a single amino acid residue in order to utilize the extended substrate-binding sites of Ht-c and Ht-d to the best advantage. This should serve to increase the specificity of these inhibitors and enhance their inhibitory potential.

This study has examined two of the less potent hemorrhagic toxins (Ht-c and Ht-d) from *C. atrox* venom. In light of their similar biochemical and kinetic characteristics, they are likely to be isoenzymes with only slight differences in their substrate-binding sites and cleavage kinetics. From the biochemical characterization and inhibition studies, these enzymes fall into the zinc metalloproteinase class and, therefore, the approaches taken toward the inhibition of various members of the class may also be relevant to these enzymes (33). Currently, we are developing inhibitors based on the substrate-specificity of these enzymes and their active site metal properties that incorporate two affinity groups. One part will serve to specifically bind the inhibitor to the substrate-binding sites in the enzymes, and the other will interact with the active site metal. Hopefully, this approach will allow for the development of more effective treatment of crotalid snake envenomation, as well as furthering our understanding of the mechanism of metalloproteinase catalysis.

REFERENCES

1. Bjarnason, J. B. and A. T. Tu. Biochemistry 17:3395 (1978).
2. Bjarnason, J. B. and J. W. Fox. Biochemistry 22:3770 (1983).
3. Civello, J. D., H. J. Duong, and C. R. Geren. Biochemistry 22:749 (1983).
4. Mori, N., T. Nikai, and H. Sugikara. Toxicon 22:451 (1984).
5. Mandelbaum, F. R., A. P. Reichel, and M. T. Assakura. Toxicon 20:955 (1982).
6. Laemmli, U. K. Nature (Lond.) 227:680 (1970).
7. Bradford, M. Anal. Biochem. 72:248 (1976).
8. Moore, S. and W. H. Stein. Methods Enzymol. 6:819 (1963).
9. Stewart, J. M. Methods Enzymol. 103:3 (1983).
10. Van Nispen, J. W., T. C. Hageman, and H. A. Scheraga. Arch. Biochem. Biophys. 182:227 (1977).
11. Johnson, M. L. Biophysical J. 44:101 (1983).
12. Castillo, M. J., K. Kurachi, N. Nishini, I. Ohkubo, and J. C. Powers. Biochemistry 22:1021 (1983).
13. Dixon, M. Biochem. J. 55:170 (1953).
14. Chen, Y. H., J. T. Yang, and K. H. Chan. Biochemistry 13: 3350 (1974).
15. Greenfield, N. and G. D. Fasman. Biochemistry 8:4108 (1969).
16. Fretto, L. and E. H. Strickland. Biochim. Biophys. Acta 235: 473 (1971).
17. Strickland, E. H. CRC Crit. Rev. Biochem. 2:113 (1974).
18. Nishino, N. and J. C. Powers. Biochemistry 17:2846 (1978).
19. Suda, H., T. Aoyagi, T. Takeuchi, and H. Umezawa. J. Antibiotics 26:621 (1973).
20. Fox, J. W. and J. B. Bjarnason. In *Toxin Reviews*, Shier, W. T. (Ed.), Marcel Dekker, New York, p. 161 (1983).
21. Hong, B. S. Toxicon 20:535 (1982).
22. Bjarnason, J. B., A. Barrish, G. S. Direnzo, R. Campbell, and J. W. Fox. J. Biol. Chem. 258:12566 (1983).
23. Pandya, B. V. and A. Z. Budzynski. Biochemistry 23:460 (1984).
24. Man, D. and J. O. Minta. Immunochemistry 14:521 (1977).

25. Bajawa, S. S., F. S. Markland, and F. E. Russell. Toxicon 18: 285 (1980).
26. Pfleiderer, G. and G. Sumyle. Biochim. Biophys. Acta 51:482 (1961).
27. Zwilling, V. R. and G. Pfleiderer. Hoppe Seylers Z. Physiol. Chem. 348:519 (1967).
28. Pfleiderer, G. and A. Krauss. Biochem. Z. 342:85 (1965).
29. Mella, K., M. Volz, and G. Pfleiderer. Anal. Biochem. 21:219 (1967).
30. Kester, W. R. and B. W. Matthews. Biochemistry 16:2506 (1977).
31. Hangauer, D. G., A. F. Monzingo, and B. W. Matthews. Biochemistry 23:5730 (1985).
32. Rasnick, and J. C. Powers. Biochemistry 17:4363 (1978).
33. Holmquist, B. and B. L. Vallee. Proc. Natl. Acad. Sci. USA 76:6216 (1979).
34. Monzingo, A. F. and B. W. Matthews. Biochemistry 23:5724 (1985).
35. Holmes, M. A. and B. W. Matthews. Biochemistry 20:6912 (1981).
36. Holmes, M. A., D. E. Tronrud, and B. W. Matthews. Biochemistry 22:286 (1983).
37. Nishino, N. and J. C. Powers. J. Biol. Chem. 255:3482 (1979).
38. Maycock, A. J., D. M. DeSousa, L. G. Payne, J. Broeke, M. T. Wu, and A. A. Patchett. Biochem. Biophys. Res. Commun. 102: 963 (1981).
39. Cawston, T. E. and G. Murphy. Methods Enzymol. 80:711 (1981).
40. Schechter, I. and A. Berger. Biochem. Biophys. Res. Commun. 27:157 (1967).
41. Civello, D. J., J. B. Moran, and C. R. Geren. Biochemistry 22:755 (1983).

VI
OTHER LABORATORY STUDIES

34

Effect of Russell's Viper Venom Infusion on Coagulation in the Rabbit

Cascade to Trickle

David L. Aronson*
Office of Biologics Research and Review, Bethesda, Maryland

INTRODUCTION

The use of snake venoms for the in vitro analysis of blood coagulation has a long history (1,2). Strangely enough, little use has been made of these specific coagulation activators in vivo, despite studies with whole venoms indicating that by certain routes, particularly the intravenous route, one of the major pathologic effects of snake venoms is the induction of thrombohemorrhagic disorders. Most of the in vivo studies with purified venom components are clinical studies of the thrombin-like venom coagulants, such as those from *Bothrops* (batroxobin) and from the Malayan pit viper (ancrod) (3,4). Schiek et al. (5) have analyzed the effects of *Echis carinatus* venom and the purified prothrombin-converting enzyme derived from this venom in rats and mice.

There are several advantages to the use of the snake venoms for in vivo studies. The absence of rapid clearance mechanisms for the snake venoms allows the introduction of a defined stimulus that will be maintained for at least several hours at an effective level. A second advantage to the use of snake venoms as compared with purified plasma coagulants is the difficulty in purification of the amounts of coagulants that are necessary for certain studies.

Along with the thrombin-like venoms, there are the venoms containing prothrombin-activating enzymes such as *Echis carinatus*, *Dispholidus typhus*, and *Oxyuranus scutellatus scutellatus*. These prothrombin-converting enzymes may be of particular interest since they have

**Present affiliation:* George Washington University Medical Center, Washington, D.C.

different converting pathways and differ in their cofactor requirements (6-9).

The venom of Russell's viper (RVV), which has been so useful because of its ability to convert factor X to activated factor X, has more recently been found to activate both factor V and factor IX. The factor V-activating enzyme is distinct from the enzyme that activates both factor IX and factor X. Further, the RVV conversion pathway for factor IX is different than the normal "bioconversion" pathway.

The studies that follow were done to determine the hierarchy of in vivo regulation of the terminal steps in coagulation (i.e., the steps between factor X activation and fibrinogen conversion to fibrin). Although these represent only a small part of the total coagulation sequence, they embody a great deal of complexity, including the activation and inhibition of factor X and prothrombin and the feedback of thrombin and factor Xa on factor V and factor VIII. A semipurified venom fraction devoid of measurable factor V-activating activity has been used to activate factor X in vivo. The preparation of RVV used presumably activated factor IX. This approach also allowed preliminary studies of the impact of such diverse thrombogenic stimuli as factor IX complex, phospholipid, and other pharmacologic agents (vide infra).

EXPERIMENTAL

The factor X-activating fraction from Russell's viper venom was purified by previously described methods, including gel filtration on ultragel and ion exchange chromatography on trisacryl-DEAE (7). The fractions eluted from the ion exchange column were aliquoted and maintained frozen at -80°C.

Rabbits (weighing 2 kg) were injected with the venom fraction into the marginal ear vein. Samples for analysis were removed from the central ear arteries at varying times and immediately anticoagulated with 1/10 volume of 4% sodium citrate. After centrifugation, the plasma was removed and either assayed immediately or frozen at -80°C until assayed. No differences were found for the assays done immediately or on frozen samples.

When 5 ml or more of this venom fraction was injected into intact, untreated rabbits, death ensued within 5 minutes. Gross autopsy findings indicated a thrombotic death. While some of the blood in the major vessels was still fluid at the time of death, that in the smaller vessels was clotted. The lethal dose could be decreased by several pretreatments, including the infusion with factor IX complex (human) prior to the injection of the RVV, or premixing the RVV with phospholipid (Inosothin, Associated Concentrates, Woodside, Long Island, New York). The pretreatment with factor IX concentrates (100 μm/kg) resulted in decreasing the lethal dose to 20 μl, and the injection of the venom mixed with the phospholipid reduced the lethal dose to about 250 µl. All deaths occurred within 5 minutes of the injection of the RVV. Table 1 summarizes these results.

Excessive thrombohemorrhagic disorders have been associated with the infusion of factor IX complex into patients. The deranged hemostasis after the infusion of factor IX complex has been ascribed to the effect of infused activated coagulants, to the accelerating effect of contaminating phospholipids on small amounts of activated coagulants (10), and to the high levels of zymogens other than factor IX in the recipients (11). The preceding experiments indicate that all of these mechanisms are possible.

When 0.5 ml of RVV was injected into the rabbits, there was neither death nor any signs of discomfort. Analysis of the plasma level of factor X showed a rapid decrease in factor X in 10 to 20 minutes to levels of about 1% of baseline. Figure 1 demonstrates the rapid and profound decrease in factor X as measured using factor X-deficient plasma. This decrease is presumed to represent activation and rapid inactivation of activated factor X. Assuming that the factor X level in rabbits approximates that in humans, this represents the activation of 15 µg/ml in 15 minutes or, on a linear basis, 1 µg/ml/min.

If this much factor X is converted, there should be rapid conversion of prothrombin, the substrate of activated factor X. Surprisingly, there is only about a 20% decrease in the plasma level of prothrombin. The temporal course of the prothrombin decrease is similar to that of factor X and was complete when factor X was depleted (Fig. 2). This implies extremely rapid inhibition of the activated factor X. Assuming that the plasma prothrombin level in rabbits is similar to that in human plasma (100 µg/ml), this would represent the activation of 20 µg/ml of prothrombin, yielding approximately 10 µg/ml of thrombin. This mass of thrombin would equal about 30 units/ml of plasma or, again, calculating on the basis of a linear reaction, the generation of 2 units/

Table 1 Lethality of Russell's Viper Venom in Rabbits

Pretreatment	Lethal dose
None	>5000 µl
Factor IX Complex	<20 µl
Factor IX (ARC)	>500 µl
Inosithin	250 µl
Pentobarbital	500 µl
Propranolol	>500 µl

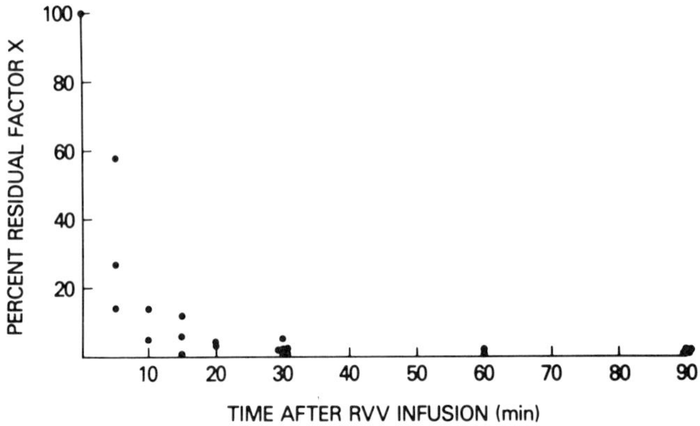

Figure 1 Decrease in factor X after the injection of Russell's viper venom.

min/ml. Acute death in rabbits was seen with the injection of 50 to 100 U/kg of thrombin or 1 to 2 u/ml of thrombin. However, in this case, no clinical signs were seen.

Ex vivo, the generation of small amounts of activated factor X results in rapid clotting and complete consumption of prothrombin. Indeed, in vitro, complete consumption of prothrombin is accomplished with the activation of less than 1% of the plasma factor X (12). In vivo in the stasis model, a minimal thrombogenic stimulus also results in complete consumption of the plasma prothrombin (13).

The substrate for thrombin, fibrinogen, shows a decrease similar to that of prothrombin (i.e., 20%) (Fig. 2). This is to be compared with the ability of 1 unit of thrombin to clot 1 ml of plasma in about 15 seconds.

The mole ratios of the substrates (i.e., prothrombin and fibrinogen) are given in Table 2. These mole ratios found in vivo are to be compared with the mole ratios found in vitro. For example, when blood is allowed to clot in vitro, less than 1% of factor X is activated, yet the majority of the prothrombin is consumed. The moles of prothrombin consumed per mole of factor X consumed is of the order of 1000:1. Kinetic analysis of the efficient conversion of prothrombin by activated factor X indicates that in the presence of all the cofactors, 1 mole of activated factor X can convert 1000 moles of prothrombin per minute.

These results may be compared with the results of Giles and coworkers (14). The administration to dogs of 0.5 μg/kg of activated factor X in the presence of phospholipid results in a decrease in the level of fibrinogen of the same order as seen in Figure 2, as well as decreases in the functional levels of factor VIII and factor V. This

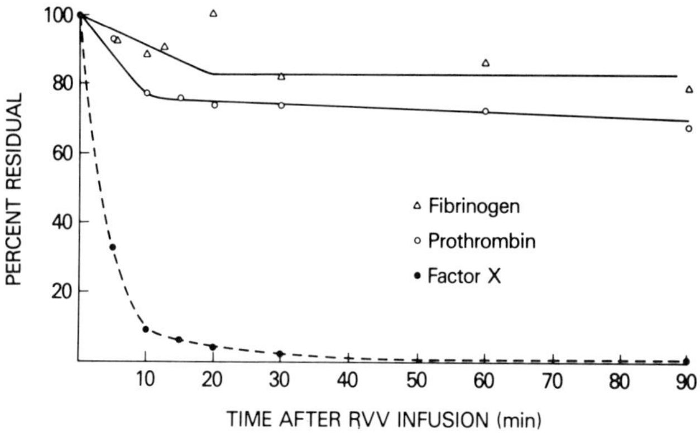

Figure 2 Effect of Russell's viper venom infusion on the level of factor X, prothrombin, and fibrinogen in the rabbit.

difference of a thousand-fold in the apparent sensitivity of the "whole animal" to the biologic effects of activated factor X could be due to species difference, to the precise effect of the phospholipid composition, or to the difference between in vivo-generated factor X and that generated factor X ex vivo.

The results in the rabbit, however, point to an extremely efficient regulatory mechanism for activated coagulants, more rapid than one would expect from the fluid phase inhibitors. One hypothesis is that the major clearance mechanism is dependent on flow and microvascular clearance. To test this hypothesis, RVV was injected into a pentothal-anesthetized rabbit and blood was immediately drawn by cardiac puncture and aliquoted into polypropylene tubes. At various times, aliquots were anticoagulated by the addition of 1/10 volume of 4% sodium citrate. Figure 3 shows that there is almost the same rate of decay of factor X as in vivo, but instead of the modest reduction in prothrombin, there is a rapid and complete consumption of the prothrombin. To rule out the possibility that this was the result of the plasma samples being

Table 2 Moles of Prothrombin and Fibrinogen Consumed per Mole of Factor X Consumed

Prothrombin/factor X	1.6
Fibrinogen/factor X	6.5
Fibrinogen/prothrombin	4

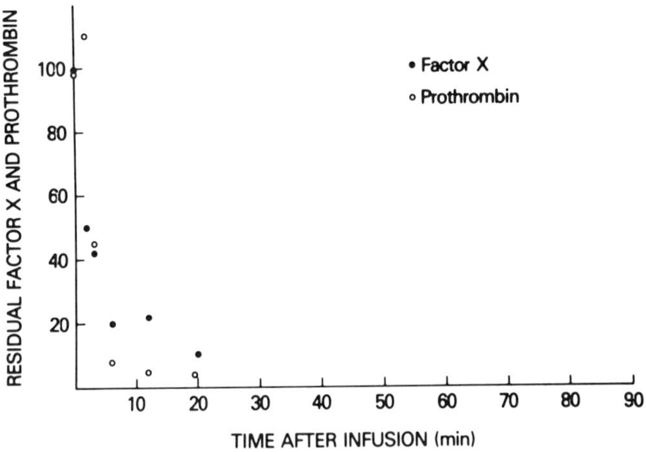

Figure 3 Ex vivo consumption of factor X and prothrombin. Russell's viper venom was injected into the rabbit. Blood was immediately removed, anticoagulated at the indicated time, and assayed immediately for prothombin and factor X.

contained in polypropylene tubes, the prothrombin consumption in an isolated venous segment was studied. Immediately after injection of 0.5 ml of RVV, a previously exposed jugular vein segment was ligated. Ten minutes later, the clotted contents of the vein were expressed into preweighed polypropylene tubes containing 0.8% citrate. The residual prothrombin level of the contents was reduced in an amount similar to that seen in the in vitro samples, in contrast to the prothrombin level in the flowing blood (Fig. 2). Thus, it is clear that the major regulation of the impact of the activated factor X in the rabbit is a flow- and probably microvasculature-dependent clearance mechanism.

Although the action of the activated factor X on prothrombin is limited, there is residual activity against the chromogenic substrate S-2222 in the flowing blood. Prior to injection of the RVV, there is no measurable S-2222 hydrolytic activity. Immediately following this, there is the development of measurable amounts of this activity, which falls off with a half-time of about 5 minutes. The maximum activity detected is only about 1% of the ex vivo-generated activity on S-2222. On storage, this activity is stable for weeks in frozen plasma. High performance liquid chromatography (HPLC) indicates that the activity is due to a high molecular weight component, which may be an activated factor X-α2-macroglobulin complex.

Other markers of in vivo coagulation are more difficult to study because of the presence of the RVV. For example, there is an apparent increase in the activity of factor VIII in the RVV-treated animals. In fact, this is the effect of the RVV on the test system. When the RVV is removed by HPLC and sieving column (TSK-3000SW), this apparent increased activity is removed, reducing the activity of the fractions eluting in the breakthrough peak by at least 50%. For this reason, we have not been able to make valid estimates of the levels of either factor VII or factor IX.

The most marked decrease in coagulant activity is that of factor V. Immediately after infusion of the RVV, there is an apparent increase in factor V and then a decrease to an apparent 25% of the initial level. These results are similar to those seen after the infusion of small amounts of activated factor X into the dog, and it may well be, as reported by Giles and co-workers (14), that this decreased factor V activity is predominantly due to the inactivation of the activated form by protein C. Table 3 summarizes the data for all of the coagulants measured.

DISCUSSION

In the present experiments utilizing Russell's viper venom (RVV) as the primary stimulus for in vivo coagulation, it is striking that in the absence of other "cofactors," the lethal dose is far above the level needed for rapid and complete activation of factor X. However, the effect of factor IX complex and extraneous phospholipid in lowering the threshold of lethality is also striking.

The thrombogenic effect of the phospholipid has two proposed components. The first is the well established accelerating effect of phos-

Table 3 Clotting Factor Activities in the Blood of Rabbits Following Intravenous Infusion of Factor X-Activating Fraction from Russell's Viper Venom

Caogulant	% Remaining activity
Factor X	1
Prothrombin	75
Factor V	25
Fibrinogen	80
Factor VIII	>50
Factor IX	Not measurable
Factor VII	Not measurable

pholipid on the conversion of prothrombin by factor Xa. As the second, there is protection of factor Xa from the fluid phase inhibitors when bound to phospholipid and increased circulation time of activated factor X in the presence of phospholipid (15).

It is not clear why factor IX complex causes such a dramatic sensitivity to the infusion of RVV. The concentrations in plasma of prothrombin and factor X are below their Michaelis-Menton constants. Therefore, any increase in concentration could cause an increase in the rate of product formation. In the case above, the levels of factor IX, factor X, and prothrombin are increased three-fold over the "normal" level. Thus, there could be a multiple-step effect. The order of magnitude of the effect shown in Table 1, however, is too great to be explained by such a simplistic mechanism.

The regulation of coagulation has been described at many levels, including fluid-phase inhibition, organ-specific clearance, and microvascular mechanisms. The most studied of these are the reactions in the fluid phase with plasma proteins, such as antithrombin-III (AT III) or α2-macroglobulin. Recent work describing the endothelial protein, thrombomodulin (16), has shown the importance of microvascular structures for the regulation of in vivo coagulation, reactions that are not independent of the fluid-phase inhibitors. Organ-specific regulation has also been proposed, with the liver being considered the primary organ.

The experiments reported here indicate that the major control for the coagulation system, once activated, is a flow-dependent system, probably dependent on the inhibition of the coagulant activity at the level of the microvasculature similar to the removal of thrombin by thrombomodulin (17). This would explain the strong association between altered flow and thrombosis. The experiments of Fuchs and associates (18) indicate that there are binding sites other than thrombomodulin for activated factors IX and X.

The limitation of the downstream effects is very striking. It seems unlikely that this is an organ-specific effect. The organ receiving the largest percentage of the blood flow during rest is the liver. Even if this were responsible for the clearance, it would take four to five circulation times for this clearance to be complete. Assuming a circulation time of about 30 seconds at the shortest, this would take several minutes for activated factor X to be eliminated, and one would then expect a more profound effect on the level of prothrombin. However, this would seem to be the ultimate clearance site for the enzyme complexes formed in the microcirculation.

The ability to form a clot and achieve quantitative prothrombin consumption in the ligated venous segment indicates that, under the conditions of our experiments, the fluid-phase inhibitors are not the major regulator and that, in the absence of flow, there is the ability to form an effective "prothrombinase" activity sufficiently long-lasting to completely consume the prothrombin. On the other hand, the presence of stable activity against the substrate S-2222 indicates that at least a small portion of the generated, activated factor X may be bound to α2-macroglobulin. In the mouse, Fuchs and associates (18) found the α2-macroglobulin complex to be the predominant complex formed in vivo. However, this complex is cleared relatively slowly, with a half-life of about 30 minutes. In our experiments, the half-life was only about 5 minutes.

SUMMARY

The purified coagulant enzymes derived from snake venoms are an unexploited tool for the evaluation of in vivo coagulant effects. In the experiments described here, Russell's viper venom was used to activate factor X in the rabbit. Despite the rapid activation of factor X, thrombotic events are not seen in the unmanipulated rabbit. Furthermore, there is very little downstream effect (i.e., there is little consumption of prothrombin or fibrinogen). This regulation of the downstream effect is flow-dependent and probably represents a regulatory mechanism of the microvasculature.

REFERENCES

1. Eagle, H. J. Exp. Med. 65:613 (1937).
2. MacFarlane, R. G. and B. Barnett. Lancet 2:985 (1934).
3. Bell, W. R., W. R. Pitney, and J. F. Goodwin. Lancet 1:490 (1968).
4. Latallo, Z. S. Thromb. Haemost. 50:604 (1983).
5. Schieck, A., E. Habermann, and F. Kornalik. Naunyn-Schmiederberg's Arch. Pharmacol. 274:7 (1972).
6. Owen, W. G. and C. M. Jackson. Thromb. Res. 3:705 (1973).
7. Esmon, C. T. Biochemistry 19:1020 (1980).
8. Franza, B. R., D. L. Aronson, and J. S. Finlayson. J. Biol. Chem. 250:7057 (1978).
9. Morita, T., S. Iwanaga, and T. Suzuki. Seikagaku 46:569 (1974).
10. Giles, A. R., M. E. Nesheim, H. Hoogendorn, P. B. Tracey, and K. G. Mann. Blood 59:401 (1982).
11. Magner, A. and D. L. Aronson. Dev. Biol. Stand. 44:185 (1979).
12. Aronson, D. L. and J. Bagley. Circulation 58:813 (1978).
13. Aronson, D. L. Thromb. Res. 1985.
14. Giles, A. R., M. E. Nesheim, and K. G. Mann. J. Clin. Invest. 74:2219 (1984).
15. Barton, P. G., E. T. Yin, and S. Wessler. J. Lipid Res. 11:87 (1970).

16. Lollar, P. and W. G. Owen. J. Clin. Invest. 66:1222 (1980).
17. Fuchs, H. E. and S. V. Pizzo. J. Clin. Invest. 72:2041 (1983).
18. Fuchs, H. E. Shifman, and S. V. Pizzo. J. Clin. Invest. 73: 1696 (1984).

35

Venom Proenzyme Activator Detection with a Novel Immunochromometric Technique

Thomas Exner
Westmead Hospital, Westmead, Sydney, Australia

INTRODUCTION

The simple binding or adsorption of functional antibodies to plastic surfaces that has been the basis of most solid-phase immunoradiometric assays, such as IRMA, and more recently of enzyme-linked immunoassays, ELISA (Engvall and Perlman, 1972) was first described by Catt and Tregear (1967) 20 years ago. Various studies have shown that proteins and enzymes retain at least some function after binding to antibodies either in solution or when immobilized on a surface or matrix (e.g., Koutts et al., 1977; Comp et al., 1983). This chapter describes a new technique that combines immunoimmobilization of specific proenzymes with activation by venom activators and subsequent assay of the activated enzymes using tripeptide-p-nitroaniline-based chromogenic substrates.

Chromogenic substrates were originally developed for specific reactivity with individual enzymes of the hemostatic mechanism by appropriate choice of constituent amino acids or analogs of amino acids (Svendsen et al., 1972). However, their specificity is not absolute and it is preferable in most cases to work with purified or isolated enzymes. This is facilitated in the present new method by specific immunoimmobilization.

This technique, which might be termed an immunochromometric assay or ICMA (by analogy with IRMA), has broad application possibilities. So far, it has been usefully applied only to assay protein C (PC) (Exner and Koutts, 1986). The latter is an important antithrombotic proenzyme that, up to the present time, has been difficult to quantitate accurately (Bertina et al., 1984). However, immunoimmobilization of PC in microtiter wells has allowed the detection of new activators of PC amidase activity in various venoms (Exner et al., 1985). In the present chapter, such studies were extended to other venoms and to include potential activators of factor X and plasminogen.

MATERIALS AND METHODS

Materials

Chromogenic substrate for plasmin, Chromozym PL (Tos-Gly-Pro-Lys-pNA), was obtained from Boehringer-Mannheim, North Ryde, Sydney. Substrate CBS-34.47 Ac-D-CHG-But-Arg-pNA used for detecting thrombin, but also sensitive for activated protein C, was purchased from Diagnostica Stago, Asnieres, France. Substrate Xa-1 (moc-CHA-Gly-Arg-pNA) for activated factor X was provided by Nyegaard and Company, Oslo, Norway.

Rabbit antisera to human protein C, plasminogen, and factor X were purchased form Diagnostica Stago. Additional antiserum to protein C was kindly provided by Dr. R. Bertina, Academisch Ziekenhis, Leiden, The Netherlands. All were monospecific and had been prepared by repeatedly immunizing rabbits with highly purified human antigens. All three antibodies were isolated using the protein A Sephrose procedure-elution, with pH 3.0, 0.1 M glycine (technical literature—Pharmacia, Uppsala, Sweden).

Normal plasma derived from ACD blood collected from healthy volunteers was obtained from the Red Cross Blood Transfusion Service, Sydney. Factor X-deficient plasma was obtained from General Diagnostics (Division of Warner-Lambert Company, Morris Plains, New Jersey).

Agkistrodon acutus, *A. bilineatus* (Tropical moccasin), *A. contortrix contortrix* (Southern copperhead), *A. contortrix laticinctus* (broadbanded copperhead), *A. piscivorus piscivorus* (Eastern cottonmouth moccasin), *A. rhodostoma*, *Bothrops atrox*, *B. neuwiedi*, *B. jararaca*, *Bitis nasicornus*, *Crotalus atrox*, *C. mollosus mollosus*, *C. scutelatus*, *Echis carinatus*, *Naja naja atra*, *Trimeresurus flavoviridis*, *Vipera aspis*, *V. russelli*, and *Walterinnaesia aegyptia* snake venoms were purchased form Sigma Chemical Company, St. Louis, Missouri.

Methods

Immunochromometric Assays

Immunochromometric assays (Exner and Koutts, 1986) were carried out in 96-well, flat-bottomed micro-ELISA plates of Immunlon II (Dynatech Laboratories, Alexandria, Virginia). Wells were "coated" with 0.2-ml volumes of 5 μg/ml antibody in 0.1 M sodium bicarobnate overnight at 4°C, washed, and then "blocked" with 5% bovine serum albumin (BSA-RIA grade, Sigma) in 0.12 M sodium chloride, 0.02 M imidazole, 0.005 M citrate, pH 7.2 buffer. The same buffer, but with 0.1% BSA present instead of 5%, was used throughout for sample dilution and well-washing procedures. Well volumes of 0.20 ml were used consistently, and all sample procedures were carried out with an eight-channel Titertek 50 to 250 μl pipetting device (Flow Laboratories, Finland). Wells were washed by a filling and evacuation method at least three times between steps, and were incubated in sealed humidified containers at 20°C.

Normal human plasma, diluted 1:50, was applied to each well for 16 hours to supply each antigen. Control wells, exposed to factor X-deficient plasma or no plasma were used as detailed in the Results section. Venom dilutions of 1:4 (0.067 ml transfers to 0.20-ml wells), usually from a starting concentration of 0.25%, were applied for 6 hours. After a final washing, the wells were incubated for 16 hours with chromogenic substrates (0.15 mM) corresponding to the proenzyme-enzyme being studied and assessed on a microtiter titer plate reader (Flow Laboratories) at 405-490 nm.

Venom Fractionation

Gel filtration of *Agkistrodon rhodostoma* venom was carried out on a 2.5- × 60-cm column of Biogel P-100 (Bio Rad Laboratories, Richmond, California) in 0.1 M Ammonium acetate, pH 7.0 buffer. Fractions containing 6.5-ml each were collected and were diluted 1:1000 to test for clotting activity on 0.2% human fibrinogen (CSL, Melbourne) in 0.12 M sodium chloride, 0.02 M imidazole/HCl, pH 7.2 buffer at 37°C. Clotting times in such assays are inversely related to thrombin-like enzyme activity (Austen and Rhymes, 1975). Fractions were tested for factor X procoagulant activator by mixing 0.02 ml of each gel filtered fraction with 0.2 ml of 0.025 M calcium chloride and 0.2 ml of normal human plasma containing 0.02% platelet-lipid substitute (Bell and Alton, 1954) and timing to a clotting endpoint at 37°C.

RESULTS

Protein C Activator in Venoms

Microtiter wells coated with immunoimmobilized protein C were incubated with dilutions of all the available venoms. After washing, the wells were loaded with chromogenic substrate CBS-34.47 and subsequently assessed on a microtiter plate reader at 405-490 nm.

Venoms that resulted in detectable hydrolysis of this substrate were tested in control experiments. These were experimentally identical to those described above, but without PC attached to the adsorbed antibody. Results obtained are shown in Figure 1. Most venoms had no apparent effect on immunoimmobilized PC chromogenic substrate activity.

The *Bothrops* venoms generally gave high absorbance, showing strong activity on the chromogenic substrate at higher concentrations. However, this effect occurred to a similar extent in control wells not containing PC, and obviously was not due to PC activation. *Agkistrodon contortrix* and *piscovorus* venoms both showed a complex effect in which less chromogenic substrate hydrolysis occurred at high concentrations of venom than at lower concentrations. This was considered to be due to reduction of the amount of attached PC at higher venom concentrations, probably due to proteolysis. At lower venom concentrations, considerable chromogenic substrate hydrolysis occurred in wells containing PC and not in control wells, suggesting that specific PC activators might occur in these venoms. A number of related moccasin and copperhead venoms were subsequently investigated. As shown in Figure 2, most of these also appeared to contain PC activators, with *Agkistrodon bilineatus* and *Agkistrodon contortrix contortrix* venoms having the most potent effects. Venoms not specifically mentioned here resulted in negligible adsorbance readings and are not shown in the figures.

Venom Activators of Factor X

Similar experiments to those just described were carried out for factor X. In this study, microtiter wells were coated with antibody to human factor X, and the chromogenic substrate Xa-1 was used. Results obtained are shown in Figure 3.

A strong factor X activator was detected in Russell's viper venom, as might be expected from previous studies with this venom. Surprisingly, however, even more activity appeared to occur in *Agkistrodon rhodostoma* venom, which is better known for its powerful thrombin-like enzyme activity (ancrod). Background activity in control wells (not containing factor X) prepared with factor X-deficient plasma in place of normal plasma was negligible with both venoms.

The factor X activator in *Agkistrodon rhodostoma* venom was further investigated after fractionating the whole venom by gel filtration on Biogel P100. Fractions obtained were tested for fibrinogen clotting activity and procoagulant activity on recalcified platelet-rich plasma, a test that is shortened in the presence of activators of factor X-clotting activity. Microtiter wells containing factor X exposed to venom fractions were further exposed to *Agkistrodon rhodostoma* venom (10^{-3} % w/v) to activate all remaining factor X. After exposure to

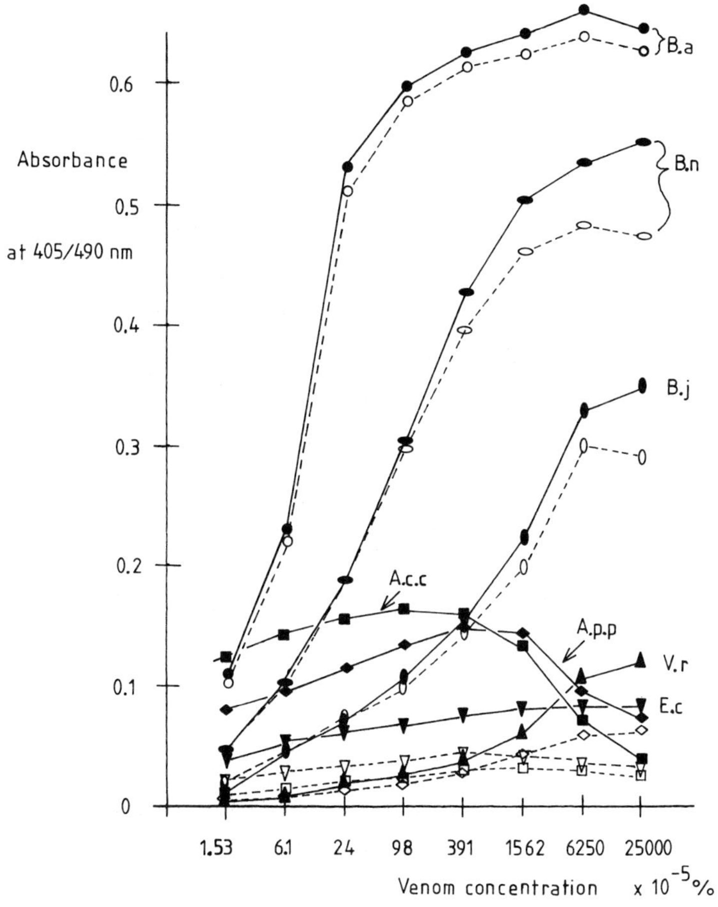

Figure 1 Effect of various venoms on the activity of immunoimmobilized protein C-containing microtiter wells on chromogenic substrate CBS-34.47. Solid symbols denote wells containing PC. Corresponding open symbols denote control wells not containing PC. *Bothrops atrox*, B.a; *B. neuwiedi*, B.n; *B. jararaca*, B.j; *Agkistrodon contortrix contortrix*, A.c.c; *A. piscivorus piscivorus*, A.p.p; *Vipera russelli*, V.r; *Echis carinatus*, E.c.

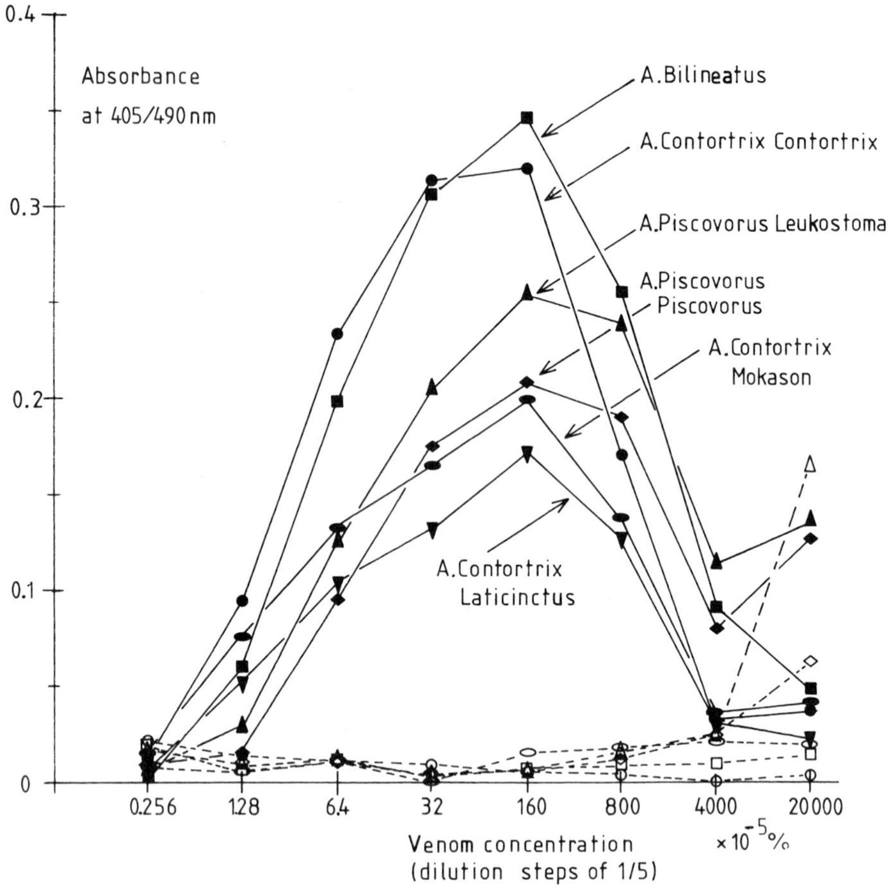

Figure 2 Effect of additional moccasin and copperhead species' venoms on immunoimmobilized protein C-containing microtiter wells. Symbols as for Figure 1.

substrate Xa-1, this procedure revealed how much of the immunoimmobilized factor X was still functional and able to be activated.

Results obtained are shown in Figure 4. The factor X amidase activator eluted predominantly with relatively high molecular weight proteins and its peak activity was distinct from the peak fibrinogen-clotting activity. Venom components that reduced the activatability of immunoimmobilized factor X eluted principally after the procoagulant activity. Interestingly, no significant procoagulant activity (expected to be associated with a factor X activator) eluted with the fac-

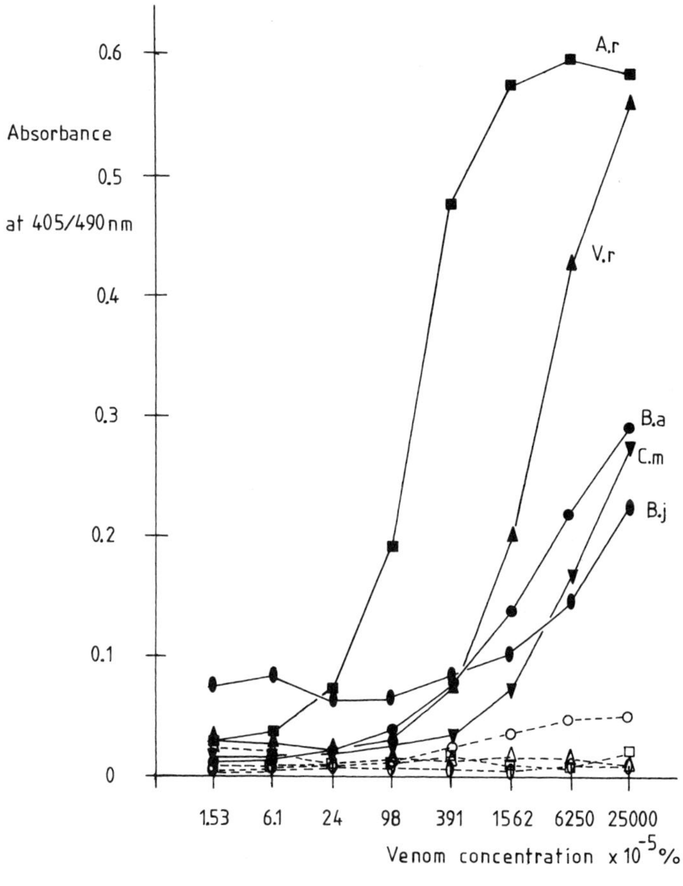

Figure 3 Effect of various venoms on the activity of factor X-containing microtiter wells on chromogenic substrate Xa-1. Factor X-containing, closed symbols; no factor X, open symbols; *Agkistrodon rhodostoma*, A.r; *Vipera russelli*, V.r; *Bothrops atrox*, B.a; *B. jararaca*, B.j; *Crotalus mollosus*, C.m.

tor X amidase activator peaks. Peak plasma procoagulant activity coeluted with the peak fibrinogen-clotting activity.

Venom Activators of Plasminogen

Immunochromometric experiments similar to those just described were carried out to detect venom activators of plasminogen. Microtiter wells containing immunoadsorbed plasminogen were exposed to all

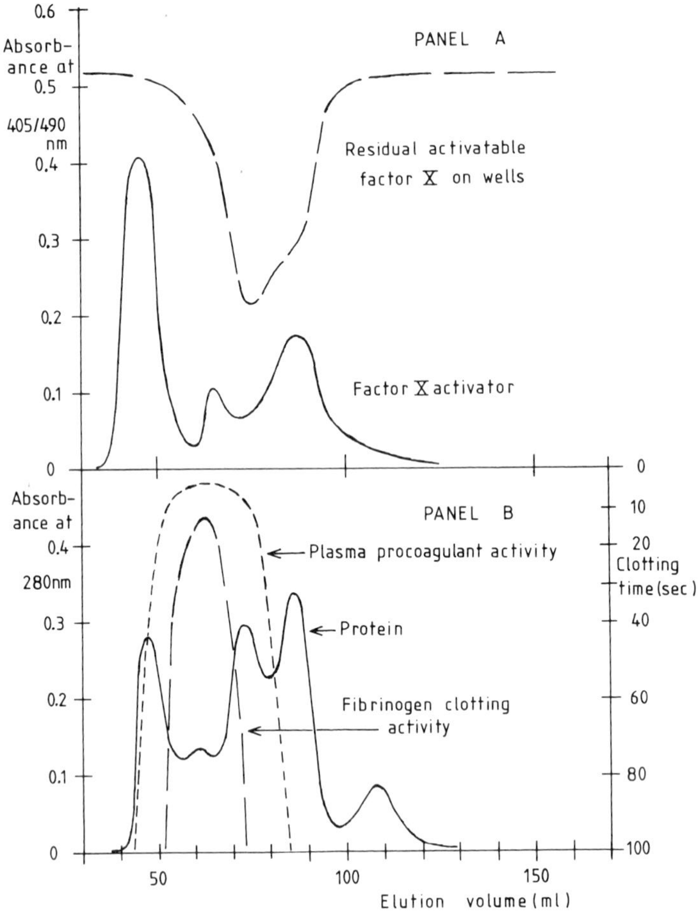

Figure 4 Elution profiles of various activities of *Agkistrodon rhodostoma* venom gel-filtered on Bio Gel P-100. Factor X chromogenic activator and residual activatable factor X after exposure to venom fractions are shown in panel A. Fibrinogen-clotting activity, plasma procoagulant activity, and protein concentration are shown in panel B.

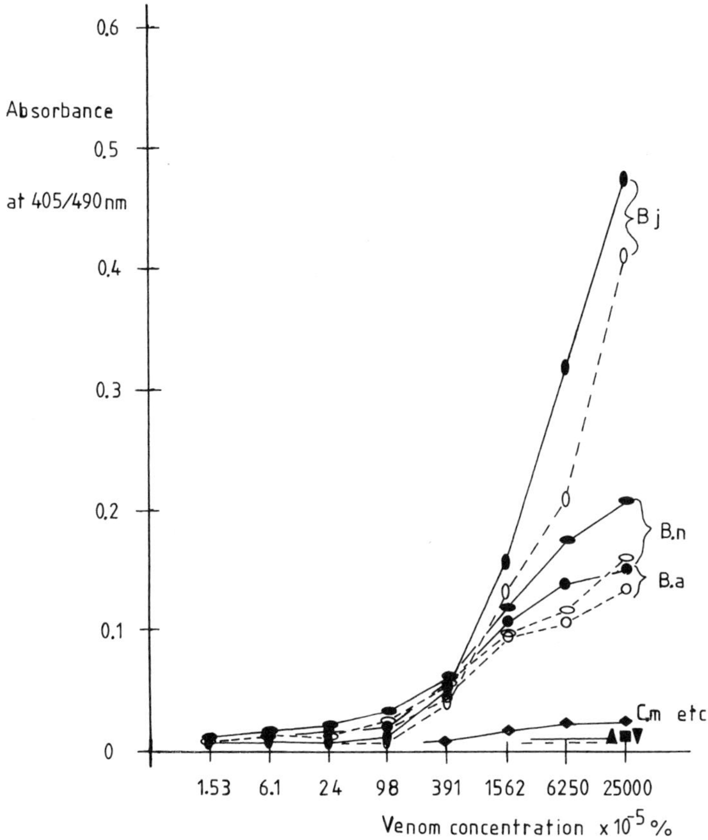

Figure 5 Effect of various venoms on the activity of plasminogen-containing microtiter wells on chromogenic substrate PL-1. Symbols and abbreviations as for Figures 1 and 3.

available venoms in various dilutions and subsequently to Chromozym PL. Results obtained are shown in Figure 5.

None of the venoms tested, except for those of the *Bothrops* species, had any significant effect in this procedure. Although wells exposed to all three *Bothrops* venoms tested showed significant activity on Chromozym PL, a similar effect was observed in control wells lacking plasminogen. Venom activators of plasmin-like chromogenic activity did not appear to be detectable by this technique or simply may not be present in the venoms tested.

DISCUSSION

The new proenzyme assay technique described here as an immunochromometric assay or ICMA combines some features of solid-phase immunoassays with the final use of chromogenic substrates for the detection of immunoimmobilized activated enzymes. Heterologous antibodies in free solution usually combine with many antigenic sties on each protein molecule and usually, but not always, inactivate the function of the target protein. However, when antibodies are "coated" or adsorbed on a surface, the stoichiometry is quite different. Each antigen molecule captured is probably bound by only one surface-associated antibody molecule. Also, due to the heterogeneity of normal heterologous antibodies, antigen molecules may be captured through a variety of epitopes and may retain various functions.

The ICMA technique has a number of potential uses. In the current work, it has been used for the screening of new activators of protein C, factor X, and plasminogen in snake venoms.

The discovery of PC activators in the venoms of *Agkistrodon contortrix contortrix* (Southern copperhead) and *A. bilineatus* (tropical moccasin) that may be more effective than Russell's viper venom (Kisiel et al., 1976) has some practical importance. It makes assays such as the ICMA for PC (Exner and Koutts, 1986) more convenient by avoiding the need for thrombin or thrombin-thrombomodulin as otherwise the only known activators of PC. One problem with thrombin is that it adsorbs nonspecifically in the ICMA wells, giving much higher "background" chromogenic activity (Exner and Koutts, 1986) than observed with these venom activators. It is interesting to note that victims of moccasin and copperhead snake bites may become depleted in PC (not measurable prior to 1980), and this may have contributed to clinical complications such as microcirculatory thrombosis leading to tissue necrosis.

The investigation of factor X activators in venoms by ICMA also yields interesting information. The demonstration of factor X activator in Russell's viper venom clearly confirms the reliability of the ICMA. Surprisingly, the generation of chromogenic amidase factor Xa-like activity by a component of *Agkistrodon rhodostoma* venom did not appear to be associated with procoagulant or anticoagulant activity. The nature of the changes in cleavages induced in factor X by this activator needs to be investigated further. According to Fujikawa and associates (1975), factor X is normally activated by the cleavage of a 51 amino acid peptide from the N-terminal region of the heavy chain. However, it is possible that cleavages occur in other sections of the proenzyme, or that binding changes might generate an active site having chromogenic activity but devoid of procoagulant activity. As a further control on the specificity of the ICMA, chromogenic activity on the factor Xa-sensitive chromogenic substrate was found to be generated by certain *A. rhodostoma* venom fractions only when factor X was present, immunoadsorbed from normal plasma, but not when human plasma congenitally deficient in factor X was used in control wells.

The presence of factor X activator in both the void volume material and also as later-eluting gel-filtered material from *Agkistrodon rhodostoma* venom suggests that the activator is of low molecular weight, but probably occurs as a high molecular weight complex.

The plasminogen experiments were unsuccessful in that no activators were detectable in the venoms tested. Unlike protein C and factor X, plasminogen loses a considerable part of its structure on its conversion to plasmin. It is possible that plasmin is dissociated from the solid phase during the activation stage of the ICMA. Furthermore, it may be necessary to test venoms fractionated (e.g., by column chromatography) to eliminate the interesting nonspecific, solid-phase binding enzymes showing such strong amidase activity.

CONCLUSION

Antibodies to plasminogen, factor X, and protein C were applied in separate experiments to microtiter wells in trays such as are commonly used in enzyme-linked immunoassays (ELISA). After incubation with dilute human plasma as a source of the corresponding proenzyme, the wells, now containing immunoimmobilized antigens, were exposed to dilutions of various snake venoms. After thorough washing, these were tested overnight with various chromogenic tripeptide substrates, known to be relatively specific for the activated forms of each proenzyme (i.e., plasmin, factor Xa, and activated protein C).

Two new activators of protein C in *Agkistrodon bilineatus* and *Agkistrodon contortrix contortrix* venoms and a new factor X activator in *Agkistrodon rhodostoma* venom were detected using the new immunochromometric assay (ICMA). Gel filtration of *A. rhodostoma* venom indicated that the factor X activator eluted with high molecular weight, was separable from the peak fibrinogen clotting activity (ancrod), and appeared to have no procoagulant or anticoagulant activity. Several *Bothrops* venoms gave strong chromogenic activity in microtiter wells independently of adsorbed proenzymes, indicating the presence of surface-binding proteinases.

REFERENCES

Austen, D. E. G. and I. L. Rhymes. (1975). In *A Laboratory Manual of Blood Coagulation*, Blackwell Scientific Publications, Oxford, pp. 81-83.

Bell, W. M. and H. G. Alton. (1954). A brain extract as a substitute for platelet suspensions in the thromboplastin generation test. Nature 174:880-881.

Bertina, R. M., A. W. Broekmans, E. C. van Krommenhoek, and A. Van Wijngaarden. (1984). The use of a functional and immunologic assay for plasma protein C in the study of the heterogeneity of congenital protein C deficiency. Thromb. Haemost. 57:1-5.

Catt, K. and G. W. Tregear. (1967). Solid phase radioimmunoassay in antibody-coated tubes. Science 158:1570-1573.

Comp, P. C., R. R. Nixon, and C. T. Esmon. (1983). Determination of functional levels of protein C, an antithrombotic protein, using thrombin-thrombomodulin complex. Blood 63:115-1158.

Engvall, E. and P. Perlmann. (1972). Enzyme-linked immunosorbent assay, ELISA. J. Immunol. 109:129-135.

Exner, T., B. Cotton, and M. Howden. (1985). Detection of specific proenzyme activators in snake venoms by a new immunoadsorbant-chromogenic substrate method. Biochim. Biophys. Acta 832:351-356.

Exner, T. and J. Koutts (1986). A simple immunochromometric assay for protein C activity. J. Lab. Clin. Med. 107:405-411.

Fujikawa, K., K. Titani, and E. W. Davie. (1975). Activation of bovine factor X (Stuart factor): conversion of factor Xaα to factor Xaβ. Proc. Natl. Acad. Sci. USA 72:3359-3363.

Kisiel, W., L. H. Ericson, and E. W. Davie. (1976). Proteolytic activation of protein C from bovine plasma. Biochemistry 15:4893-4900.

Koutts, J., J. M. Lavergne, and D. Meyer. (1977). Immunologic evidence that human factor VIII is composed of two linked moieties. Br. J. Haematol. 37:415-428.

Svendsen, L., B. Blombäck, M. Blombäck, and P. Olsson. (1972). Synthetic chromogenic substrates for determination of trypsin, thrombin and thrombin-like enzymes. Thromb. Res. 1:267-278.

36

Intraspecies Variability in the Composition of the Coagulant Active Snake Venoms

František Kornalík and Eva Táborská
Charles University, Prague, Czechoslovakia

INTRODUCTION

Echis carinatus is very likely the snake with the highest mortality rate from bite in man. A thorough epidemiological survey made by Pugh and Theakston (1980) in the West African countries showed that in these regions alone, over 20,000 people die each year from the bite of this snake, the majority of the victims several days after envenomation with a severe hemorrhagic diathesis. One of the reasons for such high mortality rate is the low potency of the specific antivenoms to neutralize those venom compounds responsible for the hemorrhagic diathesis. The venom of *Echis carinatus* was, therefore, included along with venoms from the six most dangerous species (e.g., *Vipera russelli* and *Bothrops asper*) as the reference venom to be tested in the scheme of the World Health Organization Collaborative Center for Snake Venoms Standardization and Antivenom Control.

The natural habitat of *Echis carinatus* is spread throughout the tropical and subtropical zones of the Northern hemisphere from West Africa to East Asia. It was suspected that variability in the composition of *Echis* venoms originating from different geographical regions may exist.

Individual variability in the composition of snake venoms is a known phenomenon, explained by geographical conditions, age of the snake, its feeding habits, temperature, seasons, and other conditions (Glenn et al., 1972; Kornalík and Táborská, 1972; Aragon-Ortiz and Gubenšek, 1981; Latifi, 1984). As a source for isolation of coagulant enzymes, as well as for antivenom production, it is important to select the most suitable venom sample. Pooled venom samples obtained by extracting venom from hundreds of snakes are produced by commercial firms. Some of them have already started to breed snakes in captivity in order to get a homogenous convenient material of a standard quality.

Two main questions have to be answered in this respect: (a) Is the composition of venoms dependent on the geographical origin of the snakes? (b) Is it genetically conditioned? In order to contribute to the question of individual variability of venoms within a species, the following venom samples were tested: 15 samples of pooled *Echis carinatus* venom of different geographical origins and various sources, 7 *Echis carinatus* venoms from specimens originating from a small geographical area, 7 *Vipera russelli* siblings, 8 adults unrelated, 9 *Bothrops asper* parents, and 7 *Bothrops asper* siblings.

MATERIALS AND METHODS

Fifteen samples of *Echis carinatus* venoms from different sources and various geographical regions are listed in Table 1.

Clotting Activity in Vitro

The system contained 0.1 ml citrated human platelet-poor plasma (PPP) + 0.1 ml venom + 0.1 ml 0.025 M $CaCl_2$. The venom concentration was adjusted to give a clotting time of 20 ± 2 seconds, the activity of one clotting unit (CU). Clotting activity was expressed as the number of CU/mg of venom.

Defibrinating Activity in Vivo

Two hours after intravenous administration of the venom in white rats, the plasma fibrinogen level was assayed according to the procedure of Blombäck and Blombäck (1956). A decrease in fibrinogen was expressed as a percentage as compared with 100% of the control values. The minimum defibrinating dose (MDD) was equal to the amount of venom that decreased fibrinogen under these conditions to 10 to 20% of normal values.

Hemorrhagic Activity

^{51}Cr-labeled homologous erythrocytes (red blood cells) were administered intravenously to white rats, and the specific activity of blood was measured. A total of 0.2 ml of venom was injected intradermally into the shaven part of the rats' backs. After 24 hours, animals were sacrificed, skinned, and the amount of blood at the site of injection, calculated from radioactivity, was expressed in ml $\times 10^{-4}$.

Table 1 List of *Echis carinatus* Venoms Tested

No.	Place of origin (name of batch)	No. of snakes
1.	Pakistan (Tatta District) A	4
2.	Pakistan (Tatta District) B	8
3.	Iran	50
4.	Basel-Tropeninstitut[a]	5
5.	South Algeria	3
6.	Kenya	?
7.	Nigeria	?
8.	Saudi Arabia	?
9.	Tashkhent	20
10.	Miami Serpentarium (1970)	?
11.	Pool 44 K	?
12.	Miami Serpentarium (1983 41K[b])	?
13.	Workney and Nadir	?
14.	Hartge-Hamburg 42 K	?
15.	Mali-Latoxan	?

[a]Five snakes from one litter: male from Iran (father), female from N. India (mother).
[b]Since 1979-1980, *Echis carinatus* purchased in Iran.
Samples No. 6, 7, and 8 were provided by the World Health Organization Collaborative Center for Standardization of Venoms and Antivenom Control. Samples No. 11 to 14 were provided by Behringwerke, Marburg, FRG. Sample No. 15 was provided by the Pasteur Institute.

Lethality

Mortality of mice (batches of 10 animals) was determined 24 hours after intravenous injection of different amounts of venom in 0.5 ml of saline solution. The LD_{50} was calculated by Roth's probit method. Lethality was expressed as number of LD_{50}/mg of venom.

RESULTS AND DISCUSSION

The toxicity of the venom in the intravenous, 24-hour mouse test did not disclose the potency of the venom to induce a hemorrhagic diathesis fatal to man, for which the defibrinating syndrome together with the action of hemorrhagins are responsible. In Figure 1, three activities (in vitro procoagulation and fibrinogenolysis, and in vivo defibrination) are compared among the 15 samples of venoms from *Echis carinatus*.

With some exceptions, the defibrination occurs more or less parallel with the clotting activity. Fibrinogenolysis seems to play a minor role in the defibrination. (Compare, for example, activities in venom samples 6 and 12.)

In vivo defibrinating and hemorrhagic activity of the tested venoms was further correlated with their lethality, as assayed by the intravenous 24-hour mouse test. Results are presented in Figure 2. They seem to indicate that the lethality is most probably correlated with both activities. The stronger the hemorrhagic and defibrinating activity, the greater the lethality of the venoms.

Although the results are not supported by a statistical evaluation, it seems to be evident from both analytical approaches that, by analyzing their potency to impair hemostasis in vivo, "strong and weak" venoms can be singled out from the test group. This is evident from Figure 3, where five pools of venom from different regions selected according to their ability to induce the bleeding tendency in vivo are presented. Two have apparently low and two have a high concentration of the active substances. (They were used in additional experiments, not described here, for testing the capacity of commercial antivenoms to neutralize these activities.)

It should be stressed, however, that defibrinating and hemorrhagic activities were tested in rats, whereas lethality was tested in mice. In the latter test, the majority of mice died within 2 to 10 hours after the injection of the venom. Signs of hemorrhagic diathesis, such as petechiae on the serous membranes, were observed on autopsy in only about 40% of surviving animals sacrificed 1 to 4 days after envenomation.

No life-endangering systemic bleeding from the intestines and urinary or respiratory organs was observed in mice after the LD_{50} of *Echis carinatus* venom was administered. In rats, however, massive hematuria was observed for 2 to 3 days, as were hemorrhages in respiratory and intestinal organs on autopsy after intravenous administration of 2 to 3 MDD. The cause of death after envenomation with *Echis carinatus* venom seems to be different in mice, rats, and man.

The results seemed to indicate that the quality of *Echis carinatus* venoms might be determined by the geographical origin of the snakes. This proved, however, to be a false assumption. We have analyzed 26 adult specimens of this species originating from a small geographic

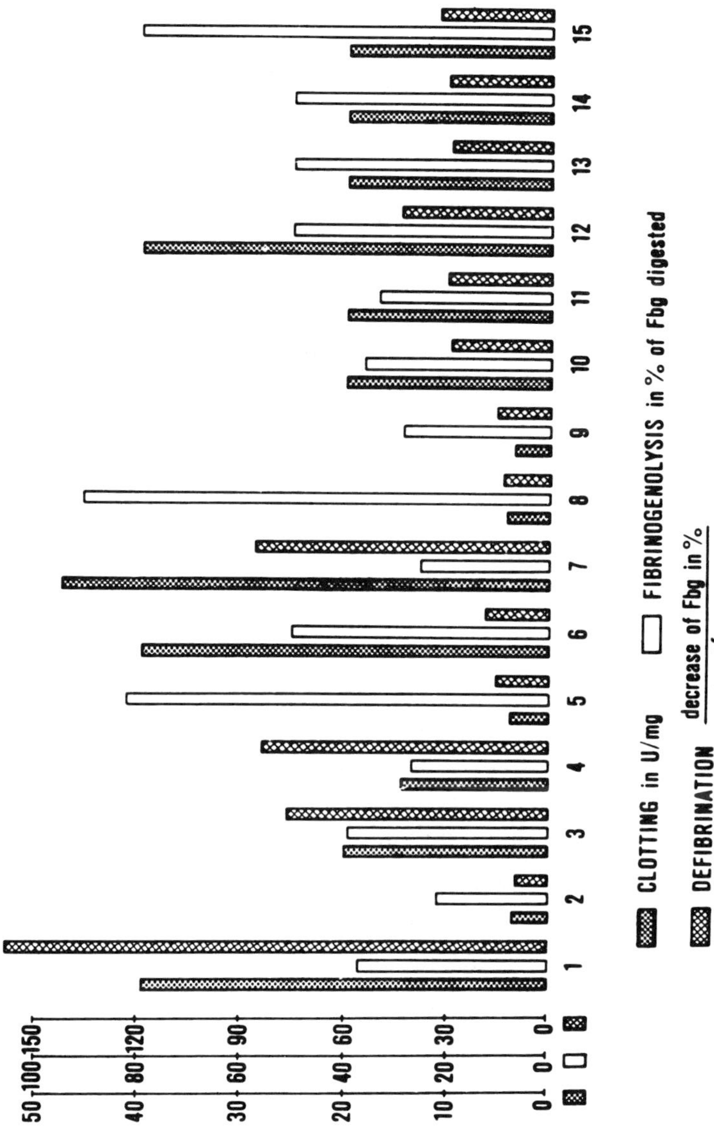

Figure 1 Clotting, fibrinogenolytic, and defibrinating activity of 15 samples of pooled Echis venoms.

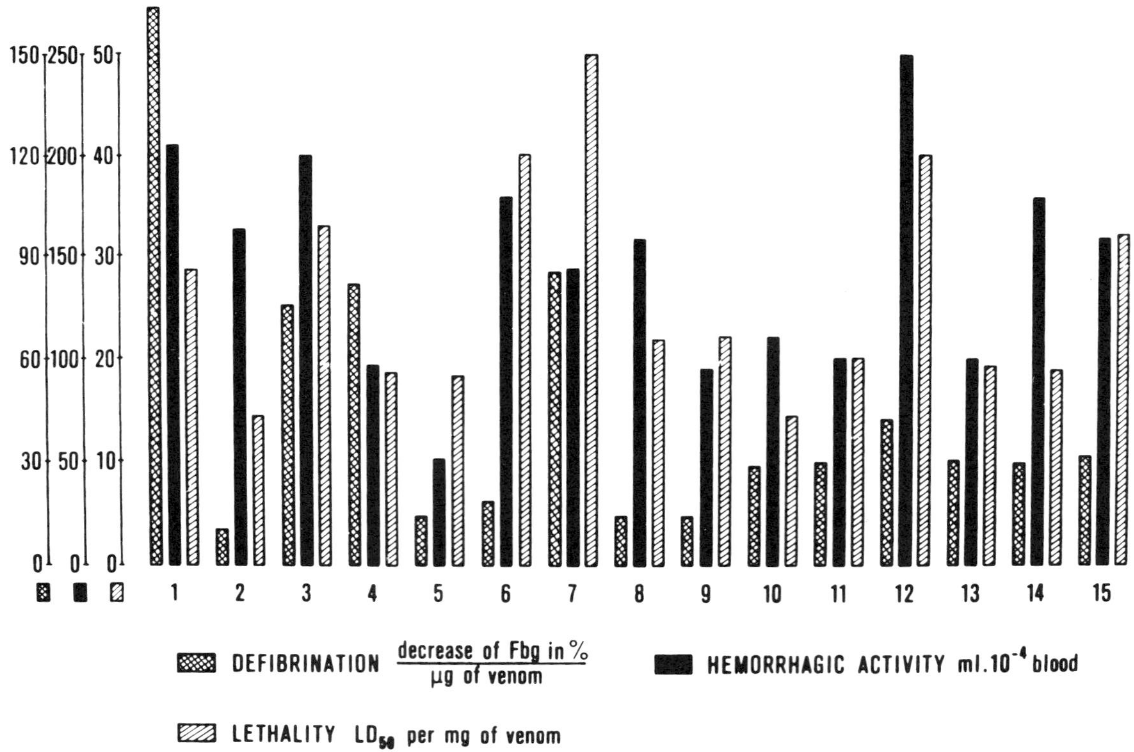

Figure 2 Defibrinating and hemorrhagic activity and lethality of 15 samples of pooled Echis venoms.

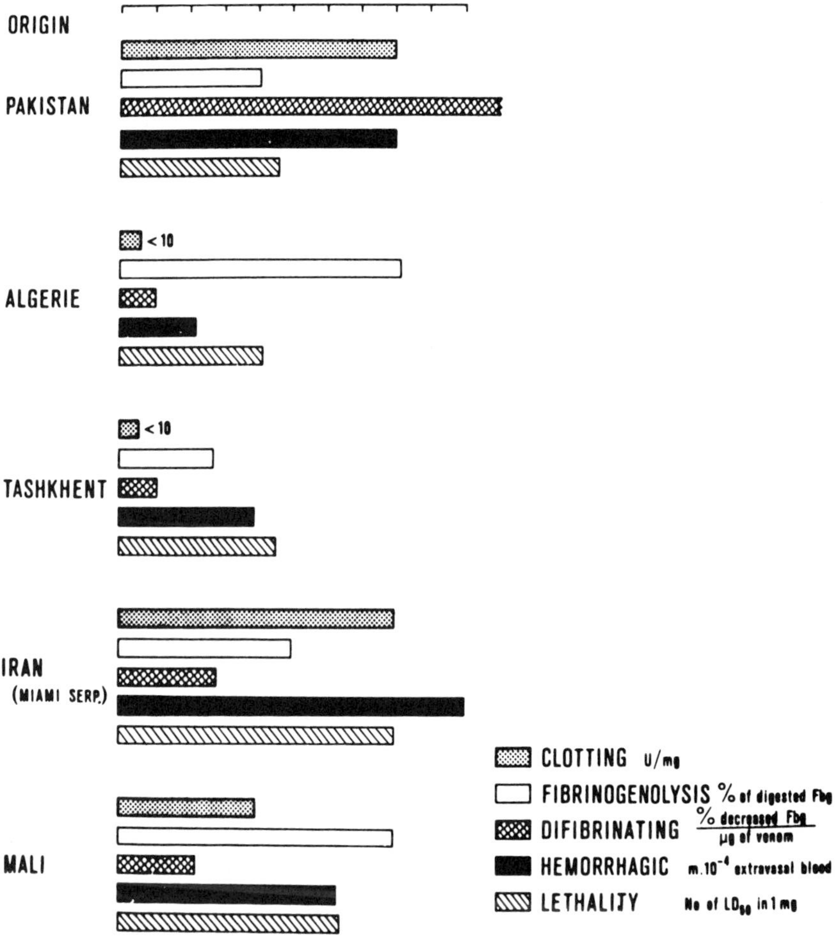

Figure 3 Five samples of pooled Echis venoms from different regions with various biological activities.

area (the Tatta District in Pakistan). The snakes were kept under identical conditions; venom was extracted every 3 to 4 weeks for 14 months, and each sample was individually analyzed for its biological, coagulant, and enzymatic qualities. The properties of the individual venoms remained constant during the experiment. The differences are depicted on Figure 4.

As can be observed, the variability of the measured parameters of these individual specimens is about the same as that observed with the pooled venoms from different geographical habitats.

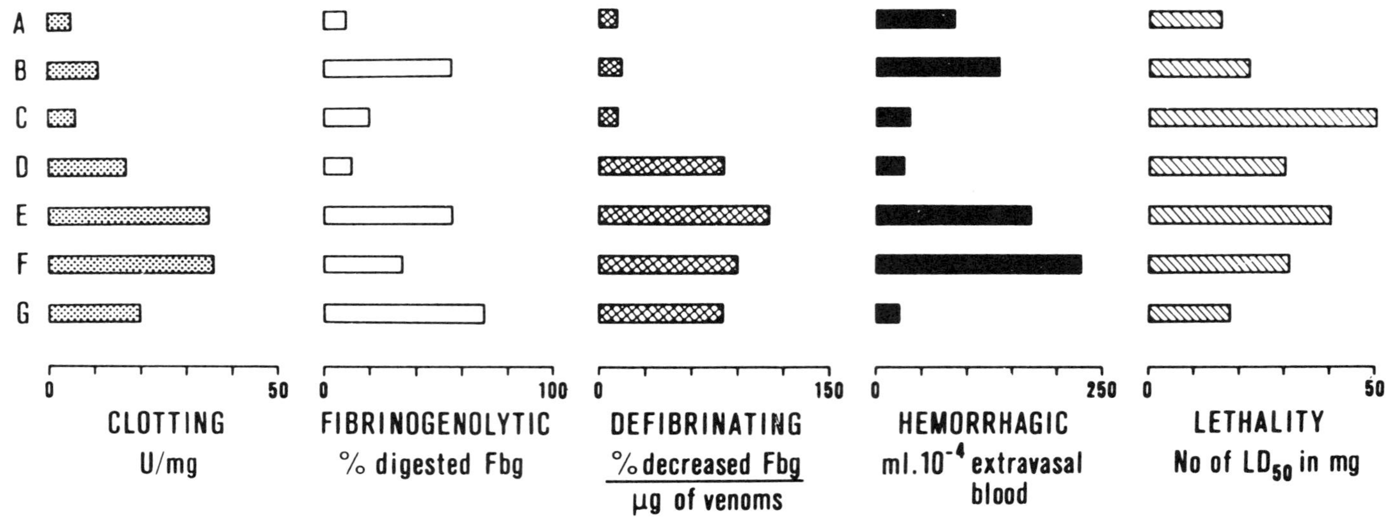

Figure 4 Biological activities of venoms collected separately from seven *Echis carinatus* originating from the Tatta District in Pakistan.

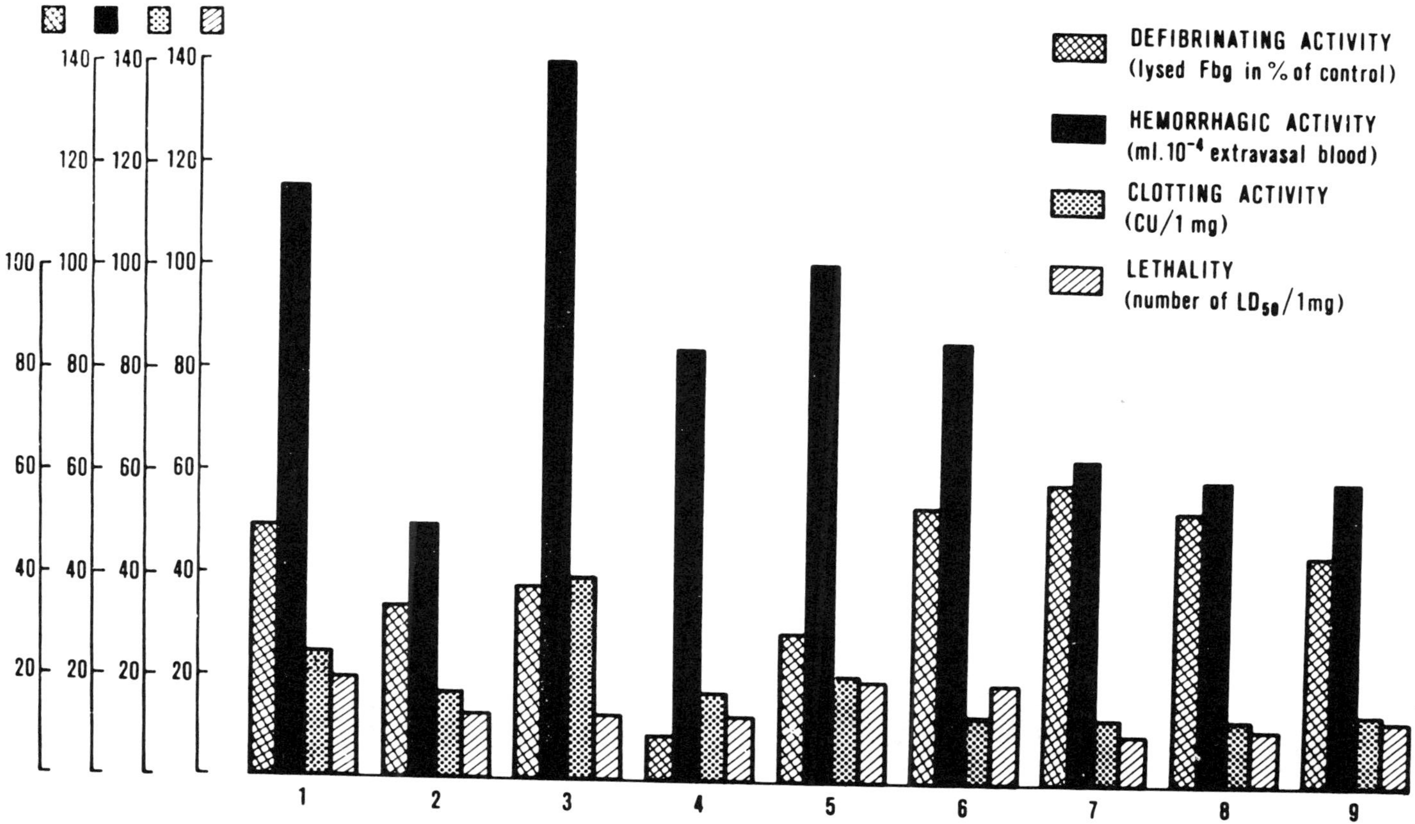

Figure 5 Biological activities of venoms collected separately from nine *Bothrops aspers*. Numbers 1 to 7, siblings; numbers 8 and 9, parents.

From a pair of *Bothrops asper*, three generations were born in captivity during the last 9 years. In order to find if any genetically determined regularity could be detected, the venoms from parents (Nos. 8 and 9 in Fig. 5), age unknown, and of the first descendent generation, age 3 years, were analyzed. The same biological activities as those tested in *Echis carinatus* venoms were determined (Fig. 5).

Similar results were obtained by testing the in vitro procoagulant activity of venom samples from seven adult *Vipera Russelli* siblings that were compared with samples from eight unrelated specimens. Here also the variability was greater in the sibling group than in the samples from unrelated snakes (Fig. 6).

These results can be summarized as follows: (a) Differences in the properties of *Echis carinatus* venoms are found in individual specimens, independent of the geographical habitat of the snakes. (b) The composition of the venom in an adult snake is a permanent, very likely inherent characteristic of specific protein synthesis. (c) No genetic pattern could be traced in the composition of snake venoms. (d) The quality of a pooled venom sample depends on the proportional concentration of active substances present in the single venoms. (e) The in vivo defibrinating capacity of a procoagulant venom is propor-

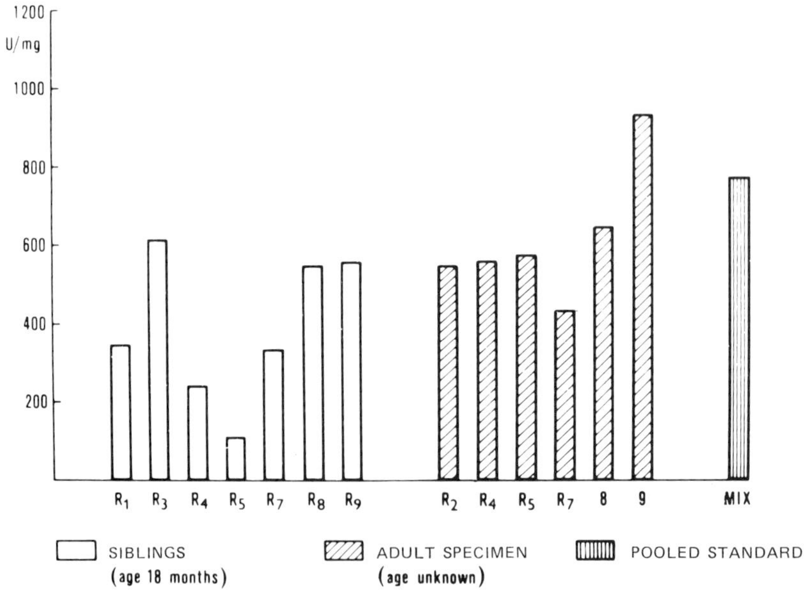

Figure 6 In vitro clotting activity of venoms collected from seven siblings and six unrelated specimens of *Vipera Russelli*.

tional to its clotting activity assayed in vitro. (f) For the isolation of clotting enzymes, pooled venoms should be selected according to their activity as assayed by in vitro and in vivo tests. These results do not exclude changes in the quality and yield of venoms described during maturation and aging of the snakes.

REFERENCES

Aragon-Ortiz, F. and F. Gubenšek. (1981). *Bothrosp asper* venom from the Atlantic and Pacific zones of Costa Rica. Toxicon 19: 797-805.

Blombäck, B. and M. Blombäck. (1956). Purification of human and bovine fibrinogen. Arkiv Kemi 10:415-443.

Glenn, J. L., R. Straight, and C. C. Snyder. (1972). Yield of venom obtained from *Crotalus atrox* by electrical stimulation. Toxicon 10:575-581.

Kornalík, F. and E. Táborská. (1972). In *Animal and Plant Toxins*, Kaiser, E. (Ed.), Wilhelm Goldmann, München, pp. 99-105.

Latifi, M. (1984). Variation in yield and lethality of venoms from Iranian snakes. Toxicon 22:373-390.

Pugh, R. N. H. and R. D. G. Theakston. (1980). Incidence and mortality of snake bite in savanna Nigeria. Lancet Nov. 29:1181.

37

Coagulation Studies on Habu (*Trimeresurus okinavensis*) Venom

Nobuo Sakuragawa and Kaoru Takahashi
Toyama Medical and Pharmaceutical University, Toyama City, Japan

INTRODUCTION

For several decades, snake venoms have been studied for their procoagulant, anticoagulant, fibrinolytic and platelet-aggregating characteristics. We have developed assay methods using *Echis carinatus* venom for prethrombin 1 (1) and *Bothrops marajoensis* venom for antithrombin III (2). This chapter deals with the activity of habu (*Trimeresurus okinavensis*) and *Echis carinatus* venoms from the viewpoint of coagulation and fibrinolysis.

MATERIALS AND METHODS

Habu (*Trimeresurus okinavensis*) venom (crude habu venom powder, obtained from Wako Pure Chemical Industry Ltd.) and *Echis carinatus* venom (Sigma Chemical Company, St. Louis, Missouri) were used. Purified bovine prothrombin was made by the method of Seegers and Smith (3).

A one-stage coagulation assay method was used with several coagulation factor-deficient plasmas. Thromboplastin-$CaCl_2$ solution (Dade Company) was used. Prothrombin and thrombin were assayed as described previously (3,4). Several chromogenic substrates were used for assaying factor Xa and thrombin. Prothrombin was assayed simultaneously by the single radial immunodiffusion method. Fibrinolytic activity was assayed using both plasminogen-poor and plasminogen-rich fibrin plates. Platelet aggregation rates were calculated from platelet aggregation tracings.

Test plasma was prepared by adding 3.8% sodium citrate solution to the blood in a ratio 1:9, and centrifuging at 3000 rpm for 15 minutes at 4°C. All test tubes were silicone-coated.

Habu venom was purified on a Sephadex G-200 column (1.8 × 90 cm) equilibrated with 0.05 M phosphate buffer (pH 7.2). One ml of 15 mg/ml of habu venom in 0.05 M phosphate buffer (pH 7.2) was applied to the column at 4°C, eluted at a flow rate of 10 ml/hour, and 2 ml/tube was collected.

Activation of the prothrombin by the venom was carried out by adding 1 ml of 1 mg/ml (120 units/ml prothrombin) of purified bovine prothrombin to 1 ml of 1 mg/ml venom containing 0.05 ml of 0.25 M $CaCl_2$ solution at 37°C, pH 7.2 (4). Experimental DIC was produced by infusing 0.5 mg of habu venom in 5 ml of saline solution into the heart of a rabbit for 10 minutes via catheter by the method of Erdmann (5). Test plasmas were collected at 0, 0.5, 1, 2, 3, and 5 hours via the catheter.

RESULTS

Influence of the Habu Venom on Coagulation, Fibrinolysis, and Platelet Function

A normal one-stage clotting time was obtained when habu venom was added to plasmas deficient in factors II, V, VII, VIII, IX, X, XI, and XII. The control study was performed using normal citrated plasma, and gave a clotting time of 24.6 seconds (Table 1). Prothrombin-deficient plasma, containing only a small percent of the normal amount of prothrombin gave a clotting time of 42.6 seconds with habu venom. Factor XI-deficient plasma also showed a longer clotting time of 41.8 seconds due to the lower amount of prothrombin in it. When fibrinogen solution (600 mg/dl) was used instead of the deficient plasma, no clot was formed in the reaction mixture. From these results, it is clear that habu venom can activate prothrombin directly.

Amidolytic activities of the habu venom (0.05 mg/ml) were assayed by using chromogenic substrates that have specificity towards kallikrein, factor Xa, thrombin, and plasmin. The results were as follows: thrombin, 20 units/ml; factor Xa, 10 units/ml; plasmin, 1 unit/ml, and kallikrein, 240% (compared with 100% for normal plasma). The influence of calcium ions on the activation of prothrombin by habu venom was investigated by adding 0.1 ml of 1 to 10 diluted plasma to 0.1 ml of fibrinogen solution (900 mg/dl) along with 0.1 ml of 0.25 M $CaCl_2$ solution and 0.1 ml of habu venom solution (0.1 mg/ml). The clotting time was 20.4 seconds. In another reaction mixture, saline was used instead of 0.25 M $CaCl_2$ solution, and the clotting time was prolonged to 39.4 seconds. Therefore, calcium ions may accelerate prothrombin activation by habu venom.

Table 1 Activity of Venom on the Coagulation Factor-Deficient Plasmas

Deficient plasma	Clotting time[a] (sec)
II	42.6
V	25.6
VII	24.0
VIII	24.2
IX	23.6
X	23.6
XI	41.8
XII	24.6
Normal plasma	24.6

[a]0.1 ml of 0.05 mg/ml habu venom solution was added to 0.1 ml of deficient plasma with 0.1 ml of 0.25 M $CaCl_2$ solution. The clotting time was examined at 37°C.

Fibrinolytic activity of habu venom was assayed by the plasminogen-rich fibrin plate method for plasminogen-activator activity, and by the plasminogen-poor fibrin plate method for plasmin activity. One mg/ml of habu venom contained the equivalent of 200 units/ml of urokinase, 2 units/ml of plasmin (fibrin plate methods), and 1 unit/ml of plasmin (amidolytic method). The platelet-aggregating potency of habu venom is shown in Figure 1.

Experimental DIC by Habu Venom

Experimental disseminated intravascular coagulation (DIC) was induced by infusion of habu venom solutions. The activated partial thromboplastin time (aPTT) was prolonged due to the decrease in coagulation factors (Fig.2). Euglobulin lysis time was shortened due to the decrease in fibrinogen. Platelets and platelet retention rates were decreased as shown in Figure 2. Many microthrombi were observed in the lungs and kidneys by autopsy studies. Thus, experimental DIC was induced by the infusion of habu venom into the rabbit.

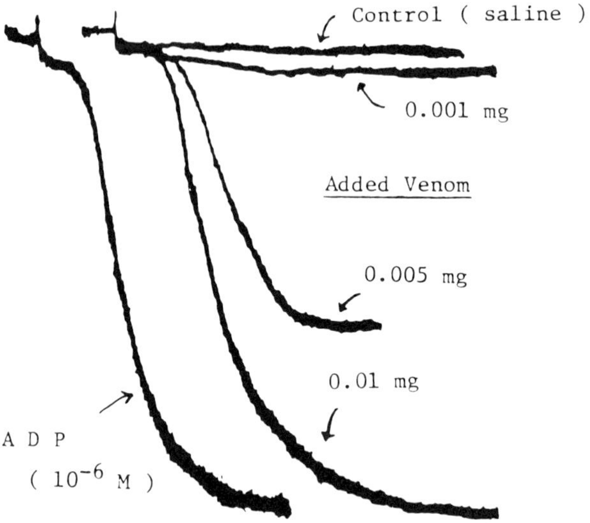

Figure 1 Platelet aggregation by habu venom. Reaction mixture: 0.2 ml of platelet-rich plasma added to 0.05 ml of saline solution or the venom solution.

Prothrombin Activation by Habu Venom

Habu venom was purified by column chromatography as described above. Only one peak with procoagulant activity was observed; it was freeze-dried.

Purified bovine prothrombin was activated by the purified habu venom. The activation profile is shown in Figure 3; full activation was obtained in 90 minutes. The activation process was also examined by disc electrophoresis, and showed thrombin, prethrombin 1, and fragments 1 and 2.

Prothrombin Assay Method Using Habu Venom

The prothrombin assay method employed a reaction mixture as follows: 0.1 ml of fibrinogen solution (900 mg/dl), 0.1 ml of 0.25 M $CaCl_2$ solution, 0.1 ml of the 1 to 10 diluted test plasma, and 0.1 ml of 0.1 mg/ml habu venom solution. The reaction was performed at 37°C and pH 7.2.

To make a calibration curve, normal plasma was diluted stepwise with saline solution to make 50, 25, and 12.5% compared with the starting plasma. The clotting times of these plasmas were determined. The clotting times were plotted on the vertical axis and the

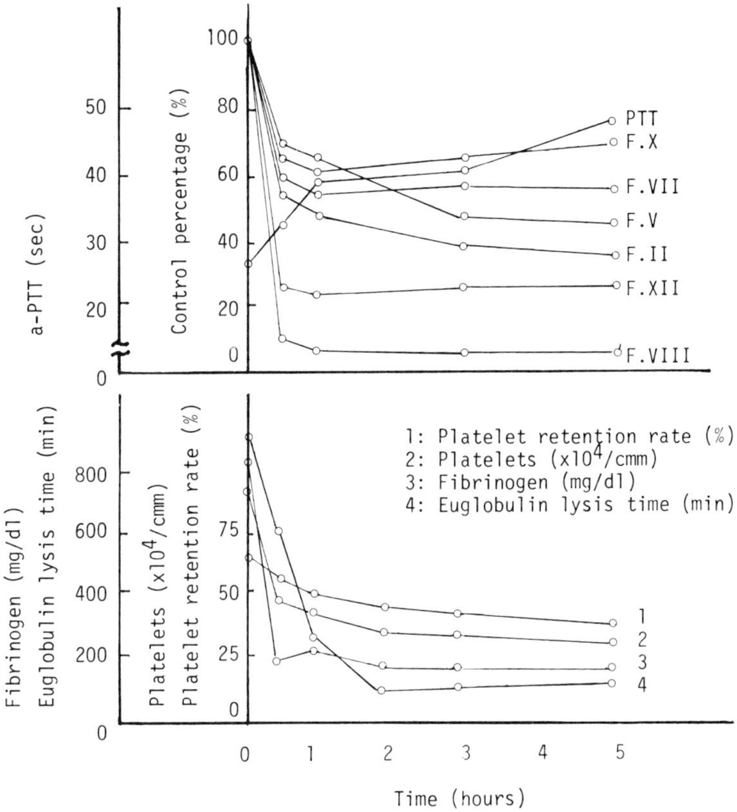

Figure 2 Experimental DIC by habu venom, in which 0.5 mg of the venom in 5 ml of saline solution was infused into a rabbit (weighing 2.5 kg) for 10 minutes according to the method of Rodriguez-Erdmann (5).

percentages on the abscissa. A straight line was obtained on a double log graph as shown in Figure 4. Normal values obtained from 20 subjects were 102 ± 12.6% (mean ± standard deviation).

To assay for PIVKA-prothrombin by the habu venom method, warfarin-treatment plasma showing well controlled thrombotest values, containing PIVKA-prothrombin, was used as well as DIC plasma containing prethrombin. Prothrombin values assayed by the method using tissue thromboplastin showed rather lower percentages than those assayed by the method using habu venom. We tested the same specimens using the *Echis carinatus* venom method and the single radial immunodiffusion method. *Echis carinatus* venom activated PIVKA-

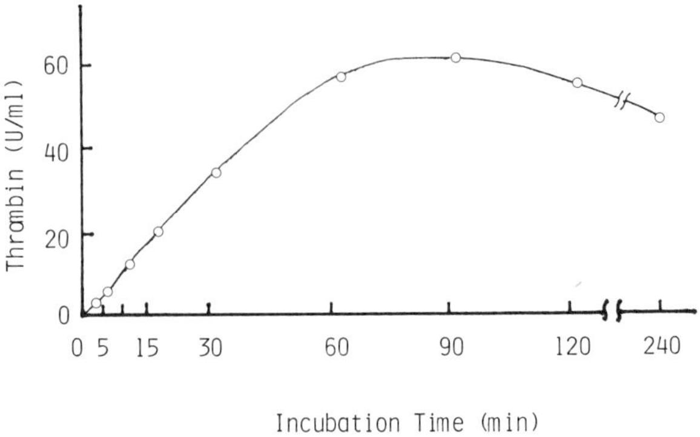

Figure 3 Prothrombin activation by habu venom. Reaction mixture: 0.1 ml of purified bovine prothrombin (1 mg/ml; 120 units/ml), 0.1 ml of habu venom solution (1 mg/ml), and 0.05 ml of 0.25 M $CaCl_2$ solution, at 37°C and pH 7.2.

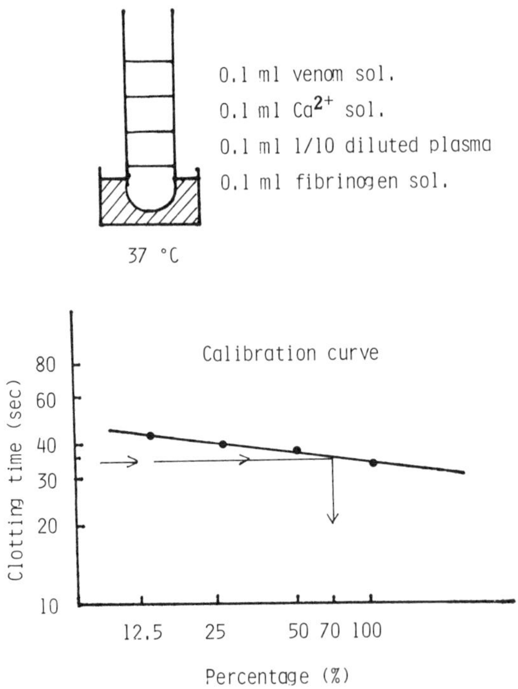

Figure 4 Prothrombin assay method using habu venom. Reaction mixture was as given in the text.

prothrombin and prethrombin 1, as we already reported (1). The prothrombin values assayed by the habu venom method agreed with the values by the method using *Echis carinatus* venom and the immunoassay method. Therefore, the assay method using habu venom appears to detect PIVKA-prothrombin and prethrombin 1.

PIVKA-prothrombin and prethrombin 1 can be calculated as follows: A value of the assay method using habu venom minus the value of the assay method using thromboplastin-$CaCl_2$ solution of over 10% indicated the presence of PIVKA-prothrombin, dependent on vitamin K deficiency or an abnormality in the production of prothrombin by the liver. When we detect prethrombin 1, we can diagnose DIC (Table 2) (1).

DISCUSSION

When a person is bitten by a venomous snake, clinical manifestations may result from activating coagulation, fibrinolysis, and platelets.

Bothrops marajoensis venom contains thrombin-like activity and is used as a defibrinating agent in patients with thrombosis. Russell's viper venom can activate factor X and is used for the assay of factor X. We studied many venoms from the viewpoint of coagulation system,

Table 2 Prothrombin Assay Method in Warfarin-Treatment Plasma

Thrombotest	TPLN (A)	Habu (B)	Echis (C)	Immuno A (D)
14%	48%	80%	90%	88%
14	35	85	86	78
13	35	80	86	78
18	40	65	80	82
19	36	78	80	82
23	46	80	76	78
17	48	85	80	88
25	52	82	98	100
18	42	86	88	90
18	46	82	86	84
14	42	82	88	90
19	48	84	90	84

and have used *Bothrops marajoensis* venom for antithrombin III assay (2) and *Echis carinatus* venom for prethrombin 1 and PIVKA-prothrombin assays (1,6). Our results suggest that we can also use habu venom to assay information for the diagnosis of DIC, liver dysfunction, or vitamin K deficiency.

SUMMARY

The characteristics of habu venom are studied form the viewpoint of coagulation, fibrinolysis, and platelet function. The results are as follows:

(a) The venom was a potent activator of coagulation, fibrinolysis, and platelet aggregation.
(b) When the venom was infused into rabbits, DIC was induced.
(c) A prothrombin assay method using the venom was devised. It can be used to diagnose DIC when there is a discrepancy of more than 10% between the method using the venom and that using thromboplastin $CaCl_2$ solution. We can also diagnose liver dysfunction or vitamin K deficiency based upon these results and determine whether PIVKA-prothrombin is present.

REFERENCES

1. Sakuragawa, N., K. Takahashi, M. Hoshiyama, C. Imbo, M. Matsuoke, and Y. Onishi. Significance of prothrombin assay method using *Echis carinatus* venom for diagnostic information in disseminated intravascular coagulation syndrome. Thromb. Res. 7:643 (1975).
2. Sakuragawa, N. and K. Takahashi. Studies on the snake (*Bothrops Marajoensis*) venom. Int. Med. 32:484 (1975) (in Japanese).
3. Seegers, W. H. and H. P. Smith. Factors which influence the activity of purified thrombin. Am. J. Physiol. 137:348 (1942).
4. Ware, A. G. and W. H. Seegers. Two-stage procedure for the quantitative determination of prothrombin concentration. Am. J. Clin. Pathol. 19:471 (1949).
5. Rodriguez-Erdmann, F. Über einfache Methode zur wiederholen Blutgewinnung beim Kaninchen. Pflüger Arch. Ges. Physiol. 269:306 (1959).
6. Sakuragawa, N. Prothrombin abnormality in the cases of Warfarin treatment. Jpn. J. Clin. Pathol. 24:279 (1976).
7. McCoy, L. E. and W. H. Seegers. Preparation of bovine prothrombin complex and purified prothrombin. Thromb. Res. 1:461 (1972).

VII
CLINICAL APPLICATIONS

38

Clinical Trials with Batroxobin

Kurt F. Stocker
Pentapharm AG, Basel, Switzerland

HUMAN PHARMACOLOGY

Based on animal experiments that demonstrated the possibility of achieving a safe, specific, dose-dependent reduction in the plasma fibrinogen level by infusion of batroxobin, and based on its thrombosis-preventing effect in various animal models, the action of batroxobin on as many as 27 parameters was monitored in more than 50 healthy human volunteers as well as in over 100 patients with thrombotic diseases requiring anticoagulant treatment.

As shown in Table 1, the results obtained in these studies confirmed the highly specific action of batroxobin, which caused a reduction in plasma fibrinogen concentration and blood viscosity and which led, as a consequence of a fibrinolytic reaction induced by this treatment, to a consumption of plasminogen and α_2-antiplasmin. Other clotting factors as well as platelets remained unaffected. The bleeding time (Duke's method) was normal during the treatment period.

The following conditions have been proposed as indications for batroxobin: (a) deep venous thrombosis, (b) pulmonary embolism, (c) central retinal vein thrombosis, (d) myocardial infarction, (e) cerebral thrombosis, (f) priapism, (g) sickle cell crisis, (h) chronic pulmonary hypertension, (i) prosthetic cardiac valve thrombosis, (j) extracorporeal dialysis, (k) extracorporeal oxygenation, (l) peripheral arterial occlusive diseases, (m) scleroderma, (n) frost bite, (o) thrombosis prevention in hip surgery and vascular surgery, (p) varicose ulcer, (q) rheumatoid arthritis, (r) renal transplant rejection, (s) prevention of metastasis in malignancy, and (t) sudden loss of hearing.

Table 1 Hemostatic Parameters in Healthy Volunteers and Patients with Thrombosis Receiving Treatment with Defibrase

	Fukutake et al. (1981) 24 volunteers	Egberg (1973) 33 patients	Martin et al. (1976) 58 patients	Yasunaga et al. (1979) 8 patients	Ishimaru et al. (1979) 18 patients
Bleeding time (Duke)	Normal	ND	ND	Normal	ND
Blood viscosity	Reduced	ND	Reduced	Reduced	ND
Platelet count	Normal	ND	Normal	Normal	Normal
Platelet adhesivity	ND	ND	Normal	Normal	Normal
Platelet aggregation	Normal	ND	ND	Normal	Normal
Fibrinogen	Reduced	Reduced	Reduced	Reduced	Reduced
Rebound effect	No	No	No	ND	No
FDP	Increased	Increased	Increased	Increased	ND
Fibrinopeptide A	Increased	ND	ND	ND	ND
PT	Prolonged	Prolonged	Prolonged	Prolonged	Prolonged
PTT	Prolonged	Prolonged	Prolonged	Prolonged	Normal
Factor II	Normal	Normal	ND	Slightly reduced	Normal

Factor V	Normal	Normal	ND	Normal	Normal
Factor VII	Normal	ND	ND	Slightly reduced	Normal
Factor VIII	Normal	Normal	ND	Normal	Normal
Factor IX	Normal	ND	ND	Normal	ND
Factor X	Normal	Normal	ND	Normal	Normal
Factor XI	ND	ND	ND	Normal	ND
Factor XII	ND	Normal	ND	Normal	ND
Factor XIII	ND	Reduced	ND	ND	ND
Euglobulin lysis time	ND	ND	ND	Shortened	ND
Clot lysis time	ND	ND	ND	Shortened	ND
Plasminogen	Reduced	Reduced	Reduced	Reduced	ND
α2-Antiplasmin	Reduced	ND	ND	ND	ND
α2-Macroglobulin	Normal	Slightly reduced	ND	Normal	ND
α1-Antitrypsin	Normal	ND	ND	ND	ND
Antithrombin III	Normal	Normal	ND	ND	ND

ND = not determined.

This potential field of indications is still being explored based on respective preclinical findings by means of pilot studies and extended clinical trials. A review on studies performed so far is provided here and in the following series of tables.

TRIALS ON EFFICACY

Deep Vein Thrombosis

Batroxobin has been given as an alternative to heparin to patients with deep venous thrombosis to inhibit thrombus propagation (Table 2). In several cases, this treatment led to complete or partial thrombolysis as verified by phlebography. Several additional studies report on improvements in patients who were not submitted for phlebography. When these trials were performed, the mechanism of action of batroxobin was not well understood. Nothing about batroxobin-induced release of tissue plasminogen activator (TPA) and stimulation of plasminogen activation by fibrin I was known, and thrombolysis was considered as a spontaneous phenomenon rather than as a benefit of batroxobin treatment.

Central Retinal Vein Thrombosis

Two pilot studies in central retinal vein thrombosis (Table 3) initiated a more extended clinical trial, the result of which is presented by Conard in this volume.

Peripheral Arterial Disease

Decrease in whole blood viscosity and improvement in blood flow and thereby in tissue oxygenation were the rationale of defibrinogenation therapy in peripheral arterial disease (Table 4). Initiated by these open trials, a controlled multicenter study was performed in Japan (see Ishimaru's report in this volume).

Table 2 Deep Venous Thrombosis in 38 Patients

n	Route	Results	Reference
10	IV	3 thrombolyses (phlebogr.)	Egberg et al. (1971)
5	IV	2 thrombolyses (phlebogr.)	Blombäck et al. (1971)
12	IV	2 thrombolyses (phlebogr.)	Blombäck and Egberg (1975)
11	IV	9 clinically improved	Donner et al. (1980)

Table 3 Central Retinal Vein Thrombosis in 20 Patients

n	Route	Results	Reference
5	IV	Visual acuity: 3 improved Retinal circulation time: 5 improved	Egberg and Nilsson (1975)
15	SC	Visual acuity: group improved	Gaudric et al. (1982)

Batroxobin and ancrod do not cross-react immunologically. Patients resistant to ancrod may be treated with batroxobin and vice versa.

Ancrod-resistant patients with peripheral arterial disease have been treated in various centers (see legend to Table 5) with batroxobin, and questionnaires requesting details on patients, treatment, and results were returned by these centers (Table 5). The effect of the treatment was assessed by the walking test, evaluation of pain at rest, consumption of analgesics, state of necrosis, flow measurements, and so on. In a patient suffering from scleroderma, the specific symptoms were evaluated. The clinical effect was considered as good or very good in 33 of the 50 cases.

Table 4 Peripheral Arterial Disease in 56 Patients

n	Route	Results	Reference
12	IV	Xe^{133} clearance, perfusion rate increased	Meissner et al. (1977)
10	IV	Femoral venous pO_2 increased	Koehler (1975)
11	IV	Blood viscosity and red cell aggregability reduced	Voelker (1975)
2	SC	Necroses reduced, clinical improvement (ancrod-resistant patients)	Sharp (personal communication)
11	IV	Dye dilution, plethysmography, skin temperature, walking test improved, skin ulcers reduced	Ohara and Sasaki (1983)
10	SC	Walking test: no significant improvement comp. with placebo	Martin et al. (1976)

Table 5 Defibrase-Treatment of 50 Ancrod-Resistant Patients with Peripheral Arterial Disease

Number of patients	50
Age	37 to 82 years
Previous ancrod therapy	41
Defibrase	
SC	49
IV	1
Duration of treatment	
Less than 21 days	16
21 to 35 days	27
More than 35 days	7
Treatment	
In-patient	43
Out-patient	7
Daily dose for 50-100 mg% plasma fibrinogen	
0.2 to 0.5 BU/kg	41
Less than 0.2 BU/kg	5
More than 0.5 BU/kg	4
Compatibility	
Good	45
Satisfactory	4
Bad	1
Undesired effects	
Bleeding	1
Thrombic reactions	1
Allergic reactions	1
Resistance against Defibrase	22
Assessment of the effect	
Walking distance	23
Pain at rest	36
Consumption of analgetics	18
State of necrosis	31
Sclerodermic status	1
Flow measurements	2
Training with expander	1

Table 5 *(continued)*

Clinical effect	
Very good	9
Good	24
Weak	10
No effect	6
Worse	1

Participating centers: Stein, Heidelberg, FRG; Pries, Hannover, FRG; Cachovan, Bad Bevensen, FRG; Kriessmann, Munich, FRG; Nissen, Bad Oeynhausen, FRG; Harch, Hannover, FRG; Gillmann, Ludwigshafen, FRG; Creutzig, Hannover, FRG; Reinöhl, Aachen, FRG; Voigt, Miltenberg, FRG; Van den Berg, Hannover, FRG; Hielmcrone, Niebüll, FRG; Fischer, Hannover, FRG; Rinck, Schwenningen, FRG; Wenzel, Homburg, FRG; Koch, Graz, Austria; Richter, Darmstadt, FRG; Hess, Munich, FRG.

Vibration Disease

Table 6 provides data on two open trials with a total of 92 patients suffering from the so-called "vibration disease." This is a peripheral circulatory disturbance in the upper extremities stimulated by vibrations emitted from hand-operated tools. The treated persons were mostly forest laborers operating chain-saws over a prolonged period of time. Based on the encouraging results of these two pilot studies, a prospective multicenter trial has been carried out (Agishi et al., 1985a, 1985b, see also the report of Fukutake et al. in this volume).

Comparison of values of the cold-loading skin temperature test and measurement of the blood flow velocity in digital arteries before and after therapy, as well as functional tests, indicated significant improvements in 79.1% of the patients, whereby 29.2% were classified as "excellent" and "good" and 62.1% were considered as "fair" cases.

Vascular Surgery

As presented in Table 7, batroxobin has been used as an anticoagulant to prevent thrombosis in vascular surgery since defibrinogenation seemed to affect hemostasis much less than heparin anticoagulation. The following conclusions have been drawn from the trials: First, the postoperative blood loss is proportional to the concentration in fibrin(ogen) degradation products (FDP). Therefore, defibrinogenation must be started either several days preoperatively to allow FDP excretion, or it should be performed as postoperative treatment. Second, the postoperative blood loss is inversely proportional to the fibrinogen concentration.

Table 6 Vibration Disease in 316 Patients

n	Route	Type of trial	Assessment	Result	Reference
78	IV	Open	Skin temperature	Improved	Nasu (1982)
14	IV	Open	Skin temperature, thermography, digital arterial pO_2	Improved	Habu et al. (1982)
224	IV	Open 9 centers	Skin temperature, digital arterial flow	50.3% excellent and good	Agishi et al. (1985a)

Table 7 Thrombosis Prophylaxis in Vascular Surgery in 43 Patients

n	Type of operation	Results	Severe bleeding complications	Reference
7	Thrombectomy	Reoccluded after 1 year: 4	3	Olsson et al. (1976)
1	Venous bypass	Patent after 2 years	0	Olsson et al. (1976)
3	Thrombectomy	2 improved	0	Ishimaru et al. (1981)
15	Venous surgery	Good/very good: 13	2	Furukawa et al. (1983)
17	Reconstructive arterial surgery	Patent: 10 Occluded: 4	2	Furukawa et al. (1983)

Combined Applications

Batroxobin has been applied in conjunction with the fibrinolytic agents streptokinase and urokinase. Moreover, a sequential treatment beginning with Defibrase (Pentapharm Ltd., Basel, Switzerland) until fibrinogen and α_2-antiplasmin levels were low, followed by Lys-plasminogen (Lab. Choay, Paris, France) infusion and plasminogen activation by urokinase infusion, has been performed in patients with pulmonary embolism.

Following an open trial, a randomized study was performed and demonstrated a significantly higher revascularization rate in the Defibrase/plasminogen/urokinase group as compared with patients treated with heparin/plasminogen/urokinase. A similar treatment was applied with success in patients with deep venous thrombosis (DVT) following thrombectomy (Table 8).

Sudden Deafness

Asai (1985) reported on an open comparative multicenter trial in which 73 patients with idiopathic deafness, suffering from a hearing loss of more than 30 dB for less than 30 days, were treated in 13 centers either with batroxobin or with a conventional treatment consisting of dextran, ATP, and corticosteroids. Markedly improved hearing ability was observed in 13 of 33 batroxobin-treated patients, with no significant difference in the 40 patients receiving the conventional treatment. Only 16 patients in the latter group showed a marked improvement. On the other hand, the symptoms remained unchanged in 32.5% of the conventional treatment group in contrast to 18.2% of the batroxobin group.

UNTOWARD EFFECTS

A list of the side effects and incidents observed during batroxobin treatment in 667 patients or volunteers is presented in Table 9. The data on 232 patients from the retrospective study of Latallo (1983) are included in Table 9.

Subarachnoidal bleeding was observed in a patient receiving combined streptokinase (SK)/Defibrase treatment with a preexisting cyst; the fibrinogen level was 0 at the moment of the incident.

Apoplexy was observed at the same center (Martin et al., 1976) again in a patient under combined SK/Defibrase treatment for deep venous thrombosis. Symptoms indicated multiple arterial emboli. Phlebography performed 3 weeks later revealed complete lysis of the venous thrombus. It was concluded that embolization occurred as a consequence of thrombolysis.

A transient decrease in platelet count observed after rapid intravenous injection of batroxobin was probably due to entrapment of

Table 8 Combined Applications in 77 Patients

n	Type of trial	Indication	Medication	Results	Reference
14	Open	DVT	Defibrase/SK	8 thrombolyses 1 bleeding complication	Martin et al. (1976)
4	Open	DVT	Defibrase/UK	3 thrombolyses	Ishimaru et al. (1981)
33	Open	Pulm. emb.	Defibrase/Lys-plgn/UK	25 revascularizations	Brochier et al. (1981)
20	Randomized	Pulm. emb.	Heparin/Lys-plgn/UK	Miller index: before 20 ± 4/34 after 15.6 ± 5/34 revascularization rate 29%	Griguer (1983)
			Defibrase/Lys-plgn/UK	Miller index: before 20 ± 5/34 after 8.2 ± 7/34 revascularization rate 54%	
6	Open	DVT following thrombectomy	Defibrase/Lys-plgn/UK	Complete or partial patency after 6 months	Ishimaru et al. (1984)

DVT = deep venous thrombosis; Lys-plgn = lysyl-plasminogen; pulm. emb. = pulmonary embolism; SK = streptokinase; UK = urokinase

Table 9 Untoward Effects During Treatment

Number of cases	Route of administration	Side effects	No.	Reference
15	IV	0	-	Harder et al. (1972a, 1972b)
15	IV	0	-	Blättler et al. (1974)
24	IV	Heaviness of the head, asthenic feeling, bleeding at the site of injection, ekchymosis	6 1 3 3	Fukutake et al. (1981)
33	IV	Allergic reaction	1	Egberg (1973)
5	IV	—	-	Straub and Harder (1971)
58	IV + streptokinase	Subarachnoidal bleeding (cyst), apoplexia	1 1	Martin et al. (1976)
8	IV	—	-	Yasunaga et al. (1979)
18	IV	0	-	Ishimaru et al. (1979)
12	SC	0	-	Martin and Auel (1976)
1	IV	Vomiting	1	Gent et al. (1979)
1	IV	Proteinuria	1	Pomianowska et al. (1973)
6	IV	Transient decrease in platelet count	2	Follana et al. (1975)
232	IV	Bleeding, symptoms of shock, transient blindness, angioneurotic edema, fever,	13 1 7	Latallo (1983)

		urticaria,	2	
		migraine, headache, sleepiness	1	
39	SC	0	-	Latallo (1983)
5	IV	—	-	Egberg (1973)
15	SC	—	-	Gaudric et al. (1982)
10	IV	0	-	Egberg et al. (1971)
5	IV	—	-	Blombäck et al. (1971)
12	IV	—	-	Blombäck and Egberg (1975)
11	IV	No bleeding	-	Donner (1980)
50	IV + plasminogen + urokinase + heparin + dipyridamol	Urticaria, fever, bleeding, death	1 17 2	Brochier et al. (1981)
12	IV, SC	Minor haematoma	-	Meissner et al. (1977)
10	IV	—	-	Koehler (1975)
11	IV	—	-	Voelker (1975)
2	IV	—	-	Sharp (personal comm.)
50	SC	Bleeding thrombotic reactions, allergic reactions	1 1 1	Pentapharm, inquiry
7	IV + surgery	Severe bleeding	3	Olsson et al. (1976)

0 = no side effect observed; — = no side effect reported.

platelets into fibrin I aggregates and subsequent release due to fibrinolytic events.

Two patients undergoing the sequential treatment with Defibrase, plasminogen, and urokinase died. One was a 63-year-old woman with a right-sided hemiplegia less than 2 months old and with a bilateral pulmonary embolism leading to a perfusion deficit of 67%. Following urokinase treatment, the perfusion deficit was still 59%. A sequential treatment with Defibrase, plasminogen, and urokinase had to be stopped on the third day of Defibrase administration because of a second cerebrovascular incident which led to death 18 days later.

The second patient who died suffered from pulmonary embolism with a perfusion deficit of 73% and from venous thrombosis. Seven days after embolism, the patient received urokinase alone and subsequently plasminogen and urokinase. Both treatments did not improve the pulmonary perfusion rate. A sequential treatment of Defibrase, plasminogen, and urokinase was instituted, but had to be stopped on the second day because of irreversible cardiac collapse. Autopsy revealed pulmonary infarction, deep venous thrombosis in the right leg, pathological changes of the aortic valves, and an old fibrous cicatrization of the myocardium.

REFERENCES

Agishi, Y., T. Uchiumi, H. Shirakawa, T. Oku, H. Kishi, K. Saito, S. Yoshimatsu, M. Kato, T. Fukushima, K. Habu, Y. Nasu, M. Fujimaki, K. Furukawa, and Y. Mishima. (1985a). Clinical investigation of defibrinating therapy on peripheral circulation disturbance of vibration syndrome patients. Clin. Med. (Japan) 1(8):1145-1168.

Agishi, Y., H. Ide, Y. Asanuma, S. Fujiya, T. Uchiumi, and M. Kondo. (1985b). Effects of batroxobin on the clinicophysiological functions and the blood coagulation and fibrinolysis in the vibration syndrome patients. Clin. Med. (Japan) 1(8):1169-1187.

Asai, H. (1985). A comparative study on defibrinogenation and conventional therapies for sudden hearing loss. Clin. Med. (Japan) 8:1189-1199.

Blättler, W., P. W. Straub, and A. Peyer. (1974). Effect of in vivo produced fibrinogen-fibrin intermediates on viscosity of human blood. Thromb. Res. 4:787.

Blombäck, M., and N. Egberg. (1975). In *Defibrinierung mit Thrombinähnlichen Schlangengiftenzymen*, Martin, M. and W. Schoop (Eds.). Huber, Bern, Stuttgart, Wien, p. 206.

Blombäck, M., N. Egberg, E. Gruder, S. A. Johansson, H. Johnsson, S. E. Nilsson, and B. Blombäck. (1971). Treatment of thrombotic disorders with Reptilase. Thromb. Diath. Haematol. Suppl. 45:51.

Brochier, M., P. Griguer, Ph. Raynaud, J. P. Fauchier, and B. Charbonnier. (1981). Augmentation de l'activité fibrinolytique par le traitement séquentiel Défibrase®-plasminogène-urokinase dans les embolies pulmonaires et les thromboses veineuses profondes. Arch. Mal. Coeur 74:761-769.

Donner, L., P. Klener, and O. Setkova. (1980). Vliv Hadiho Jedu, Defibrasy, na hemostazu. Sborn. Lék. 82:203-211.

Egberg, N. (1973). Coagulation studies in patients treated with Defibrase. Acta Med. Scand. 194:1.

Egberg, N., M. Blombäck, H. Johnsson, U. Abildgaard, B. Blombäck, G. Diener, S. Ekeström, L. Goransson, S. A. Johansson, J. McDonagh, R. McDonagh, S. E. Nilsson, S. Nordström, P. Olsson, and B. Wiman. (1971). Clinical and experimental studies on Reptilase. Thromb. Diath. Haematol. Suppl. 47:379.

Egberg, N., and S. E. G. Nilsson. (1975). In *Defibrinierung mit Thrombinähnlichen Schlangengiftenzymen*, Martin, M. and W. Schoop (Eds.). Huber, Bern, Stuttgart, Wien, p. 231.

Follana, R., J. Sampol, J. M. Salvadori, and M. Olmer. (1975). In *Defibrinierung mit Thrombinähnlichen Schlangengiftenzymen*, Martin, M. and W. Schoop (Eds.). Huber, Bern, Stuttgart, Wien, p. 233.

Fukutake, K., M. Fujimaki, H. Nagasawa, and M. Kato. (1981). Clinico-pharmacological observations of batroxobin (Defibrase®) administered to normal human adults. Acta Haem. Jpn. 44:1178-1194.

Furukawa, K., Y. Mishima, T. Tanabe, Y. Shionoya, T. Oshiro, and H. Gotoh. (1983). Effects of defibrinogenation therapy in constructive surgery of the peripheral vessel. Surgery 45:633-639.

Gaudric, A., O. Tolub, G. Coscas, J. Conard, M. Samama, and F. Lelong. (1982). Traitement médical des occlusions veineuses rétiniennes par un nouveau défibrinant (Défibrase®). Etude préliminaire. Bull. Soc. Opht. France 1:79-82.

Gent, A., G. I. C. Ingram, C. L. Arocha-Pinango, P. Fenton, and R. J. Buckley. (1979). Central retinal vein thrombosis. Serial treatment with defibrination, aspirin and plasminotropic drugs. Thromb. Res. 14:61-66.

Griguer, P. (1983). Expertise. Tours (unpublished).

Habu, K., J. Iwa, Y. Hara, T. Ueno, and K. Yangun. (1982). Treatment of patients with vibratory disease. Clinical hemorheological results of therapy with Defibrase (DF-521). Jpn. J. Med. Sci. and Pharmacol. 8:1019-1027.

Harder, A. J., H. Stadelmann, and P. W. Straub. (1972a). Reptilase-induced shortening of coagulation times in normal and hemophilic individuals. Thromb. Diath. Haematol. 27:349.

Harder, A. J. and P. W. Straub. (1972b). In vitro and in vivo induction of cryofibrinogen and "paracoagulation" by Reptilase. Thromb. Diath. Haematol. 27:337.

Ishimaru, S., M. Konishi, K. Furukawa, M. Takahashi, and M. Fujimaki. (1979). Studies on blood coagulation system during defibrinogenation therapy with batroxobin (Bothrops atrox venom). Jpn. J. Clin. Haem. 20:500-509.

Ishimaru, S., M. Konishi, K. Furukawa, and M. Takahashi. (1984). Treatment of deep vein thrombosis with combination of batroxobin, lysyl-plasminogen and urokinase. Blood and Vwssel 15:435-441.

Ishimaru, S., M. Konishi, T. Hirayama, K. Furukawa, and M. Takahashi. (1981). Application of the defibrinogenation therapy for surgery. Acta Haem. Jpn. 44:732-738.

Koehler, M. (1975). In *Defibrinierung mit Thrombinähnlichen Schlangengiftenzymen*, Martin, M. and W. Schoop (Eds.). Huber, Bern, Stuttgart, Wien, P. 132.

Latallo, Z. S. (1983). Retrospective study on complications and adverse effects of treatment with thrombinlike enzymes. Thromb. Haemost. 50:604-069.

Martin, M., and H. Auel. (1976). Technique and laboratory control of subcutaneously administered Defibrase. Thromb. Res. 9:59.

Martin, M., H. J. Engelken, F. Tambert, U. Martin, E. Hirdes, and H. Auel. (1976). Die Defibrinierungstherapie mit einem thrombinähnlichen Schlangengiftenzym (Defibrase®). Vasa Suppl. 5:1.

Meissner, J. A., M. Karpowicz, and L. Konopka. (1977). Changes in the muscle rest blood flow associated with therapeutic defibrinogenation. Eur. Surg. Res. Suppl. 1:111.

Nasu, Y. (1982). Vibratory disease resulting from hand-operated vibrating tools. J. Industr. Med. 5:47-53.

Ohara, I. and H. Sasaki. (1983). Effects of batroxobin (Defibrase®) on peripheral circulatory disturbance. Med. Sci. Pharmacol. 9:1643.

Olsson, P., M. Blombäck, N. Egberg, and S. Ekeström. (1976). Experience of extensive vascular surgery on Defibrase-defibrinogenated patients. Thromb. Res. 9:277.

Pomianowska, L., S. Luft, and M. Kopec. (1973). Massive transient proteinuria during treatment with Defibrase in a patient with amyloid nephropathy. Thromb. Res. 2:177.

Straub, P. W. and A. J. Harder. (1971). Verhalten von I^{125}-Fibrinogen bei therapeutischer Defibrinierung mit hochgereinigter Reptilase (Defibrase). Schw. Med. Wschr. 101:1802.

Voelker, D. (1975). In *Defibrinierung mit Schlangengiftenzymen*, Martin, M. and W. Schoop (Eds.). Huber, Bern, Stuttgart, Wien, p. 171.

Yasunaga, K., K. Kumada, and K. Matsuda. (1979). Coagulation studies in patients treated with Defibrase®, snake venom batroxobin. Jpn. Arch. Int. Med. 26:465-474.

39
Clinical Trials with Ancrod

William R. Bell, Jr.
The Johns Hopkins University School of Medicine, Baltimore, Maryland

INTRODUCTION

Utilization of a fraction of the crude venom of the Malayan pit viper, *Agkistrodon rhodostoma*, for therapeutic purposes occurred in light of the observations of Reid and associates in the 1960s (1,2). Recognition by these investigators that the whole blood of victims bitten by this viper was incoagulable prompted further investigation of the mechanisms involved. Thorough studies identified that the incoagulability of the blood resulted from complete and selective removal of fibrinogen by the crude venom (3-7). The active component of the crude venom responsible for the process of defibrinogenation was identified by Esnouf and Tunnah (6). The properties of the active fraction (designated ancrod*) were identified in detail, and initial clinical studies were implemented.

---

*Ancrod is the nonproprietary name given to this enzyme by the British Pharmacopeia Commission. Arvin is the registered trade name of this agent (Twyford Laboratories, Ltd., TCL Park Royal LTD Park Royal Brewery, London NW10 7RR, England). This agent is also known as Arwin (the registered trade name for Twyford Pharmaceutical Services Ltd., AG Knoll AG 67, Ludwigshafen, Postfach 21 0805, West Germany. Knoll Pharmaceutical Company, 30 N. Jefferson Road, Whippany, New Jersey 07981).

PROPERTIES OF ANCROD

The known physicochemical properties of ancrod are given in Table 1. Highly purified preparations of ancrod possess coagulant, proteolytic, and esterolytic properties. Unlike thrombin, which cleaves both fibrinopeptides A and B from fibrinogen, ancrod releases only the A fibrinopeptides (A, AP, AY) and not fibrinopeptide B. In contrast with thrombin, ancrod progressively digests the α-chain of human fibrinogen. The sites cleaved on the α-chain by ancrod are different from those degraded by plasmin. The degradation of the α-chain by ancrod can be markedly or even completely inhibited if the α-chain has been cross-linked by factor XIII via transamidation. With respect to the esterase and protease properties of this agent, serine and histidine have been identified as essential amino acids for these activities. It has been demonstrated that ancrod has methylesterase specificity limited to esters of L-arginine and the N-substituted acyl-derivatives of arginine. The Kcat/Km value for hydrolysis of benzoyl-arginine-ethyl ester was 0.9 M^{-1} sec^{-1}(x 10^{-5}), with hydrolysis occurring by means of acyl enzymes. Ancrod has only minimal amidolytic activity. The thrombin-like properties of ancrod are not inhibited by polyvalent proteinase inhibitors such as aprotinin, soybean trypsin inhibitor, heparin, antithrombin III, iodoacetamide, 0.01% sodium dodecyl sulfate, concanavalin-A, N-dansyl-L-arginine-4-ethyl-piperidine amide, and benzamidine. Ancrod is inhibited by tosyl-arginine-methyl ester.

Ancrod elicits an antigenic response in humans. The antigenic response is greater in magnitude and occurs earlier when ancrod is administered by the intramuscular versus the intravenous route.

Table 1 Physicochemical Properties of Ancrod

Property	Ancrod
Source	*Agkistrodon rhodostoma*
Mr	37,000
% carbohydrate	25-30
N-terminal residue	Valine
Partial specific volume (ml/g)	0.69
S	3.35 (at 4.9 mg/ml)
E_{280} nm	14.8
Stability	Excellent

Antisera form a complex with the antigen, neutralizing the enzymatic activity with a resultant positive precipitin reaction.

Assay methods employed to assign the relative potency of ancrod are based on coagulation capacity of ancrod on whole human plasma or purified human fibrinogen as substrates, with thrombin as a comparison, and defined in units.

MECHANISM OF ACTION

The primary mechanism of action of ancrod in vivo is its proteolytic effect on circulating fibrinogen. The release of fibrinopeptide A, but not fibrinopeptide B, which results in the conversion of soluble fibrinogen to a fibrin polymer, occurs in the absence of calcium, prothrombin, or thrombin. The resultant polymer contains only end-to-end anastomatic bonding, has a unique electrophoretic migration pattern, a reduced gel-tensile strength and elastic modulus, and appears as thin filaments on electron microscopy. The in vitro fibrin clot resulting from the action of thrombin is insoluble in 5 M of urea or 1% monochloroacetic acid in contrast with ancrod, which yields a soluble clot under the same conditions, since ancrod does not activate factor XIII (Fig. 1). Ancrod does not alter the activity of any of the coagulation factors except fibrinogen. In vitro ancrod does not convert plasminogen to plasmin. In vivo, however, there is some evidence to suggest that ancrod may induce release of tissue plasminogen activator from vascular endothelial cells, which then will convert plasminogen to plasmin.

Within minutes after the administration of ancrod, there is a significant reduction in plasma fibrinogen concentration. Within 2 to 3 hours of the institution of ancrod, the concentration of fibrinogen is less than 50 mg/dl. The plasma fibrinogen will remain at exceedingly low levels if ancrod is administered once or twice daily. When ancrod is discontinued in patients who are in a normal physiologic state, the plasma fibrinogen slowly returns (within 3 to 5 days) to a normal pretreatment value.

CLINICAL TRIALS

Since the introduction of ancrod into clinical medicine, a number of different disease entities have been studied (3-56) including deep venous thrombosis, myocardial infarction, peripheral arterial thrombosis, priapism, sickle cell crisis, and renal transplant rejection. Interpretation of these studies is difficult because of the lack of precise diagnostic criteria, appropriate controls, double-blind statistical design, and comparative studies. However, from these early studies

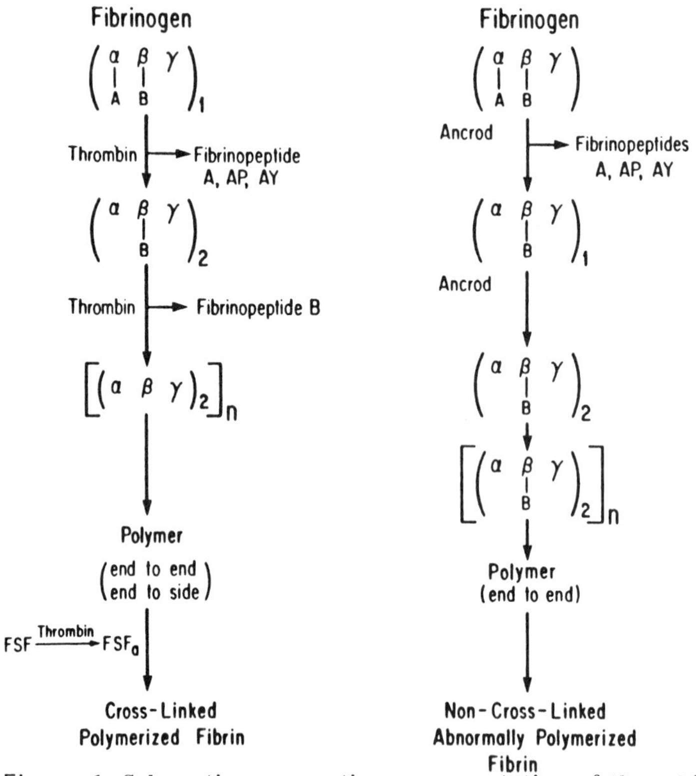

Figure 1 Schematic comparative representation of thrombin (left) and ancrod (right) and their proteolytic actions on fibrinogen and factor XIII (FSF).

it was apparent that this agent could be employed safely in man (hemorrhage was not a problem in any of these studies).

During the past 5 to 6 years, studies of adequate sample size, randomized, blinded, prospective, and appropriately controlled have been completed (57-69). Several thoroughly performed studies in an appreciable number of patients with active nephritis associated with systemic lupus erythematosus treated with defibrinogenation with ancrod have resulted in improved renal function, disappearance of glomerular thrombi, and increased patient survival (64-66,69). The results in these studies are well documented with serial measurements of renal function and histology before and at the completion of therapy. In these studies, it was observed that, during and after treatment with ancrod, reduced blood levels of prostacyclin (PGI_2)

returned to normal and remained normal for several weeks. The mechanism facilitating PGI_2 generation in this situation was not identified. In a single patient (a 35-year-old male) with hyperacute renal rejection and allograft failure, the use of ancrod reversed this problem and induced complete recovery of renal function, disappearance of glomerular fibrin, and restoration of normal glomerular architectural structure (69).

A double-blinded trial (67) compared ancrod (in 35 patients) with saline (in 38 patients) in the prophylaxis of deep venous thrombosis (DVT) in patients undergoing total hip replacement. Results revealed that DVT diagnosed by venography was reduced from 18 episodes in the control group to 5 in the ancrod-treated group ($p < 0.01$). In this study, the frequency of bleeding was not different between the two treatment groups.

In a randomized, single-blind study of 30 patients with acute ischemic stroke, ancrod was demonstrated to be more effective than mannitol plus low molecular weight dextran (68). In this study, 15 patients (mean age 63.3 ± 11.8 years) received low molecular weight dextran and mannitol. Fifteen additional patients received this combination plus ancrod. In the ancrod-treated group, the plasma fibrinogen was reduced to 100 to 130 mg/dl, with a concomitant reduction in blood viscosity to 37% of normal at a shear rate of 0.03 sec^{-1}. The mannitol/dextran-treated groups experienced a viscosity reduction of only 7%. In the ancrod-treated patients, fibrinogen-fibrin degradation products increased from 3.1 ± 0.4 mg/l to 154.3 ± 31.6 mg/l and plasminogen decreased from 98.2 ± 2.0% of normal to 79.8 ± 2.9% of normal. These changes were not evident in the mannitol/dextran-treated group. Global coagulation and platelet function tests were not influenced by the treatment regimens. Neurological score and survival were significantly better in the ancrod-treated group.

Conflicting results have been observed in the treatment of peripheral vascular disease. In one study of 23 patients with DVT treated with ancrod, all demonstrated clinical resolution and no recurrences were identified in a 2-year follow-up (61). In a double-blinded randomized prospective study of 27 patients with ischemic arterial rest pain, no treatment preference was noted after 8 days of therapy with ancrod versus saline (63). In a 6-month follow-up of these patients, there were no significant differences between treatment groups in analgesic consumption, Doppler systolic pressure ratio, or frequency of surgical intervention. In a smaller study, 15 patients with chronic occlusive arterial disease with and without ischemic necrosis were treated with ancrod (70). In this group, 86% of patients had evidence of significant improvement as identified by a reduction in Fontaine stages of severity from stage IV to stages I and II. This improvement was seen in 76% of patients as identified by acral electronic oscillography at a 12-month follow-up.

Most recently, there has been considerable enthusiasm for the use of ancrod in the treatment of acute myocardial infarction. It has been recognized for several years that ancrod induces fibrinolysis in man. This probably occurs indirectly by ancrod inducing the release of tissue plasminogen activator from vascular endothelial cells. In a study by Apprill and co-workers (71), ancrod was employed in two canine experimental models of acute ischemic heart disease to investigate its potential role as a thrombolytic agent. Ancrod (8 units/kg in saline solution, intravenously over 60 minutes) abolished cyclic flow variations in an experimentally stenosed coronary artery produced by a plastic constrictor in six of eight dogs; the frequency of cyclic flow variations per 30 minutes decreased from 5.9 ± 0.6 during control to 2.4 ± 0.9 after ancrod had been infused (mean ± SEM; $p < 0.05$). A separate group of 23 dogs had copper coils placed into their left anterior descending coronary artery to induce acute coronary thrombosis. All dogs received an intravenous infusion of lidocaine; evidence of thrombolysis following ancrod administration was assessed by repeat coronary angiography. Four dogs died suddenly of ventricular fibrillation (VF); 4 dogs received saline solution (30 cc over 30 minutes) with no evidence of thrombolysis during the subsequent 5 hours; 3 dogs died of VF during ancrod infusion; and 6 of 12 surviving dogs had thrombolysis as determined by coronary angiography in 66 ± 20 minutes (mean ± SEM) with 15 units/kg ancrod intravenously over 30 minutes. Therefore, ancrod administration reduces the frequency of cyclic flow variations in dogs with partial coronary stenosis and can induce coronary thrombolysis when acute thrombosis is induced by copper coil insertion in this experimental model.

Since these important preclinical observations by Apprill and co-workers (71), additional studies are needed to explore the thrombolytic capacity of this agent. If this agent can induce thrombolysis, the concomitant reduction in plasma fibrinogen would provide two additional benefits. First, when fibrinogen is appreciably lowered, there is a striking reduction in whole blood viscosity and a significant improvement in the fluid flow properties of the blood. Second, once the blood is depleted of fibrinogen, thrombus formation will not occur. In the absence of potential thrombus formation, endothelialization of any damaged vascular lesions will prevent rethrombosis. These potentials can only be realized when appropriately designed studies are performed.

ACKNOWLEDGMENT

This work was supported in part by Research Grants HL 24898 and HL 07377 from the National Heart, Lung, and Blood Institute, National Institutes of Health, Bethesda, Maryland, the Whitehall Foundation, and Hubert E. and Anne E. Rogers Scholars in Academic Medicine Fund.

REFERENCES

1. Reid, A., K. E. Chan, and P. C. Thean. Prolonged coagulation defect (defibrination syndrome) in Malayan viper bite. Lancet 1:621 (1963).
2. Reid, H. A. and K. E. Chan. The paradox in therapeutic defibrination. Lancel 1:485 (1968).
3. Sharp, A. A., B. A. Warren, A. M. Paxton, and M. J. Allington. Anticoagulant therapy with a purified fraction from Malayan pit viper venom. Lancet 1:493 (1968).
4. Bell, W. R. and .W. R. Pitney. The concept of therapeutic defibrination in the treatment of thrombotic disease. Lancet 1:490 (1968).
5. Bell, W. R. and W. R. Pitney. The concept of therapeutic defibrination in the treatment of thrombotic disease. Lancet 1:490 (1968).
6. Esnouf, M. P. and G. W. Tunnah. The isolation and properties of the thrombin-like activity from Agkistrodon rhodostoma venom. Br. J. Haematol. 13:581 (1967).
7. Bell, W. R., G. Bolton, and W. R. Pitney. The effect of Arvin on blood coagulation factors. Br. J. Haematol. 15:589 (1968).
8. Bell, W. R., S. S. Shaprio, J. Martinex, and H. L. Nossel. The effects of ancrod the coagulating enzyme from the venom of Malayan pit viper. A. Rhodostoma, on prothrombin and fibrinogen metaoblism and fibrinopeptide-A release in man. J. Lab. Clin. Med. 91:592 (1978).
9. Pitney, W. R., W. R. Bell, and G. Bolton. Blood fibrinolytic activity during Arvin therapy. Br. J. Haematol. 16:165 (1969).
10. Bell, W. R. and E. Regoeczi. Isotopic studies of therapeutic anticoagulation with a coagulating enzyme. J. Clin. Invest. 49:1872 (1970).
11. Regoeczi, E. and W. R. Bell. In vivo behavior of the coagulant enzyme from Agkistrodon rhodostoma venom: Studies using I^{-125} Arvin. Br. J. Haematol. 16:573 (1969).
12. Alving, B. M., B. Evatt, and W. R. Bell. Stimulation of fibrinogen synthesis by thrombin in rabbits with ancrod-induced defibrinogenemia. Am. J. Physiol. 223:H562 (1977).
13. Stocker, K. In *Handbook of Experimental Pharmacology, Volume 46,* Markwardt, F. (Ed.), Springer-Verlag, Berlin, pp. 451-484 (1978).
14 Egberg, N., M. Blomback, J. Johnson, U. Abildgaard, B. Blomback, G. Diener, S. Ekestrom, L. Goransson, S. A. Johansson, J. McDonagh, R. McDonagh, S. E. Nilsson, S. Borastrom, P. Olsson, and B. Wilman. Clinical and experimental studies on reptilase. Thromb. Diath. Haemorrh. 47(Suppl):589 (1968).

15. Martin, D. L. and G. Knollman. In *Aktuelle Probleme in der Angiologie: Defibrinierung mit Thrombinahnlichen Schlangengiftenzymen, Volume 26*, Martin, M. and W. Schoop (Eds.), Hans Huber Verlag, Stuttgart, p. 166 (1975).
16. Bell, W. R. Defibrinogenation with Arvin in thrombotic disorders. In *Platelets and Thrombosis*, Sherry, S. and A. Scriabine (Eds.), University Park Press, Baltimore, p. 273 (1974).
17. Prentice, C. R. M., W. Edgar, and G. P. McNinol. Characterization of fibrin degradation products in patients on ancrod therapy; Comparison with fibrinogen derivatives produced by plasmin. Br. J. Haematol. 27:77 (1974).
18. Harder, A. J. and P. W. Straub. In vitro and in vivo induction of cryofibrinogen and "paracoagulation" by Reptilase. Thromb. Diath. Haemorrh. 27:337 (1972).
19. Kwaan, H. C. Current status of "ancrod" and "reptilase" (Abstract). Int. Soc. Thromb. Haemost. 34:50 (1972).
20. Kwaan, H. C. Use of defibrinating agents ancrod and Reptilase in the treatment of thromboembolism. Thromb. Diath. Haemorrh. 54(Suppl):377 (1973).
21. Esbert, N. Coagulation studies in patients treated with defibrase. Acta Med. Scand. 194:291 (1973).
22. Bell, W. R. Defibrinogenation with arvin. In *Thrombosis: Mechanisms and Control*, Brinkhous, K. M. and S. Hinnom (Eds.). Verlag, Stuttgart, p. 371 (1973).
23. Olsson, P., C. Blombak, N. Egberg, S. Ekestroen, L. Goeransson, and H. Johnsson. Studies on the bleeding tendency and on the possibility of surgery in states of Reptilase-induced defibrinogenation. Thromb. Diath. Haemorrh. 47(Suppl):398 (1971).
24. Bell, W. R. Further experience in the treatment of thrombotic disorders by therapeutic defibrination. Q. J. Med. 37:658 (1968).
25. Pitney, W. R. Clinical experience with Arvin. Thromb. Diath. Haemorrh. 38(Suppl):81 (1969).
26. Gilles, H. M., H. A. Reid, A. Odutola, O. Ransome-Kuti, and S. Ransome-Kuti. Arvin treatment for sickle-cell crisis. Lancet 2:542 (1968).
27. Matsuda, J., K. Hideno, M. Ogawara, N. Kodama, and M. Murakami. Therapeutic defibrination by Bothrops marajoensis venom. Acta Haematol. Japonic. 38:299 (1975).
28. Browse, N. C. Vein surgery during defibrination. In *Aktuelle Probleme in der Angiologie: Defi brinierung mit thrombinahnlichen Schlangengiftenzymen, Volume 26*, Martin, M. and W. Schoop (Eds.), Hans Huber Verlag, Stuttgart, p. 152 (1975).
29. Hauck, G. and H. Schroer. The defibrination effect of Arvin on microcirculation and capillary permeability. Bibl. Anat. 7:229 (1975).

30. Voss, D. Therapeutic defibrination: a new concept in anticoagulant therapy. Internist 9:389 (1968).
31. Bell, W. R. Management of pamprism by therapeutic defibrination. N. Engl. J. Med. 280:649 (1969).
32. Powell, R. E., V. J. Marmion, and C. F. McCarthy. Treatment of central retinal vein thrombosis with ancrod. Lancet 1:173 (1970).
33. Hall, R. J. C., C. Young, G. C. Sutton, and S. Campbell. Anticoagulation by ancrod for haemodialysis. Br. Med. J. 4:591 (1970).
34. Bell, W. R. Current status of therapy with Arvin. Thromb. Diath. Haemorrh. 47(Suppl):371 (1971).
35. Blombäck, M., N. Egberg, E. Gruden, S. A. Johansson, H. Johnsson, S. E. G. Nilsson, and B. Blombäck. Treatment of thrombotic disorders with reptilase. Thromb. Diath. Haemorrh. 45(Suppl):51 (1971).
36. Ehringer, H. E., R. Kleinberger, K. Lechner, and W. Reiterer. Arvin: Schlangengift als neue Therapie, moglichkeit bir Durchblutungs-storungen. Wein. Lin. Wochenschr. 83:441 (1971).
37. Sharp, A. A. Clinical use of arvin. Thromb. Diath. Haemorrh. 45(Suppl):69 (1971).
38. Vinazzer, H. Zur Wirkung von Arvin auf die Blutgerinnung. Wein. Z. Inn. Med. 52:378 (1971).
39. Davies, J. A., M. V. Merrick, A. A. Sharp, and J. M. Holt. Controlled trial of ancrod and heparin in treatment of deep-vein thrombosis of lower limb. Lancel 1:113 (1972).
40. Donahoe, J. F., G. Barlow, F. G. McMahon, J. R. Ryan, and H. C. Kwaan. A study of Abbot-38414 (ancrod) in the normal adult human. In *Third Congress of the International Society on Thrombosis and Haemostasis, 82.* Washington, D.C. (1972).
41. Ehringer, H. R., K. Lechner, and F. Widhalm. Therapeutische Defibrinierung mit Schlangengift (Arvin) bei arteriellen gerinnungsfordernden. Verh. Disch. Ges. Inn. Med. 78:624 (1972).
42. Leube, G., J. J. Keuhn, and H. Hartert. Erste Erfahrungen mit Arvin, det gerinnungsforden Fraktion cines Schlangengiftes. Med. Welt. 23:601 (1972).
43. Williams, J. R. B. and E. Maughan. Treatment of tumour metastases by defibrination. Br. Med. J. 3:174 (1972).
44. Armendia, M. and C. Steuart. Ancrod therapy in acute thrombophlebitis. Clin. Res. 21:466 (1973).
45. Ehringer, R., II, R. Dudczak, and K. Lechner. Therapeutische defibrinierung mit ancord. Deutsch. Med. Wochenschr. 98:2298 (1973).
46. Forbes, C. D., C. R. M. Prentice, J. Barbanell, and C. Thomson. Sequence therapy with ancrod followed by streptokinase for thrombotic disorders. Br. J. Haematol. 24:663 (1973).

47. Caprini, J. A., H. C. Kwaan, L. Zuckerman, and R. Verduin. Thrombonelastographic patterns of ancrod and thrombin fibrin formation and dissolution. Thromb. Res. 4:199 (1974).
48. Ehringer, H., R. Dudczak, and K. Lechner. A new approach in the treatment of peripheral arterial occlusion: defibrination with arvin. Angiology 25:279 (1974).
49. Tibbett, D. A., E. W. Williams, M. W. Walker, C. N. Chesterman, J. M. Holt and A. A. Sharp. Controlled trial of ancrod and streptokinase in the treatment of deep vein thrombosis of lower limb. Br. J. Haematol. 27:407 (1974).
50. Bourgain, R. H. and F. Six. The effect of defibrase on arterial thrombus formation. Thromb. Res. 6:195 (1975).
51. Egberg. N. and S. F. G. Nilsson. Treatment of central retinal vein thrombosis with defibrase-induced defibrination. In *Aktuelle Probleme in der Angiologie: Defibrinierung mit Thrombinahn-lichen Schlangengiftenzymen, Volume 26*, Martin, M. and W. Schoop (Eds.), Hans Huber Verlag, Stuttgart, p. 231 (1975).
52. Kohler, M. Untersuchunger zur Wirkung einer defibrinierung mittels Scholangengiftenzym auf den Stoffwechsel bei chronischer peripherer arteriopathie. In *Aktuelle Problem in der Angiologie: Defibrinierung mit Thrombinahnlichen Schlangengiftenzymen, Volume 26*, Hans Huber Verlag, Stuttgart, p. 132 (1975).
53. Vinazzer, H. Coagulation studies during therapeutic application of arvin (abstract). In *Symposium on Thrombin-like Enzymes*, Trier (1975).
54. Forbes, C. E., J. Barbenell, and C. R. M. Prentice. Treatment of thrombosis by sequential therapy with ancrod followed by streptokinase clinical pharmacology and rheology. Haemostasis 5:348 (1976).
55. Martin, M., E. Hirdes, and H. Auel. Defibrinogenation treatment in patients suffering from severe intermittent claudication—a controlled study. Thromb. Res. 9:47 (1976).
56. Olsson, P., M. Blombäck, N. Egberg, and S. Ekestrom. Experience of extensive vascular surgery on defibrase-defibrinogenated patients. Thromb. Res. 9:227 (1976).
57. Roy, J., R. Guidoin, L. Martin, H. Lephat, P. Blais, M. Marois, D. Gagnon, and J. Awad. Defibrinogenation as an alternative to heparinization in prolonged extracorporeal circulation. Res. Exp. Med. 176:219 (1980).
58. Dohl, K., M. Fujioka, and Y. Nakamoto. Comparative effects of heparin, urokinase, and ancrod on intraglomerular coagulation induced in progressive masugi nephritis. Acta Pathol. Jpn. 32:1047 (1982).
59. Tomita, K., E. Muso, T. Tamura, and C. Kawai. The effect of hypfibrinogenemia with batroxobin on rat masugi nephritis. Jpn. Cir. J. 46:577 (1982).

60. Zhang, J., R. Munda, P. Glas-Greenwalt, M. A. Weiss, V. E. Pollak, and J. W. Alexander. Prolongation of survival of a heart xenograft by defibrination with ancrod. Transplantation 35:620 (1983).
61. Kounis, N. G. and W. H. Evans. Thromboembolic disease treated with anticoagulants and defibrinating drugs. The Practitioner 222:420 (1979).
62. Lowe, G. D. O. Defibrinating agents: Effects on blood rheology, blood flow and vascular diseases in controlled studies. Bibl. Haemat. 47:247 (1981).
63. Lowe, G. D. O., D. J. Dunlop, D. H. Lawson, J. G. Pollock, J. K. Watt, C. D. Forbes, C. R. M. Prentice, and M. M. Drummond. Double-blind controlled clinical trial of ancrod for ischemic rest pain of the leg. Angiology 33:46 (1982).
64. Kant, K. S., A. K. Dosekun, K. G. P. Chandran, P. Glas-Greenwalt, M. A. Weiss, and V. E. Pollak. Deficiency of a plasma factor stimulating vascular prostacyclin generation in patients with lupus nephritis and glomerular thrombi and its correction by ancrod: in-vivo and in-vitro observations. Thromb. Res. 27:651 (1982).
65. Pollak, V. E., H. I. Glueck, M. A. Weiss, A. Lebron-Berges, and M. A. Miller. Defibrination with ancrod in glomerulonephritis: effects on clinical and histologic findings and on blood coagulation. Am. J. Nephrol. 2:195 (1982).
66. Dosekun, A. K., M. R. First, K. G. Chandran, K. S. Kant, P. Glas-Greenwalt, M. A. Weiss, J. W. Alexander, and V. E. Pollak. Successful treatment by defibrination with ancrod in a patient with hyperactute renal allograft failure and a deficiency of plasma prostacylin stimulating factor. Clin. Nephrology 18:101 (1982).
67. Belch, J. J. F., D. R. Meek, G. D. O. Lowe, A. F. Campbell, A. B. Young, C. D. Forbes, and C. R. M. Prentice. Subcutaneous ancrod in prevention of deep vein thrombosis after hip replacement surgery. Thromb. Res. 25:23 (1982).
68. Hossmann, V., W-D. Heiss, H. Bewermeyer, and G. Wiedemann. Controlled trial of ancrod in ischemic stroke. Arch. Neurol. 40:803 (1983).
69. Dosekun, A. K., V. E. Pollak, P. G. Greenwalt, K. S. Kant, P. Penovich, A. L. Berges, M. A. Weiss, and J. E. Levinson. Ancrod and systemic lupus erythematosus with thrombosis clinical and fibrinolysis effects. Arch. Intern. Med. 144:37 (1984).
70. Staiger, J., S. K. Graff, and G. Keil. Long-term results after defibrinogenation by arwin in chronic occulsive arterial disease. Bibl. Haemat. 47:252 (1981).
71. Apprill, P., J. Ashton, J. Guerrero, L. M. Bija, and J. Willerson. A new intervention for producing coronary thrombolysis. J. Am. Coll. Cardiol. 5:413 (1985).

40

Reversal of Fibrinolytic Abnormalities by the Snake Venom Ancrod (Arvin) in Systemic Lupus Erythematosus (SLE) and a Thrombotic Thrombocytopenic Purpura-like Syndrome ("TTP")

Pia Glas-Greenwalt, K. Shashi Kant, and Victor E. Pollak
University of Cincinnati Medical Center, Cincinnati, Ohio

INTRODUCTION

Ancrod is a thrombin-like enzyme from the Malayan pit viper (*Agkistrodon rhodostoma*) venom with clinical potential as an anticoagulant and/or thrombolytic agent. In general, its action has been considered a consequence of the fact that it splits the fibrinogen molecule. This results in an anticoagulant effect and a decrease in blood viscosity with consequent augmentation of blood flow (1). These actions have provided the basis for the clinical use fo ancrod in venous thrombosis and vascular occlusive disease (1).

From an experimental point of view, ancrod reduces the incidence of end organ fibrin deposition in the endotoxin-induced generalized Schwartzmann reaction (1). It also has a protective effect against and therapeutic effect on renal damage in nephrotoxic serum and chronic serum sickness models of glomerulonephritis in the rabbit. Strikingly, fibrin deposition cleared, epithelial crescents decreased, and renal insufficiency improved even when ancrod was given to animals with established nephritis and renal failure (2,3).

Unlike thrombin, ancrod splits fibrinopeptide A only from fibrinogen, converting it to a fibrin-like material. It exerts a selective enzyme substrate-specificity for fibrinogen only; it does not disturb the number and turnover of platelets, or the level of other clotting factors including fibrin-stabilizing factor (FSF-factor XIII). This results in conversion of fibrinogen to soluble, circulating fibrin monomers that polymerize end-to-end without transamidated cross-links, rendering these "clots" more susceptible to lysis by plasmin (1). Simultaneously with hypofibrinogenemia, there is a reduction in plasma plasminogen and an increase in fibrin degradation products (FDP), suggesting secondary recruitment of the fibrinolytic enzyme system; the stimulus for this activation has not yet been identified (1).

In systemic lupus erythematosus (SLE), we reported thrombosis in about 50% of renal biopsy specimens from those patients with proliferative glomerulonephritis; the thrombotic lesion, rather than the necrotic or inflammatory lesions, was the forerunner of increased glomerular sclerosis (4). A decrease in fibrinolytic capacity has been reported in SLE (5); in studies on 55 patients with SLE, we found low levels of vascular plasminogen activator (VPA) and elevated levels of two endogenous fibrinolysis inhibitors—an inhibitor of plasminogen activation (IPA) and an inhibitor of plasmin (PI) (6). A subset of 18 patients with SLE and severe diffuse proliferative lupus glomerulonephritis with microvascular thrombosis was treated with ancrod (7-9). In this chapter, we report the effects of ancrod on fibrinolytic abnormalities in patients with lupus nephritis and microvascular thrombosis, and offer a potential explanation for the action of ancrod.

MATERIAL AND METHODS

Protocol for Ancrod Administration

All 18 patients studied had diffuse proliferative glomerulonephritis and fibrin deposits in glomeruli and/or vessels, graded semiquantitatively greater than 1+ on a scale of 0 to 4+. Most had cellular or fibrocellular crescents as well. All patients were ill; 13 had moderate to very marked renal insufficiency, 14 had the nephrotic syndrome, and the diastolic blood pressure level was less than 105 mmHg in 11.

Ancrod was administered intravenously. The initial dose, approximately 1 unit/kg body weight in 500 ml of 5% dextrose in water, was infused slowly over 8 hours to ensure gradual defibrinogenation (8). The estimated dose required to keep plasma fibrinogen levels, measured daily, in the 20 to 50 mg/dl range, was then given once or twice daily in 50 to 100 ml of 5% dextrose in water over 30 to 60 minutes. The duration of defibrinogenation was set arbitrarily at 2 weeks, based on experimental animal studies (2,3). Two patients were studied during each of two separate courses of treatment.

Measurement of Plasma Fibrinolytic Activity, Inhibition of Plasminogen Activation, and Inhibition of Plasmin

Plasma fibrinolytic activity was determined by a highly standardized fibrin plate method (10). Plasminogen-rich fibrinogen was used to assay vascular plasminogen activator (VPA), and an inhibitor of urokinase-induced plasminogen activation (IPA), plasminogen-free fibrinogen was used to assay plasmin inhibitor (PI). Total fibrinolytic activity, comprising both intrinsic and extrinsic (vascular) activities, and VPA were measured as described by Kluft (11). The methods, including those used to measure the inhibitors in healthy subjects and patients with SLE, and the results of these and other assays have been described in detail elsewhere (6).

Statistical Analysis

Mean values of VPA, IPA, and PI were calculated after log transformation of the values. Values during ancrod treatment were arrayed in four groups around the means and 95% confidence limits of data from 240 healthy subjects. Chi-square (χ^2) analysis was used to measure the differences in the proportion of abnormal values.

RESULTS

Ancrod Infusion

The first infusion of ancrod was given slowly over 8 hours. The mean initial dose was 1.19 (range 0.93 to 1.58) units/kg. The geometric mean plasma fibrinogen level decreased from 359 mg/dl to 32.1 mg/dl, a decrease of 86.5%, by the end of the initial infusion. Plasminogen levels decreased from a geometric mean value of 3.3 (95% confidence limits 1.9, 5.5) to 1.6 (0.8, 3.3) CTA units/ml within 24 hours of starting treatment, and serum FDP levels increased from less than 10 mg/ml to a geometric mean value of 924 (551, 1548) mg/ml.

Most patients received ancrod twice daily. The dose needed to maintain the intended fibrinogen level of 20 to 50 mg/dl fluctuated during the first few days of treatment; in most patients, a stable dose was reached by the 5th to 7th day. This dose was higher than in the first days of treatment.

Fibrinolytic Abnormalities and Effect of Ancrod—Definition of Response

Prior to the 20 courses of treatment, VPA was unmeasurable or present only in trace amounts in 15 instances; IPA was significantly increased in 19, and slightly above the upper range of normal in 1. The PI was elevated significantly in seven. In 13 patients, 15 courses of ancrod treatment resulted in prompt and persistent normalization of IPA levels. These 13 patients had an initial normal or very slightly elevated PI level, which varied little during ancrod treatment. In the other five patients, in whom the PI level was greatly elevated prior to the

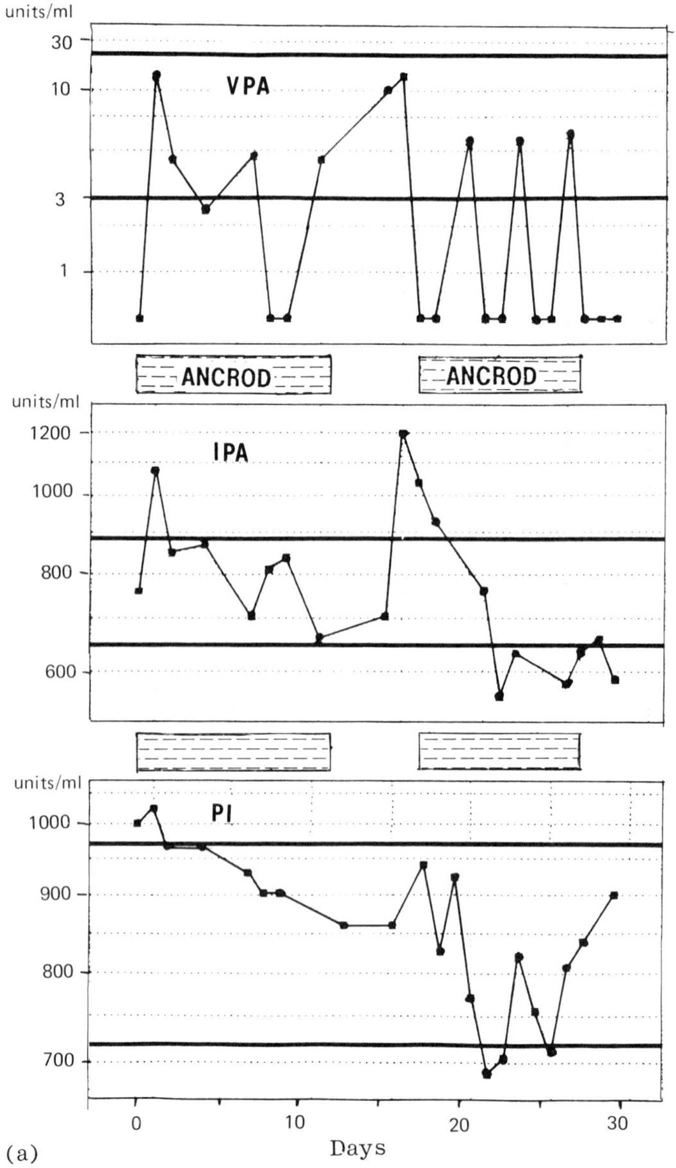

Figure 1 Serial observations on VPA, IPA, and PI levels in two patients. The range of normal values is between the horizontal lines. (a) A patient with SLE and lupus nephritis treated previously with ancrod (not shown). A second episode of severe kidney disease was associated with a "TTP"-like syndrome and severe azotemia. Ancrod was given on two occasions separated by 5 days; there was striking resolution of the thrombotic lesions in glomeruli and arterioles. The IPA level normalized rapidly, became abnormal again promptly after ancrod treatment was interrupted, and again normalized when ancrod was restarted. The VPA level fluctuated within and below the normal

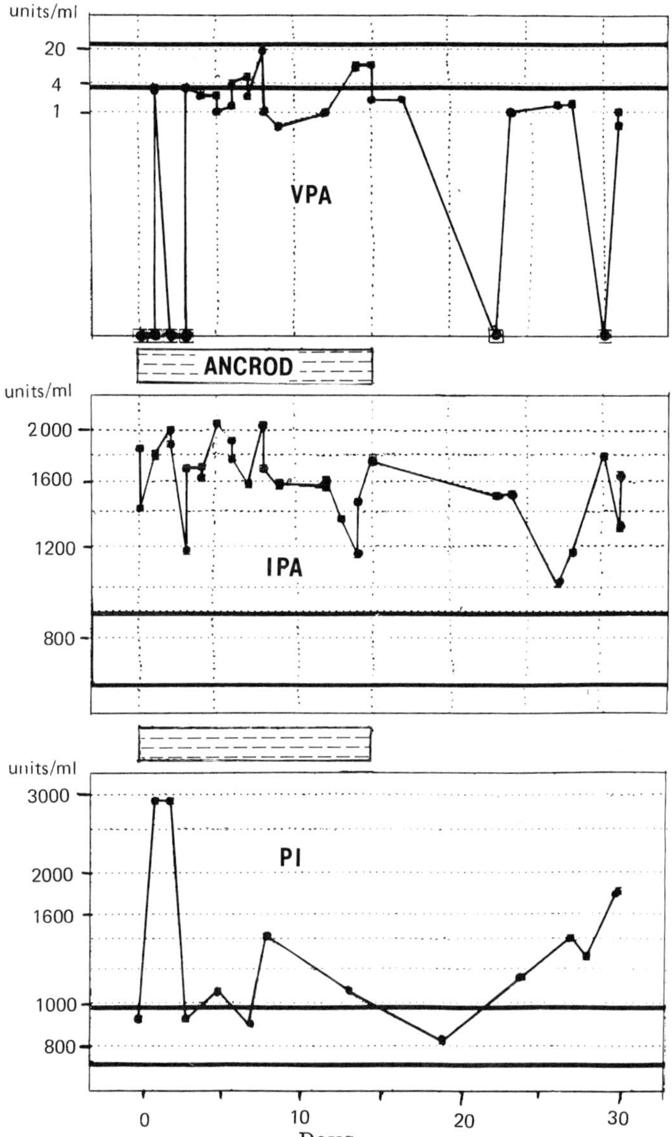

(b)

range. The PI level was increased initially, but thereafter remained within the normal range. (b) A patient with SLE, severe lupus nephritis, and striking hypertriglyceridemia. Ancrod treatment was associated with no improvement in glomerular thrombotic lesions or renal function. Note that the IPA level was high and at no time in the normal range. The VPA level was, with a few exceptions, unmeasurable or very low.

initial course of treatment, the PI level decreased gradually into the upper range of normal, but only after prolonged treatment, accompanied by fluctuating and relatively low VPA levels. Before treatment, IPA levels in these five patients were somewhat higher than in the other 13; they remained elevated throughout the period of treatment. Of 32 VPA values measured before all 20 courses of treatment, only 8 were within the normal range. During treatment, there were considerable day-to-day fluctuations; 70 of 149 were within or above the noraml range. The difference between pretreatment and treatment values was significant ($\chi^2_{[1]} = 12.2$; $p < 0.001$). Figure 1a illustrates a responder (a patient with "TTP") and Figure 1b a nonresponder (a patient with hypertriglyceridemia).

Theoretically, normalization of VPA might be expected as IPA levels returned to the normal range. The day-to-day fluctuation of individual VPA values in normal subjects (6) and the patients makes classification on the basis of VPA response difficult to judge. Patients were, therefore, classified first into two groups according to the IPA response. Those 13 patients in whom there was complete correction of the IPA level were classified as fibrinolysis responders; those 5 in whom the abnormal IPA level was not corrected were classified as fibrinolysis nonresponders. The effects of ancrod were reexamined in the light of this classification.

Fibrinogen

The action of ancrod in lowering plasma fibrinogen following the first infusion in each patient was examined separately in fibrinolysis responders and nonresponders. The percent decrease in the fibrinogen level following the initial ancrod infusion was similar in both groups. In the first few days, the actual fibrinogen level was slightly higher in the nonresponder group (Fig. 2). An increase in dose was required from the first to the last days of treatment in both fibrinolysis responders and nonresponders. To produce the same degree of defibrinogenation in the later stages of ancrod administration, more ancrod was required in nonresponders than in responders (Fig. 2). Following treatment, plasma fibrinogen levels returned to the normal range by the 4th day in both groups.

VPA

Prior to the 15 courses of ancrod treatment in the 13 responders, the VPA level was measurable in only 4. Within the first 48 hours of treatment, VPA had increased in most responders to a geometric mean value of 15.7 activator units; the response occurred later in the others. The highest VPA levels attained are summarized in Figure 3. During the 14 days of ancrod administration, a higher proportion of VAP levels was within or above the normal range in responders (62 of 103)

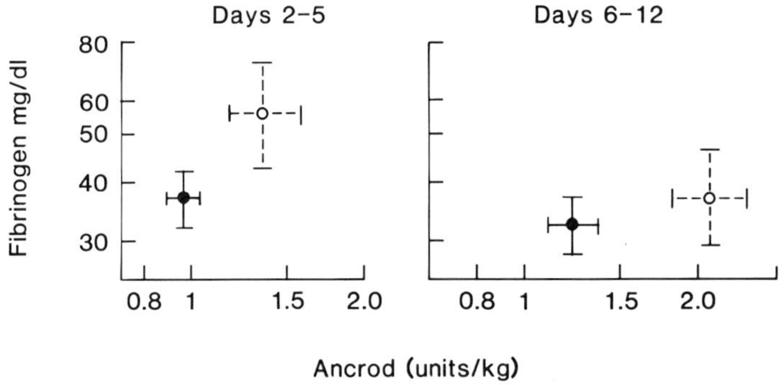

Figure 2 The relation between the fibrinogen level and the amount of ancrod infused in fibrinolysis responders (closed symbols) and nonresponders (open symbols). Values are geometric means ± 1 SE.

than in nonresponders (17 of 46); the difference was significant ($\chi^2_{[1]} = 6.11$, $p < 0.02$). The VPA returned to abnormally low levels within 2 weeks of discontinuing ancrod. In the nonresponders, the increase in VPA values on days 1 to 14 was not significant.

IPA

Before treatment, the geometric mean IPA level was 1340 inhibitor units/ml in responders and 1833 units/ml in nonresponders. The difference was not significant. In the responders, an example of which is illustrated in Figure 1a, the IPA level decreased significantly within the first 24 hours of treatment, and was within the normal range during ancrod administration. During the 14 days that ancrod was given, 72 of 93 IPA levels were in the normal range. The lowest levels attained are illustrated in Figure 3. The IPA levels were again abnormally high within 2 weeks of stopping treatment. In nonresponders, the mean IPA levels remained significantly elevated throughout treatment (Figs. 1b and 3); only 6 of 38 individual values were within the normal range. The difference between the two groups was highly significant ($\chi^2_{[1]} = 42.5$, $p < 0.001$).

The pretreatment VPA and IPA levels in individual patients and the maximum change observed during treatment with ancrod are summarized in Figure 4. In all the fibrinolysis responders, there was a clear inverse relationship between the VPA and IPA levels, and clear normalization of both VPA and IPA. In the fibrinolysis nonresponders, a slight inverse relationship between VPA and IPA levels was observed, but in no patient did both values normalize.

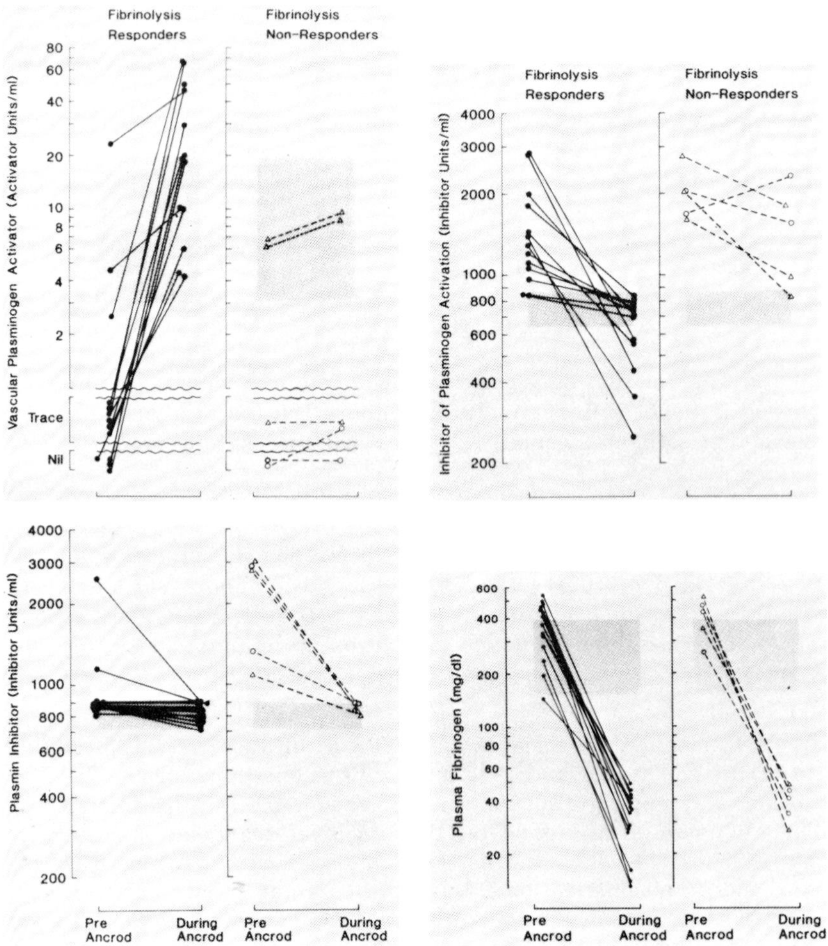

Figure 3 The VPA, IPA, PI, and fibrinogen levels in fibrinolysis responders and nonresponders. The values shown are prior to ancrod treatment and the maximum changes observed during ancrod treatment.

PI

In the 13 responders, most PI levels were all within the normal range before treatment, and changed little during treatment. In the two patients treated on a second occasion, the PI level was elevated before treatment, and rapidly became normal after ancrod was started. Only 4 of 19 individual values were elevated before ancrod was given; only

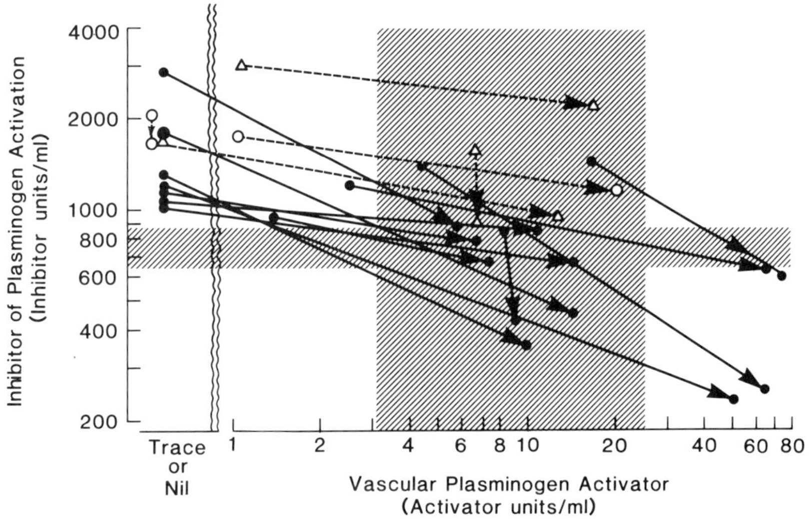

Figure 4 Levels of VPA and IPA in individual patients before treatment and the maximum change observed during ancrod treatment in fibrinolysis responders (closed symbols) and nonresponders (open symbols). The normal range is represented in the hatched area.

3 of 62 were elevated during treatment. The lowest values during treatment are summarized in Figure 3. In nonresponders, the mean PI level was 1914 inhibitor units/ml before treatment, a value significantly higher than that in responders. All eight individual values were elevated before ancrod. During ancrod treatment, the PI value decreased (Fig. 1b) in the first 4 days, fluctuated thereafter, and reached the normal range during the second week of treatment. Seventeen of 31 values during treatment were elevated. The difference between the responder and nonresponder groups was highly significant ($\chi^2_{[1]} = 22.5$, $p < 0.001$).

VPA Capacity

Since a functional assay was used to measure VPA levels, these levels reflect free VPA only; VPA complexed to IPA could be assessed only indirectly by the IPA levels. The significant decrease in IPA levels in the responders during treatment with ancrod may be an indication that relatively large amounts of VPA were complexed to IPA following VPA release from the endothelium. The capacity of the endothelium to synthesize and release VPA is, therefore, expressed by both free VPA levels and the degree to which the IPA levels decreased during ancrod infusion. On this basis, two subgroups of responders were defined; in five patients, arbitrarily called responders with a relatively low capacity, the VPA levels were within the normal range during ancrod administration in only 11 of 35 instances; in eight patients, treated on 10 occasions, arbitrarily called responders with a relatively high capacity, the VPA levels were within or above the normal range in most (50 of 66) instances. The proportion of plasma samples with VPA values within the normal range on days 1 through 14 was significantly different ($\chi^2_{[1]} = 18.8$, $p < 0.001$) in responders with relatively low and relatively high capacities to secrete VPA. In contrast, the proportion of both IPA ($\chi^2_{[1]} = 2.02$, $p > 0.10$) and PI ($\chi^2_{[1]} = 0.33$, $p > 0.70$) values within the normal range was similar in responders with a relatively high and a relatively low capacity.

Nonspecific Serine Proteinase Inhibitors and Other Findings

The actions of ancrod in increasing the levels of VPA and lowering those of the fibrinolytic inhibitors IPA and PI might be influenced by the catabolic rate and/or urinary loss of the drug in patients with proteinuria. There was no relationship between fibrinolytic response or nonresponse and the degree of proteinuria, hypoalbuminemia, or severity of renal failure.

The action of ancrod might also be influenced by the serine proteinase inhibitors. Plasma levels of the nonspecific serine proteinase inhibitors α_2-macroglobulin, α_1-antitrypsin, antithrombin III, and Cl esterase inhibitor were mostly within the normal range before ancrod treatment and were similar in fibrinolysis responders and nonresponders; there was no significant change during treatment (8).

Ancrod therapy was not associated with a significant change in platelet count. Platelet aggregation was measured in vitro in five patients with and without the aggregating agents epinephrine, ADP, and collagen (7). Prior to treatment with ancrod, platelet aggregation was normal in one patient; spontaneous aggregation was present in two, and increased sensitivity to small doses of the aggregating agents occurred in the other two. Platelet aggregation in vitro normalized in all four and become abnormal again after ancrod treatment was stopped. Elevated factor VIIIc levels decreased in four of five patients during treatment with ancrod (7).

Microvascular Thrombosis

Clinically, treatment with ancrod was associated with resolution of microvascular thrombosis, demonstrated in renal biopsy specimens 3 to 6 weeks after the start of treatment (Fig. 5). This was clearly evident only in the responders, in whom the mean score for the micro-

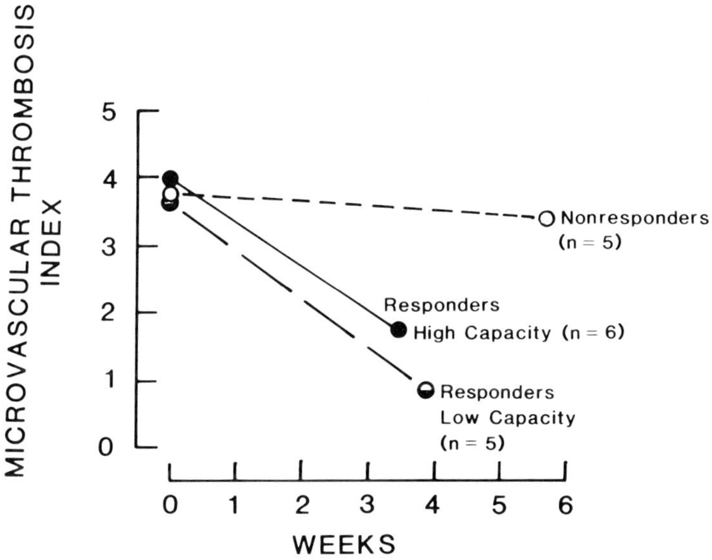

Figure 5 Mean microvascular thrombosis index before and after treatment with ancrod. The microvascular thrombosis index was the sum of the scores (each graded 0 to 4) for subendothelial deposits, for intraglomerular thrombi, and for arteriolar fibrinoid. (Reproduced with permission from Ref. 8.)

vascular thrombosis index decreased from 3.4 to 1.0; it decreased to a similar degree in those with a high (4.0 to 1.7) and those with a low (3.7 to 0.9) VPA capacity. In the nonresponders, thrombi persisted in glomeruli and/or renal veins (9); the mean score was 3.7 before and 3.5 after treatment.

DISCUSSION

The action of ancrod has hitherto been attributed to its specific effect on fibrinogen and the resulting anticoagulation and reduction in blood viscosity, which augments blood flow (1). Previous studies have provided indirect evidence to suggest recruitment of the fibrinolytic system (1), but the mechanism of this effect has been unclear.

We provide direct evidence of activation of fibrinolysis in all patients treated with ancrod (8,9). Prior to 20 courses of treatment in 18 patients with SLE, VPA was unmeasurable or present only in trace amounts in 15 of the patients, IPA was significantly elevated in all 18, and α_2-antiplasmin was significantly elevated in 5. The VPA levels were decreased and the IPA levels increased in the patient with SLE who had a "TTP"-like syndrome, and in patients with TTP who did not have SLE (12,13). In these patients, the level of α_2-AP was within the normal range. Treatment with ancrod resulted in prompt normalization of IPA in 15 of the 20 courses of treatment in SLE patients, including the patient with "TTP." The VPA levels were subject to daily fluctuations, and normalized rapidly in responders with a relatively high fibrinolysis capacity. All five nonresponders were patients with elevated levels of α_2-AP. Responsiveness to treatment was associated with effective thrombolysis in the microcirculation.

Synonymous with tissue-type plasminogen activator (t-PA), VPA is a serine proteinase synthesized and released by vascular endothelial cells (14,15). Its physiologic importance has been inferred from a close association between low plasma levels of VPA/t-PA and a propensity to thrombosis (8,12,13,16,17). The VPA/t-PA has a high affinity for fibrin (18,19), in both soluble and insoluble (20) forms, and its action is strikingly enhanced by fibrin (21-25). The presence of ancrod-fibrin in the circulation may, therefore, be an important factor triggering the release of functionally intact VPA/t-PA and its subsequent lytic effect. Rapid demand and consumption may lead to increased endothelial production de novo.

At the present time, the nature, origin, and physiologic importance of IPA are only speculative. It has been defined functionally by us as a substrate that inhibits urokinase-mediated plasminogen activation in vitro (6). The rapid normalization of elevated IPA levels that occurred concurrently with the increase in VPA levels is consistent with the view that VPA binds to IPA. We showed recently that IPA is directed against both urokinase and t-PA (13). Thus, it is most likely that IPA is identical with the novel fast-acting t-PA inhibitor described recently (26-31), which shares these properties.

α_2-Antiplasmin (α_2-AP), referred to as PI, is the most reactive and, physiologically, the major inhibitor of plasmin (32-34). In a dose-dependent fashion, α_2-AP interferes with the binding of plasminogen to fibrin (35-37) and becomes enzymatically cross-linked to fibrin (38), thereby rendering thrombi less susceptible to lysis by plasmin.

The sequence of events occurring during ancrod treatment may be analyzed tentatively as follows: in the presence of circulating ancrod-fibrin, VPA/t-PA, by virtue of its high affinity for fibrin, is released from its vascular stores and interacts with plasminogen bound to both the circulating, soluble ancrod-fibrin and fibrin precipitated pathologically in the form of thrombi in the microcirculation, converting plasminogen to plasmin, which in turn degrades fibrin. Degradation by plasmin of noncross-linked ancrod-fibrin occurred in both responders and nonresponders, as evidenced by an equal increase in fibrin degradation products in the two groups. Thus, ancrod-triggered release of VPA/t-PA from endothelial sites occurred effectively even in the nonresponders. However, their low levels of free VPA/t-PA with the resultant inability to neutralize IPA strongly suggest a limited endothelial capacity. In addition to this limiting factor, this group of patients was characterized by high levels of α_2-AP which, due to its interference with forming fibrin, results in clots resistant to lysis.

Although these studies are confined to 20 courses of treatment in 18 patients, the observations suggest that the effects of ancrod can be considered in at least three groups. That normalization of fibrinolysis was accompanied by striking evidence of resolution fo microvascular thrombosis provides insights into the action of ancrod and to the underlying mechanisms of damage in disease; this was reinforced by the absence of resolution of microvascular thrombosis associated with failure of normalization of fibrinolysis.

Any hypothesis for ancrod action must take into account the presence and extent of the clot, the amount of VPA released, the amount of VPA complexed to IPA (which is present in excess), the amount of VPA interacting with fibrin-bound plasminogen in clot lysis, and the level of PI (α_2-AP). This is illustrated schematically in Table 1. In most of the responders, the α_2-AP level was normal; it is assumed that VPA can be consumed in clot lysis. In responders, including the patient with "TTP," there was clot in glomerular capillaries and in arterioles; the volume of the clot was presumably relatively small. Responders with a relatively high capacity were considered to release 20 arbitrary units of VPA: 5 were consumed in clot lysis, 5 complexed to IPA, and 10 units were free and measurable in plasma. Responders with a relatively low capacity released 10 units of VPA: 5 were consumed in clot lysis, 5 complexed to IPA, and no free VPA could be measured.

In the nonresponders, the high levels of α_2-AP seemed to be an important determining factor. Although significant amounts of VPA may have been released, its action was restricted or limited because of less availability of plasminogen bound to fibrin. Three nonresponders also had renal vein thrombosis and, thus, a far larger volume of clot than did responders. Little VPA (2 arbitrary units) was consumed in clot lysis per unit of clot, but the total VPA consumed was high (10 units) because of the extent of the clot. The schema in Table 1 shows that a relatively low responder would consume all the VPA in clot lysis, there being no VPA to complex with IPA and no free measurable VPA. We cannot explain fully why certain nonresponders had very high IPA levels and measurable free VPA, but two had unusally severe hypertriglyceridemia, which might have interfered with the VPA-IPA interaction. This simplified hypothesis appears consistent with most of the observations. Clearly, the systems involved are extremely complex, and further studies will be required to clarify the many factors that may affect the manner in which ancrod acts.

Table 1 Hypothetical Schema to Explain the Observed Levels of Free VPA During Ancrod Treatment[a]

				VPA Units				
					Consumed in clot lysis			
Group	Capacity to release VPA	Extent of clotting (+ units)	PI α_2-AP	Released after Ancrod	Per unit of clot	Total	Complexed to IPA	New "free" in plasma
		[1]		[2]	[3]	[4=(1x3)]	[5]	[6] = [2-(4+5)]
Responder	Relatively high	+	Normal	20	5	5	5	10
Responder	Relatively low	+	Normal	10	5	5	5	0
Nonresponder	Relatively high	+++++	High	20	2	10	5	5
Nonresponder	Relatively low	+++++	High	10	2	10	0	0

[a]All units are arbitrary.

Source: Reproduced with permission form Ref. 8.

ACKNOWLEDGMENTS

We wish to thank Peggy Hunt, R.N., and the nursing staff of the General Clinical Research Center for their unfailing help with and excellent patient care; Mark Weiss, M.D., for the analysis of the renal histologic findings; Cheryl Allen, B.S., Jeff Agnew, B.S., Larry Healey, B.S., and Davis Stroop, B.S., for excellent technical help, and Violet Henry who kindly typed the manuscript.

This work was supported in part by Grants AM 17196, AM 20711, and RR 00068 from the National Institutes of Health, Bethesda, Maryland, by a grant from the Research and Education Funds of Dialysis Clinics, Inc., and by funds from Twyford Pharmaceuticals GmBH and Knoll Pharmaceutical Company.

REFERENCES

1. Bell, W. R. Defibrinogenating enzymes. In *Hemostasis and Thrombosis*, Colman, R. W., J. Hirsh, V. J. Marder, and E. W. Salzman (Eds.), J. B. Lippincott, Philadelphia, p. 1013 (1982).
2. Thomson, N. M., I. J. Simpson, and D. K. Peters. Clin. Exp. Immunol. 19:301 (1975).
3. Thomson, N. M., J. Moran, I. J. Simpson, and D. K. Peters. Kidney Int. 10:343 (1976).
4. Kant, K. S., V. E. Pollak, M. A. Weiss, H. I. Glueck, and M. A. Miller. Medicine 60:71 (1981).
5. Angles-Cano, E., Y. Sultan, and J. P. Clauvel. J. Lab. Clin. Med. 94:312 (1979).
6. Glas-Greenwalt, P., K. S. Kant, C. Allen, and V. E. Pollak. J. Lab. Clin. Med. 104:962 (1984).
7. Pollak, V. E., H. I. Glueck, M. A. Weiss, A. Lebron-Berges, and M. A. Miller. Am. J. Nephrol. 2:195 (1982).
8. Glas-Greenwalt, P., K. S. Kant, A. Dosekun, J. Frazier, C. Allen, and V. E. Pollak. J. Lab. Clin. Med. 105:99 (1985).
9. Kant, K. S., V. E. Pollak, A. Dosekun, P. Glas-Greenwalt, M. A. Weiss, and H. I. Glueck. J. Lab. Clin. Med. 105:77 (1985).
10. Haverkate, F. and P. Brakman. Fibrin plate assay. In *Progress in Chemical Fibrinolysis and Thrombolysis, Volume 1*, Davidson, J. F., M. M. Samama, and P. C. Desnoyers (Eds.), Raven Press, New York, p. 151 (1975).
11. Kluft, C. Thromb. Haemost. 41:365 (1979).
12. Glas-Greenwalt, P., K. S. Kant, A. K. Dosekun, and V. E. Pollak. Blood 62:243a (1983).
13. Glas-Greenwalt, P., J. M. Hall, T. W. Panke, K. S. Kant, C. M. Allen, and V. E. Pollak. Fibrinolysis in health and disease: the roles of tissue-type (vascular) plasminogen activator,

plasminogen activator inhibitor and protein C in thrombotic thrombocytopenic purpura. J. Lab. Clin. Med. 108:415 (1986).
14. Rijken, D. C., G. Wijngaards, and J. Welbergen. Biochemical and immunological characterization of plasminogen activator from human tissue. In *Progress in Chemical Fibrinolysis and Thrombolysis, Volume 4*, Davidson, J. E., V. Cepelak, M. M. Samama, and P. C. Desnoyers (Eds.), Churchill Livingstone, Edinburgh, London, New York, p. 349 (1980).
15. Rijken, D. C., G. Wijngaards, and J. Welbergen. Thromb. Res. 18:815 (1980).
16. Greenwalt, T. J., P. Glas-Greenwalt, A. K. Dosekun, K. S. Kant, and V. E. Pollak. Thromb. Haemost. 50:382 (1983).
17. Du, X-H., P. Glas-Greenwalt, K. S. Kant, C. M. Allen, S. Hayes, and V. E. Pollak. Nephrotic syndrome with renal vein thrombosis: Pathogenetic importance of a plasmin inhibitor (α_2-antiplasmin). Clin. Nephrol. 24:186 (1985).
18. Thorsen. S., P. Glas-Greenwalt, and T. Astrup. Thromb. Diath. Haemorrh. 28:65 (1972).
19. Wallen, P. Activation of plasminogen with urokinase and tissue activator. In *Thrombosis and Urokinase*, Paoletti, R. and S. Sherry (Eds.), Academic Press, London, p. 91 (1977).
20. Allen, R. A. and D. S. Pepper. Thromb. Haemost. 45:43 (1981).
21. Loskutoff, D. J. and L. Mussoni. Blood 62:62 (1983).
22. Camiolo, S. M., S. Thorsen, and T. Astrup. Proc. Soc. Exp. Biol. Med. 38:277 (1971).
23. Wallen, P. Chemistry of plasminogen and plasminogen activation. In *Progress in Chemical Fibrinolysis and Thrombolysis, Volume 3*, Davidson, J. F., R. M. Rowan, M. M. Samama, and P. C. Desnoyers (Eds.), Raven Press, New York, p. 167 (1978).
24. Hoylaerts, M., D. C. Rijken, H. R. Lijnen, and D. Collen. J. Biol. Chem. 257:2912 (1982).
25. Collen, D. Thromb. Haemost. 50:95 (1983).
26. Chmielewska, J., M. Ranby, and B. Wiman. Thromb. Res. 31:427 (1983).
27. Jahan-Vague, B., B. Moerman, F. De Cock, M. F. Aillaud, and D. Collen. Thromb. Res. 33:523 (1984).
28. Brommer, J. P., H. J. Verheijen, G. T. G. Chang, and D. C. Rijken. Thromb. Haemost. 52:154 (1984).
29. Wiman, B., J. Chmielewska, and M. Ranby. J. Biol. Chem. 259:3644 (1984).
30. Keuithof, E. K. O., C. Tran-Thang, A. Ransijn, and F. Bachmann. Blood 64:907 (1984).
31. Verheijen, H. J., G. T. G. Chang, and C. Kluft. Thromb. Haemost. 51:392 (1984).
32. Collen, D. Eur. J. Biochem. 69:209 (1976).

33. Moroi, M. and N. Aoki. J. Biol. Chem. 251:5956 (1976).
34. Mullertz, S. and I. Clemmensen. Biochem. J. 159:545 (1976).
35. Moroi, M. and N. Aoki. Thromb. Res. 10:851 (1977).
36. Aoki, N., M. Moroi, and K. Tachiya. Thromb. Haemost. 39:22 (1978).
37. Aoki, N., M. Moroi, Y. Sakata, M. Matsuda, and T. Tamaki. α_2-plasmin inhibitor: primary natural inhibitor of fibrinolysis. In *Progress in Fibrinolysis, Volume 5*, Davidson, J. F., I. M. Nilsson, and B. Astedt (Eds.), Churchill Livingstone, Edinburgh, London, New York, p. 144 (1981).
38. Sakata, Y. and N. Aoki. J. Clin. Invest. 65:290 (1980).

41

Effects of Subcutaneous Administration of Batroxobin in 62 Patients with Central Retinal Vein Occlusion

Jacqueline Conard, Annie Derlon,* and Meyer Michel Samama
Laboratoire Central d'Hématologie, Hospital Hôtel Dieu, Paris, France
Alain Gaudric, F. Lelong, and Gabriel J. Coscas
Hospital of Creteil-University of Paris XII, Creteil, France

INTRODUCTION

The medical treatment of retinal vein thrombosis is controversial and no drug—heparin (1), oral anticoagulants (2), or thrombolytic agents (3,4)—has really proven its efficacy for this purpose. Since blood viscosity may play a role in retinal vein occlusion (5,6), treatment with defibrinating agents could be beneficial. It has been shown that batroxobin (Defibrase) and ancrod (Arvin) induce a decrease in fibrinogen (7,8) and, consequently, in blood viscosity (9). In addition, an increase in fibrinogen-fibrin degradation products and a decrease in plasminogen have also been observed, but there was no increased fibrinolytic activity on fibrin plates (7). Ancrod has been previously used in a short series of eight patients with retinal vein occlusion, and the results were considered encouraging (10).

We have studied the effect of batroxobin (Defibrase Stago) on different parameters of coagulation and fibrinolysis in a group of 62 patients with various forms of central vein occlusion (ischemic or non-ischemic). A second group of 20 patients was included in a prospective randomized study comparing batroxobin and heparin; the clinical results of this study will be presented here.

---

*Present affiliation: Centre Hospitals-Universitaire, Caen, France

MATERIAL AND METHODS

Patients

All patients had a central retinal vein occlusion confirmed by fluorescein angiography. Duration of occlusion was less than 2 months. Patients older than 75 years of age were excluded, as were those with contraindications to anticoagulant treatment.

The functional result of the treatment was judged by the comparison of the visual acuity before and after a 3-week treatment and by fluorescein angiographic findings. A difference in visual acuity of 0.2 or more was considered a clinical improvement. Patients included in the open study had any type of capillaropathy, whereas in the randomized study, all patients had a nonischemic capillaropathy (11).

Treatments

Batroxobin (Defibrase Stago) was administered subcutaneously at a daily dose of 0.5 μg/kg/day for 2 days. The dose was then adjusted to maintain the fibrinogen level below 1 gm/l. Doses of Defibrase required were varied from patient to patient; in the prospective randomized study, total doses of 258 to 671 units were infused during the 21-day treatment, corresponding to 2.8 to 8.0 units/kg body weight.

Heparin was given subcutaneously twice a day at a dose to obtain a patient APTT twice that of the control value. Patients received a 3-week treatment.

Tests Performed

Tests were performed on 10 different days in each patient (before treatment and on days 1, 2, 3, 4, 5, 8, 11, 14, and 21). Blood was drawn on 3.8% trisodium citrate (1 vol/9 vol). For measurement of plasma viscosity, EDTA was used as an anticoagulant.

The following tests were done: fibrinogen by the Clauss method (12); plasma viscosity using a Fica capillary viscometer; fibrinogen-fibrin degradation products (FDP) by the method of Merskey and coworkers (13); soluble complexes by the method developed by Largo and associates (14) using FS Prest Stago; α_2-antiplasmin by amidolytic (Coatest antiplasmin Kabi) or by the Laurell method; histidine-rich glycoprotein (HRGP) antigen by the Laurell method using an antiserum from Behringwerke; plasminogen by the amidolytic method (S-2251 Kabi); plasmin-antiplasmin complexes detection by crossed-immunoelectrophoresis; and antithrombin III (AT III) by the amidolytic method (Coatest AT III Kabi).

In addition, in a small group of patients, amidolytic α_2-macroglobulin assay (15), cross-linked fibrin derivative determination (Dimertest Mab Co.), and tissue plasminogen activator (t-PA) antigen measurement (kit from Biopool) were performed.

RESULTS

Laboratory Results

Fibrinogen

Before treatment, the fibrinogen level was normal in 56 patients and above the normal limit (>4.50 gm/l) in 6 patients (10%). As expected, there was a significant decrease after initiation of Defibrase treatment, and on day 3, 56 of 62 patients had a fibrinogen level below 1 gm/l (mean 0.76 gm/l). Results are shown in Table 1. It should be pointed out that after day 5, blood samples were drawn just before the injection of Defibrase, so that the results on days 8, 11, 14, and 21 do not reflect the maximal defibrinating effect of the treatment. On day 21, the fibrinogen level was still below 1 gm/l in 14 patients, indicating that there was no acquired resistance to Defibrase after a 3-week treatment in these patients.

FDP and Cross-Linked Fibrin Degradation Products

Before treatment, a borderline FDP level was observed in four patients. During treatment, FDP levels were above normal limits in all patients on days 1 to 5, and normal or elevated on other days. The FDP levels reached 1024 μg/ml or more in several patients, the largest number on day 2 (16).

Cross-linked degradation products of fibrin were moderately increased in three patients studied on five different days each (200 to 500 ng/ml in most cases). Variations were parallel to those of FDP.

Soluble Complexes

Before treatment the test was negative in all patients, but two tests were slightly positive, one associated with increased FDP, and one with normal FDP. During Defibrase, a positive test was frequent on the first days, occurring in 71% to 92% of the patients between days 1 and 4.

Plasma Viscosity

Before treatment, plasma viscosity ranged between 1.13 and 1.64 cs. A significant ($p < 0.05$) decrease was observed during treatment (on day 2) and was associated with the decrease in fibrinogen (correlation coefficient 0.47 on day 2). Plasma viscosity was below 1.20 cs when fibrinogen level decreased below 1 gm/l.

Antithrombin III and α_2-Macroglobulin

These parameters were normal before and during treatment (Table 1).

Table 1 Results During Defibrase Treatment

	Before treatment	Day 1	Day 2	Day 3	Day 4	Day 5	Day 8	Day 21
Fibrinogen (gm/l)	3.2 ± 0.8	2.0 ± 0.7	0.8 ± 0.6	0.5 ± 0.4	0.7 ± 0.3	0.9 ± 0.4	1.0 ± 0.5	1.2 ± 0.7
(n = 62)	(2.0-5.6)	(0.4-3.9)	(0-2.5)	(0-1.4)	(0-1.6)	(0-2.5)	(0-2.5)	(0.3-3.2)
Viscosity (cs)	1.29 ± 0.11	1.22 ± 0.07	1.20 ± 0.07	1.18 ± 0.07	1.18 ± 0.07	1.18 ± 0.07	1.21 ± 0.07	1.24 ± 0.08
(n = 62)	(1.15-1.64)	(1.11-1.38)	(1.07-1.38)	(1.07-1.45)	(1.07-1.36)	(1.06-1.36)	(1.06-1.33)	(1.13-1.45)
Plasminogen (%)	102 ± 18	87 ± 22	63 ± 18	53 ± 14	55 ± 13	61 ± 15	64 ± 16	72 ± 20
activity (n = 62)	(60-162)	(53-160)	(34-113)	(31-87)	(32-81)	(28-96)	(28-103)	(38-134)
α2-Antiplasmin (%)		72 ± 23	39 ± 21	38 ± 24	42 ± 20	48 ± 21	50 ± 20	57 ± 27
activity (n = 62)	105 ± 18							
antigen (n = 62)	102 ± 21	72 ± 21	45 ± 14	37 ± 13	39 ± 14	40 ± 16	41 ± 18	46 ± 19
α2-M activity (%)	92 ± 18	95 ± 22	98 ± 27	97 ± 22	97 ± 29	90 ± 17	91 ± 19	87 ± 20
(n = 19)								
HRGP antigen (%)	131 ± 34	129 ± 35	136 ± 35	144 ± 42	153 ± 43	151 ± 38		
(n = 15)								
AT III activity (%)	100 ± 14	94 ± 17	95 ± 14	97 ± 14	97 ± 15	97 ± 12	100 ± 14	103 ± 13
(n = 42)								
FDP < 16 μ/ml	58	0	0	0	0	1	5	12
16 to 128	4	11	5	9	24	29	28	22
>128	0	42	52	53	35	23	26	15
Soluble complexes								
negative ±	48	6	4	10	14	14	20	16
positive	2	40	45	40	35	26	28	19

Data are presented as mean values ± SD.
AT III = antithrombin III; FDP = fibrinogen-fibrin degradation products; HRGP = histidine-rich glycoprotein.

Fibrinolysis Parameters

The decrease in α_2-antiplasmin was more pronounced than the decrease in plasminogen, with mean levels of 38.4% and 52.9%, respectively, on day 3 and 57.1% and 72.1% on day 21.

Cross-immunoelectrophoresis using an antiserum against α_2-antiplasmin was done in six patients. Two peaks were present during treatment, one corresponding to free antiplasmin and the other to plasmin-antiplasmin complex.

The HRGP was measured in 15 patients only. Before treatment, HRGP levels were higher than normal laboratory values (97 ± 22%) and remained high during Defibrase. Treatment did not modify t-PA antigen in the patients studied.

Clinical Results

Clinical results of the prospective study, including 10 patients treated with Defibrase and 10 patients who received heparin, are summarized in Table 2. No significant difference in visual acuity was observed between the two series.

Fluorescein angiography remained unchanged in eight patients treated with Defibrase. In another patient, venous occlusion deteriorated with an increase in retinal hemorrhages. In the last patient, macular bleeding resorbed, leading to a marked improvement of visual acuity (0.1 to 0.8). In heparin-treated patients, fluorescein angiography was unchanged in nine patients and, in the other patient, macular bleeding resorbed (visual acuity <0.05 before treatment and 0.8 after heparin). No side effects were observed in either group, except itching at the sites of infection in a few instances.

Doses of Defibrase required to decrease the fibrinogen level below 1 gm/l in this series of 10 patients showed some variation. The total dose infused during the 21-day treatment ranged between 258 and 671 units (i.e., 2.8 to 8.0 units/kg body weight).

DISCUSSION

Retinal vein occlusion was first described by Michel in 1878 (in reference 5), but its pathophysiology is still unclear and the presence of definite venous thrombosis in retinal vein occlusion has been questioned (16). It is suggested that the pathological process may be due to changes in the vessel walls, blood constituents, and blood flow. An increase in viscosity has been reported by different groups (5,17). Trope and co-workers (6) found that viscosity, hematocrit, fibrinogen, fibrinopeptide A, and β-thromboglobulin were increased in patients with capillary nonperfusion or new vessels, but not in patients without ischemic complications. In this latter group, the only

Table 2 Clinical Results of the Prospective Study

	Batroxobin (n = 10)	Heparin (n = 10)
Mean age (yr) (range)	59.0 (44-72)	53.9 (40-70)
Duration of central retinal vein occlusion	2.8 weeks (1-8)	2.1 weeks (1-8)
Visual acuity		
before treatment	0.25 (<0.05 to 0.6)	0.33 (<0.05 to 0.7)
end of treatment	0.31 (<0.05 to 0.8)	0.46 (<0.05 to 10/10)
No change	5	4
Improvement (⩾0.2)	3	4
Deterioration (⩾0.2)	2	2
Fluorescein angiography	No change in 8	No change in 9

abnormality was a lower level of antithrombin III. Abnormal fibrinolysis has been found in patients with retinal vein occlusion, with a decreased fibrinolytic response to venous stasis (18,19). Antithrombin III, fibrinogen, and plasminogen antigen were normal (19).

The rationale for using Defibrase in the treatment of retinal vein occlusion was to decrease blood viscosity (20). In our prospective randomized study comparing Defibrase and heparin, no clinical benefit of Defibrase was observed in the short series of patients with non-ischemic capillaropathy. These results may be due to insufficient defibrination, to the type of occlusion selected, or to inefficacy of the drug. However, treatment was associated with markedly altered laboratory parameters.

Defibrase is a defibrinating venom with thrombin-like activity. In the 62 patients studied, we obtained a decrease in fibrinogen associated with a decrease in plasma viscosity. A fibrinogen level below 1 gm/l was required to induce a decreased viscosity below 1.20 cs. All patients responded to the injection of Defibrase with a defibrinating effect. A resistance could be expected after several days of subcutaneous treatment; however, after 21 days, 14 patients had still a fibrinogen level below 1 gm/l. Development of resistance has been reported to occur after a mean of 33 days (21).

Plasminogen and α_2-antiplasmin decreased during treatment, the latter being the lower. The levels of α_2-antiplasmin antigen and activity were rather similar before treatment and on days 1 through 4, while α_2-antiplasmin activity tended to be higher than the antigen on the following days. Plasmin-antiplasmin complexes, FDP, cross-linked fibrin derivatives, and soluble complexes were also found in various amounts on different days of treatment. In contrast, AT III, α_2-macroglobulin, and HRGP remained unchanged.

Defibrase, like other thrombin-like enzymes, such as ancrod, splits fibrinopeptide A (not B) from fibrinogen. It results in removal of fibrinogen from circulating blood. Simultaneously, soluble fibrin complexes (detected by the Largo test) appear, as do cross-linked fibrin derivatives (detected by Dimertest). The subsequent steps in the action of Defibrase are not completely understood. Soluble fibrin monomers may facilitate fibrinolysis by activation of plasminogen to plasmin. This could explain the observed decrease in plasminogen and in α_2-antiplasma in formation of plasmin-antiplasmin complexes, and the increase in FDP. However, in a study with Arvin, no increased systemic fibrinolytic activity could be shown on fibrin plates (7). Different mechanisms of plasminogen activation have been reported. Activation of plasminogen may be induced by release of tissue plasminogen activator (t-PA), but plasma t-PA antigen was not increased in our patients. One explanation could be a t-PA-independent activation. Factor XII pathway activation was not investigated in this study.

Whatever their mechanism, modifications induced by Defibrase can be compared with those observed during thrombolytic treatments with streptokinase (SK) or urokinase (UK). Treatments with the three drugs are associated with a decrease in fibrinogen, α_2-antiplasmin, and plasminogen, and an increase in FDP. However, the frequency of bleeding is different. No minor or severe bleeding was observed during subcutaneous administration of Defibrase for 21 days in 62 patients. These results are in good agreement with those of the multicenter trial reported by Latallo (21), who observed that incidence of bleeding was low, the subcutaneous route of administration being even less hemorrhagic than the intravenous one. Thus, one may wonder why SK or UK are associated with an increased risk of bleeding. As previously reported for SK (22), the degree of fibrinogen-lowering and most tests are of little help in predicting bleeding. Recently, the influence of plasmin on platelets has been suggested (23). During Defibrase treatments, platelets may not be altered and systemic fibrinolytic activity is not increased, as it is in thrombolytic treatments.

CONCLUSION

We have confirmed the defibrinating effect of Defibrase after subcutaneous administration in 62 patients with venous retinal vein occlusion. The tolerance to the drug was good. Coagulation and fibrinolytic parameters were studied on 10 different days during the 21-day treatment. The expected defibrination was obtained, and an increase in FDP, soluble complexes, and cross-linked fibrin derivatives was observed. In contrast to thrombolytic substances like SK or UK, there was no bleeding. Greater clinical efficacy of Defibrase as compared with heparin in retinal vein occlusion, characterized by a non-ischemic capillaropathy, was not shown in our short series of patients.

REFERENCES

1. Vanna, S. and H. Orma. Experience of treating retinal venous occlusion with anticoagulants and antisclerosis therapy. Arch. Ophthal. Chicago 58:812-828 (1957).
2. Thorne, B. and M. B. Thorne. Anticoagulants in the treatment of retinal veins thrombosis. Proc. Roy. Soc. Med. 54:856-859 (1961).
3. Rossmann, H. Treatment of retinal vascular occlusions by means of fibrinolysis. Postgrad. Med. J. Suppl. 5:105-108 (1973).
4. Kwann, H. C., J. G. Dorme, C. L. Fetkenhour, and R. D. Levin. Thrombolytic therapy of central retinal vein occlusions. In *New Concepts in Streptokinase Dosimetry*, Martin, M., W. Schoop, and J. Hirsch (Eds.), Hans Huber Publishers, Bern, 37:221 (1978).
5. Ring, C. P., T. C. Pearson, M. D. Sanders, and G. Wetherley-Mein. Viscosity and retinal vein thrombosis. Br. J. Ophthalmol. 60:397-410 (1976).
6. Trope, G. E., G. D. O. Low, B. M. Mc Ardle, J. T. Douglas, C. D. Forbes, C. W. Prentice, and W. S. Foulds. Abnormal viscosity and haemostasis in long-standing retinal vein occlusion. Br. J. Ophthalmol. 67:137-142 (1983).
7. Bell, W. R., W. R. Pitney, and J. F. Goodwin. Therapeutic defibrination in the treatment of thrombotic disease. Lancet 1:490-493 (1968).
8. Latallo, Z. S. Report of the task force on clinical use of snake venom enzymes. Thromb. Haemost. 39:768-774 (1978).
9. Weaver, J. P. A., A. Evans, and D. N. Walder. The effect of increased fibrinogen content on the viscosity of blood. Clin. Sci. 3:1-10 (1969).
10. Bowell, R. E., V. J. Marmion, and C. F. McCarthy. Treatment of central retinal vein thrombosis with ancrod. Lancet 1:173-174 (1979).
11. Coscas, G. and P. Dhermy. *Occlusions Veineuses Rétiniennes. Rapport de la Société Francaise d'Ophtalmologie*, Masson, Paris (1978).
12. Clauss. Gerinnungsphysiolofische schnellmethode zur bestimmung des fibrinogens. Acta Haematol. (Basel) 17:237-246 (1957).

13. Merskey, C., P. Lalezari, and A. J. Johnson. A rapid, simple, sensitive method for measuring fibrinolytic products in human serum. Proc. Soc. Exp. Med. 131:871-875 (1969).
14. Largo, R., V. Heller, and P. W. Straub. Detection of soluble intermediates of the fibrinogen-fibrin conversion using erythrocytes coated with fibrin monomers. Blood 47:991-1002 (1976).
15. Latallo, Z. S., E. Teisseyre, and S. Lopaciuk. Assessment of plasma fibrinolytic system with use of chromogenic substrate. Haemostasis 7:150-154 (1978).
16. Kohner, E. M. and J. S. Shilling. Retinal vein occlusion. In *Medical Ophthalmology*, Rose, F. C. (Ed.), Chapman & Hall, London (1976).
17. McGrath, M. A., F. Wechsler, A. B. L. Hunyor, and R. Penny. Systemic factors contributing to retinal vein occlusion. Arch. Intern. Med. 138:216-220 (1978).
18. Pedruzzi, M., V. Derosa, S. Fonda, and S. Coccheri. Haemostatic studies in retinal vien occlusion. Fibrinolytic response to venostasis as a prognostic factor for spontaneous recanalization. Thrombos. Res. 24:105-118 (1981).
19. Lorient-Roudaut, M. F., C. Vergnes, J. Bonnet, M. J. Le Rebeller, M. Boisseau, and H. Bricaud. Hémostase et occlusions veineuses de la rétine. A propos de 40 patients à la phase aiguë. Méd. Hyg. 42:4055-4058 (1984).
20. Gaudric, A., O. Tolub, G. Coscas, J. Conard, M. Samama, and F. Lelong. Traitement médical des occlusions veineuses rétiniennes par un nouveau défibrinant (Defibrase). Etude préliminaire. Bull Soc. Ophtal. France 82:79-82 (1982).
21. Latallo, Z. S. Retrospective study on complications and adverse effects of treatment with thrombin-like enzymes. A multicentre trial. Thromb. Haemostas. 50:604-609 (1983).
22. Conard, J., M. Samama, P. Milochevitch, M. H. Horellou, B. Chabrun, and J. Prestat. Complications hémorragiques au cours de 98 traitements par la streptokinase. Place de la surveillance biologique. Nouv. Presse Méd. 8:1319-1325 (1979).
23. Adelman, B., J. Loscalzo, A. D. Michelson, and R. I. Handin. Fibrinolytic therapy inhibits platelet, von Willebrand factor interactions. Blood 62(Suppl. 1):296 (1983).

42

Therapeutic Efficacy of Thrombin-like Snake Venom Enzyme (Batroxobin) in Peripheral Vascular Disease

Shin Ishimaru, Kinichi Furukawa, and Michio Fujimaki
Tokyo Medical College, Tokyo, Japan

INTRODUCTION

There is still no unanimity of opinion on the relative advantages of thrombectomy using Fogarty catheters or fibrinolytic therapy for deep vein thrombosis. Treatment for ischemic ulcer of the foot caused by chronic occlusion of the peripheral arteries is also difficult. In this chapter, the efficacy of a thrombin-like enzyme purified from *Bothrops atrox* venom (batroxobin) as a postoperative cotherapy for deep vein thrombosis and on the healing of ischemic ulcers was investigated from basic and clinical points of view.

METHODS

Ischemic Ulcer of the Foot

From 1982 to 1983, a multicenter study was carried out in Japan on 85 cases of ischemic ulcer in which surgery was not indicated because of peripheral arterial occlusion from arteriosclerosis obliterans (ASO) or thromboangiitis obliterans (Buerger's disease). The efficacy of batroxobin (Defibrase, Pentapharm) in ulcer healing after 4 weeks' treatment was evaluated by photographs taken before and every two weeks during treatment. No laboratory data except plasma fibrinogen level were investigated.

Deep Vein Thrombosis

Revascularization of the deep vein by thrombectomy using a Fogarty catheter was performed in 72 of 104 patients suffering from deep vein thrombosis in our hospital. Ten of 72 patients were treated with a combination of batroxobin (Defibrase, Pentapharm) and urokinase after operation. Seventeen patients were treated with a combination of batroxobin, urokinase, and human placental lys-plasminogen after operation. These patients were divided into two groups as follows:

Group I: 10 BU (batroxobin units) of batroxobin was given immediately after operation. Beginning on the next day, 60 mg of lys-plasminogen, 120,000 IU of urokinase, and 10 BU of batroxobin were given in that order every day for 5 to 7 days. Conventional anticoagulant therapy with warfarin (coumarin) was followed for 6 months after the combination therapy.

Group II: Same doses of batroxobin, lys-plasminogen, and urokinase as in Group I were given in this order every day after operation for 5 to 7 days. Warfarin was given for 6 months. In the other 45 patients, urokinase was given as cotherapy after operation.

Changes in the fibrinolytic system were evaluated by plasminogen level (single radial immunoassay), α_2-plasmin inhibitor level (chromogenic substrate S-2251), and plasmin (plasminogen-free fibrin plate method). The efficacy of treatment 6 months after operation was graded in three groups (excellent, good, and poor). The patients having a patent deep vein were graded excellent. The patients graded good had a rethrombosed deep vein, but good collaterals and almost no clinical symptoms related to venous occlusion. In the patients graded poor, their clinical symptoms still remained and the deep vein was completely occluded.

RESULTS

Ischemic Ulcer of the Foot

Improved ulcer healing was demonstrated by serial photographic observation in 66 (78%) of 85 patients. In 53 patients in whom the fibrinogen level was reduced below 120 mg/100 ml, 93% of ulcers were improved. In contrast, ulcer healing was observed in only 53% of patients when the minimum level of fibrinogen was above 120 mg/100 ml. Bleeding complications were observed in two patients from gastric ulcer and hemorrhoids (Table 1).

Deep Vein Thrombosis

Thirty-three of 45 patients were operated on within 7 days after onset of their clinical symptoms, and were treated with urokinase after operation. In 21 (64%) of these patients, the postoperative course was

Table 1 Efficacy of Defibrinogenation in the Treatment of Ischemic Ulcer of the Food

Ulcer healing	Fibrinogen >120 mg/100 ml	Fibrinogen ≦120 ml/100 ml
Excellent	4 (13%)	32 (60%)
Good	7 (22%)	13 (25%)
Fair	6 (19%)	4 (8%)
(Subtotal)	(17 [53%][a])	(49 [93%][a])
Poor	15 (47%)	4 (7%)
(Total)	(32)	(53)

[a] $p < 0.01$.

There were two bleeding complications—gastric ulcer in one patient and hemorrhoids in one other patient. Pronounced efficacy of ulcer healing was demonstrated when fibrinogen level was below 120 mg/100 ml.

graded as excellent. The remaining 12 patients were operated on within 8 to 14 days after onset, and also treated with urokinase postoperatively. An excellent postoperative course was obtained in 5 (42%) of these patients. Three patients were graded as poor (Table 2). The postoperative course of 17 patients treated with a combination of batroxobin, lys-plasminogen, and urokinase was monitored for 6 months. There were no patients graded as poor in this series, which had been operated on within 14 days of onset. Twelve (71%) of 17 patients were graded as excellent, and the remaining 5 as good. There was no significant difference in the postoperative course between those treated within 7 days and those treated from 8 to 14 days (Table 3).

Phlebograms taken before and 6 months after operation showed complete patency of the iliofemoral vein (Figs. 1A and B). Although thrombectomy was not completely undertaken because of organized thrombus, mainly located at the iliac vein bifurcation, partial patency of the reconstructed vein was maintained and blood flow increased through the collateral veins branching from it (Figs. 2A and B).

The plasma fibrinogen level was markedly reduced in patients given batroxobin, whether combined with lys-plasminogen or not (Fig. 3).

After infusion of batroxobin and urokinase, plasminogen level decreased by approximately 40% compared with the pretreatment levels. Reduction in the plasminogen level after infusion of batroxobin, lys-plasminogen and urokinase was not statistically significant.

Table 2 Postoperative Prognosis of Deep Vein Thrombosis

Onset[a] (days)	No. of patients	Postoperative prognosis		
		Excellent	Good	Poor
0-7	33	21	12	0
8-14	12	5	4	3
(Total)	(45)	(26 [58%])	(15)	(3)

[a]Onset = period from onset of symptoms until operation.

Also after infusion of batroxobin and urokinase, the α_2-plasmin inhibitor level decreased by 50% of the pretreatment levels. Marked reduction in α_2-plasmin inhibitor was detected in patients treated with batroxobin, lys-plasminogen, and urokinase as compared with batroxobin and urokinase (Fig. 4). Plasmin (fibrinolytic activity) measured by plasminogen-free fibrin plate was detected in 10% of the blood samples taken after urokinase infusion, and 18% of samples taken after batroxobin and urokinase administration. In the patients treated with batroxobin, lys-plasminogen, and urokinase in that order every day (group II), plasmin was detected in 10% of the blood samples after urokinase administration. In group I in which batroxobin was given more than 12 hours prior to lys-plasminogen and urokinase, positive fibrinolytic activity was detected in 90% of the blood samples after urokinase administration (Fig. 5).

Table 3 Postoperative Prognosis of Deep Vein Thrombosis Treated by Thrombectomy Combined with Batroxobin, Lys-Plasminogen, and Urokinase

Onset (days)	No. of patients	Postoperative prognosis		
		Excellent	Good	Poor
0-7	13	9	4	0
8-14	4	3	1	0
(Total)	(17)	(12 [71%])	(5)	(0)

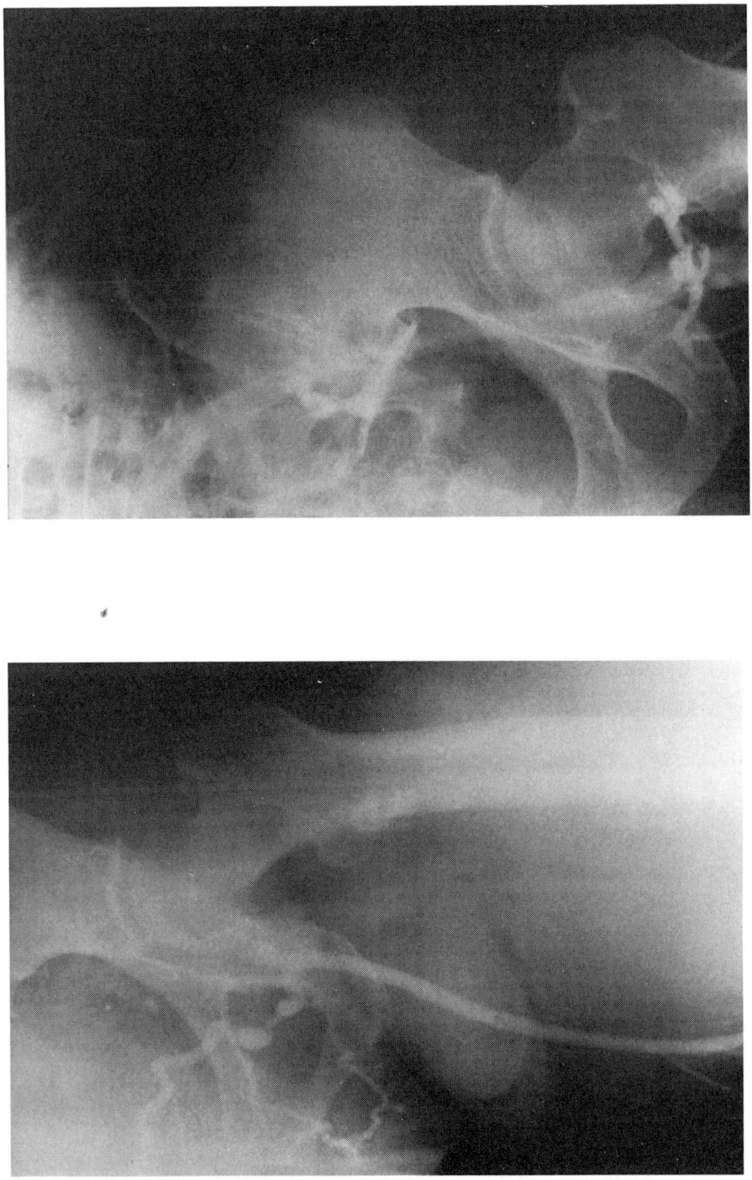

A B

Figure 1 Phlebograms of a patient with iliofemoral vein thrombosis. (A) Before operation. (B) Complete patency of the reconstructed vein 6 months after operation.

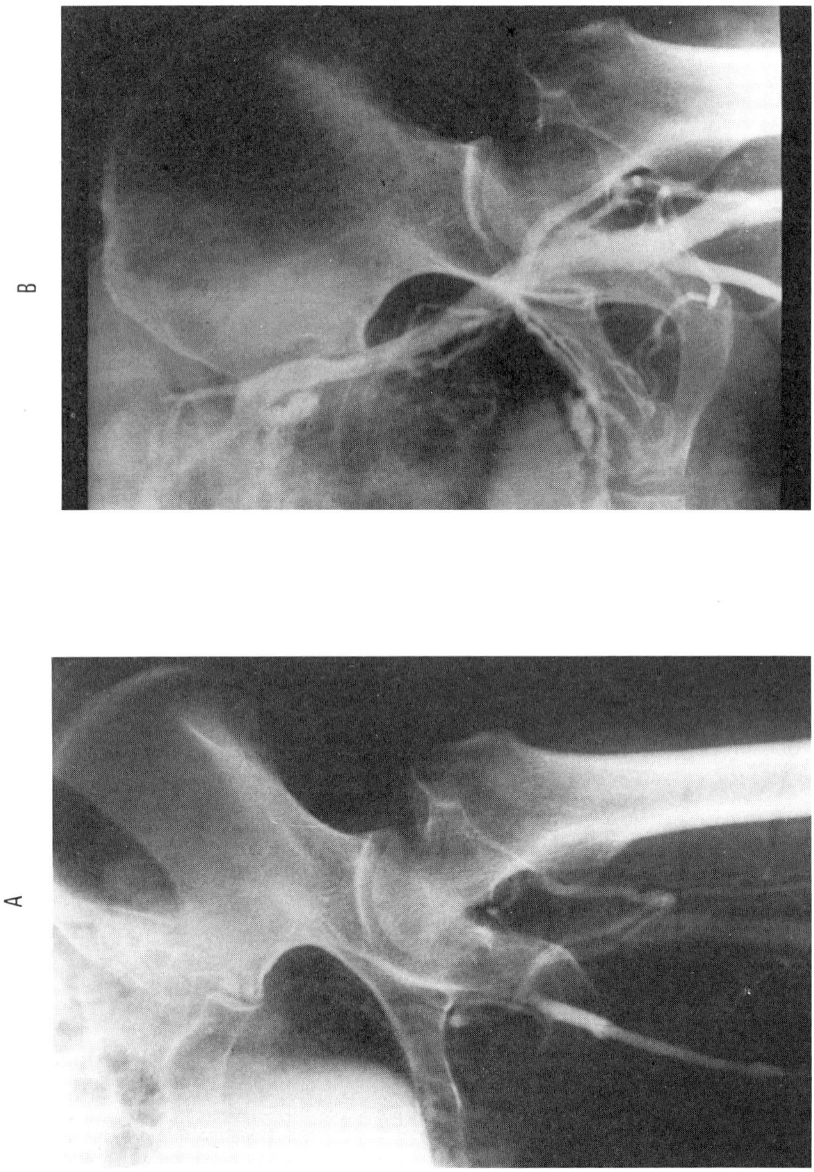

Figure 2 Phlebograms of a patient with iliofemoral vein thrombosis. (A) Before operation. (B) Partial patency of the reconstructed vein and good collateral vessels 6 months after operation.

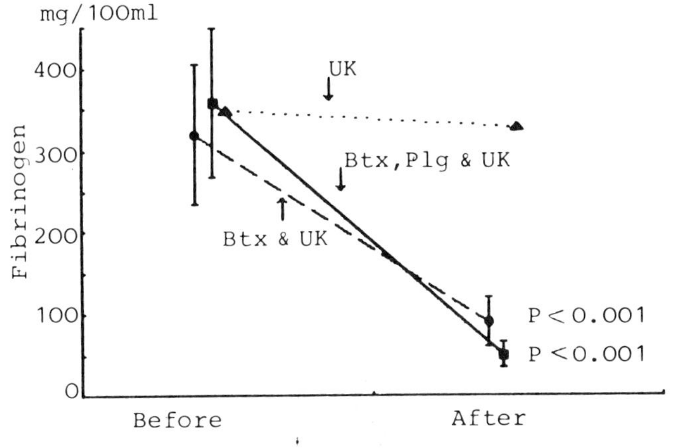

Figure 3 There was a marked decrease in fibrinogen level in patients given batroxobin, whether combined with lys-plasminogen or not.

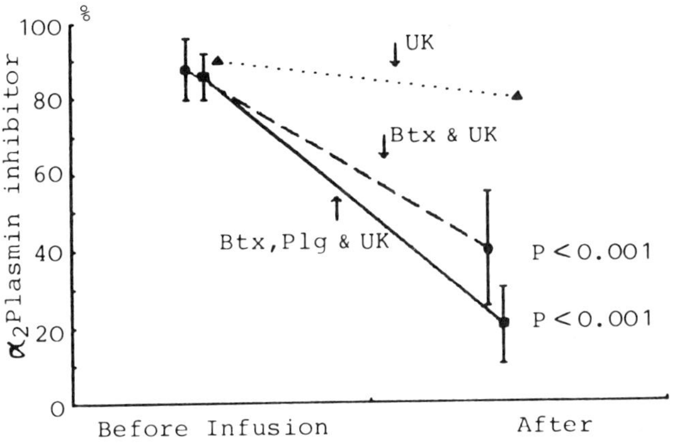

Figure 4 There was a marked decrease in α_2-plasmin inhibitor in patients given batroxobin, lys-plasminogen, and urokinase.

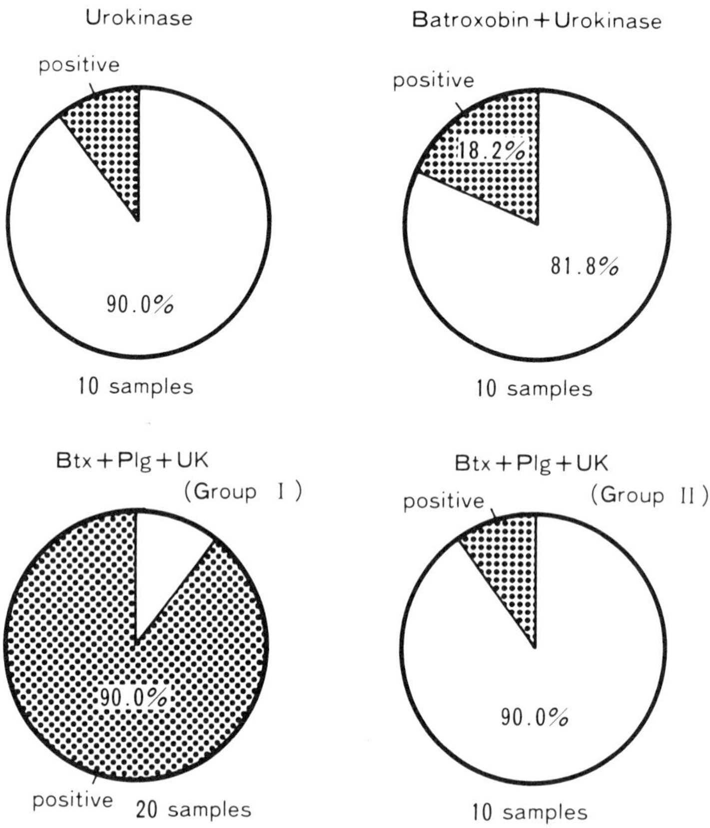

Figure 5 Positive fibrinolytic activity was detected by the plasminogen-free fibrin plate method in 90% of the blood samples after urokinase infusion in patients treated with batroxobin more than 12 hours before treatment with lys-plasminogen and urokinase (Group I).

DISCUSSION

Batroxobin is a defibrinogenating enzyme purified from *Bothrops atrox* venom (Stocker, 1978). It is noteworthy that microcirculation can be improved with this enzyme, inducing low blood viscosity by reducing plasma fibrinogen level (Dormandy and Rein, 1978). A multicenter trial in Japan on batroxobin applied to ischemic ulcers of the lower extremities indicated that a reduction in the diameter of ulcer in 78% of the patients after 4 weeks of treatment. Moreover, pronounced efficacy of ulcer healing was demonstrated when the fibrinogen level was below 120 mg/100 ml. These results suggest that defibrinogenation with batroxobin can be a favorable treatment for chronic peripheral arterial disease without major complications by reducing the fibrinogen level to about 100 mg/100 ml. Our animal experiments, previously reported, indicated the importance of platelets to ensure hemostasis (Ishimaru et al., 1981). The batroxobin-induced abnormal fibrin (des A fibrin) is unstable and more susceptible to human plasmin, thus it promptly degraded (Laurent and Blombäck, 1958). Consequently, plasma fibrinogen is selectively removed from the blood circulation, leading to a decrease in plasminogen and α_2-plasmin inhibitor. The fibrinolytic system might be activated by abnormal fibrin monomers that activated plasminogen in the vessel wall (Katou, 1982). Concerning the reason why thrombolysis can be easily activated under conditions of low α_2-plasmin inhibitor (Aoki et al., 1977), it is conceivable that fibrinolytic effect of urokinase can be enhanced by the combination of batroxobin with urokinase. Moreover, additional lys-plasminogen prior to urokinase administration may possibly correct the low plasminogen level caused by batroxobin. Because of disagreement among various reports concerning the thrombolytic effect of urokinase, combined therapy using batroxobin, human placental lys-plasminogen (Lormeau et al., 1976) and urokinase was investigated in this chapter to evaluate whether the fibrinolytic effect of urokinase could be enhanced by the cotherapy.

It is obvious that α_2-plasmin inhibitor was markedly decreased after batroxobin, lys-plasminogen, and urokinase therapy as compared with batroxobin and urokinase therapy. This indicated that the fibrinolytic system was strongly activated by administration of lys-plasminogen, rather than urokinase alone.

In the patients treated with batroxobin, lys-plasminogen, and urokinase in that order, fibrinolytic activity measured by the plasminogen-free fibrin plate method was detected in only 10% of the blood samples taken after urokinase infusion. In contrast, fibrinolytic activity was consistently noted after urokinase infusion in patients given batroxobin more than 12 hours prior to lys-plasminogen and urokinase administration. These results indicated that dynamic changes in the fibrinolytic system caused by urokinase might be concealed when batroxobin is given immediately prior to the administration of lys-plasminogen and urokinase because the fibrinolytic system is affected, especially in the initial period of batroxobin infusion.

It is concluded that even moderate dosages of urokinase (such as 120,000 IU) can induce fibrinolytic activity under the condition of low antiplasmin levels caused by batroxobin given several hours previously. Additional lys-plasminogen given before urokinase infusion might also be effective in activating fibrinolysis.

Even if complete removal of the thrombus was impossible due to delayed surgery or organic iliac compression, clinical manifestations can be improved by potent fibrinolytic therapy to maintain the partial patency of the vein over the long-term and ensure effective collateral blood flow. Our clinical experiences indicated that thrombectomy was preferable for treatment of deep vein thrombosis within 7 days of onset. The patency of incompletely reconstructed veins can be maintained by postoperative combined therapy in patients operated on within at least 14 days of onset.

REFERENCES

Aoki, N., M. Moroi, M. Matsuda, and K. Tachiya. (1977). Behavior of $\alpha_2$plasmin inhibitor in fibrinolytic states. J. Clin. Invest. 60: 361-369.

Dormandy, J. A. and H. Reid. (1978). Controlled defibrination in the treatment of peripheral vascular disease. Angiology 29:80-85.

Ishimaru, S., E. Berglin, H. A. Hansson, A. C. Teger-Nisson, and G. William-Olsson. (1981). Haemostatic clot formation at anastomosis of synthetic venous graft in defibrinogenated dogs: A scanning electron microscopic study. Thromb. Haemost. 45:276-281.

Katou, M. (1982). Studies on fibrinolysis induced by batroxobin infusion. J. Tokyo. Med. Coll. 40:605-618.

Laurent, T. C. and B. Blombäck. (1958). On the significance of the release of two different peptides from fibrinogen during clotting. Acta Chem. Scand. 12:1875-1881.

Lormeau, J. C., J. Choay, J. Goulay, E. Sache, and E. G. Vaurel. (1976). Purification et propietes des plasminogenes extracts du placenta human. Ann. Pharm. Franc. 34:287-296.

Stocker, K. (1978). Defibrinogenation with thrombin-like snake venom enzymes. In *Handbook of Experimental Pharmacology*, Markwardt, F. (Ed.), Springer-Verlag, Berlin, pp. 451-484.

43

Effects of Defibrase on Fibrinogen and Its Derivatives in Patients with Vibration Syndrome

Katsuhiro Fukutake and Minoru Ukita
Tokyo Medical College, Tokyo, Japan
Yuko Agishi
Hokkaido University School of Medicine, Medical Hospital, Noboribetsu Branch Hospital, Noboribetsu City, Hokkaido, Japan

INTRODUCTION

This study presents clinical experiments with Defibrase in 224 patients and 5 normal volunteers. These experiments yield additional findings on the biological effects of Defibrase due to defibrinogenation and secondary induced fibrinolysis, which have been reported by several authors (1-4). The defibrase-induced changes in the blood formed the basis for treatment of circulation disturbances associated with the vibration syndrome resulting from chain-saw handling by professional woodcutters.

MATERIALS AND METHODS

Des-A fibrin monomer was prepared by clotting human fibrinogen (10 mg/ml) with 6 units of Defibrase in 0.3 ml; the clot was lysed in a 7-M urea solution after washing four times with saline solution. Thrombomodulin was prepared from human placenta by the method of Salem and co-workers (5). Protein C was prepared from human pooled plasma by the method described by Suzuki and co-workers (6). Fibrinogen-fibrin degradation product-E (FDP-E) was assayed by the LPIA-E method devised by the Mitsubishi group (7) using special assay apparatus. Quantitative detection of tissue plasminogen activator in the plasma was performed by the enzyme immunoassay (EIA) method of Bergsdorf and co-workers (8). Potentiation of tissue plasminogen activator by des-A fibrin monomer was determined in a mixture of tissue plasminogen activator (2 ng, American Diagnosis Inc.) in 100 μl of Tris buffer and 100 μl of glu-plasminogen (85 μg/ml). The plasmin activity was measured by a chromogenic method using S-2251 with or without the addition of 10 μl of des-A fibrin monomer to the reaction mixture. Activated protein C was measured by the chromogenic method of Bertina and associates (9). Defibrase in plasma and urine was determined by the sandwich EIA method of Imamura and co-workers (10). Blood viscosity was measured with a Silenus Viscosimeter (Silenus Instruments Fty. Ltd., Australia). The blood flow rate in the hands was determined with an ultrasonic Doppler-type current meter (Hayashi Electronic Co. Ltd., Japan). Assay of D-dimer FDP was performed by the immunoassay of Rylatt and associates (11), using monoclonal antibodies for human D-dimer.

RESULTS AND DISCUSSION

To analyze the mechanism of the fibrinolytic phenomenon induced by the administration of Defibrase, studies were conducted on five normal volunteers. As shown in Figure 1, the administration of Defibrase induced secondary fibrinolysis with the production of FDP and soluble des-A fibrin monomer measured by the LPIA method (7) and FM test (12). At the same time, the consumption of tissue plasminogen activator (t-PA) in the plasma, probably caused by the complex formation of t-PA and des-A fibrin, was found as shown in Figure 2. Furthermore, as shown in Figure 3, human des-A fibrin monomer potentiated t-PA activation of plasminogen in comparison with fibrinogen or buffer control. Therefore, the Defibrase-induced fibrinolytic state may entail the intervention of the extrinsic pathway of fibrinolysis through t-PA.

We also studied the clinical effects of Defibrase therapy on 224 patients with blood vessel disturbances and Raynaud's phenomenon resulting from vibration associated with long-term handling of chain-saws by professional woodcutters. This study was performed by the collaboration of 14 institutes (Table 1). These patients each received Defibrase by drip-infusion of 10 Batroxobin units every 2 days for 4 weeks. As shown in Figure 4, plasma fibrinogen levels decreased to about 50 to 100 mg/dl within 4 hours after the drip-infusion and remained low. The survival time ($t_{1/2}$) of Defibrase, measured by the enzyme immunoassay of Imamura and co-workers (10) was 6.2 hours.

As shown in Figure 5, the patients showed a gradual decrease in blood viscosity, which could be related to the improvement of circulation at disturbed areas, mainly in the hands. In addition, an increase in local temperature in the hands could be detected by thermistor with cold water loading tests after 4 weeks of Defibrase therapy. This was especially apparent in severe cases, as shown in Figure 6. There was also an increase in local blood flow in the hands shown in Figure 7.

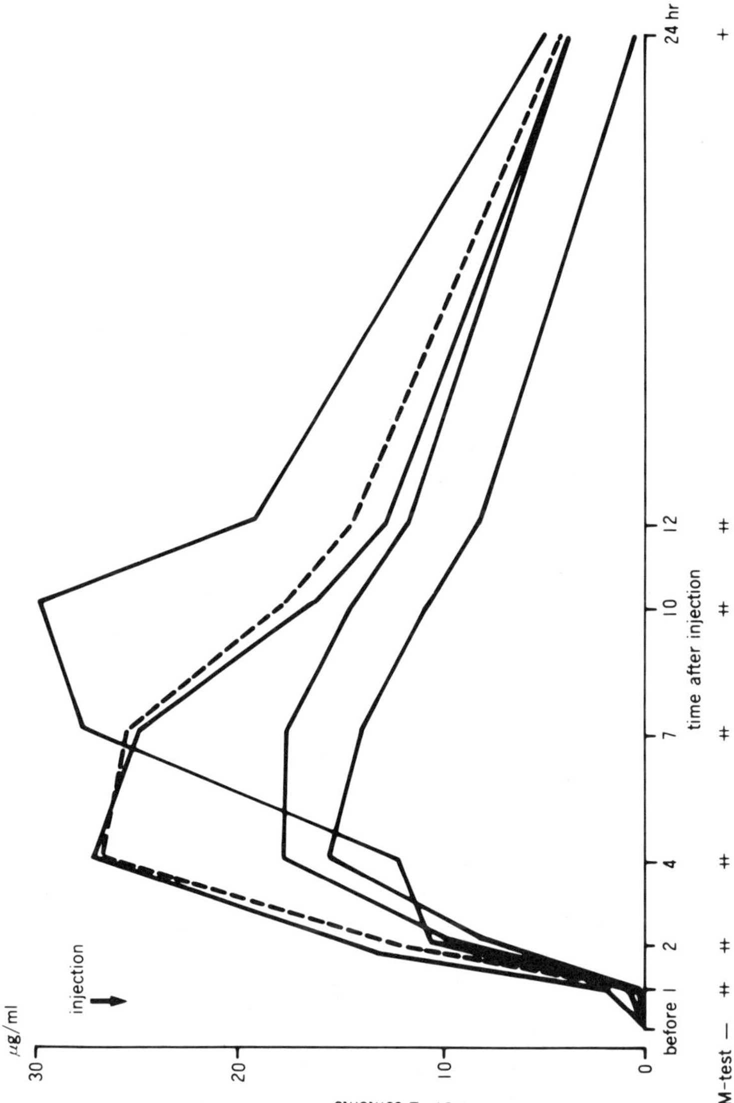

Figure 1 FDP-E contents in serum after Defibrase injection (20 BU).

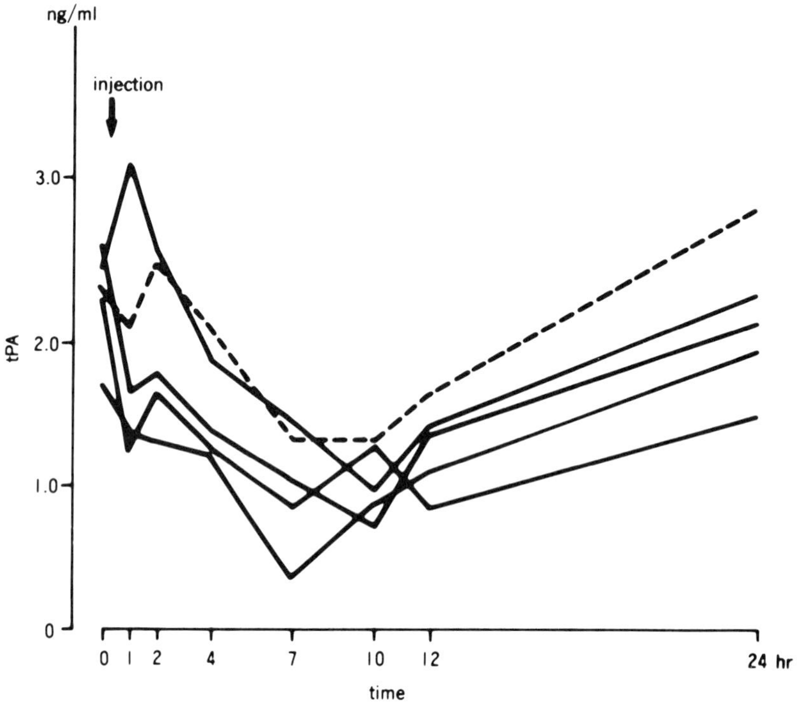

Figure 2 Consumption of tissue plasminogen activator after Defibrase injection.

It is presumed that the cause of hypoviscosity in the patients' plasma was defibrinogenation resulting from Defibrase, a consequence of the lower viscosity of fibrinolytic derivatives of fibrinogen as compared with fibrinogen itself.

In 77.6% of the cases, the patients reported a decrease in the frequency of cold or paralyzed feeling in the fingers. As evaluated by the doctors of the 14 institutes collaborating in this study, 76.8% of the patients were improved and without adverse effects, except for four cases with headache and an unpleasant feeling in the chest. Among general laboratory findings were statistically significant increases in serum GPT, total cholesterol, and leukocyte counts, although these values were still within normal ragnes.

In three patients with microthrombosis in the retina, we detected a smaller amount of D-dimer in the plasma after injection of Defibrase (Fig. 8) than would be expected from their FDP levels. Since D-dimer as measured by the immunoassay of Rylatt and associates (11) is based

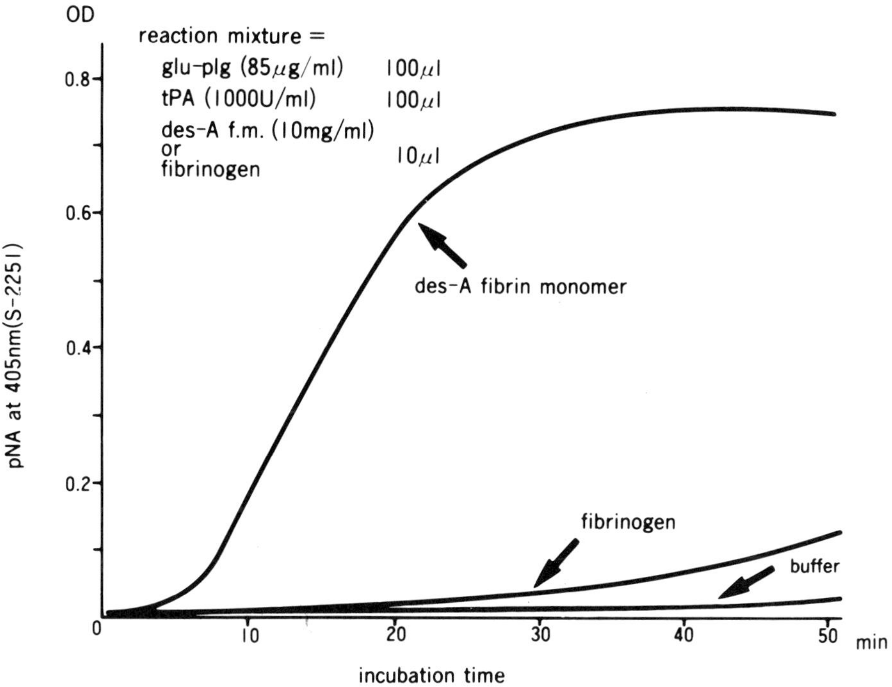

Figure 3 Potentiation of t-PA activity by coexistence of des-A fibrin monomer in vitro.

on completely cross-linked fibrin, this finding suggests that treatment with Defibrase promptly produces a large amount of noncross-linked des-A fibrin in a soluble state and, thereafter, simple fibrinogenolysis or fibrinolysis of des-A fibrin monomer and its noncross-linked polymerization complex (1,13) could be induced by the extrinsic pathway of fibrinolysis. This might be one reason for the absence of thrombogenic effects of Defibrase.

SUMMARY

The data presented in this chapter suggest that the in vivo development of fibrinolysis seen in Defibrase therapy could be a secondary phenomenon related to the extrinsic pathway of plasmin formation through tissue plasminogen activator that was consumed in the plasma after the injection of Defibrase. Furthermore, the action of tissue

Table 1 Collaborating Institutes on Vibration Syndrome Treatment with Defibrase

Institutes	Doctors
Department of Internal Medicine, Hokkaido University School of Medicine, Noboribetsu Branch Hospital	Yuko Agishi
Department of Internal Medicine, Nakaizu SPA Hospital	Kikujiro Saito
The Second Department of Surgery, Tokyo Medical and Dental University, School of Medicine	Yoshio Mitshima
Department of Surgery, Tokyo Medical College	Kin-ichi Furukawa
Department of Clinical Pathology, Tokyo Medical College	Michio Fujimaki
Department of Orthopedic Surgery, Noboribetsu Welfare Pension Hospital	Toshihiko Uchiumi
Department of Internal Medicine, National Noboribetsu Hospital	Hisashi Nishihara
Department of Surgery, Kushiro Rosai Labour Accident Hospital, Labour Welfare Projects Corporation	Tetsuo Oku
Department of Orthopedic Surgery, Shirakawa Orthopedic Surgery Clinic	Hisanari Shirakawa
Department of Internal Medicine, Izunirayama Rehabilitation Hospital	Hisahiro Kishi
Department of Surgery, Gifu Prefectural Gero Hot Spring Hospital	Masao Kato
Department of Orthopedic Surgery, Toyama Rosai Labour Accident Hospital, Labour Welfare Projects Corporation	Tsuyoshi Fukushima
Department of Surgery, Oyodo Town Hospital	Kyusaku Habu
Department of Orthopedic Surgery San-in Rosai Labour Accident Hospital, Labour Welfare Projects Corporation	Yoshiro Nasu

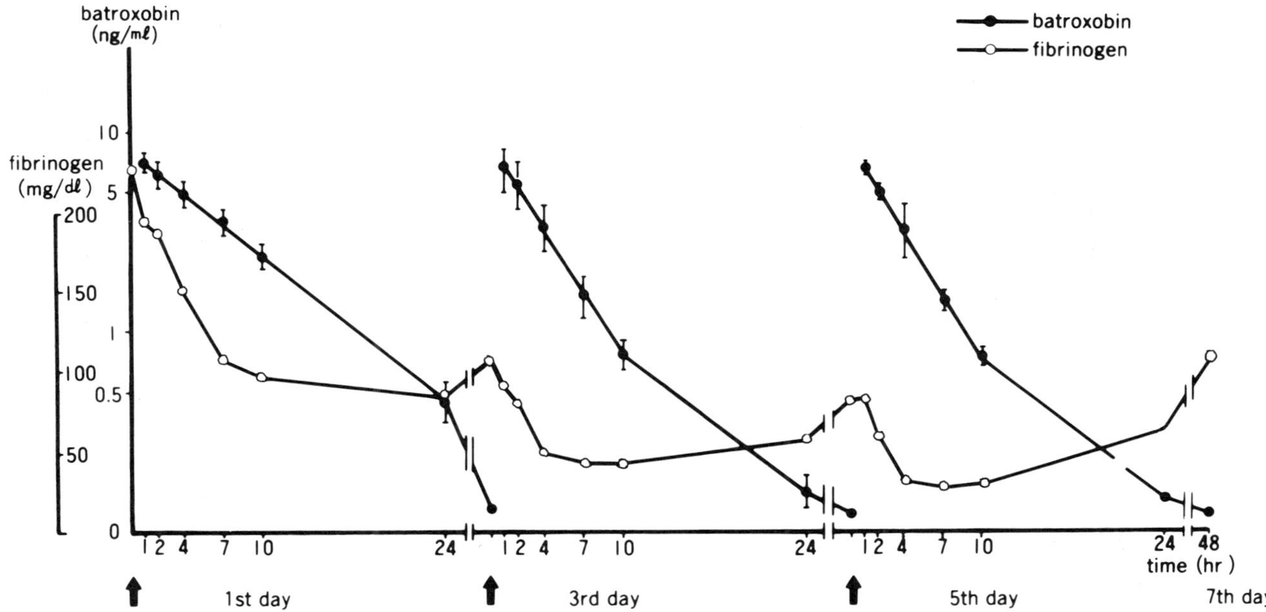

Figure 4 Relationship between Defibrase units and fibrinogen levels during repeated injections of Defibrase (10 BU/body, n = 4).

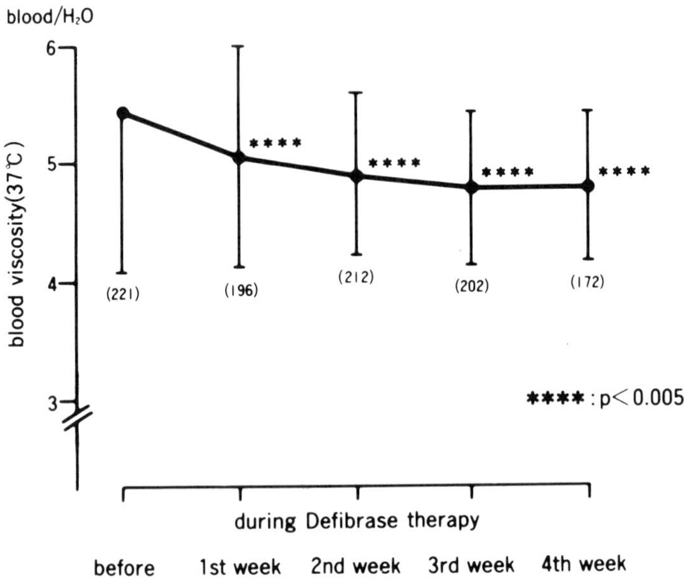

Figure 5 Changes in blood viscosity during Defibrase therapy.

plasminogen activator was enhanced by the presence of des-A fibrin monomer.

Clinical data suggested beneficial effects of Defibrase therapy in 224 patients with vibration syndrome with no adverse effects of the agent except for mild symptoms in four cases. These results may point to a new therapeutic application for Defibrase.

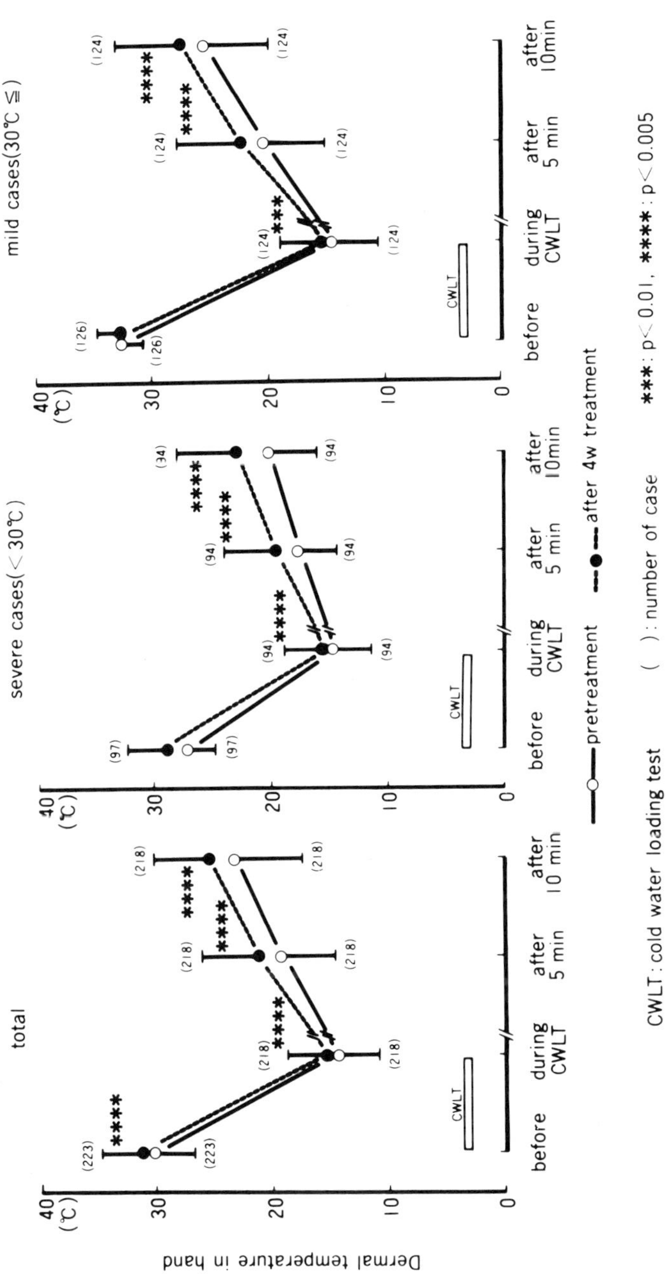

Figure 6 Improving dermal temperature tendencies in the hands, detected by cold water loading tests.

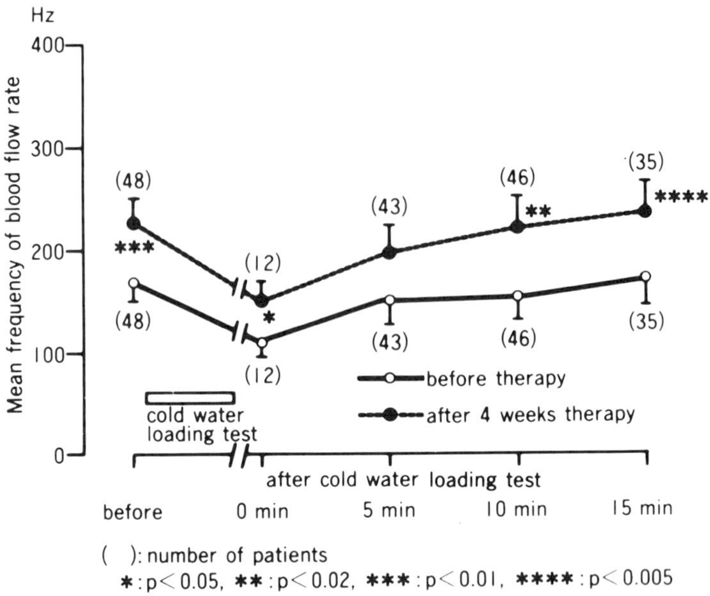

Figure 7 Comparison of local blood flow before and after Defibrase therapy in cold water loading tests in patients, indicating below 30°C cutaneous temperature in the hands.

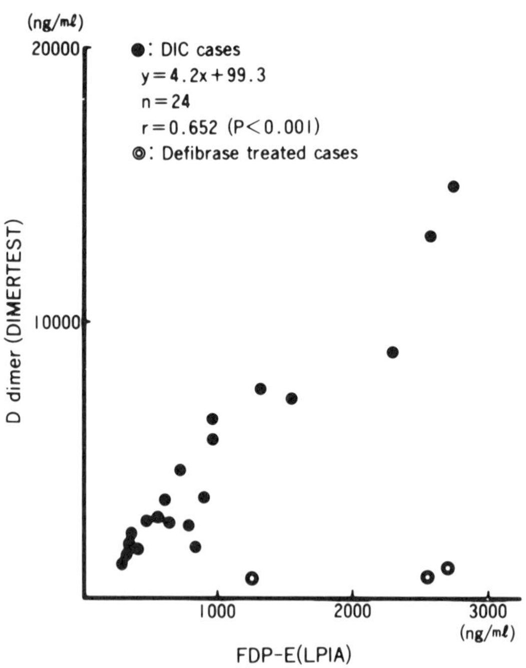

Figure 8 Correlation of FDP values between FDP-E (LPIA) and D-dimer FDP (Dimertest).

REFERENCES

1. Kato, M. J. Tokyo Med. Coll. 40:605 (1982).
2. Habu, K., H. Iwa, I. Hara, T. Ueno, and R. Kim. Med. Pharmacol. 8:1019 (1982).
3. Fukutake, K., M. Fujimaki, H. Nagasawa, and M. Kato. Acta Haematol. Jpn. 44:1178 (1981).
4. Ishimaru, S., S. Fujiwara, K. Furukawa, M. Takahashi, I. Takeuchi, S. Ikematsu, and M. Fujimaki. Blood Vessel 8:545 (1977).
5. Salem, H. H., I. Maruyama, H. Ishii, and P. W. Majerus. J. Biol. Chem. 259:12246 (1984).
6. Suzuki, K., J. Stenflo, B. Dahlbäck, and B. Teodorsson. J. Biol. Chem. 258:1914 (1983).
7. Sawai, M., T. Sudo, and S. Enomoto. U.S. Pat. No. 4:118, 192 (1978).
8. Bergsdorf, N., T. Nilsson, and P. Wallen. Thromb. Haemost. 50:740 (1983).
9. Bertina, R. M., A. W. Broekmans, C. Krommenhoek-van Es, and A. van Wijngaarden. Thromb. Haemost. 51:1 (1984).
10. Imamura, Y., Y. Ohkubo, F. Ohzeki, K. Sugai, and N. Mori. *Reports on Progress in Polymer Physics in Japan* 26:759 (1983).
11. Rylatt, D. B., A. S. Blake, L. E. Cottis, S. A. Massingham, W. A. Fletcher, P. P. Masci, A. N. Whitaker, M. Elms, I. Bunce, A. J. Webber, D. Wyatt, and P. G. Bundesen. Thromb. Res. 31:767 (1983).
12. Largo, R., V. Heller, and P. W. Straub. Blood 47:991 (1976).
13. Takeuchi, I. J. Tokyo Med. Coll. 35:825 (1977).

44

Use of Ecarin in Diagnosis of Coagulant Disorders

Frantisek Kornalík
Charles University, Prague, Czechoslovakia

INTRODUCTION

The capacity to convert prothrombin to thrombin can be detected in a small group of venoms including venom from most of the Australian Elapidae (Marshall and Herrmann, 1983), two African opistoglyphous snakes (Guillin et al., 1978; Kornalík et al., 1978), and two Afro Asian Viperidae species (*Echis carinatus* and *Echis coloratus*). The responsible enzymes from four venoms (*Echis carinatus* [Kornalík, 1971], *Oxyuranus scutellatus*-Taipan [Pirkle et al., 1972], *Notechis scutatus scutatus* [Tans et al., 1985], and *Thelotornis kirtlandi* [Kornalík et al., 1978]) were isolated. Of these, ecarin, obtained from *Echis carinatus*, has found widest use in coagulation laboratory techniques. It was isolated by various combinations of column chromatography procedures and obtained in a highly purified form that enabled a detailed biochemical and biophysical characterization (Morita and Iwanaga, 1978; Rheer et al., 1982; Dyr et al., 1983).

Besides the main prothrombin-converting activity, ecarin has a weak fibrinogenolytic but no general proteolytic effect and it does not affect other clotting proteins. It is a metalloproteinase with a high specificity for prothrombin, its main target being the Arg_{323}-Ile_{324} bond between the A- and B-chains. By incubation of ecarin with purified prothrombin, thrombin activity is produced. Analysis of the fragments ensuing during the activation, without addition of any thrombin inhibitors, indicated that fragment 1 is split off first, but almost simultaneously the Arg_{323}-Ile_{324} bond in prethrombin is cleaved. This causes a rapidly increasing thrombin activity due to ecarin-thrombin, which is composed of fragment 2 and the A- and B-chains of thrombin (Franza, et al., 1975; Kornalík and Blombäck, 1975).

From the results obtained by using various thrombin inhibitors, several investigators came to the conclusions that ecarin acts on prothrombin by limited proteolysis, whereby the first product is meizothrombin. Subsequent autocatalytic reactions give rise to two products: meizothrombin 1 (des-fragment 1) and α-thrombin (Novoa and Seegers, 1980; Briet et al., 1982; Rheer et al., 1982). The action of ecarin on prothrombin is schematically outlined in Figure 1. No cofactors, such as factor Va, phospholipids, or Ca^{2+}, are necessary. Since this activation is independent of any form of the N-terminal segment (fragment 1) of the prothrombin molecule, it is understandable that ecarin converts all forms of the dicoumarol-induced precursors of prothrombin, the so-called PIVKA II (Suttie, 1973; Franza et al., 1975; Malhotra, 1979). This highly specific and unique action of ecarin qualifies it for use in several laboratory techniques. Franza and Aronson (1976) recommended the use of ecarin for the detection of minute amounts of prothrombin in human plasma protein preparations. Prothrombin could be detected in concentrations as low as 0.02 U per ml in the test system. Sakuragawa and co-workers (1976) elaborated on a method for a more precise diagnosis of the early procoagulant stages of disseminated intravascular coagulation (DIC) using ecarin. The ecarin test, using

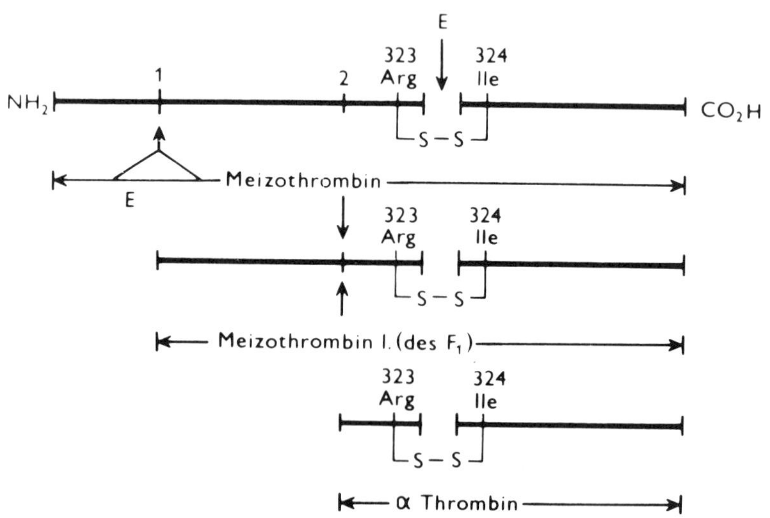

Figure 1 Action of ecarin on prothrombin molecule.

ecarin instead of tissue thromboplastin, proved to be a valuable tool in the differential diagnosis of liver disease and vitamin K deficiency (Corrigan and Earnest, 1980), as well as in a few cases of dysprothrombinemia (Weinger et al., 1980).

We have concentrated our attention on the possible use of ecarin for the laboratory control of the dicoumarol therapy, the differential diagnosis of liver disease and vitamin K deficiency, and the diagnosis of DIC.

MATERIALS AND METHODS

A group of more than 340 patients who were receiving dicoumarol treatment for several weeks to several years was followed for 1 year. Each patient was tested several times. In samples of oxalated plasma, prothrombin was assayed by the routine Quick method, using tissue thromboplastin (Imuna Michalany) and 0.025 M $CaCl_2$. The level of prothrombin was expressed as a percentage of normal values. The PIVKA was assayed in $BaSO_4$-adsorbed plasma (prothrombin time >5 minutes) by adding to equal volumes of plasma and saline solution (pH 7.3) ecarin in a concentration to give a clotting time of 20 seconds with normal plasma, equivalent to 1 ecarin unit (i.e., 0.5 to 1.0 μg of coagulant protein/ml). Clotting time was inversely related to PIVKA concentration. This was established by using mixtures of $BaSO_4$-adsorbed plasmas from a normal person and from a patient with a prothrombin level lower than 10%. Shortening of the clotting time after addition of ecarin was parallel to the increasing amount of PIVKA (Table 1).

In another group of patients (liver disease, DIC, or vitamin K deficiency), the Quick prothrombin time was assayed in citrated plasma using either thromboplastin plus $CaCl_2$ or ecarin with saline as prothrombin (or PIVKA) activators. Thrombin activity was expressed as a percentage of normal values.

RESULTS AND DISCUSSION

Prothrombin and PIVKA estimations in a group representative of the total of 347 patients are presented in Figure 2. As can be seen, PIVKA could be detected only in cases in which the prothrombin level was lower than 50%. With prothrombin levels between 50 and 100%, PIVKA is very likely accumulated in hepatocytes and does not appear in the circulation. Maximal PIVKA concentration was found in patients with the therapeutically optimal range of prothrombin (i.e., between 10 and 20%). The increase in PIVKA concentration, however, did not correspond to the decreased prothrombin level.

Table 1 Ecarin Clotting Time (ECT or PIVKA Time) in Mixtures of Normal and Dicoumarolized Deprothrombinized Plasmas

Ads. DP (%)	Ads. NP (%)	Clotting time (sec)
0	100	>180
10	90	56
25	75	49
50	50	40
75	25	32
90	10	28
100	0	25

Ads. DP = dicoumarolized platelet-poor plasma adsorbed with $BaSO_4$.
Ads. NP = normal platelet-poor plasma (PPP) adsorbed with $BaSO_4$.

In about 3% of cases, patients receiving dicoumarol for longer than 2 years, no PIVKA was detected, even when the prothrombin concentration was below 20% (open triangles, in Fig. 2). This discrepancy between the decrease in prothrombin and the PIVKA concentration can be hypothetically explained by formation of partially carboxylated populations of dicoumarol-induced prothrombin molecules, as they were described in dicoumarolized bovine plasma (Malhotra, 1982). Partially carboxylated prothrombin precursors cannot be converted to thrombin by factor Xa, but are still adsorbable on $BaSO_4$ and, thus, can escape detection by ecarin in $BaSO_4$-treated plasma.

In a few cases, PIVKA could be detected in patients with a prothrombin level well above 50% (closed triangles in Fig. 2). All of them have received 10,000 to 15,000 U of heparin because of an acute thromboembolism 1 to 4 hours before the ECT was assayed in the adsorbed plasma. Thromboembolic disease is always accompanied by DIC of various intensity. It is assumed that in the early procoagulant stages of DIC, prethrombin 1 or prethrombin 2 can be produced autocatalytically by an excess of thrombin (Seegers et al., 1974). In these cases, the ECT positive test can be explained by formation of meizothrombin 1 by ecarin. This intermediate is not adsorbed on $BaSO_4$ because it lacks fragment 1, containing γ-carboxylated groups, enabling the binding to the barium salt, and is insensitive to heparin (see Fig. 1).

Determination of prothrombin and its precursors in untreated platelet-poor plasma, using tissue thromboplastin plus Ca^{2+} or ecarin as activators was performed in four groups of patients: (a) normal

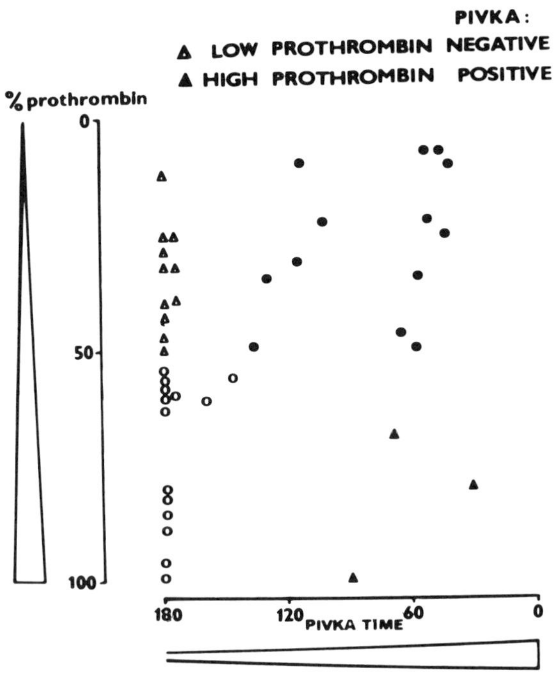

Figure 2 Estimation of prothrombin and PIVKA in the plasma of patients treated with dicoumarol. Vertical axis: prothrombin level expressed as a percentage of average normal values, increasing downwards; horizontal axis: PIVKA concentration (inversely related to PIVKA time in seconds) increasing from left to right; open circles and open triangles: PIVKA concentration ECT negative; closed circles and closed triangles: PIVKA concentration ECT positive.

persons for control, (b) patients with liver disease with manifest clinical symptoms, (c) patients with vitamin K deficiency, and (d) patients with DIC in the early hypercoagulable state. Results are summarized in Table 2.

The nearly identical decrease in prothrombin levels as assayed using thromboplastin as well as ecarin in patients with liver disease indicated that a dysfunction of the hepatocytes is at a more basic level of protein synthesis than is the last step of prothrombin formation. Carboxylation of the glutamic acid residues is not involved in the functional defect as long as vitamin K is supplied.

There is a highly significant difference between the low levels of prothrombin assayed using tissue thromboplastin and the high levels

Table 1 Comparison of Prothrombin Assay in Patients

Activator	Thromboplastin	Ecarin
Normal	99 (85-100)	100 (90-110)
Liver disease	43 (21-88)	44 (10-75)
Vitamin K deficiency	19.3 (2-29)	98 (60-110)
DIC	67 (55-100)	100 (>100)

Prothrombin and ecarin values are expressed as thrombin activity as a percentage of normal values.

of thrombin activity obtained using ecarin. High values in the ecarin test are due to PIVKA present in the plasma of patients treated with dicoumarol.

A group of patients with DIC, corroborated by routine laboratory tests (e.g., ethanol gelation, FDP immunoassay, thrombocytopenia, and euglobulin lysis time) was also tested. The higher level of thrombin activity found with the ecarin test indicates the presence of prothrombin derivatives that are not convertible by tissue thromboplastin, but can be activated by ecarin. As explained earlier, prethrombin 1 and prethrombin 2, occurring in the earlier procoagulant stages of DIC, can be converted to meizothrombin 1 and α-thrombin by ecarin. In the later stages of DIC, an increased synthesis of prothrombin is triggered by its decrease due to consumption coagulopathy. In this case, acarboxylated precursors of prothrombin can be released into plasma and activated by ecarin.

All these findings indicate that the ecarin test can be a useful tool in the differential diagnosis of some hemostatic disorders.

REFERENCES

Briet, E., C. M. Noyes, H. R. Roberts, and M. J. Griffith. (1982). Cleavage and activation of human prothrombin by *Echis carinatus* venom. Thromb. Res. 27:591-600.

Corrigan, J. J. and D. L. Earnest. (1980). Factor II antigen in liver disease and warfarin-induced vitamin K deficiency correlation with coagulant activity using Echis venom. Am. J. Hematol. 8:249-255.

Dyr, J. E., B. Blombäck, and F. Kornalík. (1983). The action of prothrombin activated by ecarin on fibrinogen. Thromb. Res. 30:225-234.

Franza, B. R. and D. L. Aronson. (1976). Detection and measurement of low levels of prothrombin. Use of a procoagulant from *Echis carinatus* venom. Thromb. Res. 8:329-336.

Franza, R. B., D. L. Aronson, and J. S. Finlayson. (1975). Activation of human prothrombin by a procoagulant factor from venom of *Echis carinatus*. J. Biol. Chem. 250:7057-7068.

Guillin, M. C., A. Bezeaud, and D. Mènachè. (1978). The mechanism of activation of human prothrombin by an activator isolated from *Dispholidus typus* venom. Biochim. Biophys. Acta 537:160-168.

Kornalík, F. (1971). Fibrinolytische Proteasen aus Schlangengiften. Folia Haemat. 95:193-208.

Kornalík, F. and B. Blombäck. (1975). Prothrombin activation induced by ecarin—a prothrombin converting enzyme from *Echis carinatus* venom. Thromb. Res. 6:53-63.

Kornalík, F., E. Táborská, and D. Mebs. (1978). Pharmacological and biochemical properties of a venom gland extract from the snake *Thelotornis kirtlandi*. Toxicon 16:535-542.

Malhotra, O. P. (1979). Purification and characterization of dicoumarol-induced prothrombins. Thromb. Res. 15:427-437.

Malhotra, O. P. (1982). Partially carboxylated prothrombins. Biochim. Biophys. Acta 702:178-192.

Marshall, L. R. and R. P. Herrmann. (1983). Coagulant and anticoagulant actions of Australian snake venoms. Thromb. Haemost. 50:707-711.

Morita, T. and S. Iwanaga. (1978). Purification and properties of prothrombin activator from the venom of *Echis carinatus*. J. Biochem. (Tokyo) 83:559-570.

Novoa, E. and W. H. Seegers. (1980). Mechanisms of α thrombin and β thrombin-E formation: use of ecarin for isolation of meizothrombin I. Thromb. Res. 18:657-668.

Pirkle, H., M. McIntosh, I. Theodor, and S. Vernon. (1972). Activation of prothrombin with Taipan snake venom. Thromb. Res. 1:559-567.

Rheer, M. J., S. Morris, and D. P. Kosow. (1982). Role of meizothrombin and meizothrombin-(des F1) in the conversion of prothrombin to thrombin by the *Echis carinatus* venom coagulant. Biochemistry 21:3437-3443.

Sakuragawa, N., K. Takahashi, M. Hoshiyama, C. Jimbo, K. Niiya, M. Matsuoka, and J. Ohnishi. (1976). Changes of prothrombin assayed by the method using *Echis carinatus* venom in experimental D.I.C. Thromb. Res. 8:697-701.

Seegers, W. H., D. A. Walz, J. Reuterby, and L. E. Mc Coy. (1974). Isolation and some properties of thrombin E and other prothrombin derivatives. Thromb. Res. 4:829-859.

Suttie, J. W. (1973). Mechanism of action of vitamin K. Demonstration of a liver precursor of prothrombin. Science 179:192-194.

Tans, G., J. W. P. Govers-Riemslag, J. L. M. L. van Rijn, and J. Rosing. (1985). Purification and properties of a prothrombin activator from the venom of *Notechis scutatus scutatus*. J. Biol. Chem. 260:9366-9372.

Weinger, R. S., C. Rudy, J. L. Moake, J. D. Olson, and P. L. Cimo. (1980). Prothrombin Houston: a dysprothrombin identifiable by crossed immunoelectrofocusing and abnormal *Echis carinatus* venom activation. Blood 55:811-815.

Taxonomic Index

(See also Subject Index, pp. 617-628)

Subject Index

(See also Taxonomic Index, pp. 611-616)

U

V